Bioceramics with Clinical Applications

Bioceramics with Clinical Applications

Edited by

María Vallet-Regí

Library of Congress Cataloging-in-Publication Data

Bioceramics with clinical applications / edited by Maria Vallet-Regi.
    pages cm
  Includes bibliographical references and index.
  ISBN 978-1-118-40675-5 (cloth)
 1. Ceramics in medicine. 2. Biomedical materials. I. Vallet-Regi, Maria, editor of compilation.
  R857.C4B5534 2014
  610.28'4–dc23
                                        2013049091

A catalogue record for this book is available from the British Library.

ISBN: 9781118406755 (Cloth)

Set in 10/12pt TimesLTStd by Laserwords Private Limited, Chennai, India
Printed and bound in Malaysia by Vivar Printing Sdn Bhd

1   2014

Contents

5. Silica-based Ceramics: Mesoporous Silica

Montserrat Colilla

6. Alumina, Zirconia, and Other Non-oxide Inert Bioceramics

Juan Peña López

List of Contributors

Daniel Arcos, Departamento de Química Inorgánica y Bioinorgánica, Facultad de Farmacia, Universidad Complutense de Madrid, Spain **and** CIBER de Bioingeniería, Biomateriales y Nanomedicina (CIBER-BBN), Spain

Alejandro Baeza, Departamento de Química Inorgánica y Bioinorgánica, Facultad de Farmacia,Universidad Complutense de Madrid, Spain **and** CIBER de Bioingeniería, Biomateriales y Nanomedicina (CIBER-BBN), Spain

M. Victoria Cabañas, Departamento de Química Inorgánica y Bioinorgánica, Facultad de Farmacia, Universidad Complutense de Madrid, Spain

Oscar Castaño, Biomaterials for Regenerative Therapies, Institute for Bioengineering of Catalonia (IBEC), Spain **and** CIBER de Bioingeniería, Biomateriales y Nanomedicina (CIBER-BBN), Spain **and** Department of Materials Science and Metallurgy, Polytechnic University of Catalonia (UPC), Spain

Montserrat Colilla, Departamento de Química Inorgánica y Bioinorgánica, Facultad de Farmacia, Universidad Complutense de Madrid, Spain **and** CIBER de Bioingeniería, Biomateriales y Nanomedicina (CIBER-BBN), Spain

Elisabeth Engel, Biomaterials for Regenerative Therapies, Institute for Bioengineering of Catalonia, Spain **and** Department of Materials Science, Technical University of Catalonia, Spain **and** Biomedical Research Networking center in Bioengineering, Biomaterials and Nanomedicine (CIBER-BBN), Spain

Blanca González, Departamento de Química Inorgánica y Bioinorgánica, Facultad de Farmacia, Universidad Complutense de Madrid, Spain **and** CIBER de Bioingeniería, Biomateriales y Nanomedicina (CIBER-BBN), Spain

Isabel Izquierdo-Barba, Departamento de Química Inorgánica y Bioinorgánica, Facultad de Farmacia, Universidad Complutense de Madrid, Spain **and** CIBER de Bioingeniería, Biomateriales y Nanomedicina (CIBER-BBN), Spain

Miguel Manzano, Departamento de Química Inorgánica y Bioinorgánica, Facultad de Farmacia, Universidad Complutense de Madrid, Spain **and** CIBER de Bioingeniería, Biomateriales y Nanomedicina (CIBER-BBN), Spain

Juan Peña López, Departamento de Química Inorgánica y Bioinorgánica, Facultad de Farmacia, Universidad Complutense de Madrid, Spain

Soledad Pérez-Amodio, Biomaterials for Regenerative Therapies, Institute for Bioengineering of Catalonia, Spain **and** Biomedical Research Networking center in Bioengineering, Biomaterials and Nanomedicine (CIBER-BBN), Spain

Josep A. Planell, Biomaterials for Regenerative Therapies, Institute for Bioengineering of Catalonia (IBEC), Spain **and** CIBER de Bioingeniería, Biomateriales y Nanomedicina (CIBER-BBN), Spain **and** Open University of Catalonia (UOC), Spain

Antonio J. Salinas, Departamento de Química Inorgánica y Bioinorgánica, Facultad de Farmacia, Universidad Complutense de Madrid, Spain **and** CIBER de Bioingeniería, Biomateriales y Nanomedicina (CIBER-BBN), Spain

María Vallet-Regí, Departamento de Química Inorgánica y Bioinorgánica, Facultad de Farmacia, Universidad Complutense de Madrid, Spain **and** CIBER de Bioingeniería, Biomateriales y Nanomedicina (CIBER-BBN), Spain

Mercedes Vila, Departamento de Química Inorgánica y Bioinorgánica, Facultad de Farmacia, Universidad Complutense de Madrid, Spain **and** CIBER de Bioingeniería, Biomateriales y Nanomedicina (CIBER-BBN)

Preface

What do I intend with this book? What do I want to convey to the readers? Who is going to be interested? Why have I chosen this organization of the contents?

All these questions have arisen before and during the writing of this book. In fact, the initial contents list has been modified a few times during the almost two years of this adventure.

One of my first targets is to stimulate interest in bioceramics and provide the tools for their development and commercialization with biomedical applications. This book aims to give a clear view of bioceramics, and tries to do it in such a way that the reader realizes that there is plenty still to be learned, as reflected in the book.

This book was born as a consequence of a kind invitation from Rebecca Stubbs, Senior Commissioning Editor at Wiley-Blackwell. At the start I had many doubts about editing a book and I let Rebecca know about it. However, her proposal made my head think about it, because undoubtedly it was a proposition that my heart could not reject, although common sense was telling me to stop. As you can see, finally, I became fully involved in this adventure of editing a book about bioceramics, which has been the central topic of my research and academic activities during the last 20 years. I wanted to write it with the senior members of my research team and a couple of friends who have also been working for a long time in this area, coordinated closely by myself to produce a textbook accessible to those uninitiated in this area, not simply a collection of chapters that are not integrated and coordinated in a single literary work. Of course, I also wanted to produce a book in which experts could go straight to the chapter in which they are interested without the necessity of reading the whole book. You, my friends the readers, will let me know if I have succeeded.

In the last two decades, Biomaterials has been incorporated into the studies of several universities. Beyond research activities, Master's courses, and PhD studies, this subject is also included in different degrees, such as Materials Engineering, Pharmacy, and so on. Bioceramics is one of the main topics in Biomaterials. In the last decade, the efforts of many groups involved in this new research area have led to new bioceramics and, in some recent cases, clinical applications. However, there is no text book dealing with the subjects proposed above in a complete way. This book will provide important information concerning the synthesis and characterization of bioceramics, aimed to be helpful for all those researchers involved in this field. I will pay special attention to explaining the relationship between the synthesis processes and the subsequent clinical applications, in an attempt to help students in the development of the new ideas framed in these topics.

The book has been divided into four parts:

I. Introduction
II. Materials
III. Material Shaping
IV. Research on Future Ceramics

In the first part there is a general view of bioceramics, highlighting their reactivity and possibilities to regenerate bone, within a multidisciplinary context and especially nowadays in biomedical engineering. In the second chapter I give general and basic ideas of biomimetics. In both chapters there is a list of recommended reading to increase the reader's knowledge of these ideas.

Part II deals with the study of materials to fabricate bioceramics: from calcium phosphates, treated in Chapter 3; glasses, Chapter 4; and silica mesoporous materials, Chapter 5; up to inert bioceramics, described in Chapters 6 and 7.

Part III describes diverse possibilities of materials shaping. Thus, Chapter 8 deals with cements, Chapter 9 with coatings, and Chapter 10 with scaffolds fabricated as fibers, foams, or three-dimensional pieces.

Finally, Part IV tries to give a vision of new bioceramics still under research, some with future applications that are almost ready for commercialization, others needing a long time to come to market. It starts with an in-depth review of bone biology to be able to relate it to bone regeneration. This is followed by a chapter dedicated to ceramics for drug delivery, Chapter 12; another for gene transfection, Chapter 13; and, finally, Chapter 14 is dedicated to ceramic nanoparticles for cancer treatment.

This book will probably be read by a great community of researchers involved both in the academic and commercial world, such as materials engineers, chemical engineers, chemists, biologists, physicists, medical doctors, and so on. Amongst people from the academic community, undergraduates will have the chance of a first contact with top class research in the world of bioceramics and they might be motivated to start their scientific career in this topic. Postgraduates will have the opportunity to discover the important milestones in bioceramics and their application to biomedicine. Post-docs and senior researchers will use this book as the basis to start building their research. In this way, this book can be both an introductory and an advanced tool for academic researchers involved in the topic of bioceramics. As a second audience we could find medical doctors with further interest in implantable materials, and different companies interested in commercializing bioceramics with clinical applications.

I want to finish this preface by acknowledging Wiley editorial and Sarah Tilley, Senior Project Editor, for giving me the opportunity to editing this book, which collects my work in academia and research during the last 20 years, and also for giving their comprehensive technical support. Likewise, I want to express my greatest thanks to Dr. Fernando Conde, Pilar Cabañas, and Jose Manuel Moreno for their assistance during the elaboration of this manuscript. And, of course, I want to thank you, the reader of this book. I would love it to be useful to you to go forward in this passionate world of bioceramics, which undoubtedly can be of great usefulness in the society in which we live, helping in the health area and giving a better quality of life to everybody who could benefit from our research.

Now, my work as editor ends, in a book that I was always temped to write, but would never have had the guts to do it without the motivation of Rebecca. I am fully aware that some of the content of this book will be old in a few years, even before this book sees the light for the first time. This is only a small indication of the intensive and good research work carried out in the area of bioceramics, where the advance is constant and systematic. I wish that important solutions will be found and I also wish there will be companies brave enough to put them into practice in clinical applications.

And now, it is time for you, my dear reader, to enjoy the book, wishing from my heart that you like it and that it will be useful for you.

María Vallet-Regí

Part I
Introduction

1

Bioceramics

María Vallet-Regí
Departamento de Química Inorgánica y Bioinorgánica, Facultad de Farmacia,
Universidad Complutense de Madrid, CIBER de Bioingeniería, Biomateriales y
Nanomedicina (CIBER-BBN), Spain

1.1 Introduction

Ceramic materials are important sources of biomaterials for applications in biomedical engineering. Those ceramics intended to be in contact with living tissues are called *bioceramics*, and have experienced great development in the last 50 years. The medical needs of an increasingly aging population have driven a great deal of research work looking for new materials for the manufacture of implants. These are used to regenerate and repair living tissues damaged by disease or trauma. For specific clinical applications, mainly in orthopedics and dentistry, bioceramics are playing a key role.

In general, ceramics are inorganic materials with a combination of ionic and covalent bonding. The use of new ceramic materials represents an evolution of many aspects of mankind history. Many millennia ago, the possibility to store grains in ceramic receptacles allowed man to become a settler instead of a nomad hunter. Some centuries ago, the use of structural ceramics also brought great advances in the quality of life of man with the possibility of making clay bricks and tiles. Decades ago, ceramics produced a new revolution in the human way of life, with the development of functional ceramics in dielectrics, semiconductors, magnets, piezoelectrics, high temperature superconductors, and so on. In addition, ceramics have played an important role in improving the quality and length of human life through their use in biomaterials and medical devices.

As observed, the investigation of bioceramics has also evolved when, as will be explained later, more restrictive properties for the new ceramics were required. Thus, alumina, zirconia, calcium phosphates, and certain glasses and glass-ceramics are genuine examples of bioceramics. Figure 1.1 shows a classification of bioceramics according to their reactivity and their main clinical applications. Carbon is an element, not a compound, and conducts

Bioceramics with Clinical Applications, First Edition. Edited by María Vallet-Regí.
© 2014 John Wiley & Sons, Ltd. Published 2014 by John Wiley & Sons, Ltd.

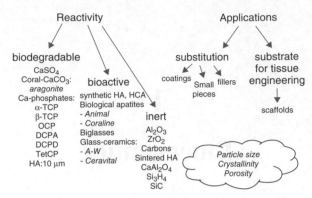

Figure 1.1 *Classification of bioceramics according to their reactivity. Particle size, crystallinity, and porosity are important factors to classify certain bioceramics, like apatites, in one group or the other. HA: hydroxyapatite, HCA: hydroxycarbonate apatite, A-W: apatite–wollastonite, TCP: tricalcium phosphate, OCP: octacalcium phosphate, DCPA: dicalcium phosphate anhydrous, DCPD: dicalcium phosphate dihydrate, TetCP: tetracalcium phosphate monoxide (See insert for color representation of the figure)*

electricity in its graphite form, but it is considered a ceramic because of its many ceramic-like properties. Nowadays, new advanced bioceramics are under study, including ordered mesoporous silica materials or specific compositions of organic–inorganic hybrids.

Ceramic materials have high melting temperatures, low conduction of electricity and heat and relatively high hardness. With regards to their mechanical behavior, ceramic materials exhibit great compression strengths and very much lower tensile strengths. Moreover, they are stiff materials, with high Young's modulus, and brittle because failure takes place without plastic deformation.

In relation to their surface properties, ceramics show high wetting degrees and surface tensions which favor the adhesion of proteins, cells, and other biological moieties. Furthermore, a ceramic surface can be treated to reach very high polish limits. Currently, as will be explained latter, much research effort is devoted to ceramics with interconnected porosity and in these cases the mechanical properties will change drastically.

Nowadays, it is possible to manufacture implants to replace any part of our body, except the brain.

Obviously, different types of materials are in use depending on the tissue to be replaced. Regarding the materials to be used, it is critical to bear in mind that a group of biomaterials will be applied in body reconstruction functions, hence they must perform their duty for an undefined period of time, that is, for the rest of the patient's life. Another group of biomaterials will be used in temporary body support functions. This "permanent" or "temporary" feature allows for a larger and better choice of materials for implant manufacture.

1.2   Reactivity of the Bioceramics

Many different factors affect the reactivity of any chemical substance and greatly determine its reaction kinetics. Figure 1.2 shows some of these. If we take into account the almost inert or bioactive nature of the different ceramics for medical applications, as well as kinetic

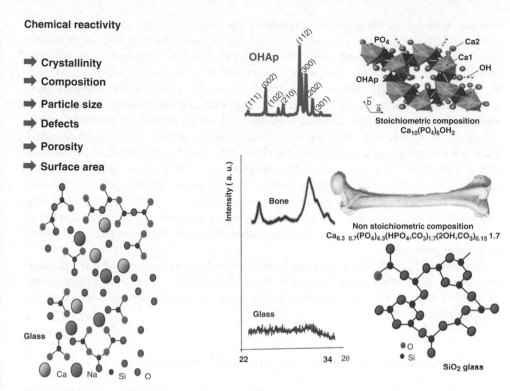

Figure 1.2 *Governing factors in chemical reactivity of bioceramics. Composition in between glasses (disordered) and crystals (ordered) (See insert for color representation of the figure)*

factors such as particle size and porosity, three groups of bioceramics in use nowadays may be distinguished, *inert, bioactive*, and *biodegradable*, as we can see in Figure 1.1. The final purpose of the artificial synthesis of ceramics for bone replacement (hard tissue) is to implant a ceramic material able to regenerate the damaged bone. This is feasible if the ceramic is bioactive. Otherwise, if the ceramic is inert, the bone will be replaced by a material that the organism can tolerate, but which cannot substitute it by means of bone regeneration.

Reactivity, rather than the type of bioceramic is a suitable criterion to classify bioceramics. For instance, in the field of amorphous ceramics it is possible to obtain glasses that, in the same chemical system, behave as bioinert, bioactive, or resorbable because they have somewhat different compositions. It is also possible to find glasses with identical composition behaving as bioinert when obtained by melting, or bioactive when synthesized by a sol−gel method. Moreover, some glass compositions considered bioactive can be completely resorbed when used as particulates under a certain size limit, for instance, 90 μm for Bioglass® 45S5 (all this will be dealt with in Chapter 4). Analogous examples can be found among crystalline ceramics. For instance, the *in vivo* reactivity of hydroxyapatite (HA) can range from almost bioinert, when highly sintered as dense monoliths, to resorbable, when used in particle size, omitting the bioactive character generally attributed to HA (to be discussed in Chapter 3).

When some glass compositions presenting the highest levels of bioactivity were investigated, it was found that they were able to bond to hard and soft tissues, whereas other bioactive materials only bond to hard tissues. To explain these differences in reactivity Hench defined in 1994 two classes of bioactivity: class A, osteoproductive and class B, osteoconductive. The first occurs when the material elicits extracellular and intracellular responses whereas in the second only an extracellular response is obtained. It was explained that the ions released from these bioactive glasses, in particulate form, stimulated a regeneration of living tissues mediated by genes. These osteoproductive glasses were considered as third generation bioceramics and are the basis of the research efforts looking for new biomaterials that stimulate the cellular response. Nowadays the research efforts are concentrated on porous second generation bioceramics and new advanced bioceramics. In these materials the ceramic plays the role of a scaffold of cells and substances with biological activity (growth factors, hormones ...) which are released to the medium in a controlled way. Thus, they are starting to be used in applications related to tissue engineering.

On the other hand, bioceramics must be biocompatible and functional for the required implantation time. They must also not be toxic, carcinogenic, allergic, or inflammatory. In general, because of their ionic bonds and chemical stability ceramic materials are biocompatible.

1.3   First, Second, and Third Generations of Bioceramics

The study of bioceramics can be divided into first, second, and third generations. The study of first generation bioceramics started in the 1960s, when the goal was reactivity as low as possible. The more representative examples of this kind of bioceramics are alumina, Al_2O_3 and zirconia, ZrO_2. They are widely used as biomaterials because of their high strength, excellent corrosion and wear resistances, stability, non-toxicity, and *in vivo* biocompatibility. Around the 1980s the objective changed to obtain favorable interactions with the living body, namely a bioactive response or degradation. Specific compositions of calcium phosphates or sulfates, bioactive glasses, and glass-ceramics are examples of second generation bioceramics used clinically for bone tissue augmentation, as bone cements, or for metallic implants coating. As was indicated, in the last decade bioceramics with more demanding properties were required. The studies in third generation bioceramics are more based in biology and follow the purpose of substituting "replacement" tissues by "regenerating" tissues. This category includes bioceramics based on porous second generation bioceramics, loaded with biologically active substances, and new advanced bioceramics like silica mesoporous materials, mesoporous ordered glasses, or organic–inorganic hybrids.

The situation described above summarizes the trends in bioceramics between the year 1950 – when the first hip prosthesis was implanted in Oxford, by Charnley – and the early twenty-first century, showing the enormous evolution that has taken place in the field of bioceramics, with significant advancements. By the middle of the twentieth century, inert ceramics began to be used as replacements for damaged parts of the human skeleton. Only a few types of ceramics, such as alumina and zirconia, not specifically developed for biomedical tasks, were used back then. Nowadays, in the early stages of the twenty-first century, the bioceramics applied in the clinical field are specifically designed to repair and regenerate human bone; there are several manufacturing companies providing different bioceramics to trauma and maxillofacial surgeons, among other specialities. All these commercially available bioceramics, commonly used in the clinical field, could be termed as "traditional," that is, used according to all regulatory and legal restrictions required for such bone substitutes and prostheses (this will be described in detail in Chapter 3).

1.4 Multidisciplinary Field

At this point, it becomes clear that the field of biomaterials requires the input of knowledge from very different areas so that the implanted material in a living body performs adequately. In Figure 1.3 are shown the interactions between these areas. The biomaterials discipline is founded in the knowledge of Materials Science and Biological Clinical Science. In this sense, biomaterials are an excellent example of a multidisciplinary field where the material, developed by material scientists and engineers, has to be validated and must perform its task inside the human body, under the expertise of physicians and biologists; the final outcome must be analyzed and coordinated by all the intervening scientists.

The process itself is very long because it starts when a specific need is identified, then the idea of a potential implant is developed, and it ends with the final insertion of the implant in a patient. Several stages have to be verified: material synthesis, design, and manufacture of the prosthesis, combined with multiple material tests. It must also fulfill all regulatory requirements before its application to patients. The complete process is depicted schematically in Figure 1.4.

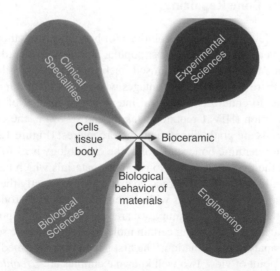

Figure 1.3 *Interactions between the different disciplines contributing in the field of bioceramics*

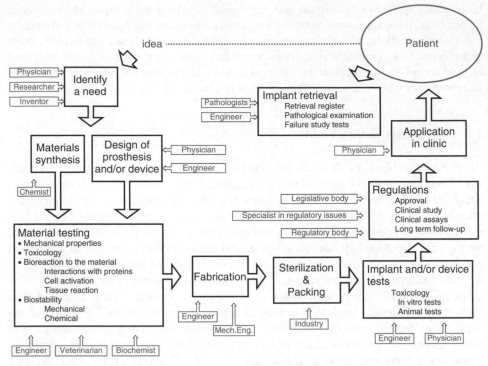

Figure 1.4   *Schematic view of the different stages to from the idea for a new bioceramic to the point where said material can be used in humans*

1.5   Solutions for Bone Repairing

We may begin by reviewing the present solutions for bone repairing, those which are being applied today. Not very long ago, the most popular solutions involved the use of natural materials, using bone from the patient himself, or from a bone donor bank, or even from animals. However, there are some disadvantages when using natural materials: in the first case, the patient has to endure two surgical interventions instead of one, and there are general risks of infection (HIV, Creutzfeldt-Jakob, and so on) in the other cases. This is why artificial materials are gradually attracting more interest (Figure 1.5).

When searching for ceramic bone substitutes, the chronology is as follows: It all started in the 1950s, when the first attempt was to use inert materials which had no reaction with living tissues. Later, in the 1980s, the trend changed toward exactly the opposite; the idea was to implant ceramics able to react with the environment and produce newly formed bone. And now, in the twenty-first century, we are searching for new porous ceramics that act as scaffolds for cells while hosting certain molecules able to drive self-regeneration of tissues. Let us now analyze this situation: The first generation is formed by *inert ceramics*. From the chemical point of view, two well known examples are *zirconia* and *alumina* (see Chapters 6 and 7). However, these ceramics, in a similar way to what happens with metallic and polymeric biomaterials, provoke foreign body reactions. Therefore, and although they

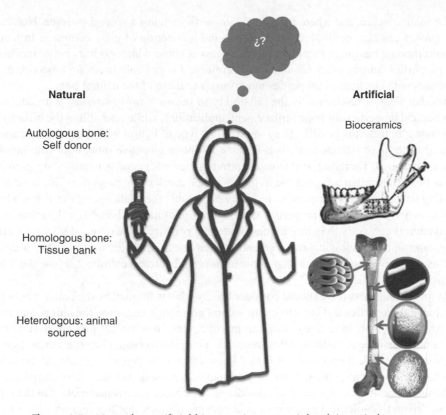

Natural

Autologous bone:
Self donor

Homologous bone:
Tissue bank

Heterologous: animal
sourced

Artificial

Bioceramics

Figure 1.5 *Natural or artificial bioceramics to repair hard tissue in humans*

are biocompatible, the body will react against them due to their foreign nature; the implant will be then surrounded by an acellular collagen capsule which isolates it from the body. In this way, the material will never transform itself into bone, and its artificial nature prevails. The search for *bioactive ceramics* yielded promising results in the 1980s. These ceramics can react with the physiological fluids forming biological-type apatite as a by-product; in the presence of living cells, this apatite can form new bone. Among these ceramics we can mention calcium phosphates, and several compositions of glasses and glass ceramics. For medical applications, these materials are provided in the following forms: *powder, porous pieces, dense pieces, injectable mixtures and cements, and coatings*. They have excellent features in terms of biocompatibility and bioactivity, but their mechanical properties are very poor (see Chapters 3–5, 8–10).

The *bioactive glasses*, when in contact with a simulated acellular physiological fluid, react to form an apatite that closely resembles the biological apatite found in bones. But if we perform the same process under *in vivo* conditions, hence in the presence of living cells, what is actually obtained is newly formed bone. Although their bioactivity is excellent, the great problem of glasses is that their mechanical properties are very poor, rendering it impossible to use them in the repair of large osseous defects. However, these glasses have an excellent field of application in the filling of small defects, where the rate of regeneration is the main concern, and where mechanical properties are just a secondary issue. However, the glasses can also be used as precursors in the production of glass ceramics. In fact, a certain thermal treatment to a glass yields a glass ceramic which exhibits better mechanical properties, among other advantages. Therefore, it is possible to obtain bioactive glass ceramics with mechanical properties much closer to those of the natural bone.

Another field of discussion is the possibility to obtain *hybrid materials* with adequate mechanical properties for bone replacement applications, while resembling the bioactivity of glasses. In fact, it is possible to synthesize this type of hybrid with mechanical features similar to those of natural bone. It is feasible to obtain bioactive monoliths with molded shapes and sizes. Therefore, still from the point of view of second generation bioceramics, it has been shown that it is possible to improve their mechanical properties. We have been dealing so far with dense materials. In fact, we have met the challenge of the 1980s which was to achieve mechanical properties similar to those of natural bone. But the needs have evolved; it is necessary to induce in these materials porosity in the range of microns so that they can fulfill physiological requirements in their use as scaffolds for tissue engineering. In this sense, we need methods that allow us to obtain porous scaffolds keeping the small particle size of the ceramics.

At this point, perhaps we should consider briefly what is implied by the balance between mechanical properties and bioactivity. In second generation ceramics, the aim was primarily to improve their bioactivity, while trying to achieve mechanical properties similar to those of natural bone. Following this route, it is possible to obtain bioactive ceramics with improved mechanical properties. It is also known that more dense materials exhibit better mechanical properties, although the transformation into bone becomes less complete. As a consequence, it was necessary to consider the option of porous materials. On the other hand, driven by biological requirements, it became obvious that the ceramics have to be porous; and such porosity has to exhibit a hierarchical structure. All these ideas will be discussed in detail throughout this volume. We may recall here the concept of porosity and its range of order. Those materials with mesoporosity between 2 and 50 nm are of interest for applications where drugs or biologically active molecules are loaded, and later released to help in the bone regeneration process. Macroporous materials, where the pore sizes are on the order of microns, are adequate as scaffolds for tissue engineering. In Figure 1.6 it can be observed how the fabrication of scaffolds for tissue engineering requires choosing a method that yields pieces with interconnected porosity and pores in the 2–1000 μm range.

Nowadays, reviewing the research on porous inorganic materials for medical applications two large subject areas can be highlighted: *design and fabrication of scaffolds*, and development of *substrate materials eligible for loading and releasing drugs or biologically active molecules* in a controlled manner, within the pharmacology field. Both fields interact deeply in many situations, which indicates the interdisciplinary research necessary in this area. However, even if these are not the only research lines currently in use for these kinds of materials with biomedical applications, their relevance is very remarkable. We know that the reactivity of solids begins on their surface. This general statement is of particular importance in the field of bioceramics, since they will be in contact with a wet medium and in the presence of cells and proteins. Depending on the function to be performed, bioceramics can be manufactured from very different materials.

This is in short the path followed to reach third generation bioceramics. The main aim now is to obtain porous ceramics that act as scaffolds for cells and inducting molecules,

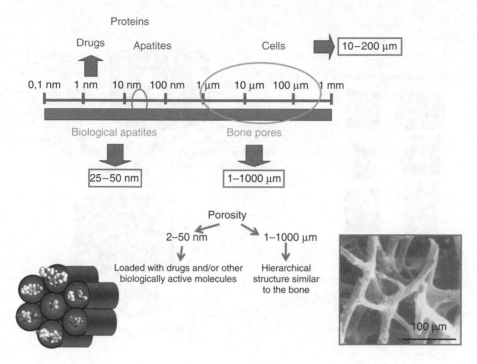

Figure 1.6 *Layout of orders of magnitude involved in bioceramics*

able to drive self-regeneration of tissues. With these requirements listed in Figure 1.7 it would be possible to keep using second generation bioceramics with added porosity. However, next generation ceramics could also be devised, with porosity values in agreement with biological requirements (see Chapters 10 and 11). As starting materials nanometric apatites could be used, shaped in the form of pieces with interconnected and hierarchical porosity, within the micron range. At present, the aim is to find bioceramics which induce the regeneration of hard tissues stimulating the response of the cells involved. Figure 1.8 shows the three basic pillars in tissue engineering. The requirements for these ceramics are to act as a scaffold and also to be porous so that the cells can do their job. This porosity implies a certain sacrifice of their mechanical properties. Some sort of "smart" behavior is also required, so that they can modify their properties in response to certain stimuli.

In some cases it is also required to allow the loading of biologically active molecules onto such ceramics. Therefore, the first step would be to find methods that yield pieces with interconnected porosity, with porosity in the range of microns, and this must be possible with all the bioceramics previously discussed. Nowadays, there are several methods which allow one to obtain pieces at room temperature. Moreover, working at room temperature allows one to include biomolecules of interest in many cases to treat different diseases, or to improve the treatment of various bone pathologies.

An important challenge is to design materials that can help the human body to improve its regeneration features, not only recovering the structure of the damaged tissue, but also its function (this will be covered in Chapters 10 and 11) (Figure 1.8).

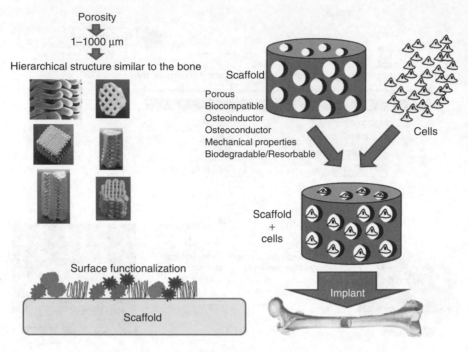

Figure 1.7   *Scaffolds and the requirements to be met*

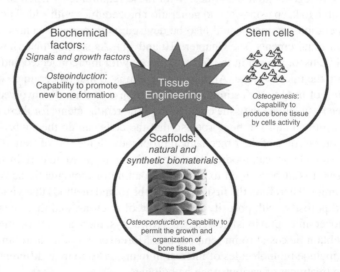

Figure 1.8   *The three cornerstones in tissue engineering*

1.6 Biomedical Engineering

Biomedical engineering is a recently created discipline. By general agreement, it is commonly defined as the application of engineering knowledge to life sciences. It combines design criteria in engineering with analysis tools from mathematics, physics, and chemistry in order to solve issues in medicine, biology, biotechnology, pharmacy, and so on.

Biomedical engineering has been growing as a multidisciplinary field among well-established sectors, such as engineering, physics, or mathematics, although it is currently acknowledged as a field of knowledge per se, hence its increasing importance from both academic, scientific, and professional points of view.

We find three related disciplines under the term of biomedical engineering, with specific differences according to their engineering origin:

- *Chemical engineering*: This section of biomedical engineering is focused on the design of biomaterials, tissue engineering, genetic engineering, pharmaceutical engineering, nanoparticle design and applications, design of novel drug transport and activation systems, and so on.
- *Mechanical engineering*: This is focused on biomechanics, prosthesis design and analysis, fluid mechanics in medical and laboratory devices, rheology, tribology, and so on.
- *Information technology and communications engineering*: This is focused on the relation between biomedical engineering and the design of electronic laboratory and clinical devices, software development for research and clinical applications, signal analysis, image diagnostics, statistical data analysis, design of telemedical systems, applications of artificial intelligence systems, and so on.

Biomedical engineering applies concepts, principles, and methods of engineering to the resolution of challenges in biology and medicine. It is a strategic socioeconomic activity with a great impact in modern medicine. It encompasses many fields one of which is biomaterials (Figure 1.9), which includes the bioceramics.

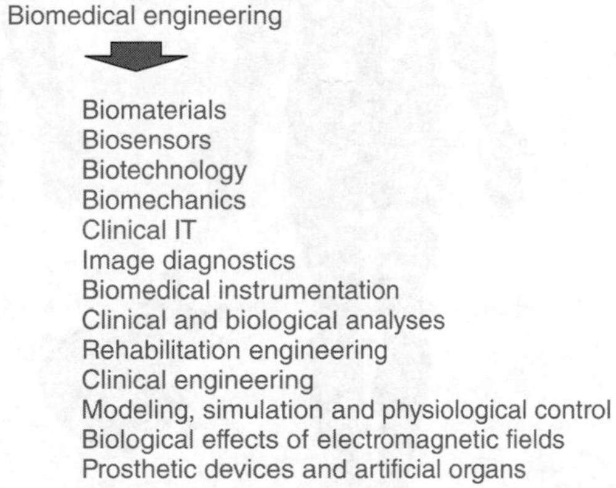

Biomedical engineering

Biomaterials
Biosensors
Biotechnology
Biomechanics
Clinical IT
Image diagnostics
Biomedical instrumentation
Clinical and biological analyses
Rehabilitation engineering
Clinical engineering
Modeling, simulation and physiological control
Biological effects of electromagnetic fields
Prosthetic devices and artificial organs

Figure 1.9   *Fields of interest in biomedical engineering*

Besides these bioceramics, new ones are being developed, each with specific design and manufacture requirements aimed at addressing a particular issue. These new bioceramics, still at a preliminary research stage, are not yet commercially available; there is still a long road to travel before a potential application in the human body. So far, they have been designed, manufactured, and tested *in vitro*, and some of them have also been tested *in vivo* in animals, but the stages to pass before any human application are still numerous. Many will not meet the standard requirements, or may not solve adequately the problem at hand. But others will succeed and, certainly, will represent a breakthrough in patients' health, both in terms of quality and quantity. Research, development and regulatory tests are absolutely crucial to obtain these new bioceramics with real applications in the clinical field, as for any other biomaterial, and this path requires perfectly defined stages and long deadlines.

This book deals with both types of bioceramics; the so called "*traditional*", with real and well-proven applications in healthcare already, which will be described both in composition and structure, as well as through their applications; and, in contrast, the *new* bioceramics (described in Chapters 12–14), currently at the forefront of knowledge, specifically designed for particular missions, which will also be described in terms of design, manufacture, *in vitro* and *vivo* assays, while the assessment of their real clinical applications cannot be performed yet (Figure 1.10).

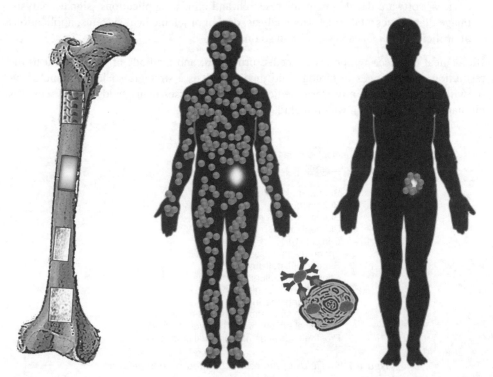

Figure 1.10   *Bioceramics to replace, repair, or regenerate bone nanoparticles for cancer treatment*

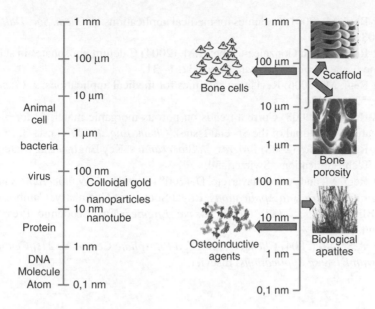

Figure 1.11   *The size scales of the various agents present throughout this book*

Before continuing with this book, it is important to bear in mind the sizes of the different components involved in the various chemical and biological processes present in the replacement, repair, or regeneration of bones. Figure 1.11 might be of help in this regard.

Recommended Reading

1. Frieb, W. and Werner, J. (2000) Biomedical applications, in *Handbook of Porous Solids* (eds F. Shüth, S. Kienneth, W. Sing and W. Kamps), Wiley-VCH Verlag GmbH, Weinheim, p. 2923.
2. Guelcher, S.A. and Hollinger, J.O. (2006) *An Introduction to Biomaterials*, CRC Press, Taylor & Francis.
3. Hulbert, S.F. (1993) *An Introduction to Bioceramics* (eds L.L. Hench and J. Wilson), World Scientific, Singapore.
4. Lanza, R.P., Langer, R. and Vacanti, J. (eds) (2000) *Principles of Tissue Engineering*, 2nd edn, Academic Press, New York.
5. LeGeros, R.Z. (1991) Calcium phosphates in oral biology and medicine, in *Monographs in Oral Science*, vol. **15** (ed H. Myers), Karger, Basel.
6. Park, J.B. and Lakes, R.S. (2007) *Biomaterials an Introduction*, 3rd edn, Springer, New York.
7. Planell, J. (ed) (2009) *Bone Repair Biomaterials*, Woodhead Publishing Ltd, Oxford, U.K.
8. Ratner, B.D., Hoffman, A.S., Schoen, F.J. and Lemons, J.E. (2013) *Biomaterials Science. An Introduction to Materials in Medicine*, 3rd edn, Academic Press, New York.
9. Castner, D.G. and Ratner, B.D. (2002) Biomedical surface science: foundations to frontiers. *Surf. Sci*, **500**, 28.

10. Vallet-Regí, M. (2001) Ceramics for medical applications. *J. Chem. Soc. Dalton Trans*, 97–108.
11. Vallet-Regí, M. and Gonzalez-Calbet, J.M. (2004) Calcium phosphates in substitution of bone tissue. *Prog. Solid State Chem*, **32**, 1–31.
12. Vallet-Regí, M. (2006) Revisiting ceramics for medical applications. *J. Chem. Dalton Trans*, 5211–5220.
13. Vallet-Regí, M. (2008) Current trends on porous inorganic materials for biomedical applications (Editorial of the Special Issue). *Chem. Eng. J*, **137** (1), 1–3.
14. Vallet-Regí, M. (ed) (2008) *Progress in Bioceramics*, Key Engineering Materials, **377**, Trans Tech Publications, Switzerland.
15. Vallet-Regí, M. and Arcos-Navarrete, D. (2008) *Biomimetic Nanoceramics in Clinical Use: From Materials to Applications*, Royal Society of Chemistry, Cambridge.
16. Van Blitterswijk, C. (ed) (2008) *Tissue Engineering*, Academic Press/Elsevier, London.
17. Wagh, A.S. (ed) (2004) *Chemically Bonded Phosphate Ceramics. 21st Century Materials with Diverse Applications*, Elsevier.

2

Biomimetics

María Vallet-Regí

*Departamento de Química Inorgánica y Bioinorgánica, Facultad de Farmacia,
Universidad Complutense de Madrid, CIBER de Bioingeniería, Biomateriales y
Nanomedicina (CIBER-BBN), Spain*

2.1 Biomimetics

For millions of years, living creatures have used and perfected biomineral-based materials with outstanding properties.

Nature has designed its materials and structures, which generate variability through mutation and recombination processes, and then the selection mechanisms favored the optimal solutions for each biological environment. Some examples of such processes would be *microskeletons, biomagnets, teeth, shells*, and *bones*.

The term *biomimetic* is generally used to describe the understanding of those solutions found by Nature to solve each challenge, and our use of them as a source of inspiration to solve in turn our technological challenges (Figure 2.1).

Biomimetics could therefore be considered as the technology transfer between Nature and the man-made artificial world.

We can also find many examples where man copied from Nature, such as the Lotus effect based on the self-cleaning ability of *Nelumbo nucifera* leaves, which allowed the design and development of self-cleaning surfaces which are nowadays present in many building materials.

Another example of biomimetics can be found in the study of particular features of certain insects and other creatures in terms of their nanostructure. For instance, lizards rely on nanostructure to move freely on vertical surfaces or to walk upside down. This is possible because the lizard's feet are coated with countless thin hairs, each of them ramified in spatula-ended filaments with dimensions in the nanometer range. These structures follow the surface roughness; hence the total contact surface of a lizard's foot is much larger than the equivalent surface on other species without said nanostructures.

Bioceramics with Clinical Applications, First Edition. Edited by María Vallet-Regí.
© 2014 John Wiley & Sons, Ltd. Published 2014 by John Wiley & Sons, Ltd.

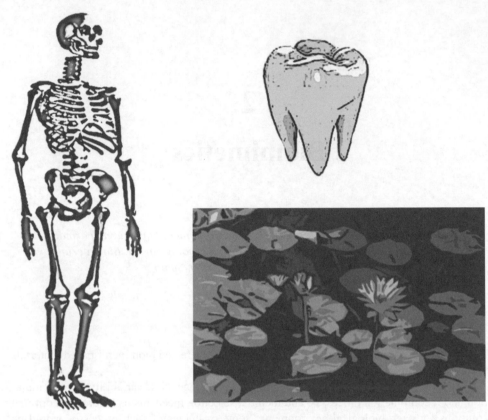

Figure 2.1   *In the kingdom of vertebrate animals, nature fabricates hard tissues (bones and teeth). In the botanical kingdom, we find instead non-adherent systems*

Homing pigeons or bees find their way over the earth's crust thanks to magnetite nanoparticles embedded in their tissues; Nature provides a beautiful model for nanobiosensors.

Additional examples can be found in butterfly wings and in the shells of certain beetles. The colors of those wings and shells are not due to pigmentation. The color stems from the vast amount of tiny nanostructured scales which reflect the incident light in different ways.

And finally, our very own *bones*, nanocomposite materials with such amazing mechanical properties that made them the source of inspiration in the fields of engineering and cement production for the building industry.

2.2   Formation of Hard Tissues

If we focus on how Nature deals with the production of hard tissue, the first conclusion would be that biomineralization processes mainly use calcium and silicon combined with carbonates, phosphates, and oxides. Bone, for instance, is formed by such biomineralization processes, natural sequences of physical-chemical reactions that yield the formation

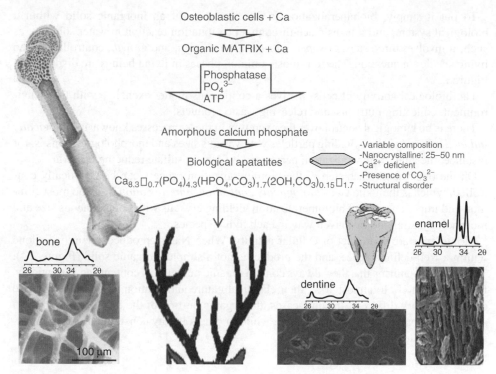

Osteoblastic cells + Ca

Organic MATRIX + Ca

Phosphatase
PO_4^{3-}
ATP

Amorphous calcium phosphate

Biological apatatites

-Variable composition
-Nanocrystalline: 25–50 nm
-Ca^{2+} deficient
-Presence of CO_3^{2-}
-Structural disorder

$Ca_{8.3}\square_{0.7}(PO_4)_{4.3}(HPO_4,CO_3)_{1.7}(2OH,CO_3)_{0.15}\square_{1.7}$

bone

26   30 34   2θ

enamel

26   30   34   2θ

dentine

26   30   34   2θ

100 µm

Figure 2.2   *Hard tissues in vertebrates are mainly formed by biological apatites (See insert for color representation of the figure)*

of hard tissues in vertebrates or protective tissues in invertebrates and inferior zoological species (see Figure 2.2). As a result, natural composites are formed, which often feature exceptional mechanical properties impossible to obtain from pure materials otherwise.

The inorganic phase of our bones is an apatite-like phase. Its structure has the special ability to accommodate several different ions in its three sublattices. Bone apatite can be considered basic calcium phosphate. Bones of vertebrate animals are organic–inorganic composite materials which can be simply described as follows: the inorganic component is carbonated and calcium deficient nonstoichiometric hydroxyapatite; these biological apatite crystals are nanometric in size, ranging from 25 to 50 nm. Such crystals grow at the mineralization sites of the collagen molecules, which are grouped together forming collagen fibers. Furthermore, a certain hierarchical bone porosity is necessary for several physiological functions performed by the bone.

2.3   Biominerals versus Biomaterials

A deep understanding of biomineralization and biomimetics is essential to design and obtain artificial biomaterials aimed at human body repair. The rules of Nature to produce biominerals are extremely different from those governing mineral growth on Earth's crust.

To put it simply, biomineralization is the formation of an inorganic solid within a biological system, and it takes place through a precipitation reaction in water solution; as such, it involves three steps of *supersaturation, nucleation*, and *growth*, controlled by the living species in question. The four most common phases in living beings are displayed in Figure 2.3.

The biological activity of cells involves a continuous matter exchange with their environment, collecting nutrients, and releasing waste products.

There is no biological control over the excretion in these processes known as *biologically induced mineralization*, yielding particles with various sizes and morphologies. This is, for instance, the kind of mineralization performed by certain sulfate reducing bacteria.

On the other hand, each step of the biomineralization process can be genetically controlled, which is then known as *biologically controlled mineralization*. It is the most common and important type of biomineralization yielding crystals with homogeneous size and morphology, arranged in a given way in each living species.

Biominerals are a product of cellular activity. When Nature produces them, it is done within a well-defined space, and the product is not a simple inorganic solid (Figure 2.4). The analysis confirms that they always contain organic matter; this content can vary widely, from less than 1% to almost 40%; their chemical nature and relation to the mineral phase can also be very different. In some cases, the organic portion of the biomineral may be a simple film coating an inorganic crystal, while in others it may constitute a matrix intrinsically linked to the mineral phase.

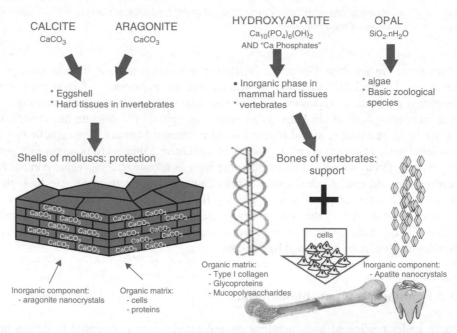

Figure 2.3 *The four most common inorganic phases in biomineralization processes. In all cases, they appear embedded in an organic matrix to form protection systems (molluscs, shells) or structural systems (vertebrates) (See insert for color representation of the figure)*

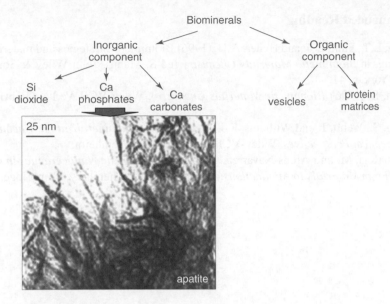

Figure 2.4 *Biominerals exhibit mainly two components, organic and inorganic, respectively*

Using common terms in materials science and engineering, most biomaterials can be defined as true composites. This feature is responsible for their optimal mechanical properties, unequaled by synthetic materials. The study of the structure of biomaterials and biomineralization processes not only provides important biochemical information, it can also lead to new strategies for scientists and engineers in the development of novel industrial materials (Figure 2.5).

For all of the above, it is crucial to understand the mechanism of mineralization in order to mimic it in material production processes and to obtain artificial devices able to replace, repair, or even better, regenerate damaged tissues in the human body.

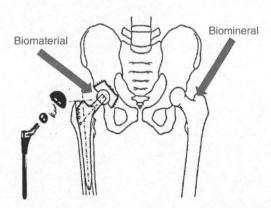

Figure 2.5 *Biomaterial versus biomineral. The artificial and the natural routes*

Recommended Reading

1. Kuhn, L.T., Fink, D.J. and Heuer, A.H. (1996) Biomimetic strategies and materials processing, in *Biomimetic Materials Chemistry* (ed S. Mann), John Wiley & Sons, Inc., New York, p. 41.
2. Mann, S. (1996) *Biomimetic Materials Chemistry*, Wiley-VCH Verlag GmbH, Weinheim.
3. Mann, S., Webb, J. and Williams, R.J.P. (eds) (1989) *Biomineralization, Chemical and Biochemical Perspectives*, Wiley-VCH Verlag GmbH, Weinheim.
4. Vallet-Regí, M. and Arcos-Navarrete, D. (2008) *Biomimetic Nanoceramics in Clinical Use: From Materials to Applications*, Royal Society of Chemistry, Cambridge.

Part II
Materials

3

Calcium Phosphate Bioceramics

Daniel Arcos
Departamento de Química Inorgánica y Bioinorgánica, Facultad de Farmacia,
Universidad Complutense de Madrid, CIBER de Bioingeniería, Biomateriales y
Nanomedicina (CIBER-BBN), Spain

3.1 History of Calcium Phosphate Biomaterials

Humans have always felt the necessity of repairing their aged and damaged tissues. From ancient times, the substitution and grafting of teeth and bones were a clinical practice registered in historic records as well as confirmed in archeological findings [1, 2]. The "ancient" bone and dental implants were commonly made of metals, wood, and/or xenografts, that is, manufactured from natural tissues of other species. Many of the calcium phosphate (CaP) bioceramics currently used for bone and dental grafting are of natural bone origin, so we could consider that this kind of implant has been around for a long time. Laying aside the use of natural teeth and bone, calcium phosphate materials with biomedical application are more recent. The first report on the implantation of an artificial calcium phosphate was in 1920. Albee [3] tested the stimulatory effect on osteogenesis, when a tricalcium phosphate (TCP)-based bioceramic was implanted into surgically created bone defects in rabbits. Many years later, in 1975, calcium phosphate materials broke into the field of periodontal surgery research [4] and, a few years later, Denissen and de Groot published their results in implementing hydroxyapatite (HA) cylinders as tooth root substitutes [5].

The 1980s can be considered as the golden age for the expansion of the use of calcium phosphates in clinical practice. The marketing of synthetic bioceramics based on calcium phosphates was consummated in this decade, thanks to the efforts of different research groups, mainly located in the Netherlands (de Groot *et al.* [6, 7]), the USA (Jarcho *et al.* [8]), and Japan (Aoki *et al.*) [9]. Moreover, HA implants of natural origin were also developed. Currently, these materials are reference bioceramics for dental and orthopedic applications, being marketed under different brand names by different companies. During the 1980s,

Bioceramics with Clinical Applications, First Edition. Edited by María Vallet-Regí.
© 2014 John Wiley & Sons, Ltd. Published 2014 by John Wiley & Sons, Ltd.

other CaP-based bone substitutes were incorporated into the field of bioceramics. These new materials offered the potential to be injected and molded *in situ* and were marketed as *calcium phosphate bone cements*. These biomaterials will be studied in depth in Chapter 8.

3.2   Generalities of Calcium Phosphates

The application of calcium phosphates for repair and reconstruction of hard tissues has been motivated by their chemical similarity to the mineral component of bones and teeth [10, 11]. As explained in Chapter 2, the mineral component of calcified tissues in mammals is made of nanocrystalline calcium deficient carbonate HA. This helps to understand the good biocompatibility of calcium phosphates without the mediation of foreign body response. Some calcium phosphates can even be integrated with the living organism, following a process very similar to living bone remodeling, that is, they *osteointegrate* with bone [12]. They provide a suitable surface to guide the bone remodeling and direct the bone growth by a process named *osteoconduction*. Finally, the possibility of producing *osteoinductive* calcium phosphates by means of specific ionic substitutions, such as silicon, has been also considered. *Osteoinduction* is the property exhibited by those materials that stimulate the mesenchymal stem cells (MSCs) along the osteoblastic pathway and lead to these cells in the formation of new bone [13]. In this sense, calcium phosphate bioceramics have broken into the field of regenerative medicine and into tissue engineering strategies.

The main limitation posed by calcium phosphates is common to almost all bioceramics: their poor mechanical properties. They are generally brittle, rigid, and have low wear resistance [14]. Therefore, their main applications are limited to filling small bone defects and as coatings of metallic prostheses and periodontal implants. To date, their application for the repair of large bone defects is very difficult, metallic materials being the best choice in these cases.

From the viewpoint of the anion phosphate, calcium phosphates can be classified into ortho (PO_4^{3-}), meta (PO_3^-), pyro ($P_2O_7^{4-}$), and poly ($PO_3)_n^{n-}$. However, since most of the calcium phosphate bioceramics are used as the ortho type, this chapter will focus on this group. Calcium phosphates are formed primarily by three major elements: calcium (in oxidation state +2), phosphorus (+5), and oxygen (−2) as a part of the orthophosphate anion. Furthermore, the chemical composition of many of them includes H, as in acid orthophosphate anions (HPO_4^{2-}, $H_2PO_4^-$), when incorporated in H_2O, or forming part of hydroxyl anions (OH^-) in the case of HA, $Ca_{10}(PO_4)_6(OH)_2$.

The general structure of calcium orthophosphates can be described as a network of PO_4 tetrahedral units, which give stability to the whole structure. Mostly they are fundamentally ionic compounds, sparingly soluble in water, soluble in acids, and insoluble in basic medium. The crystallography and chemistry of calcium orthophosphates allows different substitutions in both anionic and cationic positions. As a representative example, Figure 3.1 shows the crystalline structure of HA and some possible ionic substitutions. For example, Ca^{2+} ions can be easily replaced by cations of Sr, Ba, Mg, K, Na, Fe, and so on. Moreover phosphate anions can be partially replaced by carbonate groups (CO_3^{2-}), silicate (SiO_4^{4-}), and so on. In the case of HA, the OH^- groups may be replaced quite frequently by F^-, Cl^-, CO_3^{2-}, O^{2-}, and so on. Finally, it is noteworthy that the substitutions are closely linked to many nonstoichiometric calcium phosphates. Frequently, crystalline network positions

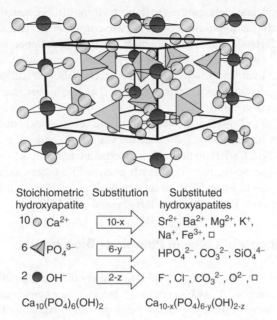

Stoichiometric   Substitution   Substituted
hydroxyapatite   hydroxyapatites

10 ○ Ca^{2+} [10-x] Sr^{2+}, Ba^{2+}, Mg^{2+}, K^+,
                            Na^+, Fe^{3+}, □

6 ◁ PO_4^{3-} [6-y] HPO_4^{2-}, CO_3^{2-}, SiO_4^{4-}

2 ● OH^- [2-z] F^-, Cl^-, CO_3^{2-}, O^{2-}, □

$Ca_{10}(PO_4)_6(OH)_2$ $Ca_{10-x}(PO_4)_{6-y}(OH)_{2-z}$

Figure 3.1 *Crystalline structure of hydroxyapatite and some of the most common ionic substitution occurring in the unit cell*

Table 3.1 *Calcium orthophosphates ordered by the Ca/P molar ratio*

Compound name	Formula	Ca/P molar ratio	Solubility 25 °C (g·l^{-1})
Tetracalcium phosphate (TTCP)	$Ca_4(PO_4)_2O$	2.0	0.0007
Hydroxyapatite (HA or OHAp)	$Ca_{10}(PO_4)_6(OH)_2$	1.67	0.0003
Calcium-deficient hydroxyapatite (CDHA)	$Ca_{10-x}(PO_4)_{6-x}(HPO_4)_x(OH)_{2-x}$	1.67 to 1.5	0.0094
β-Tricalcium phosphate (β-TCP)	β-$Ca_3(PO_4)_2$	1.5	0.0005
α-Tricalcium phosphate (α-TCP)	α-$Ca_3(PO_4)_2$	1.5	0.0025
Octacalcium phosphate (OCP)	$Ca_8(HPO_4)_2(PO_4)_4 \cdot 5H_2O$	1.33	0.0081
Dicalcium phosphate anhydrous (DCPA)	$CaHPO_4$	1.0	0.048
Dicalcium phosphate dihydrate (DCPD)	$CaHPO_4 \cdot 2H_2O$	1.0	0.088
Monocalcium phosphate anhydrous (MCPA)	$Ca(H_2PO_4)_2$	0.5	17
Monocalcium phosphate monohydrate (MCPM)	$Ca(H_2PO_4)_2 \cdot H_2O$	0.5	18
Amorphous calcium phosphate (ACP)	$Ca_xH_y(PO_4)_z \cdot nH_2O$, $n = 3–4.5$	1.2–2.2	–

are not fully occupied, leaving vacancies in those positions (represented as □ in Figure 3.1) and resulting in the formation of nonstoichiometric compounds, such as calcium deficient hydroxyapatites (CDHAs). In these compounds the ionic imbalance is balanced with the substitution of PO_4^{3-} groups by HPO_4^{2-} or CO_3^{2-}, or by generating additional vacancies at the position of the hydroxyl groups (OH^-). These crystal-chemical defects often lead to materials with enhanced biological responses, based on the higher reactivity derived from the presence of defects in solids.

There are many CaPs (Table 3.1 shows the most important ones) but only a few are used in medicine. As can be seen from Table 3.1, the Ca/P ratio and the solubility are closely correlated. Those with a Ca/P ratio below 1 are too acidic and soluble to be implanted [15]. HA and β-TCP are among the most commonly used, while other calcium phosphates can be part of biphasic mixtures (α-TCP) or as components of bone cements like dicalcium phosphate anhydrous (DCPA) or tetracalcium phosphate (TTCP).

3.3   *In vivo* Response of Calcium Phosphate Bioceramics

Calcium phosphates can be classified considering many different parameters. However, if we consider these compounds as bioceramics, that is, *as ceramic materials intended to be in contact with living tissues for health purposes*, then the classification in terms of their *in vivo* response seems to be most appropriate. Those CaP used for clinical applications are classified into *bioactive* and *resorbable* bioceramics. The reactivity of a CaP is dependent on its composition and structure [16]. One of the mechanisms underlying the phenomena of *in vitro* bioactivity is that dissolution from the ceramic produces solution-mediated events leading to mineral precipitation. Under *in vivo* conditions, the process involves more complicated biological reaction affecting cellular activity and organic matrix deposition [17, 18].

Hulbert *et al.* defined a *bioactive material* as "*one that elicits a specific biological response at the interface of the material which results in the formation of a bond between the tissues and the material*" [19]. The most representative bioactive calcium phosphate is HA. This dissolves slightly at the surface under physiological conditions, to a degree that depends on the chemical composition, porosity, and microstructure. Subsequently a new biological apatite is formed at the surface, followed by direct bonding with the bone tissue at the atomic level. Bioactive CaPs commonly exhibit osteoconduction, giving rise to good stabilization, providing a bioactive surface on which the bone can grow without implant resorption [20, 21]. The ability to bond to bone tissue is a unique property of *bioactive materials*. During this process, dissolution and precipitation reactions occur. Figure 3.2 shows schematically these phenomena, indicating the events occurring during the bioactive process onto a bioactive calcium phosphate coating. These events are summarized as follows:

1. Ionic dissolution from the ceramic, increasing the concentration of Ca^{2+} and orthophosphate species in the surrounding fluids.
2. Precipitation from solution of nanocrystalline and/or amorphous calcium phosphate (ACP) onto the ceramic.
3. Ion exchange and structural rearrangement at the ceramic/tissue interface. Nanocrystalline calcium-deficient carbonate apatite is formed.

4. Interdiffusion from the surface boundary layer into the ceramic.
5. Solution-mediated effects on cellular activity.
6. Deposition of either the mineral phase or the organic phase, without integration into the ceramic.
7. Deposition with integration into the ceramic.
8. Chemotaxis to the ceramic surface.
9. Cell attachment and proliferation.
10. Cell differentiation to osteoblastic phenotype.
11. Extracellular matrix formation.

The scheme displayed in Figure 3.2 does not represent a mechanism by itself, but only a description of observable events occurring at the interface after implantation. An important aspect of the overall reaction sequence between these materials and tissues is that dissolution, precipitation, and ion exchange reactions lead to a biologically equivalent apatite surface on the implanted material: *the in vivo bioactivity is only strongly expressed if this new calcium deficient carbonate apatite is formed* [22]. The events that constitute the bioactive process are commonly overlapped or occurring simultaneously, and the scheme displayed in Figure 3.2 should not be considered in terms of a time sequence.

Bioresorbable materials are those that dissolve after implantation, thus allowing newly formed tissue to grow into any surface irregularities but may not necessarily interface directly with the material. The most representative bioresorbable CaPs are TCPs, mainly the β-TCP polymorph. Resorbable CaP bioceramics are based on biological principles of bone repair, as natural bones repair and replace themselves throughout life. The main complications in the development of resorbable calcium phosphates are maintenance of the stability at the interface during the degradation period, as well as matching the resorption rate to the repairing one of the natural host tissues. Resorbable biomaterials must consist only of metabolically acceptable substances. In this sense, TCPs fulfill the requirements for resorbable hard tissue replacement when low mechanical solicitation is required [11].

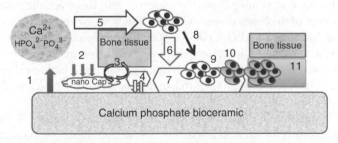

Figure 3.2 *Events occurring during bone formation onto bioactive CaP ceramics. (1) Dissolution from the ceramic. (2) Precipitation from solution onto the ceramic. (3) Ion exchange and structural rearrangement at the ceramic/tissue interface. (4) Interdiffusion from the surface boundary layer into the ceramic. (5) Solution mediated effects on cellular activity. (6) Deposition of either the mineral phase or the organic phase, without integration into the ceramic. (7) Deposition with integration into the ceramic. (8) Chemotaxis to the ceramic surface. (9) Cell attachment and proliferation. (10) Cell differentiation. (11) Extracellular matrix formation*

The factors that rule the *in vivo* degradation of CaP bioceramics can be summarized as *crystal-chemical, microstructural*, and *biological*. The *crystalline structure* and *chemical composition* of ionic compounds determine the solubility product (Kps) for a specific environmental pH. Thus, CaP bioceramics are more soluble under acid conditions. The dissolution rate increases as the Ca/P ratio decreases, as can be observed in Table 3.1. Thus, TCP (both α or β) with a Ca/P ratio of 1.5 are fully resorbed *in vivo* in a few weeks (depending on the implant size and location), whereas stoichiometric HA with Ca/P of 1.67 can remain in the tissue for years. The *microstructure* of any solid greatly influences its reactivity. In fact, any solid starts to react from the microstructural defects. Grain boundaries, pores, and surfaces are the best attack points for the biological environment to start with. Those CaP-based implants with high pore volume, small grain sizes, and sintering defects are likely to degrade faster than highly crystalline dense pieces. Finally, macrophage-mediated phagocytosis always occurs as part of the non-specific response to foreign body invasion. This inflammatory response is associated with a pH decrease, as a consequence of the macrophage enzymatic payload drop over the material. This pH decrease facilitates the CaP dissolution during the first stages after implantation.

In general terms, *the implant solubility enhances the bone repair process* in the case of CaP bioceramics [23, 24]. This does not mean that only highly soluble CaP are useful for bone repair; calcium phosphate with higher solubility should be implanted in those scenarios where the implant resorption is desired, followed by bone colonization, whereas less soluble calcium phosphates are intended as osteoconductive materials, providing a bioactive surface that supports bone growth without dissolving, with better mechanical stability during the first stages of the repair process.

3.4   Calcium Hydroxyapatite-Based Bioceramics

HA, $Ca_{10}(PO_4)_6(OH)_2$ is the most widely used calcium phosphate for dental and skeletal implantology [25–28]. As mentioned above, the main reason for its biocompatibility and bioactive behavior is its similarity to the mineral component of bone and teeth. Therefore, the use of HA somehow responds to a biomimetic concept of biomaterials application: let us provide to the bone something similar to what was lost. However, the description of the mineral component of the bones given in Chapter 2 does not correspond to a stoichiometric compound with formula $Ca_{10}(PO_4)_6(OH)_2$. Actually, biological apatites in mammals are calcium-deficient, carbonated, ionically substituted and, consequently, nonstoichiometric. In addition, in contrast to that exhibited by ceramic compounds, biological apatites exhibit a nanocrystalline structure consisting of rod-shaped or elongated platelets, 20–30 Å thick and 200–400 Å long.

The aim of mimicking biological apatites has resulted in different calcium HAs, with chemical and microstructural characteristics resembling those of biological ones. In addition to apatites of animal origin, CDHA, carbonated hydroxyapatites (CHAs) and nanocrystalline apatites have been incorporated to the market, mainly due to the development of low temperature synthesis methods and new methods to manufacture biological derived calcium phosphates. Finally, silicon-substituted hydroxyapatites (Si-HAs) have been recently marketed with better bioactive behavior than that exhibited by non-substituted ones. The next sections study these kinds of HA from a crystal-chemical perspective and their relevant properties as bioceramics.

3.4.1 Stoichiometric Hydroxyapatite (HA)

3.4.1.1 *Chemical and Structural Properties*

Stoichiometric HA with composition $Ca_{10}(PO_4)_6(OH)_2$, is a low soluble basic calcium phosphate with a Ca/P ratio of 1.67. The structure of HA is depicted in Figure 3.1 and can be described as a hexagonal unit cell with space group P_63/m and lattice parameters $a = 9.432$ Å and $c = 6.881$ Å, having one formula unit $Ca_{10}(PO_4)_6(OH)_2$ per unit cell [29]. The OH^- anions and four Ca^{2+} cations at Ca(I) sites lie along columns parallel to the c axis. The hydroxyls are sited along the c axis and the O-H bond direction is parallel to it, without straddling the mirror planes at $z = 1/4$ and 3/4. The remaining six Ca^{2+} cations, positioned at Ca(II) sites, are associated with the two hydroxyl groups in the unit cell, where they form triangles centered on, and perpendicular to, the OH axis and lying on the mirror planes. The PO_4 tetrahedrons form the remaining basic structural unit of HA.

The physical properties and *in vivo* performance of HA depend strongly on their synthesis route. The main preparation route of HA is aqueous precipitation methods, although solid-state processing is also widely used to prepare dense HA implants. Properties such as surface area and crystallite size are very sensitive to the temperature and to the synthesis route (dry or wet), thus determining the material *bioreactivity*. The bioactive behavior can be improved by introducing some substitutions into the structure. Actually, the stoichiometric apatite can incorporate a wide variety of ions that can affect both cationic and anionic sublattices. The substitutions incorporate cations with the same oxidation state as Ca^{2+}, such as Sr^{2+}, Pb^{2+}, Mg^{2+}, and so on [30–33], and anions with the same oxidation state as OH^-, such as F^- or Cl^- [34] at the 4e Wyckoff position of the HA unit cells. Ionic substitutions with different oxidation states, such as Na^+, K^+, CO_3^{2-}, and so on, are also very common [35–37]. Small crystal-chemical changes in the hydroxyl sites have important consequences for the reactivity of the apatites, thus leading to significant biological behavior. The 4e Wyckoff position (or OH position) is placed on the c axis at 0.25 and 0.75, surrounded by Ca(II) triangles, as shown in Figure 3.3. The fluorine anion is small enough to be placed in the triangle plane, thus optimizing the ionic interactions with the Ca(II). This stronger ionic bond results in lower solubility of fluorapatite (FHA) compared with HA. On the contrary, in the case of chloroapatites (ClHA), Cl^- anions are placed out of the trigonal plane and at a greater distance than OH^- in HA. Consequently, the solubility of these compounds is ordered as FHA < HA < ClHA.

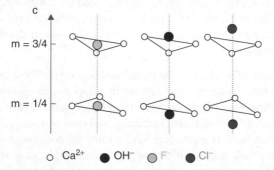

Figure 3.3 *Scheme of the ions along the c axis in fluorapatite (left), hydroxyapatite (center), and chloroapatite (right)*

3.4.1.2   Preparation Methods

Solid State Reaction.   Solid state reactions involve long treatment periods and high temperatures to improve the diffusion of the atoms throughout their respective solid precursors, reaching the interface where the reaction is occurring. Keeping the stoichiometry of pure HA requires some caution when it is prepared at high temperature by solid-state methods. For instance, small deviations in the weights of raw materials (that could shift the Ca/P ratio from 1.67) result in the occurrence of TCP (α or β polymorphs) during the thermal treatment.

The most common reactants for the solid state synthesis are $CaHPO_4 \cdot 2H_2O$ (DCPD (dicalcium phosphate dihydrate), brushite) and calcium carbonate ($CaCO_3$). Commercial DCPD contains a small and non-quantified amount of the anhydrous form DCPA (monetite, $CaHPO_4$), thus resulting in weighing errors. To avoid this, the mixture DCPD + DCPA can be treated at 1000 °C for 10 h or longer, obtaining β-calcium pyrophosphate (β-$Ca_2P_2O_7$), which is reacted with the appropriate amount of $CaCO_3$ as follows to obtain HA:

$$3\,Ca_2P_2O_7(s) + 4\,CaCO_3(s) + H_2O\,(g) \rightarrow Ca_{10}(PO_4)_6(OH)_2(s) + 4CO_2(g)$$

For HA synthesis, an appropriate thermal treatment consists of repetitive annealing at 1050 °C for 48 h with an intermediate milling process before the final sintering stage. The starting precursors commonly exhibit a particle size of several micrometers, which is the order of magnitude of the particle size of any chemical salt-type compound in commercial form. This particle size may roughly correspond to a succession of 10 000 unit cells, which is the diffusion route to be covered by ions of each one of the reactants in order to reach the interface of the other reactant. Therefore, the kinetic obstacles greatly restrict the ionic diffusion, rendering impossible in many cases the complete solid–solid chemical reaction. All cases require starting products in stoichiometric amounts, submitted to milling and homogenizing processes. The obtained starting mixture must be submitted then to generally very high temperatures and long treatment times. This procedure results in very crystalline apatites as opposed to biological ones.

The degree of hydroxylation of stoichiometric HA is another important topic that must addressed, especially when HA is prepared at high temperatures. HA loses hydroxyl groups at temperatures above 900 °C as a consequence of the loosening of constitutional water. In fact, during the synthesis by solid-state methods, HA becomes oxyapatite, $Ca_{10}(PO_4)_6O$ at 1300 °C and subsequently is partially re-hydroxylated during cooling to oxyhydroxyapatite, $Ca_{10}(PO_4)_6(OH,O,\square)_2$, where $\square$ means vacancies at the OH sites [38]. The preparation of fully hydroxylated HA can be carried out by carrying out the thermal treatment in steam at 0.1 MPa pressure [39].

Precipitation from Basic Solutions.   Wet route methods are the most usual strategy to prepare stoichiometric HA. Since the reactants are in solution, chemical reactions take place at a much lower temperature. Consequently, the yielded solids exhibit smaller crystallite sizes than those prepared by solid state reactions, and the HA microstructure can be more easily controlled. Moreover, HA powders have a more homogeneous distribution without the dehydroxylation problem of the high temperature methods. Pure stoichiometric HA is not directly obtained from aqueous systems. However, it can be prepared when apatites are precipitated from very basic solutions (pH > 11) and thereafter sintered above 900 °C [8, 40]. In this way, precipitated HA powders exhibit higher reactivity and the fabrication of the final product (grains or blocks) requires lower temperature and shorter heating periods compared with those prepared by solid state reactions.

Methods based on precipitation from aqueous solutions are most suitable for the preparation of large amounts of apatite, as needed for processing into ceramic bodies. There are different parameters that must be strictly controlled, otherwise the reproducibility of the method is seriously affected [41, 42]. Problems can arise due to the usual lack of precise control of the factors governing the precipitation, pH, temperature, Ca/P ratio of reagents, and so on, which can lead to products with slight differences in stoichiometry, crystallinity, morphology, and so on, that could then contribute to different "*in vivo/in vitro*" behaviors. Hence, it is important to develop a methodology able to produce massive and reproducible quantities of apatite, optimized for any specific application or processing requirements by controlling composition, impurities, morphology, and crystal and particle size. For quantitative reactions in solution, the reactants must be calcium and phosphate salts with ions that are unlikely to be incorporated into the apatite lattice, such as NO_3^- and NH_4^+. Therefore, a suitable reaction could be

$$10Ca(NO_3)_2.4H_2O + 6(NH_4)_2HPO_4 + 8NH_4OH \rightarrow Ca_{10}(PO_4)_6(OH)_2$$
$$+ 20NH_4NO_3 + 6H_2O.$$

The effects of varying the concentration of the reagents, the temperature, reaction time, initial pH, aging time, and the atmosphere within the reaction vessel can also be controlled with a device like that represented in Figure 3.4. The as-precipitated apatite powders can exhibit crystal sizes similar to those of human bones and enamel.

3.4.1.3   *Sintering and Mechanical Properties of HA Bioceramics*

Regardless of HA powder preparation, most of the stoichiometric HA bone grafts require a sintering process to achieve their final clinical presentation (commonly dense or porous grains and blocks). The sintering process involves temperatures above 900 °C for several hours. During this process, moisture, carbonates, and residuals such as ammonia or nitrates are removed. The sintering process also involves chemical and microstructural transformation, thus increasing the crystal size and decreasing the surface area. In the fabrication of HA blocks, sintering plays a fundamental role regarding mechanical properties since it causes toughening and increases the mechanical strength. Temperatures between 900 and 1000 °C are enough to remove residuals, to create pores by porogen decomposition and also to induce crystal growth. If HA powders are prepared by precipitation methods, this temperature interval can induce particle coalescence but with insufficient densification to get suitable mechanical properties. Pressurized powders followed by thermal treatments between 1000 and 1200 °C can result in blocks and grains with skeletal densities close to the HA theoretical one. Temperatures above 1250 °C make HA chemically unstable by dehydroxylation, while mechanical properties can be decreased because of exaggerated grain growths.

HA is a brittle ceramic compound constrained to be applied in non-bearing areas. Despite its application as a coating of metal prostheses, HA-based implants exhibit weak mechanical properties even after the thermal treatments. Table 3.2 shows the mechanical parameters of dense HA together with those of bone. As can be seen, HA behaves as a brittle ceramic with low reliability to act as a supporting material in load-bearing skeletal areas. In addition, porosity exponentially decreases the mechanical strength. As most ceramics,

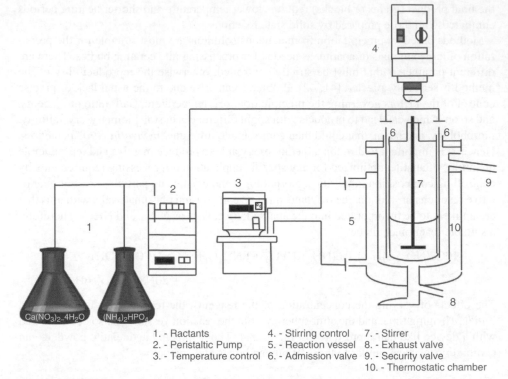

1. - Ractants   4. - Stirring control   7. - Stirrer
2. - Peristaltic Pump   5. - Reaction vessel   8. - Exhaust valve
3. - Temperature control 6. - Admission valve   9. - Security valve
                                                 10. - Thermostatic chamber

Figure 3.4 *Controlled crystallization equipment for the preparation of calcium phosphates by wet route synthesis*

Table 3.2 *Mechanical properties for sintered dense hydroxyapatite and bone*

Material	Tensile strength (MPa)	Yield strength (MPa)	Compressive strength (MPa)	Elastic modulus (GPa)	Fracture toughness (MPa m$^{1/2}$)
Hydroxyapatite	115–120	–	350–917	80–110	1
Cortical bone	78–196	30–70	160–250	4–30	2–12
Cancellous bone	10–20	–	23	0.2–0.5	–

HA has high compressive strength and comparatively similar tensile strength to cortical bone. However, the fracture toughness of HA is far below that of this tissue. Although bone contains a considerable amount of the apatite phase, the collagen fibrils greatly improve these parameters and prevent catastrophic mechanical failure. Finally, the higher stiffness of HA could also be a problem in a load bearing application, leading to stress shielding by a mismatch with the surrounding bone tissue.

3.4.1.4 *Porosity in HA Bioceramics*

Porous blocks and granules are often used for bone grafting and reconstruction. The presence of macropores enhances the contact surface and improves both the mechanical fixation and the chemical bond derived from the bioactive behavior. The pore size and the pore connectivity determine the cell migration and adhesion, as well as the bone ingrowth capability. Moreover, macropores must be large enough and interconnected to allow blood vessel formation (angiogenesis), otherwise the new bone tissue will not be viable in the medium term [43].

Porous HA is an excellent osteoconductive biomaterial. A good design of a macroporous arrangement guarantees success in many clinical scenarios that require bone grafting and reconstruction. This porosity is achieved by mixing the HA powders with volatile materials (for instance naphthalene, polymethylmethacrylate paraffin, sucrose, etc.), which act as porogens. These porogens are intended to leave interconnected macropores during a subsequent thermal treatment, with sizes larger than 100 μm and 70% in pore volume. The pore-forming additives are added to the ceramic slurry made of HA powders. Thereafter, the mixture is shaped into the block morphology by pressing into dies, slip casting, gel casting, and so on or any other of the shaping methods developed by the ceramic industry. The sintering process must be carefully tailored to achieve the required porosity without cracking, taking into account that material shrinkage occurs while volatile species tend to push out from inside the ceramic body. Although the pore size can be easily controlled, the significant challenge is to tailor the pore connectivity. The scientific literature describes several preparation procedures of HA porous blocks. A wide perspective of these methods can be consulted in reference [44]. Porogens addition commonly results in randomly porous blocks that can also be applied as scaffolds for tissue engineering methods. The incorporation of solid free forms fabrication (rapid prototyping (RP) techniques) to the biomaterials field has resulted in the development of scaffolds with designed pore arrangements, interconnected, and even hierarchically porous structures. These scaffolds intended for tissue engineering will be studied below (Section 3.8.1 and Chapter 10).

There is agreement about the significance of macroporosity to improve the osteoconduction of HA implants. However, determining the best pore size for any specific dental or orthopedic application is a difficult task. Some *in vivo* studies reveal that different pore sizes (from 150 to 1220 μm) do not involve large differences [45]. The literature on this topic indicates that pore sizes between 200 and 500 μm are appropriate for osteoconduction and bone ingrowth [46, 47]. The occurrence of smaller pores, that is, those below 10 μm in size, is strongly related with the sintering process or the powder shaping procedure. These pores also determine the *in vivo* response, as they somehow govern the implant wettability, ion dissolution, and protein adhesion. Although mechanical properties are significantly affected, the presence of these pores clearly improves the bioactive behavior of HA implants [48].

3.4.1.5   *Clinical Applications*

Porous Granules.   The *in vivo* behavior of HA relies on cell-mediated mechanisms, osteoconductivity, and dissolution, which are related to the implant surface area. Porosity, as well as particle size, will therefore affect the degradation rate of the compound and the biomechanical properties of the implant. Thus, it is crucial to differentiate between dense and macroporous forms of HA for each specific clinical application. Porous HA exhibits higher osteoconduction and resorption rates but is mechanically weaker.

Stoichiometric HA is mainly used as cancellous bone substitute in the form of porous particles. Ceramics must have pores of at least 100 μm to allow bone ingrowth. Porous HA is typically produced by isostatic compaction of calcium phosphate powder with porogens. The porogen is subsequently removed, creating a porous structure that is then sintered. Most of the different commercial grafts are presented as granules with sizes up to 2 mm, where the structure is a porous scaffold characterized by an interconnecting, open macro-porosity resembling cancellous bone, with porosity of about 80% (see Figure 3.5). HA porous particles osseointegrate over the course of several months.

Porous HA granules are applied as bone defects fillers in oral-maxillofacial surgery, dentistry, implantology, and periodontology. This form of stoichiometric HA is indicated for bone grafting in defects after removal of bone cysts, augmentation of the atrophied alveolar ridge, sinus floor elevation, filling of alveolar defects following tooth extraction, filling of extraction defects to create an implant bed, defects after surgical removal of retained teeth or corrective osteotomies and other multi-walled bone defects of the alveolar ridge.

In orthopedic surgery, porous HA is implanted in combination with metallic fixation devices. In a typical surgical procedure using porous calcium phosphate particles on a long bone, the surgeon determines the amount of bone graft he needs and shapes the block to fit into the damaged area. The graft area is stabilized with a metal plate, screws, or some other form of internal fixation. This protects the bone graft area so it can grow strong and durable. Eventually, the CaP graft is replaced by bone, leaving the bone as strong as it was before it was fractured.

Porous Blocks.   Although most of the published work and surgical procedures have been on the particulate form, stoichiometric HA is also clinically available as solid or porous blocks. The blocks are more predictable in their use than the particles, and they are adaptable and versatile. For example, bulk materials, available in dense and porous forms, are also used for alveolar ridge augmentation, immediate tooth replacement, and maxillofacial reconstruction. In the field of orthopedic applications, HA blocks can be implanted as spacers in cases of spine fusion or in pelvic osteotomy in children [49]. These implants are mechanically stable and prevent pelvic deformities. Another application of HA porous blocks is as supporting material wedges after osteotomy in patients with osteoarthritis of the knee [50, 51]. In this case, implantation of HA blocks for reconstruction has provided osteointegration and a good clinical outcome [52].

Other examples of synthetic HA implants are orbital eyes [53]. Synthetic orbital HA implants are designed with interconnecting pores of 900–1100 μm in diameter. These pores

(a)   (b)

Figure 3.5 *Commercial porous granules of hydroxyapatite bone graft (a) and a scanning electron micrograph of one of these granules (b). Reproduced with permission from Zimmer Dental*

allow fibrovascular ingrowth in four to six months. When vascularization has occurred, implants may be drilled and pegged like the coral implants. Synthetic HA orbital implants must be light in weight (commonly less than 2 g) and are available in spherical and conoid shapes. HA orbital implants are porous to the extent of 75% and are tailored to provide fibrovascular growth when used for filling the volume of an empty eye socket following evisceration/enucleation surgeries.

Dense Hydroxyapatite.   Dense HA implants can be prepared as both grains and blocks. HA grafts are considered as dense (or nonporous), when they exhibit a porosity less than 5% by volume with maximum pore sizes less than 1 μm in diameter. Dense HA are prepared in blocks, complex shapes, and particulates in irregular or spherical shapes. The last ones are obtained by milling and rolling the compacted sintered blocks. The main uses of dense HA blocks are as immediate tooth root replacement, augmentation of the alveolar ridge for better denture fit, or in orthopedic surgery. In addition, ossicular prostheses have been used successfully as shown in Figure 3.6. Dense HA in particulate form is used as a filler in bony defects in dental and orthopedic surgery, as a filler in association with the placing of metal implants, or for repair of failing metal dental implants.

3.4.2   Calcium Deficient Hydroxyapatites (CDHA)

3.4.2.1   Chemical and Structural Properties

CDHAs are nonstoichiometric HAs, which exhibit a Ca/P molar ratio lower than 1.67. In biological systems, unsubstituted CDHA does not exist, as they always appear with cationic substitutions such as sodium or magnesium together with carbonates substituting

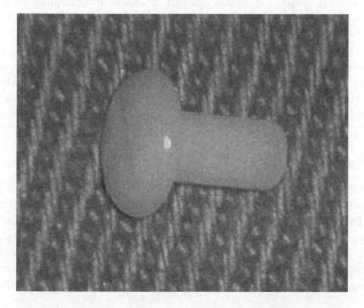

Figure 3.6   *Ossicular prosthesis made of dense HA*

for phosphates. In addition to the Ca deficiency, the low carbonate content generally found in biological CDHA contributes to the structural disorder of the CDHA.

Synthetic CDHA can be prepared from aqueous systems at temperatures between 25 and 100 °C. The crystalline structure is analogous to stoichiometric HA, with a unit cell belonging to the space group P_63/m, although CDHA shows small crystalline differences from HA, for instance a larger a-axis in the unit cell attributable to the presence of HPO_4^{2-} anions and water molecules thought to be in OH^- positions.

CDHA synthesis requires a pH below 12 in order to ensure enough amounts of HPO_4^{2-} anions, which must balance the Ca^{2+} deficiency mainly occurring at the Ca(2) sites. In this way, the simplest general formula for CDHA would be: $Ca_{10-x}(HPO_4)_{2x}(PO_4)_{6-2x}(OH)_2$, where x can vary from 0 to almost 2 [54]. However, other models of lattice substitutions in CDHAs consider vacancies at the OH^- positions. Here, general formulas $Ca_{10-x}(HPO_4)_x(PO_4)_{6-x}(OH)_{2-x}$ with $0 \leq x \leq 2$ and $Ca_{10-x-y}(HPO_4)_x(PO_4)_{6-x}(OH)_{2-x-2y}$ with $0 \leq x \leq 2$ and $y \leq (1 - x/2)$, fit better with the experimental results. This set of crystal-chemical changes leads to a higher initial dissolution rate of CDHA when compared to HA. The enhanced Ca^{2+} release would be a consequence of the weak bonding interaction of Ca^{2+} ions around the vacancies at the OH^- sites. Regarding phosphate release, CDHA also shows a higher dissolution rate compared with HA. The same factors contributing to Ca^{2+} release can explain the phosphate dissolution.

When CDHAs are treated above 700 °C, β-TCP formation occur,s resulting in β-TCP/HA biphasic materials, where the amount of β-TCP increases as the Ca/P ratio decreases. This CDHA decomposition has become one of the most used methods to obtain bioceramics based on biphasic calcium phosphates (BCPs), as will be described in Section 3.6.

3.4.2.2   Preparation Methods

Synthetic CDHA is obtained from aqueous systems, either by precipitation or hydrolysis methods at temperatures between 25 and 100 °C. As mentioned above, the initial pH must be lower than 12 and decreases as the synthesis progresses until reaching a final pH of 5. Synthesis of CDHA by the hydrolysis method can be carried out by dissolving $CaHPO_4$ in a calcium solution (calcium nitrate or calcium acetate) keeping the initial pH between 5 and 7 and stirring for several hours at 95–100 °C. CDHAs with a Ca/P molar ratio from 1.21 to 1.64 are also prepared by the titration of a saturated solution of $Ca(OH)_2$ with H_3PO_4; or by the hydrolysis of DCPD or of OCP (octacalcium phosphate) in Ca-containing solutions at 60–100 °C, with an initial pH of 7.

3.4.2.3   Clinical Applications

CDHAs have important clinical applications as they are one of the calcium phosphate-based cements (CPCs). CaP cements are low temperature bioceramics, the two major types being CDHA and brushite. Their main clinical applications are those requiring minimal invasive surgery. Taking advantage of the self-setting properties, CaP cements are applied as injectable and moldable bone substitutes. CDHA cements are commonly obtained by reaction between acid and basic CaP in the presence of aqueous solutions. These cements and those based on brushite compositions will be studied in depth in Chapter 8.

3.4.3 Carbonated Hydroxyapatites (CHA)

3.4.3.1 Chemical and Structural Properties

Biological apatites of bone and teeth contain carbonate anions, CO_3^{2-}, in their crystalline structure. This ionic substitution contributes to the higher solubility of biological apatites compared with the stoichiometric and highly crystalline synthetic ones. Thus, the presence of CO_3^{2-} in the HA structure helps to keep the constant bone regeneration through dissolution–crystallization cycles. Following this biomimetic argument, the preparation of synthetic carbonate HA has been one of the most widely studied topics. Mimicking biological apatites is a difficult task, but incorporating carbonate anions is possible by high temperature methods or by precipitation and other solution reactions. CHAs are classified as *A-type* and *B-type*, depending on the crystal site occupied by the CO_3^{2-}. In A-type CHA, carbonate anions are sited at the hydroxyl site, that is, on the sixfold screw. Since CO_3^{2-} anion is larger than OH^- the *a*-axis parameter is increased compared to stoichiometric HA.

The incorporation of CO_3^{2-} into the HA structure leads to distortions that result in microstrains around these crystal sites [55]. Differently to PO_4^{3-} and OH^- anions, which have tetrahedral and linear structures, respectively, CO_3^{2-} is a trigonal planar anion, thus stressing the crystal lattice when substituting any of these anions. Different models have been proposed to describe the orientation of CO_3^{2-} anions in PO_4^{3-} sites. As an example, polarized IR spectroscopy carried out on enamel HA has shown that the CO_3^{2-} ions have their planes oblique to the *c*-axis and are thought to occupy the sloping faces of tetrahedral sites [56]. These strains have a significant effect on the HA solubility. However, the higher solubility of CHAs can also be explained on the basis of the carbonate effect on the crystal size. Actually, from the same synthesis conditions (temperature, time, and pressure) CHAs exhibit smaller crystal size and increased surface area compared with stoichiometric HA. In addition, the $Ca^{2+}-CO_3^{2-}$ ionic interactions are weaker than $Ca^{2+}-PO_4^{3-}$, thus making the former more soluble in aqueous systems.

3.4.3.2 Preparation Methods

The best method to obtain A-type carbonate apatites is to heat HA in dry CO_2 atmosphere at $\sim 900\,^\circ C$ for some days. This reversible reaction can be described as:

$$Ca_{10}(PO_4)_6(OH)_2 + xCO_2 \leftrightarrow Ca_{10}(PO_4)_6(OH)_{2-2x}(CO_3)_x + xH_2O$$

When the CO_2 uptake is a maximum, A-type CHA is almost fully dehydroxylated. In this case the resulting A-type CHAs are given by the formula

$$Ca_{10-x}(PO_4)_6(CO_3)_{1-x}$$

In the case of B-type carbonate apatites, CO_3^{2-} substitutes for PO_4^{3-} reducing the *a*-axis parameter, whereas the *c*-axis parameter is increased [57]. The neutrality is reached by incorporation of single valence cations (Na^+ or K^+) in the Ca^{2+} positions [58, 59], represented by the general formula

$$(Ca_{10-x}Na_x)(PO_4)_{6-x}(CO_3)_x(OH)_2$$

B-type CHA can be prepared from aqueous systems by a direct precipitation method at $37-100\,°C$. As mentioned above, the $CO_3{}^{2-}$ for $PO_4{}^{3-}$ anion substitution is commonly coupled to Na^+ substitution for Ca^{2+}. The amount of $CO_3{}^{2-}$ incorporated depends on the carbonate content of the solution, which is often added as $NaHCO_3$ in concentrations ranging from 0.01 to 0.4 M (higher concentrations result in the formation of aragonite, $CaCO_3$). Hydrolysis of ACP, DCPD, DCP, or OCP in the same carbonate-containing solution is also a suitable method for the preparation of B-type CHA.

3.4.3.3   Clinical Applications

Since the discovery of the carbonate content in biological apatites, CHA has attracted the attention of several research groups. These bioceramics more closely resemble the mineral component of bones than stoichiometric ones. In addition, they are more reactive and bioresorbable due to the crystal-chemistry and microstructural characteristics explained above.

Sintered carbonate apatites have been proposed to fabricate blocks for bone grafting and reconstruction [60]. However, the high temperature significantly decreases the solubility and only under acid conditions (pH 5.5) is this material bioresorbable. In the 1980s, Impladent Ltd. marketed a synthetic bioactive resorbable graft, OsteoGen, based on CHA. This graft is intended for sinus elevations, filling tooth extractions and cyst defects, as well as repair of periodontal bone defects. Although keeping the crystalline structure of HA, this bioceramic is described as a non-ceramic crystal cluster that is highly hydrophilic, allowing the material to readily absorb liquid. The commercial production of this CHA avoids the thermal treatment and leads to a more porous and resorbable bone graft. With low temperature methods, CHA behaves as an osteoconductive material in that the highly porous crystalline clusters act as a slowly-resorbing matrix permitting the infiltration of bone-forming cells and the subsequent deposition of host bone [61].

3.4.4   Silicon-Substituted Hydroxyapatite (Si-HA)

3.4.4.1   Chemical and Structural Properties

In 1970 Carlisle demonstrated the importance of silicon in bone formation and mineralization [62, 63]. She reported the detection of silicon *in vivo* within the unmineralized osteoid region (active calcification region) of the young bone of mice and rats. Silicon levels up to 0.5 wt% were observed in these areas, suggesting that Si has an important role in the bone calcification process. On the other hand, the mechanism proposed for the bioactive behavior in SiO_2-based glasses [64, 65] involved silanol groups (Si-OH) acting as a catalyst of the apatite phase nucleation, and the silicon dissolution rate is considered to have a major role in the kinetics of this process [66, 67]. These events suggested the idea of incorporating Si or silicates into the HA structure in an attempt to enhance the *in vivo* performance [68, 69]. The term silicon-substituted means that silicon substitutes into the apatite crystal lattice and is not simply added. Silicon or silicates are supposed to substitute for phosphorus or phosphates, with the subsequent charge imbalance [70]. The amount of silicon that can be incorporated seems to be limited. The literature gives values ranging from 0.1 to 5% by weight in silicon [71, 72]. Small amounts of 0.5 and 1% are enough to yield important biomimetic improvements.

The amount of silicon incorporated also has an important influence on the thermal stability. Figure 3.7 shows the XRD patterns of SiHA with nominal formula

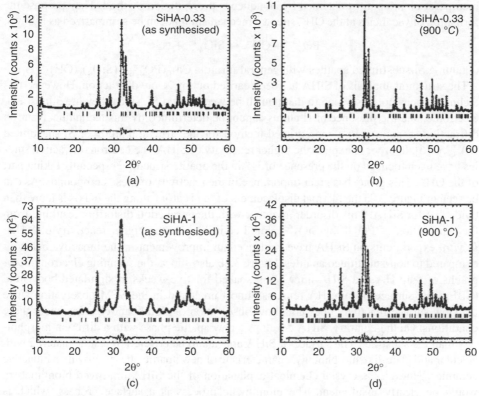

Figure 3.7 *XRD patterns collected for Ca$_{10}$(PO$_4$)$_{6-x}$(SiO$_4$)$_x$(OH)$_{2-x}$ apatites obtained by controlled precipitation method; (a) x = 0.33 without subsequent thermal treatment; (b) x = 0.33 treated at 900 °C for 2 h; (c) x = 1 without subsequent thermal treatment; and (d) x = 1 treated at 900 °C for 2 h. Reproduced with permission from Kobayashi et al, American Journal of Otolaryngology, Vol 23, No 4 (July–August), 2002: pp 222–22. Copyright © Elsevier, 2002*

Ca$_{10}$(PO$_4$)$_{6-x}$(SiO$_4$)$_x$(OH)$_{2-x}$, for x = 0.33 (Figure 3.7a,b) and x = 1 (Figure 3.7c,d). The as-precipitated samples (without thermal treatment) are always a single nanocrystalline apatite phase (Figure 3.7a,c). After heating at 900 °C, the samples with Si content of 0.33 remained as a single apatite phase, whereas higher Si content led to decomposition into HA and α-TCP [73, 74]. In fact, this is an appropriated method to obtain biphasic materials α-TCP-HA at relatively low temperature. α-TCP is a high-temperature phase that appears when HA or β-TCP is treated at over 1200 °C. The presence of silicon seems to stabilize the α-TCP at lower temperatures.

Silicon (or SiO$_4^{4-}$) for P (or PO$_4^{3-}$) is a non-isoelectronic substitution. This means that the extra negative charge introduced by SiO$_4^{4-}$ must be compensated by means of some mechanism, for example, creating new anionic vacancies. The Si, or SiO$_4^{4-}$ incorporation into the apatite structure at the P, or PO$_4^{3-}$, position has been studied by several authors. Gibson *et al.* [70] reported on the structure of aqueous precipitated SiHA. The main structural evidences reported were the decrease and increase of the *a* and *c* parameters, respectively, the absence of secondary phases, and the increase in tetrahedral distortion. They proposed a mechanism to compensate the negative charge introduced by the SiO$_4^{4-}$

incorporation in apatites obtained by an aqueous precipitation method. They propose the formation of vacancies at the OH⁻ site in a mechanism that can be summarized as follows:

$$PO_4^{3-} + OH^- \leftrightarrow SiO_4^{4-} + \square$$

obtaining Si-substituted apatites with general formula $Ca_{10}(PO_4)_{6-x}(SiO_4)_x(OH)_{2-x}\square_x$.

The structural analysis of SiHA has been carried out by X-ray diffraction. However, this technique does not allow one to distinguish between P and Si, since they are almost iso-electronic, and the presence of H atoms cannot be determined by this technique. Neutron diffraction seems to be an appropriated method for the structural study of Si-substituted HA [75, 76]. In order to explain the higher reactivity of SiHA, the neutron diffraction studies have been focused on the presence of H⁺ in the apatite structure, especially taking part of the OH⁻. This group has great importance in the reactivity of these compounds. As can be seen in Figure 3.8a the thermal displacement of the H atom along the *c*-axis is more than twice that for SiHA. This disorder, together with the tetrahedral distortion resulting from the substitution of PO_4^{3-} by SiO_4^{4-}, could contribute to the higher reactivity of SiHA. *In vitro* experiments on Si-HA have evidenced an improvement in the bioactive behavior compared to non-substituted apatites. Figure 3.8b also shows the scanning electron micrographs of pure HA and SiHA after being soaked for five weeks in simulated body fluid (SBF). The surface of pure HA remains almost unaltered in the SEM observation since the slow surface reactivity does not allow observation of significant changes under these conditions. On the contrary, SiHA develops a new apatite phase with a different morphology than the substrate. The surface of SiHA appears to be covered by a new material with acicular and plate-like morphology, characteristic of new apatite phases grown on bioactive ceramics. However, a crystal-chemical explanation of the SiHA improved biomimetism would be clearly insufficient. The biomimetic process is a surface process, which is enhanced by the material reactivity. The sum of the different factors may justify the enhanced reactivity. From the point of view of *crystalline structure*, silicon yields tetrahedral distortion and disorder at the hydroxy site, which could decrease the stability of the apatite structure and, therefore, increase the reactivity. From the point of view of the *microstructural level*, the changes are even more evident. Grain boundaries defects are the starting points of dissolution under *in vivo* conditions. There is a close relationship between the amount of silicon, the number of sintering defects at the grain boundaries and the dissolution rate. Particularly, the number of triple junctions in SiHA may have an important role in the material reactivity and, consequently, in the rate at which the ceramic reacts with the bone. Finally, the *surface charge* undergone by the ceramic due to the presence of SiO_4^{4-} would also play an important role for Ca^{2+} incorporation at the new biomimetic layer. This effect could also be responsible in part for the alteration in its biological response. Substituting silicate into HA boosts the bioactivity of the material by increasing the negative surface charge on the synthetic graft. The negative charge attracts circulating proteins, essential for bone growth, in greater numbers to the graft's surfaces, stimulating the production of more bone in less time. Summarizing, the understanding of the improved biomimetic behavior in SiHA should be considered as a sum of different factors at different levels.

3.4.4.2 *Preparation Methods*

The controlled crystallization method is, by far, the most common synthesis route to SiHA found in the scientific literature. This process comprises the reaction of a calcium

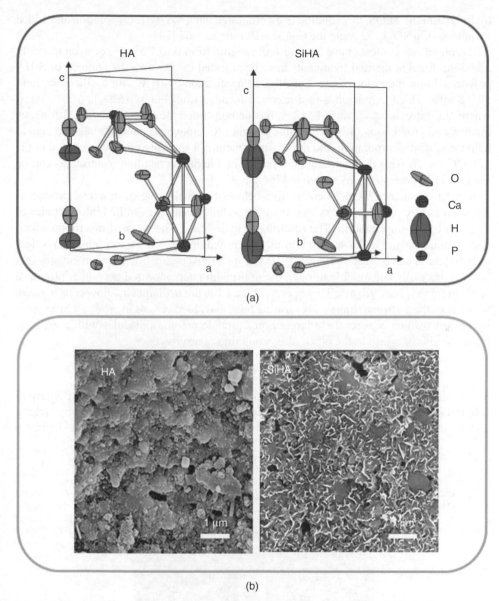

Figure 3.8 *(a) Anisotropic displacement of the atoms into HA and SiHA lattices. (b) Scanning electron micrographs of pure HA and SiHA after five weeks soaked in simulated body fluid (See insert for color representation of the figure)*

salt or calcium hydroxide $Ca(OH)_2$ with orthophosphoric acid (H_3PO_4) or a salt of orthophosphoric acid in the presence of a silicon-containing compound such as silicon acetate $(Si(CH_3COO)_4)$. In this case, the synthesis of SiHA is carried out maintaining the $Ca/(Si + P)$ ratio as 1.67. Under these conditions it is believed that the silicon-containing compound yields silicon-containing ions, such as silicate ions, which substitute in the apatite lattice. Alternative synthetic routes to incorporate Si into the HA structure use silicon alkoxides, for instance tetraethylorthosilicate (TEOS), $Si(OCH_2CH_3)_4$. After hydrolysis of the TEOS, this solution is incorporated into $(NH_4)_2HPO_4$ solution and added to aqueous $Ca(NO_3)_2$, keeping the $Ca/(Si + P)$ ratio as 1.67 [74].

Several silicon contents have been tested, ranging from 0 to 1.5 wt%, or even more. In addition, different thermal treatments have been tested to optimize the features of SiHA powders. From the point of view of clinical applications, SiHA with a silicon content of 0.8 wt%, which corresponds to a nominal formula $Ca_{10}(PO_4)_{5.7}(SiO_4)_{0.3}(OH)_{1.7}\square_{0.3}$, where $\square$ means vacancies at OH^- sites, has demonstrated the best *in vivo* level for rate, quality, and progression of bone healing. In order to remove all the residuals and also to help the silicon incorporation into the apatite structure, a sintering process is carried out at 1300 °C for 2 h. This thermal treatment results in a highly crystalline material, as can be seen by TEM observation as shown in Figure 3.9.

In order to enhance the osteoconductive behavior of SiHA, the commercial product is prepared as macroporous blocks by a foaming technique and then milled into particles of 1–2 mm before implantation. The resulting granules have a network of macropores with sizes around 100 μm and a second porous system within the walls (strut porosity) of less than 10 μm, as can be seen in Figure 3.10. The material's microstructure then ensures that bone grows easily and quickly through the graft, with interconnected pores that provide a scaffold for new bone growth. These pores, at least 100 μm in diameter, allow cells to move throughout the graft, enabling newly forming bone and blood vessels to grow. Smaller pores in the struts which connect these larger pores provide protein molecules with a pathway from one side of a strut to the other, also promoting bone growth.

3.4.4.3   Clinical Applications

SiHAs were launched to market under the trademark Actifuse and Actifuse ABX (Apatech, UK). As described above, clinically approved SiHA contains 0.8% silicon by weight, and the interconnected macro and microporous structure and enhanced surface chemistry

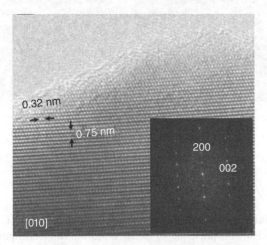

Figure 3.9 *HRTEM image and corresponding FT pattern for a silicon-substituted hydroxyapatite treated at 1300°C. Reprinted from Physica B: Condensed Matter, 350, 1–3, Arcos, Rodriguez-Carvajal and Vallet-Regi, Neutron scattering for the study of improved bone implants, E607 - E610, 2004, with permission from Elsevier*

Figure 3.10 *Scanning electron micrograph of a commercial Si-substituted HA showing (a) strut porosity and (b) a macroporous grain*

encourages the rapid formation of host bone and the growth of capillary blood vessels throughout the network of interconnecting pores. The clinical applications comprise spinal, orthopedic, periodontal, oral, and maxillocranial surgery. The macroporous SiHA particles are used as bone grafting filling bone defects (Actifuse™) or can be applied suspended in an aqueous gel carrier (Actifuse ABX™).

3.4.5 Hydroxyapatites of Natural Origin

HAs of natural origin are prepared from three kinds of sources: coral, algae, and treated bone. Commonly they are low crystalline materials with some carbonate content that keep, to some degree, the original porosity of the natural source. For these reasons they are more soluble than pure synthetic HAs and are successful bioceramics used clinically in dental, orthopedic, and ophthalmological applications.

3.4.5.1 Coralline HA

The rationale for the use of coralline HA as a bone graft is based on the porous exoskeleton of corals, which resembles the structure of the human bone [77]. Such a structure provides a relatively strong and natural matrix through which blood vessels and new bone tissue can grow. When this bone implant is placed in contact with the patient's viable bone, it creates a strong structure to support the healing tissues and stimulates new bone growth. As can be seen in Figure 3.11, coralline HA exhibits a macrostructure similar to human cancellous bone.

HAs derived from coral are prepared by hydrothermal treatment of the aragonite $CaCO_3$ phase of these species. The marine coral exoskeleton is hydrothermally converted into HA. The coral, which is primarily composed of aragonite calcium carbonate, is heated with $(NH_4)_2HPO_4$ at temperatures between 200 and 400 °C and pressures about 12 000 psi for 24–60 h to obtain a 95% HA material. This substance is processed into block or granular form and sterilized by gamma radiation. These "block" and "granular" forms allow surgeons to mold coralline HA into a compatible shape to fill defects in bones.

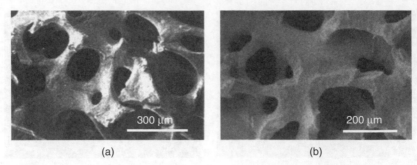

(a)   (b)

Figure 3.11   *Scanning electron micrographs of (a) cancellous bone and (b) coralline hydroxyapatite*

The final calcium phosphate ceramic often contains some Mg-substituted β-TCP as a consequence of the magnesium contained in coral. Coral-derived HA also exhibits carbonates substituting both phosphate and hydroxy groups (A and B) sites. Other products such as ProOsteon 500R consist of $CaCO_3$ coated by a thin layer of HA (2–10 μm) as a consequence of incomplete hydrothermal transformation. This biphasic composition allows the bone apposition onto the HA layer and a subsequent faster resorption of the more soluble calcium carbonate.

Coralline HA hydrothermically treated exhibits very weak mechanical properties which limits their clinical applications. For instance, coralline HA is indicated for the repair of metaphyseal defects and long bone cysts and defects. However, it should be used within one month of fracture. Coralline HA does not possess enough mechanical strength to support the reduction of a defect site prior to soft and hard tissue ingrowth. Therefore standard osteosynthesis elements are required. In a typical long bone application, an appropriately sized coralline HA block or certain amounts of particles is fitted into the damaged area and stabilized with a metal plate, screws, or some other form of internal fixation. This protects the bone graft area so it can grow strong and durable. Eventually, the bioceramic is replaced by bone, leaving the mended bone as strong as prior to the fracture.

Coralline HA is also applied for manufacturing HA orbital implants, which are commonly used after enucleation, evisceration, and as a secondary implant. Coralline HA was the pioneering material in the field of porous orbital implants. Actually, the original implants, derived from marine coral, have a uniform interconnecting porous architecture that allows them to act as a passive framework for host fibrovascular ingrowth. The first ocular implant made of HA was implanted in 1985 by Perry [78]. This implant was patented and marketed as Bio Eye, Integrated Orbital Implants San Diego, USA) and is the one most commonly used by ophthalmic plastic surgeons in the USA. The eye muscles can be attached directly to this implant, allowing it to move within the orbit – just like the natural eye. Some of this movement is automatically transferred to the artificial eye, which fits over the tissues that cover the implant. If greater movement is desired, a titanium peg is used to connect the artificial eye to the implant. In this way, even the small, darting movements of the natural eye can be transformed directly to the artificial eye. The result is a more natural-looking artificial eye that can be difficult to distinguish from the natural eye.

3.4.5.2   Algae-Derived HA

Algae-derived HA exhibits a certain similarity to the natural bone ones. It is prepared by the hydrothermal conversion of the original calcium carbonate of calcarean red algae at about 700 °C. This process preserves the porosity of the algae. The temperature of the hydrothermal treatment is significantly higher than those used to convert coralline aragonite to HA. Consequently, the crystallinity is higher and the resorption time is much longer. High resolution transmission electron microscopy (HRTEM) studies (Figure 3.12a) and XRD patterns (Figure 3.12b) of algae-derived HA indicate that it is composed of a highly crystalline apatite phase showing ordered and homogeneous regions.

From the point of view of the microstructure, algae HA particles show a characteristic and very uniform shape. Under SEM observation, algae-derived HA exhibits a certain porosity that may have a positive effect on the interaction of these materials with living tissues (Figure 3.13). This kind of bioceramic is marketed by Friadent as Frios Algipore® granules

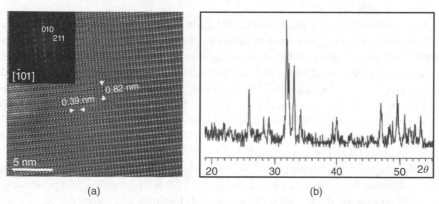

(a)   (b)

Figure 3.12 *Structural characterization of commercial algae derived HA. (a) HRTEM image and Fourier transform diagram and (b) XRD pattern. Reproduced with permission from Arthur C. Perry, MD of Integrated Orbital Implants, Inc.*

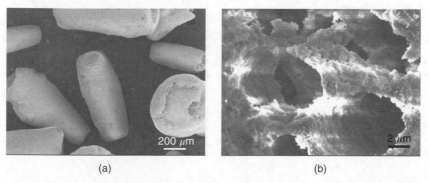

(a)   (b)

Figure 3.13 *Scanning electron micrographs of commercial algae derived HA. (a) Magnification ×50 and (b) magnification ×5000*

and is indicated for those clinical situations where keeping the volume stability for long periods is important, as a result of the slow resorption of this graft.

3.4.5.3   Treated Bone-Derived HA

At present, there are many bone substitutes based on treated bone xenograft. The animal sources are mainly bovine, porcine, and equine and have demonstrated excellent biocompatibility as well as osteoconductive behavior. Some researchers have compared naturally occurring sintered HA with the artificial ones. For instance, Matsumoto *et al.* [79] compared the naturally occurring HA and artificial HA-based materials *in vivo*, concluding that the sintered bovine HA gives better results in enhancing proliferation and differentiation of osteoblast-like cells. The properties and clinical applications of treated bone-derived HA mainly depend on the tissue treatment. The bone grafts mostly come from cancellous bone sources treated with processes that, mostly, are under intellectual property protection. However, the treated bone HA can be differentiated in high temperature and low temperature bone tissue treatments.

High temperature treated bone results in highly crystalline HA without the presence of organic components. These kind of implants perform excellently in terms of biocompatibility and osteoconduction but do not exhibit osteoinductive or osteogenic properties. A suitable procedure to fabricate high temperature treated bone is to collect cancellous bone from the femoral condyles [80]. To avoid cracking and soot formation in the material during the heat treatment process, the raw bone is boiled in distilled water. After boiling, the bone blocks are dehydrated in an alcohol series and subsequently dried at 70 °C for three days. Thereafter, the bone is calcined at 800 °C to remove the organic matrix, before proceeding to the sintering at, for instance, 1250 °C for 1 h.

Low temperature treated bone processes are aimed to neutralize the antigenic components present in animal bony tissues, while keeping the HA microstructure and preserving the collagen matrix inside the granules of biomaterial. Figure 3.14a shows the XRD patterns

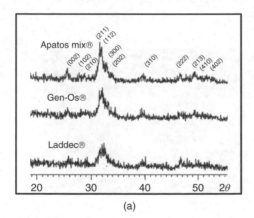

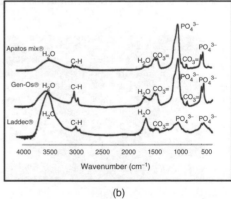

(a)   (b)

Figure 3.14   *XRD patterns (a) and FTIR spectra (b) of three commercial bone treated derived HAs*

obtained from three bone grafts treated by this kind of process. The diffraction patterns show poorly defined maxima compared to biological apatites that keep their initial small crystal size. Moreover, FTIR spectra of these products (Figure 3.14b) evidence the presence of collagen (C-H absorption band) that remains after the tissue process.

Commonly, the brochures of commercial grafts inform about the type of calcium phosphate, the origin (biological or synthetic), the sterilization procedure, and the particle size. This last parameter often determines the specific application (periodontal or skeletal grafting), with grain sizes below 1 mm indicated for periodontal grafting and those of about 2 mm limited to skeletal applications. However, information about the crystalline degree, presence of amorphous phases, textural properties, or any other microstructural aspects is absent and unknown for the implantologist. This is somehow strange, because it is widely known that the microstructure, the crystallinity, and porosity of CaP-based implants determine the solubility, surface reaction and, consequently, the *in vivo* behavior. New and fundamental microstructural characteristics are revealed when the grafts are analyzed by HRTEM. The microstructural studies at the nanometric scale show up features that play a fundamental role in the *in vivo* behavior but are not included in the manufacturer specifications provided to the clinicians. Figure 3.15 shows the HRTEM studies of a HA bone graft obtained from treated bone. Contrary to synthetic HA or other calcium phosphates, treated bone HAs exhibit heterogeneity on the nanometric scale. Figure 3.15a shows crystalline domains (clearly magnified in Figure 3.15b) with their corresponding electron diffraction pattern (Figure 3.15c), together with other non-crystalline regions in the same particle.

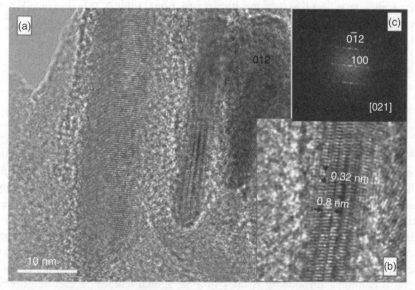

Figure 3.15 *HRTEM images (a,b) for commercial bone treated derived HA. (c) Fourier transform (FT) pattern obtained from image (b)*

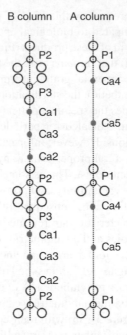

Figure 3.16   *Structural Ca and phosphate columns oriented in parallel to the* c *axis in β-TCP*

Since bone tissue is not a uniform and homogeneous compound, the derived calcined bioceramics somehow reflect these compositional variations. Table 3.3 shows some HA-based implants of different origins and commercially available.

3.5   Tricalcium Phosphate-Based Bioceramics

3.5.1   *β*-Tricalcium Phosphate (*β*-TCP)

3.5.1.1   *Chemical and Structural Properties*

β-TCP is the low temperature phase of $Ca_3(PO_4)_2$. It is a biodegradable bioceramic that dissolves in physiological media and can be replaced by bone after implantation. β-TCP crystallizes in the rhombohedral system, with a unit cell described by the space group R3c and unit cell parameters $a = 10.41$ Å, $c = 37.35$ Å, $\gamma = 120°$ [25]. The unit cell (hexagonal setting) contains 21 $Ca_3(PO_4)_2$ unit formula with a volume of 168 Å^3. The crystalline structure can be described as an arrangement of two types of columns oriented in parallel to the *c* axis. These columns, denoted as types A and B, comprise PO_4^{3-} anions and Ca^{2+}, as described in Figure 3.16. The unit cell contains three A columns and nine B columns, with type B columns surrounding A ones in a hexagonal pattern, as can be observed in Figure 3.17.

At temperatures above 1125 °C it transforms into the high temperature phase α-TCP. Being the stable phase at room temperature, β-TCP is less soluble in water than α-TCP. Pure β-TCP never occurs in biological calcifications, that is, there is no biomimetic process that

Table 3.3　*Some examples of commercially available HA-based bioceramics*

Origin	Product name	Company	Form
Synthetic HA	Apaceram	Pentax Corp., Japan	Porous (15–60%) and dense (<0.8%) blocks
	Apapore	ApaTech, UK	Porous irregular granules or chips
	Boneceram	Olympus Biomaterials, Japan	Porous blocks and grains (35–48%)
	Bonefil	Mitsubishi Materials Co, Japan	Porous blocks and grains (60–70%)
	Bonetite	Mitsubishi Materials Co, Japan	Porous blocks (70%)
	Calcitite	Zimmer, IN, USA	Dense grains
	Cerapatite	Ceraver, France	–
	IngeniOs HA	Zimmer, IN, USA	Porous particles
	Ostim	Heraeus Kulzer, Germany	Paste of nanocrystalline HA
	Synatite	SBM, France	Porous blocks (20–30%)
Calcium-deficient HA	Cementek	Teknimed, France	Cement
Carbonate HA	Osteogen	Impladent, NY, USA	Grains
Si substituted HA	Actifuse	ApaTech, UK	Porous grains
Coralline HA	Interpore	Interpore, CA, USA	Porous blocks (55–65%)
	ProOsteon	Interpore, CA, USA	Porous granules
	Bioeye	Integrated Orbital Implants, CA, USA	Orbital porous implants
Algae HA	Frios Algipore	Dentsply-Friadent, Germany	Porous granules
Treated bone unsintered	BioOss	Geitslich, Switzerland	Porous granules
	Laddec	BioHorizons, AL, USA	Porous granules
	Apatos mix	OsteoBiol, Italy	Porous granules
	GenOs	OsteoBiol, Italy	Porous granules
	Putty	OsteoBiol, Italy	Paste
Treated bone sintered	Cerabone	Aap Implantate, Germany	Porous granules and blocks
	Endobone	Merck, Germany	Porous granules and blocks
	Sinbone	Purzer, Taiwan	Porous blocks

results in β-TCP. Only the Mg-substituted form (withlockite) is found in some pathological calcifications (dental calculi, urinary stones, dentinal caries, etc.) [26].

3.5.1.2　Preparation Methods

Solid State Reaction.　β-TCP does not form in aqueous systems under normal laboratory conditions (up to 100 °C and 1 atm pressure). However, it can be prepared by heating at

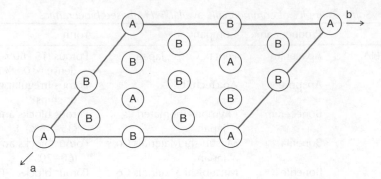

Figure 3.17 *Projection of the β-TCP unit cell containing three A columns and nine B columns*

about 1000 °C a mixture of DCPA and $CaCO_3$ keeping the Ca/P molar ratio at 1.5 as follows:

$$2\,CaHPO_4(s) + CaCO_3(s) \rightarrow Ca_3(PO_4)_2(s) + CO_2(g) + H_2O\,(g)$$

As explained in Section 3.4.1, commercial DCPA often appears mixed with DCPD and should be heated at 1000 °C for about 12 h to obtain β-calcium pyrophosphate ($\beta\text{-}Ca_2P_2O_7$). This reactant is subsequently mixed with the stoichiometric amount of $CaCO_3$ and heated between 900 and 1000 °C for about 10 h to obtain β-TCP as follows:

$$Ca_2P_2O_7(s) + CaCO_3(s) \rightarrow Ca_3(PO_4)_2(s) + CO_2(g)$$

3.5.1.3   Clinical Applications

β-TCP bone grafts are widely used in those clinical scenarios where resorption of the material is expected in four to six months. β-TCP bone grafts are indicated for filling periodontal bone defects, cysts extirpation, grafting of alveolar defects, and sinus elevations and augmentation. The bone regeneration and graft resorption can be fitted for each clinical scenario with the grain size as indicated in Table 3.4.

Currently, the most common strategy to shape β-TCP-based grafts is to prepare granules or blocks with interconnecting scaffolding that allows them to be completely resorbed while simultaneously creating new bone formation. For instance, bone grafts such as Cerasorb® M or Vitoss have between 60 and 75% of interconnecting porosity, which increases osteoconductivity within the grafted area and blood impregnation compared with dense bone grafts. These implants have a predictable resorption and new bone formation in four to six months. In order to enhance the solubility, β-TCP particles and blocks are prepared with a twofold hierarchical porosity, that is, a strut porosity to favor blood impregnation and macroporous systems arising from calcination of porogens or from foaming techniques (Figure 3.18).

Table 3.4   *Appropriated grain size of β-TCP bone grafts for different clinical scenarios*

Grain size (μm)	Clinical applications
150–500	Periodontal bone defects
500–1000	Small and intermediate-size cysts; grafting of alveolar defects
1000–2000	Large cysts, sinus elevations, and augmentation

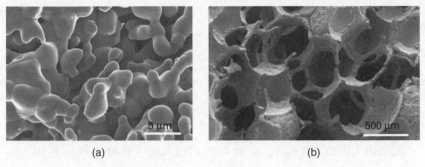

(a)   (b)

Figure 3.18 *SEM micrographs of β-TCP showing strut porosity (a). Reproduced with permission from http://www.curasan.de/imgs/fachkreise/cerasorb-dental/10_SEM_Cerasorb_M_500 .jpg Copyright (2013) curasan AG date last accessed 27/08/13. A macroporous structure (b). Reproduced with permission from http://www.zimmerdental.com/Products/ Regenerative/rg_ingeniOsB-TCPBioSynthBonePart.aspx Copyright (2013) Zimmer Dental Date last accessed 18/09/13*

3.5.2   α-Tricalcium Phosphate (α-TCP)

3.5.2.1   Chemical and Structural Properties

α-TCP might be considered as a high temperature phase of β-TCP. It crystallizes in the monoclinic system, with a unit cell described by the space group $P2_1/a$ and unit cell parameters $a = 12.89$ Å, $b = 27.28$ Å, $c = 15.21$ Å, $\beta = 126.2°$. The unit cell contains 24 unit formulas and occupies a volume of 180 Å^3. Thus α-TCP and β-TCP have exactly the same chemical composition, $Ca_3(PO_4)_2$ but they differ in crystal structure. The structure comprises columns of Ca^{2+} cations and PO_4^{3-} anions parallel to the c-axis. There are columns comprised of only Ca^{2+} cations and columns comprised of cations and anions following a PO_4 Ca PO_4 □ PO_4 Ca PO_4 □..., where □ are vacancies. Columns are distributed in a pseudohexagonal arrangement, where each cation column is surrounded by six cation–anion columns and each cation–anion column is surrounded by alternating cation and cation–anion columns (Figure 3.19). These structural differences determine that α-TCP is less stable than the β-phase. Actually, α-TCP is obtained at higher temperature,

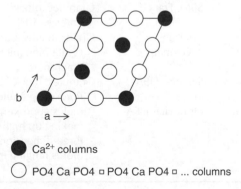

● Ca^{2+} columns

○ PO4 Ca PO4 □ PO4 Ca PO4 □ ... columns

Figure 3.19   *Pseudohexagonal arrangement of Ca and phosphate columns in α-TCP unit cell*

is more reactive in aqueous systems, has a higher specific energy and can be hydrolyzed to a mixture of other calcium phosphates.

3.5.2.2   Preparation Methods

Basically, the synthesis of α-TCP is accomplished by thermal transformation of a previously obtained precursor with molar ratio Ca/P 1.5 (CDHA, ACP, or β-TCP). For instance thermal decomposition of CDHA into α-TCP is represented as:

$$Ca_9(HPO_4)(PO_4)_5(OH)\ (s) \rightarrow 3a - Ca_3(PO_4)_2(s) + H_2O\ (g)\ T \geq 1150\,^{\circ}C$$

Another synthesis route is the solid state reaction of a mixture of solid precursors at high temperatures. The synthesis is carried out in the usual manner for solid state reactions, that is, the solid precursors are milled together to reduce particle size, increase the contact area, and mix them intimately. Wet milling is generally preferred. After milling, the mixture of powders may be directly heated above the transformation temperature or previously pressed to improve contact between the particles. The recommended reaction temperatures vary between 1250 and 1500 °C, and the dwell time is between 2 and 48 h. Quenching is highly recommended to avoid the reversion of the α-phase. The same reactants used above for β-TCP preparation can be used for α-TCP. α-TCP can also be obtained by heating Si-HA at temperatures above 1000 °C, obtaining the so-named silicon-stabilized α-TCP [81, 82].

Table 3.5   *Some examples of commercially available alpha and beta tricalcium phosphates*

Tricalcium phosphate	Product name	Company	Form
α-TCP	BioBase	Biovision, Germany	Granulate (200–500 µm). Re-entry time: 4 mo
α-TCP	ArrowBone α	BrainBase, Japan	Granulate (250 µm) with open porosity of 75%
β-TCP	Bioresorb	Sybron Implant Solutions, Germany	Granulate (500–1000 µm) Granulate (1000–2000 µm) Porosity 50%
β-TCP	Biosorb	SBM, France	Granules, blocks, wedges Porosity 50%
β-TCP	Cerasorb	Curasan, Germany	Grains, foams, pastes, and blocks. Containing micro-, meso-, and macropores
β-TCP	ChronOS	Synthes, USA	Granules, blocks, and wedges Porosity: granules 60%; preforms 70%
β-TCP	Vitoss	Orthovita, USA	Particles approximately 100 nm in size, making up highly porous, three-dimensional ß-TCP scaffolds (pores ranging from 1 to 1000 µm in size)

3.5.2.3   Clinical Applications of α-TCP

α-TCP is receiving growing attention as a raw material for several injectable hydraulic bone cements, biodegradable bioceramics, and composites for bone repair [83]. Similarly to the β-phase, α-TCP never occurs in biological calcifications, and it is occasionally used in calcium phosphate cements [84]. In recent years, α-TCP has been used not only as a component of biphasic HA-α-TCP bioresorbable scaffolds but also as pure α-TCP granulates, although there are only a few α-TCP products available in the market (see Table 3.5). These grafts are indicated in periodontology (for filling small bony defects), implantology (defect augmentation, elevation of sinus floor, etc.), and grafting after cysts removal. However, the faster dissolution rate of this polymorph compared with β-TCP reduces its clinical application when implanted alone.

3.6   Biphasic Calcium Phosphates (BCP)

3.6.1   Chemical and Structural Properties

BCPs are bioceramics consisting of two different CaP phases which are intimately mixed. BCPs used for reconstruction of bone defects are typically made of a low soluble calcium phosphate phase (commonly HA) and a more resorbable one such as TCP (typically β polymorph) [85]. In this way, the *in vitro* and *in vivo* behavior of these mixtures strongly correlates with the HA/TCP ratio.

The rationale for preparing BCPs arose when several clinical failures appeared with sintered HA base implants. For instance, some implants made of calcined HA to reconstruct mandibular ridges defects have resulted in a high failure rate in human clinical applications [86]. In order to avoid this problem, the use of granular instead of block forms of HA was suggested [87], but HA still presented low biodegradability, independently of the implant form. On the other hand, β-TCP ceramics have been developed as a biodegradable bone replacement and are commercially available as, for instance, ChronOS^{TM}, Vitoss^{TM}, and so on. However, when used as a biomaterial for bone replacement, the rate of biodegradation of TCP has been shown to be too fast. In different clinical scenarios, HA and TCP do not exhibit appropriate biodegradability kinetics, which eventually will be a disadvantage to the host tissue surrounding the implant. The presence of an optimum balance of stable HA and more soluble β-TCP should be more favorable than pure HA and β-TCP.

The excellent properties of BCPs rely on the controlled dissolution and the bone ingrowth at the expense of the ceramic [88]. The main advantage over other non-soluble calcium phosphates is that the mixture is gradually dissolved in the human body, acting as a stem for newly formed bone and releasing Ca^{2+} and PO_4^{3-} to the local environment. *In vivo* tests have confirmed the excellent behavior of BCP concerning the biodegradability rate [89, 90]. Due to the biodegradability of the β-TCP component, the reactivity increases with the β-TCP/HA ratio. The bioreactivity of BCPs, that is, biodegradation, biodissolution, and biological apatite precipitation on these compounds can be controlled through the β-TCP/HA ratio. Daculsi *et al.* [91] observed that, after six months of *in vivo* implantation, newly formed bone was present on a series of β-TCP/HA with weight ratios of 15/85, 35/65, and 85/15. Their results demonstrated that the dissolution of BCP ceramic implants and the abundance of the newly formed apatite crystals observed was influenced by the β-TCP/HA ratios: the higher the β-TCP/HA ratio, the better.

3.6.2   Preparation Methods

Biphasic HA/β-TCP material can be prepared by solid state methods, such as *physical mixing* of pure HA and β-TCP phases, or by the typical *ceramic method* through the thermal treatment of different raw materials at temperatures above 1000 °C. For instance, Yang *et al.* [92] carried out the synthesis of a series of BCPs by a process based on the solid state reaction of brushite with calcium carbonate. However, the most common strategy to prepare BCPs consists of sintering CDHAs, as indicated by the reaction:

$$Ca_{10-x}\square_x(PO_4)_{6-y}(HPO_4)_y(OH)_2 \rightarrow Ca_{10}(PO_4)_6(OH)_2 + Ca_3(PO_4)_2$$

This method results in more intimate HA/β-TCP mixtures compared with those obtained by ceramic or mechanical mixing methods. The presence of β-TCP in BCP ceramics is observed when CDHA is treated above 700 °C (see Figure 3.20), whereas thermal treatment above 1125 °C results in β-TCP transformation into α-TCP.

Silicon or silicates incorporation determines the thermal stability of the CaP phases. When SiO_2 is present during CDHA heating, α-TCP/HA biphasic materials can be obtained at temperatures as low as 700 °C. The α-TCP so obtained is denoted as silicon-stabilized α-TCP and different combinations of BCPs, [93] have been synthesized.

3.6.3   Clinical Applications

BCPs are used clinically as an alternative or as an additive to autogenous bone for dental and orthopedic applications. Implants shaped as particles, dense, or porous blocks, customized pieces, and injectable polymer-BCP mixtures are common BCP-based medical devices. Table 3.6 shows some of the commercially produced BCPs. Moreover, research is in progress to enlarge the clinical applications to the field of scaffolding for tissue engineering [94, 95] and carriers loading biotech products [96].

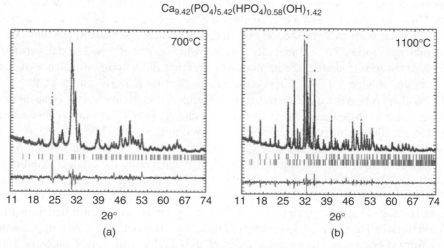

$$Ca_{9.42}(PO_4)_{5.42}(HPO_4)_{0.58}(OH)_{1.42}$$

Figure 3.20   *XRD patterns for a calcium-deficient HA. (a) Treated at 700 °C (single apatite-like phase) and (b) treated at 1100 °C (biphasic HA/β-TCP mixture) (See insert for color representation of the figure)*

Table 3.6 *Some of the commercially available BCPs*

BCP	Product name	Company	Form
HA + β-TCP	4Bone	MIS, Israel	Granulate (1000–20 000 µm); 60% HA to 40% β-TCP; Micro and macroporous
HA + β-TCP	Biosel	Depuy Bioland, France	Granulate (2000–3000 µm) 75% HA to 25% β-TCP Macropores 200–500 µm
HA + β-TCP	BoneSave	Stryker Orthopaedics, NJ, USA	Granulate (2000–3000 µm) 20% HA to 80% β-TCP
HA + β-TCP	CellCeram	Scaffdex, Finland	Macroporous blocks (100–800 µm) 60% HA to 40% β-TCP
HA + β-TCP	Ceraform	Teknimed, France	Granulate, blocks, and powder 65% HA to 35% β-TCP Macropores 300–500 µm
HA + β-TCP	Kainos	Signus, Germany	Injectable paste 60% HA to 40% β-TCP
HA + β-TCP	Triosite	Zimmer, IN, USA	Granulate 60% HA to 40% β-TCP Macroporosity 300–600 µm Micropores < 10 µm

3.7 Calcium Phosphate Nanoparticles

3.7.1 General Properties and Scope of Calcium Phosphate Nanoparticles

The development of nanotechnology has acquired great scientific interest in the biomaterials field. Nanoparticles are a bridge between bulk materials and atomic or molecular structures. When the material's size approaches the nanoscale the percentage of atoms at the surface of a material becomes significant, and the properties of materials change with their size. For bulk materials larger than 1 µm the percentage of atoms at the surface is minuscule relative to the total number of atoms of the material. The interesting and sometimes unexpected properties of nanoparticles are partially due to the aspects of the surface of the material dominating the properties in lieu of the bulk properties. The inherent nanoceramics properties allow the tackling of traditional problems in the bioceramics field, such as mechanical performance, bone regeneration kinetics, biocompatibility, and so on, as well as new challenges such as the optimization of scaffolds for bone tissue engineering and the design of nano-drug delivery systems. The nanoapatites contribution to the biomaterials field is also mainly justified by their surface features. Tissue–implant interaction is a surface event and the high surface of nanoapatites supplies new reactivity features; bioactivity, bioresorption, foreign body responses, and so on are significantly modified [97]. Nanoapatites give the chance to tailor at the nanometric scale the interactions between the material and the osteoblast adhesion proteins, with the purpose of optimizing osteoblast adhesion. The nanoceramic surfaces can be easily functionalized and can incorporate biologically active molecules [98]. Nowadays, there are experimental evidences that metallic biomaterials with nanometer grain sizes stimulate osteoblast activity, leading to more bone growth [99]. Similarly, long term functions, such as cell proliferation, synthesis of alkaline phosphatase (ALP), and concentration of calcium in the extracellular matrix are enhanced when osteoblasts are seeded on nanoceramics [100]. This enhancement can be explained because of the osteoblast selectivity, due to the types of protein adhered in the first stages after implantation. Cells do not directly attach onto the material's surface, but on the proteins previously linked to the implant. Therefore, the first physico-chemicalexplanations of the better performance of nanoceramics must be found in the events occurring between the nanoceramics and the serum proteins.

Nano-sized grains provide higher surface roughness in the range of tens of nanometers, which appears to be a critical characteristic that determines the nanoceramic biocompatibility [101]. Moreover, the nanostructure provides a higher number of grain boundaries as well as an increased surface wettability, which is also associated with enhanced protein adsorption and cell adhesion. However, the enhanced biocompatibility of nanoceramics exhibits a much more interesting selective mechanism. When considering several protein anchorage-depending cells, for instance osteoblast, fibroblast, and endothelial cells, it is possible to correlate the adsorbed protein type and concentration with the observed cell adhesion on the materials. *Vitronectin* is mainly adsorbed on nanoceramics, whereas *laminin* is preferentially adsorbed on conventional ceramics. Although the mechanism is not well established, fibroblast and endothelial cells attach better on conventional ceramics, whereas osteoblasts are adhere better on nanoceramics.

Webster *et al.* [100] explain this in terms of the inherent defect sizes of each kind of bioceramic. In addition to enhanced surface wettability, the roughness, dictated by grain and pore size, of nanoceramics influences interactions (such as adsorption and/or configuration/bioactivity) of determined serum proteins and thus affects subsequent cell adhesion. Vitronectin, which is a linear protein of 15 nm length preferentially may be adsorbed preferentially on the small pores present in nanophase ceramics, while laminin (cruciform configuration, 70 nm in length and width) would be preferentially adsorbed on the large pores present in conventional ceramics. The specificity of nanoceramics with respect to the type of cell has also beenobserved with bone marrow MSCs and osteosarcoma cells [102]. When both cultures are exposed to HAs nanoparticles, greater cell viability and proliferation of MSCs were observed on the nano-HA, especially in the case of the smallest nanoparticles (around 20 nm). On the contrary, the growth of osteosarcoma cells was inhibited by the nano-HA and the smallest particles exhibit the higher inhibitory effect.

3.7.2   Preparation Methods of CaP Nanoparticles

3.7.2.1   Sol–Gel

The chemistry of the sol–gel process is based on the hydrolysis and condensation of molecular precursors [103]. There are two different routes described in the literature, depending on whether the precursor is formed by an aqueous solution of an inorganic salt or by a metal-organic compound. In any case, this method requires the careful study of parameters such as oxidation states, pH, and concentration. The sol–gel method is a process divided into several stages where different physical and chemical phenomena are performed, such as hydrolysis, polymerization, drying, and densification. This process is known as sol–gel because the viscosity increases at a given instant during the process sequence. A sudden increase in viscosity is a common feature in all sol–gel processes, indicating the onset of gel formation.

The sol-gel process for HA synthesis usually produces fine-grain microstructures containing a mixture of nano-to-submicron crystals. The low temperature formation of the apatite crystals has been the main contribution of the sol-gel process in comparison with conventional methods for HA powder synthesis. A number of combinations of calcium and phosphorus precursors have been employed for sol–gel HA synthesis. The sol–gel synthesis of calcium phosphates finds a very interesting application in coating metallic prostheses and this topic will be further discussed in Chapter 9. As an example of sol–gel HA synthesis, Liu *et al.* [104] prepared a sol with triethyl phosphate diluted in anhydrous ethanol together with a small amount of distilled water for hydrolysis. Thereafter a stoichiometric amount of calcium nitrate is subsequently dissolved in anhydrous ethanol and dropped into the hydrolyzed phosphorus sol. The solution thus obtained is aged and dried, resulting in a white gel that can be treated at temperatures ranging between 600 and 1100 °C, depending on the particle size desired.

The major limitation of the sol-gel technique for CaP synthesis is hydrolysis of phosphates and the high cost of the raw materials. In order to overcome these problems, alternative reactants can be used, such as phosphoric pentoxide and calcium nitrate tetrahydrate [105]. This sol-gel method provides a simple route for synthesis of HA nanopowder, where the crystalline degree and morphology of the obtained nanopowder are also dependent on the sintering temperature and time.

3.7.2.2 Solidification of Liquid Solutions

The solidification of liquid solutions technique is based on the Pechini patent [106]. This patent was originally developed for the preparation of multicomponent oxides, allowing the production of massive and reproducible quantities with a precise homogeneity in both composition and particle size. This method is based on the preparation of a liquid solution that retains its homogeneity in the solid state. Its application has now been extended to the preparation of CaPs. The synthesis of CaPs by the Pechini method relies on the formation of complexes of calcium with a bi- and tridentate organic chelating agent such as citric acid, in the presence of an orthophosphate salt, such as $NH_4H_2PO_4$ [107]. Thereafter, ethylene glycol is added to create linkages between the chelates, resulting in gelation of the reaction mixture. Finally, the gel is heated to pyrolyze the organic species, thus forming submicron particles of CaPs. By modifying the synthesis conditions large amounts of single phases or biphasic mixtures can be prepared. This method makes it possible to obtain not only single phase HA, but also β-TCP, α-TCP, and BCPs. The specific content in β-TCP and HA can be precisely predicted from the Ca/P ratio in the precursor solution.

3.7.2.3 Synthesis of Apatites by an Aerosol Process

Aerosol-based processes can be considered as a type of synthesis of solids that involves the transformation from *gas* to *particle* or from *droplet* to *particle*. Commonly, calcium and phosphate precursors in the form of soluble salts are dissolved in an aqueous solution. An aerosol is then formed by means of an atomizer (Venturi effect or high frequency signal generator, etc.). When the aerosol reaches the reaction area, different decomposition phenomena may take place, depending on the precursor features and the temperature. The aerosol synthesis technique has been used to produce small particles of pure HA with different crystallinity as well as TCPs [108]. Its main advantage is that it has the potential to create particles of unique composition, for which starting materials are mixed in a solution at atomic level. A thermal treatment then allows tailoring modifications on morphology and texture.

3.7.2.4   *Synthesis of Apatites by Precipitation Methods with Controlled Morphology*

Starting from liquid solutions instead of solid grains facilitates the microstructural and morphological control of the calcium phosphate-based bioceramics. There exist several synthesis routes that allow tailoring of the shape and surface composition of the nanoparticles. Palazzo *et al.* [109] have proposed the preparation of biomimetic nano-HA with both *needle-shaped* and *plate-shaped* morphologies and different physical-chemical properties. *Needle-shaped* nanocrystals have also been prepared from an aqueous suspension of $Ca(OH)_2$ by slow addition of H_3PO_4 [110], obtaining needle-shaped nanocrystals with a granular dimension around 100 ± 20 nm. *Plate-shaped* nanocrystals can be precipitated from an aqueous solution of $(NH_4)_3PO_4$ by slow addition of an aqueous solution of $Ca(CH_3COO)_2$ keeping the pH at a constant value of 10 by addition of $(NH_4)OH$ solution. The reaction mixtures are stirred at low temperature for 72 h and then the deposited inorganic phase is commonly isolated by filtration. In this way, *plate-shaped* nano-HA is obtained having granular dimensions of 25 ± 5 nm [111]. These procedures are likely to produce CDHA, with a Ca/P ratio of 1.65 to 1.62. This deficiency is greater in the surface of the nanoparticles, as the Ca/P ratio can range between 1.30 and 1.45 for needle- and plate-shaped nanocrystals, respectively.

3.7.3   Clinical Applications

There are few commercial products based on synthetic HA nanoparticles. Commonly they are injectable bone substitutes composed of 100% HA nanoparticles. These products are totally resorbable and enhance bone formation due to their nanocrystalline structure similar to natural bone.

Injectable nanoapatites are intended as temporary osteoconductive grafts for the ingrowth of viable bone and are not intended to provide structural support. This kind of product can be used in applications such as reconstruction of posttraumatic bone defects, periodontal defects filling, cystectomy filling, alveolar bone filling, osteotomies, spinal surgery cages filling, acetabulum reconstruction, and metaphyseal fractures.

3.8   Calcium Phosphate Advanced Biomaterials

3.8.1   Scaffolds for in situ Bone Regeneration and Tissue Engineering

Calcium phosphates bioceramics, of natural or synthetic origin, are among the most appropriate biomaterials for fabrication of scaffolds for *in situ* bone regeneration and tissue engineering. Scaffolds are intended to provide a 3D environment that works as an extracellular matrix for the mesenchymal stem cells MSCs present in bone. The scaffold must facilitate the MSCs migration and adhesion, and stimulate their differentiation to osteoblast phenotype. Moreover, the porous architecture should facilitate bone colonization and angiogenesis, thus exhibiting a framework of interconnected pores between 200 and 800 μm. Otherwise the newly formed bone will not be viable in the long term.

Among natural origin CaP, coralline HA has had considerable success, as coral possess a macroporous (150–500 μm) and interconnected structure that somehow mimics human cancellous bone. Regarding synthetic CaP scaffolds, those prepared with HA, β-TCP, or BCP are by far the most widely studied. In addition, ceramic–polymer composites also constitute a very important research line in the field of scaffolds for bone regeneration and tissue engineering. The design and preparation of scaffolds intended to assist bone regeneration therapies must fulfill several conditions. These scaffolds must fit into the anatomic defect, enhance tissue ingrowth, produce non-toxic byproducts and stimulate the osteogenesis. This last point involves a set of biological requirements such as a source of MSCs able to differentiate into osteoblast cells, growth factors (osteoinductive proteins) that lead MSCs to migrate toward the bone defect and subsequently proliferate and differentiate, development of newly formed blood vessels (angiogenesis), mass transport of nutrients, and new bone formation within the scaffold. Fulfilling all these features requires a multidisciplinary approach to the scaffolds preparation, in such a way that the design must be optimized considering the macroscopic architecture, chemical composition, microstructure, and presence of osteogenic components.

Bone tissue scaffolds must exhibit anatomically shaped morphologies, control over porosity (including pore size and pore connectivity), and scaffold density (or wall thickness) that determine the mechanical properties of the implant. Rapid prototyping (RP) techniques, such as the robocasting method (Figure 3.21), allow accurate control of the scaffold architecture and permit the design of the scaffold morphology to adapt to the specific defect. The use of RP, also known as solid freeform fabrication technology, allows

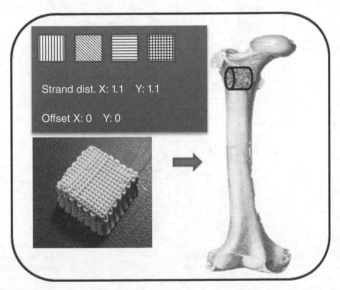

Figure 3.21 *Scheme of a scaffold preparation with an HA ink by robocasting rapid prototyping technique*

the production of scaffolds with defined and reproducible internal structures taken straight from computer data. In Chapter 10 the different strategies for scaffolds preparation will be described in detail.

3.8.2 Drug Delivery Systems

Synergy between bioceramics and pharmacology is a very powerful alternative to tackle both the bone defects and their etiology. The studies involving bioceramics and drugs association can be related to one of the following therapies: bone regeneration in critical defects, bone infection treatments, osteoporotic fractures consolidation, and bone tumor treatments [112].

3.8.2.1   CaP-Based Drug Delivery Systems: Bone Regeneration

Bones exhibit self-repairing mechanisms, but grafts are required when the defect is critical. The implant must provide a substrate that stimulates the self-healing processes of bone. The combination of bone-bonding bioceramics with osteogenic factors is a widely studied strategy and one of the most promising therapies to regenerate bone tissue in critical defects. Figure 3.22 shows some of the growth factors used to date for bone regeneration purposes. Clinical studies combining bone morphogenetic proteins (BMPs) and BCP have also shown encouraging results. For instance, Dimar *et al.* have compared the efficiency on a vertebral fusion model of an iliac crest autograft versus BCP combined with BMP-2 [113]. The

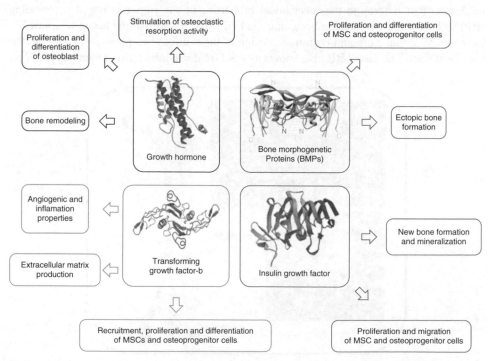

Figure 3.22   *Growth factors most used for bone regeneration and their biological activities (See insert for color representation of the figure)*

implantation of combined BCP gave a clear improvement compared to the iliac crest bone graft in terms of vertebral fusion rate (88% vs 73%), operatory conditions (duration of the intervention 2.4 h vs 2.9 h and hemorrhages 273 cm^3 vs 465 cm^3), and morbidity (leg and back pain).

The association of drugs with CPCs has also provided promising results for bone tissue treatments. Haddad *et al.* [114] evaluated the bone repair of a critical-sized calvarial vault defect in rabbits, after implantation of apatitic cement loaded with BMP-2 (25 mg/ml). After 12 weeks of implantation, they observed a bone formation of 45.8% more than control. Studies led by Seeherman *et al.* [115] confirm the efficiency of these combined systems. For example, the injection of combined apatitic cement with rh-BMP-2 in primates, in an osteotomy model of the fibula, accelerated the filling of the bone defect by 40% after 14 weeks of implantation compared to unloaded cement.

Blom *et al.* [116] have studied not only the stimulation of bone formation but also the cement resorption kinetics, when loaded with the growth factor TGF-1. For this study, calvaria critical-sized defects on adult rats were filled with TGF-β1. Eight weeks after implantation, the combined cement stimulated bone formation by 50% and improved the bone–cement contact by 65% compared with the control cement.

3.8.2.2   *CaP-Based Drug Delivery Systems: Bone Infection*

Loading antibiotic agents in bioceramics is a promising and effective procedure for delivering drugs to the implantation site. This strategy is aimed at preventing implant-associated infections by reducing the concentration of bacteria and/or impeding bacterial adherence to the implant surface. Systemic antibiotic administration does not always allow for efficient concentrations, mainly because of poor blood flow in the bone tissue. This necessitates the administration of large antibiotic doses in order to obtain acceptable concentrations in the affected region. Most of the studies consider the combination of CaP with polymers as matrices for antibiotic and other drug controlled release. However, the bioceramics alone have demonstrated very good results as an antibiotic system. The homogeneous incorporation of drugs into ceramic matrices, as well as the compatibility of the high temperature processes with active drugs, are the main challenges to overcome from the point of view of processing strategy.

Cold fabrication methods of bioactive ceramics enable any antibiotic to be placed into the ceramic implant without thermal damage to the drug. For instance, the use of dynamic compaction and isostatic compression to consolidate calcium phosphate powder loaded with a therapeutic agent avoids a sintering step that could destroy the drug [117]. Hard templating strategies have been revealed as an excellent alternative to develop CaP-based particles containing antibiotics [118]. These methods lead to bioceramics with high surface area and porosity, whereas the chemical routes are friendly toward the drug. For this purpose, mesoporous carbon is filled with the calcium phosphate precursors and subsequently removed. The bioceramic thus obtained shows much higher charging capacity than that of commercially available calcium orthophosphate which did not have any nanoporosity.

Clinical trials have demonstrated the efficacy of HA blocks used as drug delivery systems, for the treatment of infected hip arthroplasty. After implantation of HA blocks saturated with antibiotics, six of the seven patients who had contracted an infection after hip arthroplasty did not contract an infection during the five years of monitoring [119].

The conclusion extracted from this study is that antibiotic-impregnated CHA ceramic is an excellent drug delivery system for infected hip arthroplasty.

3.8.2.3   CaP-Based Drug Delivery Systems: Osteoporotic Fractures

Osteoporosis is a bone disease characterized by decrease in bone mass and density, resulting in a predisposition to fractures and bone deformities, such as the collapse of one or more vertebrae. It occurs most commonly in women after menopause as a result of estrogen deficiency. Nowadays, bisphosphonates (BPs) are the drugs of choice for osteoporosis treatment since they inhibit bone resorption mediated by osteoclasts. Bisphosphonates are analogs of pyrophosphates, with nonhydrolyzable P-C-P groups rather than P-O-P, which makes them resistant to hydrolysis and therefore showing poor intestinal absorption, typically less than 1%. The affinity of BPs for HA has been used to develop new CaP systems for the controlled release of BP and HA is envisaged as a potential vector for alendronate. HA nanocrystals loaded with alendronate at 7 wt% have demonstrated, under *in vitro* conditions, the capability to reduce the number of osteoclasts to 30% while increasing the osteoblastic activity, characterized by doubling in the synthesis of ALP, osteocalcin, and type I collagen [120].

Although a variety of synthetic CaPs are currently developed for their use as bone substitutes (i.e., BCPs, α-, or β-TCP, CDAs, and DCPD), there are only a few studies considering the combination of BPs with these compounds. This alternative can provide new release profiles because, in contrast to the case of HA, which is highly stable under physiological conditions, they can be degraded in bone defects simultaneously with the formation of new bone. CaP-based coatings are also suitable to be loaded with antiosteoporotic agents. Since the coatings are in intimate contact with the living tissues, the incorporation of drugs within or onto the surface can result in very effective drug delivery systems. When zoledronate is grafted onto HA coatings of titanium implants, a statistically significant increase in the peri-implant bone volume fraction is observed when the implants are tested in rats [121].

3.8.2.4   CaP-Based Drug Delivery Systems: Bone Tumors

Implantation of bioceramic bone grafts combined with specific local cancer treatments is an excellent alternative to restore large bone defects that occur after tumor extirpation or partial bone resection, resulting in tumor inhibition with low levels of systemic toxicity. Several antitumoral drugs have been incorporated within ceramic matrices, and subsequently tested both *in vitro* and *in vivo*. For instance, methotrexate release from HA and β-TCP has been tested [122] as well as *cis*-platinum enclosure into blocks of porous calcium HA ceramic [123]. These antitumoral-loaded bioceramics behave as drug delivery systems, which exhibit acceptable mechanical properties and allow partial surgical excision and replacement of the bone defect at the same time.

Intracellular targeting is one of the most important challenges pursued by anticancer therapies. For this reason, the association of antitumoral drugs with calcium phosphate nanoparticles has acquired great significance in recent years. The nanoceramics surfaces can be easily functionalized and can incorporate biologically active molecules. Since nanomaterials exhibit a maximum surface/volume ratio they are excellent candidates as vehicles for *drug delivery* applications. For instance, cisplatin can be easily adsorbed onto HA nanoparticles [124]. The chemical and physical characteristics of the apatite crystals, including the chemical composition, structure, porosity, particle size, and surface area, as well as the ionic composition of the equilibrating solution (pH, ionic strength, concentration of ion constituents), all play an important role in both the binding and release of the specific chemical components from calcium phosphates. However, the key to treating bone cancer with nanoparticles still remains a far from a satisfactory solution. CaP nanoparticles are likely to target bone tissue but unlikely to specifically target tumoral cells. Functionalization of CaP nanoparticles with targeting agents, such as folic acid or antibodies, can open new possibilities for bone cancer treatments.

References

1.   Ring, M.E. (1992) *Dentistry: an Illustrated History*, Harry N. Abrams Inc., New York, p. 32.
2.   Bobbio, A. (1970) The first endosseous alloplastic implant in the history of man. *Bull. Hist. Dent*, **20**, 1–6.
3.   Albee, F.H. (1920) Studies in bone growth – triple calcium phosphate as stimulus to osteogenesis. *Ann. Surg*, **71**, 32–39.
4.   Nery, E.B., Lynch, K.L., Hirthe, W.M. and Mueller, K.H. (1975) Bioceramic implants in surgically produced intrabony defects. *J. Periodontal*, **46**, 328–347.
5.   Denissen, H.W. and de Groot, K. (1979) Immediate dental root implants from synthetic dense calcium hydroxyapatite. *J. Prosthet. Dent*, **42**, 551–556.
6.   De Groot, K. (1983) *Bioceramics of Calcium Phosphate*, CRC Press, Boca Raton, FL, p. 146.
7.   De Groot, K. (1980) Bioceramics consisting of calcium phosphate salts. *Biomaterials*, **1**, 47–50.
8.   Jarcho, M. (1981) Calcium phosphate ceramics as hard tissue prosthetics. *Clin. Orthop. Relat. Res*, **157**, 259–278.
9.   Aoki, H., Kato, K.M., Ogiso, M. and Tabata, T. (1977) Studies on the application of apatite to dental materials. *J. Dent. Eng*, **18**, 86–89.
10. Lowenstam, H.A. and Weiner, S. (1989) *Biomineralization*, Oxford University Press, Oxford, p. 324.
11. Dorozhkin, S.V. (2009) Calcium orthophosphates in nature, biology and medicine. *Materials*, **2**, 399–498.
12. Ducheyne, P. and Qiu, Q. (1999) Bioactive ceramics: the effect of surface reactivity on bone formation and bone cell function. *Biomaterials*, **20**, 2287–2303.
13. Hench, L.L. and Polack, J.M. (2002) Third-generation biomedical materials. *Science*, **295**, 1014–1017.
14. Hench, L.L. and Best, S. (2004) Ceramics, Glasses and Glass-Ceramics, in *Biomaterials Science. An Introduction to Biomaterials in Medicine*, Elsevier Academic Press, San Diego, CA, p. 153.
15. Driessens, F.C.M. (1983) *Formation and Stability of Calcium Phosphate in Relation to the Phase Composition of the Mineral in Calcified Tissue*, CRC Press, Boca Raton, FL.
16. Le Geros, R.Z., Parsons, J.R., Daculsi, G. *et al*. (1988) *Bioceramics: Material Characteristics Versus in Vivo Behavior*, vol. **523**, New York Academic Science, New York.

17. Fujui, T. and Ogino, M. (1984) Difference of bond bonding behavior among surface active glasses and sintered apatite. *J. Biomed. Mater. Res*, **18**, 845–859.
18. Hench, L.L. (1988) *Bioceramics: Material Characteristics Versus in Vivo Behavior*, vol. **523**, New York Academic Science, New York.
19. Hulbert, S.F., Hench, L.L., Forbers, D. and Bowman, L.S. (1982) History of bioceramics. *Ceram. Int*, **8**, 131–140.
20. Hench, L.L. (1991) Bioceramics: from concept to clinic. *J. Am. Ceram. Soc*, **74**, 1487–1510.
21. Vallet-Regí, M. and González-Calbet, J. (2004) Calcium phosphates in the substitution of bone tissue. *Prog. Solid State Chem*, **32**, 1–31.
22. Neo, M., Nakamura, T., Yamamuro, T. *et al.* (1993) *Bone-Bonding Biomaterials*, Reed Healthcare Communications, Leiderdorp, p. 111.
23. Ducheyne, P., Beight, J., Cuckler, J. *et al.* (1990) Effect of calcium phosphate coating characteristics on early post-operative bone tissue ingrowth. *Biomaterials*, **11**, 531–540.
24. de Bruijn, J.D., Novell, Y.P. and van Blitterswijk, C.A. (1994) Structural arrangements at the interface between plasma sprayed calcium phosphates and bone. *Biomaterials*, **15**, 543–550.
25. Elliott, J.C. (1994) *Structure and Chemistry of the Apatites and Other Calcium Orthophosphates*, Studies in Inorganic Chemistry, vol. **18**, Elsevier, Amsterdam.
26. LeGeros, R.Z. (1991) *in Calcium Phosphates in Oral Biology and Medicine*, Monographs in Oral Science, vol. **15**, Karger, Basel.
27. Vallet-Regí, M. (2001) Ceramics for medical applications. *J. Chem. Soc., Dalton Trans*, **2**, 97–108.
28. Vallet-Regí, M. (2006) Revisiting ceramics for medical applications. *Dalton Trans*, 5211–5220.
29. Kay, M.I., Young, R.A. and Posner, A.S. (1964) Crystal structure of hydroxyapatite. *Nature*, **204**, 1050–1052.
30. Bigi, A., Ripamonti, A., Brückner, S. *et al.* (1989) Structure refinements of lead-substituted calcium hydroxyapatite by X-ray powder fitting. *Acta Crystallogr. Section B*, **45**, 247–251.
31. Bigi, A., Falini, E., Foresti, M. *et al.* (1996) Rietveld structure refinements of calcium hydroxyapatite containing magnesium. *Acta Crystallogr. Section B*, **52**, 87–92.
32. Ergun, C., Webster, T.J., Bizios, R. and Doremus, R.H. (2002) Hydroxyapatite with substituted magnesium, zinc, cadmium, and yttrium. I. Structure and microstructure. *J. Biomed. Mater. Res*, **59**, 305–311.
33. Webster, T.J., Ergun, C., Doremus, R.H. and Bizios, R. (2002) Hydroxyapatite with substituted magnesium, zinc, cadmium, and yttrium. II. Mechanisms of osteoblast adhesion. *J. Biomed. Mater. Res*, **5**, 312–317.
34. Mackie, P.E., Elliott, J.C. and Young, R.A. (1972) Monoclinic structure of synthetic $Ca_5(PO_4)_3Cl$, chloroapatite. *Acta Crystallogr. Section B*, **28**, 1840–1848.
35. Elliott, J.C., Bonel, G. and Trombe, J.C. (1980) Space group and lattice constants of $Ca_{10}(PO_4)_6CO_3$. *J. Appl. Crystallogr*, **13**, 618–621.
36. DeBoer, B.G. (1991) Determination of the antimony substitution site in calcium fluorapatite from powder X-ray diffraction data. *Acta Crystallogr. Section B*, **47**, 683–692.

37. Serret, A., Cabañas, M.V. and Vallet-Regí, M. (2000) Stabilization of calcium oxyapatites with lanthanum(III)-created anionic vacancies. *Chem. Mater*, **12**, 3836–3841.
38. Seuter, A.M.J.H. (1972) Existence region of calcium hydroxyapatite and the equilibrium with coexisting phases at elevated temperatures, in *Reactivity of Solids*, Chapman and Hall, London, p. 806.
39. Elliot, J.C. (1994) *Structure and Chemistry of the Apatites and Other Calcium Orthophosphates*, Elsevier Science, Amsterdam, p. 111.
40. Hayek, E. and Newesely, H. (1963) Pentacalcium monohydroxyorthophosphate. *Inorg. Synth*, **7**, 63–65.
41. Suchanek, W. and Yoshimura, M. (1998) Processing and properties of hydroxyapatite based biomaterials for use as hard tissue replacement implants. *J. Mater. Res*, **13**, 94–117.
42. Narasaraju, T.S. and Phebe, D.E. (1996) Some physico-chemical aspects of hydroxyapatite. *J. Mater. Sci*, **31**, 1–21.
43. Klawitter, J.J. and Hulbert, S.F. (1971) Application of porous ceramics for the attachment of load bearing internal orthopedic applications. *J. Biomed. Mater. Res. A*, **5**, 161–229.
44. Guda, T., Appleford, M., Oh, S. and Ong, J.L. (2008) A cellular perspective to bioceramics scaffolds for bone tissue engineering; the state of the art. *Curr. Top. Med. Chem*, **8**, 290–299.
45. Von Doernberg, M.C., von Rechenberg, B., Bohner, M. *et al.* (2006) In vivo behavior of calcium phosphate scaffolds with four different pore sizes. *Biomaterials*, **27**, 5186–5198.
46. Holmes, R.E. (1979) Bone regeneration within a coralline hydroxyapatite implant. *Plast. Reconstr. Surg*, **63**, 626–633.
47. Dorozhkin, S.V. (2010) Bioceramics of calcium orthophosphates. *Biomaterials*, **31**, 1465–1485.
48. Hing, K., Annaz, B., Saeed, S. *et al.* (2005) Microporosity enhances bioactivity of synthetic bone graft substitutes. *J. Mater. Sci. Mater. Med*, **16**, 467–475.
49. Amemiya, M., Kikkawa, I., Watanabe, H. *et al.* (2008) The use of hydroxyapatite blocks for innominate osteotomy: a report of three cases. *J. Orthop. Surg*, **16**, 237–240.
50. Koshino, T., Murase, T., Takagi, T. and Saito, T. (2001) New bone formation around porous hydroxyapatite wedge implanted in opening wedge high tibial osteotomy in patients with osteoarthritis. *Biomaterials*, **22**, 1579–1582.
51. Koshino, T., Murase, T. and Saito, T. (2003) Medial opening-wedge high tibial osteotomy with use of porous hydroxyapatite to treat medial compartment osteoarthritis of the knee. *J. Bone Joint Surg. Am*, **85-A**, 78–85.
52. Maruyama, M., Tensho, K., Wakabayashi, S. and Terayama, K. (2012) Hydroxyapatite block for reconstruction of severe dysplasia or acetabular bone defects in total hip arthroplasty. Operative technique and clinical outcome. *J. Arthroplastia*, **27**, 591–597.
53. Jordan, D.R., Munro, S.M., Brownstein, S. *et al.* (1998) A synthetic hydroxyapatite implant: the so-called counterfeit implant. *Ophthalmic. Plast. Reconstr. Surg*, **14**, 244–249.

54. van Raemdonck, W., Ducheyne, P. and de Meester, P. (1984) *Metal and Ceramic Biomaterials*, CRC Press, Boca Raton, FL, p. 149.

55. Young, R.A. and Mackie, P.E. (1980) Crystallography of human tooth enamel: initial structure refinement. *Mater. Res. Bull.*, **15**, 17–29.

56. Elliott, J.C. (1965) The interpretation of the infrared absorption spectra of some carbonate-containing apatites, in *Proceedings of the International Symposium on the Composition, Properties, and Fundamental Structure of Tooth Enamel* (eds R.W. Fearnhead and M.V. Stack), John Wright & Sons, Ltd, Bristol, p. 20 and 277.

57. LeGeros, R.Z. (1965) Effect of carbonate on the lattice parameters of apatite. *Nature*, **206**, 403–404.

58. De Maeyer, E.A.P., Verbeeck, R.M.H. and Naessens, D.E. (1993) Stoichiometry of sodium(+)- and carbonate-containing apatites obtained by hydrolysis of monetite. *Inorg. Chem.*, **32**, 5709–5714.

59. Verbeeck, R.M.H., De Maeyer, E.A.P. and Driessens, F.C.M. (1995) Stoichiometry of potassium- and carbonate-containing apatites synthesized by solid state reactions. *Inorg. Chem.*, **34**, 2084–2088.

60. Doi, Y., Shibutani, T., Moriwaki, Y. *et al.* (1998) Sintered carbonate apatites as bioresorbable bone substitutes. *J. Biomed. Mater. Res.*, **39**, 603–610.

61. Valen, M. and Ganz, S.D. (2002) A synthetic bioactive resorbable graft for predictable implant reconstruction; part one. *J. Oral Implant.*, **28**, 167–177.

62. Carlisle, E.M. (1970) Silicon: a possible factor in bone calcification. *Science*, **167**, 279–280.

63. Carlisle, E.M. (1981) Silicon: a requirement in bone formation independent of vitamin D. *Calc. Tissue Int*, **33**, 27–34.

64. Hench, L.L. and LaTorre, G.P. (1993) *Bioceramics*, vol. **5**, Kobunshi Kankokai, Inc., Kyoto, p. 67.

65. Ohtsuki, C., Kokubo, T. and Yamamuro, T. (1992) Mechanism of apatite formation on $CaO\ SiO_2\ P_2O_5$ glasses in a simulated body fluid. *J. Non-Cryst. Solids*, **143**, 84–92.

66. Arcos, D., Greenspan, D.C. and Vallet-Regí, M. (2002) Influence of the stabilization temperature on textural and structural features and ion release in $SiO_2–CaO–P_2O_5$ sol–gel glasses. *Chem. Mater*, **14**, 1515–1522.

67. Arcos, D., Greenspan, D.C. and Vallet-Regí, M. (2003) A new quantitative method to evaluate the in vitro bioactivity of melt and sol-gel-derived silicate glasses. *J. Biomed. Mater. Res*, **65A**, 344–351.

68. Gibson, I.R., Huang, J., Best, S.M. and Bonfield, W. (1999) *Bioceramics*, vol. **12**, World Scientific Publishing, Singapore, p. 191.

69. Hing, K.A., Saeed, S., Annaz, B. *et al.* (2004) *Transactions 7th World Biomaterials Congress*, Australian Society for Biomaterials, Brunswick Lower, VIC, p. 108.

70. Gibson, I.R., Best, S.M. and Bonfield, W. (1999) Chemical characterization of silicon-substituted hydroxyapatite. *J. Biomed. Mater. Res*, **44**, 422–428.

71. Gibson, I.R., Best, S.M. and Bonfield, W. (2002) Effect of silicon substitution on the sintering and microstructure of hydroxyapatite. *J. Am. Ceram. Soc*, **85**, 2771–2777.

72. Rashid, N., Harding, I. and Hing, K.A. (2004) *Transactions 7th World Biomaterials Congress*, Australian Society for Biomaterials, Brunswick Lower, VIC, p. 106.

73. Arcos, D., Rodriguez-Carvajal, J. and Vallet-Regí, M. (2004) Silicon incorporation in hydroxyapatite obtained by controlled crystallization. *Chem. Mater*, **16**, 2300–2308.

74. Vallet-Regí, M. and Arcos, D. (2005) Silicon substituted hydroxyapatites. A method to upgrade calcium phosphate based implants. *J. Mater. Chem*, **5**, 1509–1516.

75. Arcos, D., Rodriguez-Carvajal, J. and Vallet-Regí, M. (2004) The effect of the silicon incorporation on the hydroxyapatite structure. A neutron diffraction study. *Solid State Sci*, **6**, 987–994.

76. Arcos, D., Rodriguez-Carvajal, J. and Vallet-Regí, M. (2004) Neutron scattering for the study of improved bone implants. *Physica B*, **350**, e607–e610.

77. Clarke, S.A., Walsh, P., Maggs, C.A. and Buchanan, F. (2011) Designs from the deep: marine organism for bone tissue engineering. *Biotech. Adv*, **29**, 610–617.

78. Perry, A.C. (1991) Advances in enucleation. *Ophthal. Plast. Reconstr. Surg*, **4**, 173–182.

79. Matsumoto, T., Kawakami, M., Kuribayashi, K. *et al.* (1999) Effects of sintered bovine bone on cell proliferation, collagen synthesis and osteoblastic expression in MC3T3-E1 osteoblast-like cells. *J. Orthop. Res*, **17**, 586–592.

80. Tsi, W.C., Liao, C.J., Wu, C.T. *et al.* (2010) Clinical result of sintered bovine hydroxyapatite bone substitute: analysis of the interface reaction between tissue and bone substitute. *J. Orthop. Sci*, **15**, 223–232.

81. Sayer, M., Stratilatov, A., Reid, J. *et al.* (2003) Structure and composition of silicon-stabilized tricalcium phosphate. *Biomaterials*, **24**, 369–382.

82. Langstaff, S., Sayer, M., Smith, T. *et al.* (1999) Resorbable bioceramics based on stabilized calcium phosphates. Part I: rational design, sample preparation and material characterization. *Biomaterials*, **20**, 1727–1741.

83. Carrodeguas, R.G. and De Aza, S. (2011) α-Tricalcium phosphate: synthesis, properties and biomedical applications. *Acta Biomater*, **7**, 3536–3546.

84. Yamamoto, H., Niwa, S., Hori, M. *et al.* (1998) Mechanical strength of calcium phosphate cement in vivo and in vitro. *Biomaterials*, **19**, 1587–1591.

85. Dorozhkin, S.V. (2012) Biphasic, triphasic and multiphasic calcium orthophosphates. *Acta Biomater*, **8**, 963–977.

86. Hupp, J.R. and McKenna, S.J. (1988) Use of porous hydroxyapatite blocks for augmentation of atrophic mandibles *J. Oral Maxillofac. Surg*, **46**, 538–545.

87. El Deeb, M. and Roszkowski, M. (1988) Hydroxyapatite granules and blocks as an extracranial augmenting material in rhesus monkeys. *J. Oral Maxillofac. Surg*, **46**, 33–40.

88. Daculsi, G., Laboux, O., Malard, O. and Weiss, P. (2003) Current state of the art of biphasic calcium phosphate bioceramics. *J. Mater. Sci.: Mater. Med*, **14**, 195–200.

89. Daculsi, G., Passuti, N., Martin, S. *et al.* (1990) Macroporous calcium phosphate ceramic for long bone surgery in humans and dogs. Clinical and histological study. *J. Biomed. Mater. Res*, **24**, 379–396.

90. Schopper, C., Ziya-Ghazvini, F., Goriwoda, W. *et al.* (2005) HA/TCP compounding of a porous CaP biomaterial improves bone formation and scaffold degradation – A long-term histological study. *J. Biomed. Mater. Res*, **74B**, 458–467.

91. Daculsi, G., Legeros, R.Z., Nery, E. *et al.* (1989) Transformation of biphasic calcium phosphate ceramics in vivo: ultrastructural and physicochemical characterization. *J. Biomed. Mater. Res*, **23**, 883–894.

92. Yang, X. and Wang, Z. (1998) Synthesis of biphasic ceramics of hydroxyapatite and β-tricalcium phosphate with controlled phase content and porosity. *J. Mater. Chem*, **8**, 2233–2237.

93. Reid, J., Pietak, A., Sayer, M. *et al.* (2005) Phase formation and evolution in the silicon substituted tricalcium phosphate/apatite system. *Biomaterials*, **26**, 2887–2897.

94. Sendemir-Urkmez, A. and Jamison, R.D. (2007) The addition of biphasic calcium phosphate to porous chitosan scaffolds enhances bone tissue development in vitro. *J. Biomed. Mater. Res*, **81A**, 624–633.

95. Sánchez-Salcedo, S., Izquierdo-Barba, I., Arcos, D. and Vallet-Regí, M. (2006) In vitro evaluation of potential calcium phosphate scaffolds for tissue engineering. *Tissue Eng*, **12**, 279–290.

96. Alam, I., Asahina, I., Ohmamiuda, K. and Enomoto, S. (2001) Comparative study of biphasic calcium phosphate ceramics impregnated with rhBMP-2 as bone substitutes. *J. Biomed. Mater. Res*, **54**, 129–138.

97. Vallet-Regí, M. and Arcos, D. (2008) *Biomimetic Nanoceramics in Clinical Use. From Materials to Applications*, RSC Nanoscience and Nanotechnology, Cambridge.

98. Aronov, D., Rosen, R., Ron, E.Z. and Rosenman, G. (2006) Tunable hydroxyapatite wettability: effect on adhesion of biological molecules. *Process Biochem*, **41**, 2367–2372.

99. Webster, T.J. and Ejiofor, J.U. (2004) Increased osteoblast adhesion on nanophase metals: Ti, Ti6Al4V, and CoCrMo. *Biomaterials*, **25**, 4731–4739.

100. Webster, T.J., Ergun, C., Doremus, R.H. *et al.* (2000) Enhanced functions of osteoblasts on nanophase ceramics. *Biomaterials*, **21**, 1803–1810.

101. Maxian, S.H., Zawadski, J.P. and Duna, M.G. (1993) In vitro evaluation of amorphous calcium phosphate and poorly crystallized hydroxyapatite coatings on titanium implants. *J. Biomed. Mater. Res*, **27**, 111–117.

102. Cai, Y.R., Liu, Y.K., Yan, W.Q. *et al.* (2007) Role of hydroxyapatite nanoparticle size in bone cell proliferation. *J. Mater. Chem*, **17**, 3780–3787.

103. Hench, L.L. and West, J.K. (1990) The sol-gel process. *Chem. Rev*, **90**, 33–72.

104. Liu, D.M., Troczynski, T. and Tseng, W.J. (2001) Water-based sol–gel synthesis of hydroxyapatite: process development. *Biomaterials*, **22**, 1721–1730.

105. Fathi, M.H. and Hanifi, A. (2007) Evaluation and characterization of nanostructure hydroxyapatite powder prepared by simple sol–gel method. *Mater. Lett*, **61**, 3978–3983.

106. Pechini, M.P. (1967) Method of preparing lead and alkaline earth titanates and niobates and coating method using the same to form a capacitor. US Patent 3,330,697.

107. Peña, J. and Vallet-Regí, M. (2003) Hydroxyapatite, tricalcium phosphate and biphasic materials prepared by a liquid mix technique. *J. Eur. Ceram. Soc*, **23**, 1687–1696.

108. Cabañas, M.V. and Vallet-Regí, M. (2003) Calcium phosphate coatings deposited by aerosol chemical vapour deposition. *J. Mater. Chem*, **13**, 1104–1107.

109. Palazzo, B., Iafisco, M., Laforgia, M. *et al.* (2007) Biomimetic hydroxyapatite–drug nanocrystals as potential bone substitutes with antitumor drug delivery properties. *Adv. Funct. Mater*, **17**, 2180–2188.

110. Landi, E., Tampieri, A., Celotti, G. and Sprio, S. (2000) Densification behaviour and mechanisms of synthetic hydroxyapatites. *J. Eur. Ceram Soc*, **20**, 2377–2387.

111. Sz-Chian, L., San-Yuan, C., Hsin Yi, L. and Jong-Shing, B. (2004) Structural characterization of nano-sized calcium deficient apatite powders. *Biomaterials*, **25**, 189–196.

112. Arcos, D. and Vallet-Regí, M. (2013) Bioceramics for drug delivery. *Acta Mater.*, **61**, 890–911.

113. Dimar, J.R., Glassman, S.D., Burkus, K.J. and Carreon, L.Y. (2006) Clinical outcomes and fusion success at 2 years of single-level instrumented posterolateral fusions with recombinant human bone morphogenetic protein-2/compression resistant matrix versus iliac crest bone graft. *Spine*, **31**, 2534–2539.

114. Haddad, A.J., Peel, S.A.F., Clokie, C.M.L. *et al.* (2006) Closure of rabbit calvarial critical-sized defects using protective composite allogenic and alloplastic bone substitutes. *J. Craniofac. Surg*, **17**, 926–934.

115. Seeherman, H.J., Bouxsein, M., Hyun, K. *et al.* (2004) Recombinant human bone morphogenetic protein-2 delivered in an injectable calcium phosphate paste accelerates osteotomy-site healing in a nonhuman primate model. *J. Bone Joint Surg. Am*, **86**, 1961–1972.

116. Blom, E.J., Klein-Nulend, J., Burger, E.H. *et al.* (2001) Transforming growth factor-β1 incorporated in calcium phosphate cement stimulates osteotransductivity in rat calvarial bone defects. *Clin. Oral Implants Res*, **12**, 609–616.

117. Obadia, L., Amador, G., Daculsi, G. and Bouler, J.M. (2003) Calcium-deficient apatite: influence of granule size and consolidation mode on release and in vitro activity of vancomycin. *Biomaterials*, **24**, 1265–1270.

118. Fan, J., Lei, J., Yu, C. *et al.* (2007) Hard-templating synthesis of a novel rod-like nanoporous calcium phosphate bioceramics and their capacity as antibiotic carriers. *Mater. Chem. Phys*, **103**, 489–493.

119. Sudo, A., Hasegawa, M., Fukuda, A. and Uchida, A. (2008) Treatment of infected hip arthroplasty with antibiotic-impregnated calcium hydroxyapatite. *J. Arthroplasty*, **23**, 145–150.

120. Boanini, E., Torricelli, P. and Gazzano, M. (2008) Alendronate–hydroxyapatite nanocomposites and their interaction with osteoclasts and osteoblast-like cells. *Biomaterials*, **29**, 790–796.

121. Peter, B., Pioletti, D.P., Laïb, S. *et al.* (2005) Calcium phosphate drug delivery system: influence of local zoledronate release on bone implant osteointegration. *Bone*, **36**, 52–60.

122. Itokazu, M., Sugiyama, T., Ohno, T. *et al.* (1998) Development of porous apatite ceramic for local delivery of chemotherapeutic agents. *J. Biomed. Mater. Res*, **39**, 536–538.

123. Uchida, A., Shinto, Y., Araki, N. and Ono, K. (1992) Slow release of anticancer drugs from porous calcium hydroxyapatite ceramic. *J. Orthop. Res*, **10**, 440–445.

124. Barroug, A., Kuhn, L.T., Gerstenfeld, L.C. and Glimcher, M.J. (2004) Interactions of cisplatin with calcium phosphate nanoparticles: in vitro controlled adsorption and release. *J. Orthop. Res*, **22**, 703–708.

4

Silica-based Ceramics: Glasses

Antonio J. Salinas

Departamento Química Inorgánica y Bioinorgánica, Facultad de Farmacia, Universidad Complutense de Madrid, CIBER de Bioingeniería, Biomateriales y Nanomedicina (CIBER-BBN), Spain

4.1 Introduction

4.1.1 What Is a Glass?

A glass is a rigid material obtained by heating a mixture of solid materials called precursors until reaching a viscous state that is quickly cooled to avoid the formation of a crystalline structure. When the melt is quenched, the atoms remain in the disordered state characteristic of liquids. Consequently a glass is not a liquid or a solid, although it exhibits the characteristics of both. It can be considered a solid with the structure disordered. The term glass is often used to denote other amorphous solids, like some polymers or metallic alloys. However, this chapter is focused on silica-based glasses although the most relevant features as biomaterials of other covalent glasses based on phosphate or borate will also be mentioned. A glass is a metastable state of matter that tends to be transformed with time into a crystalline substance. However, its useful life in the vitreous state can reach hundreds of years which has made glasses useful for many technological applications and also as biomaterials, as will be explained in this chapter.

The main components of a glass are the network former and the network modifiers (see Figure 4.1). The network former is the essential component of the glass and is the oxide of an element with high valence, that is, one able to form three-dimensional (3D) glass structures. Silicon dioxide, SiO_2, a component of sand, containing silicon with valence (IV) is the most common network former. During the traditional synthesis of a glass, the former needs to be heated at very high temperature until it becomes viscous. For instance, the melting temperature of silica is over 1700 °C. Network modifiers are oxides of low valence metals (+1 or +2) that interrupt the $Si-O-Si$ bonding, the basis of the silicate glass network. The presence of these cations leads to the formation of non-bonding oxygen

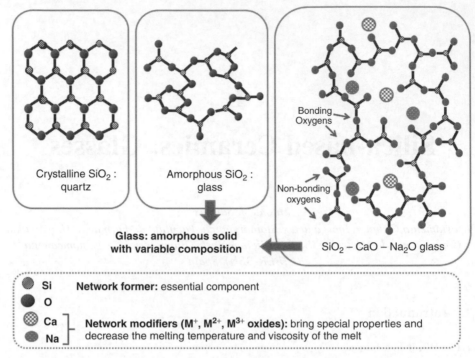

Crystalline SiO$_2$:
quartz

Amorphous SiO$_2$:
glass

Bonding
Oxygens

Non-bonding
oxygens

**Glass: amorphous solid
with variable composition**

SiO$_2$ – CaO – Na$_2$O glass

Si   **Network former:** essential component
O

Ca
Na   **Network modifiers (M$^+$, M^{2+}, M^{3+} oxides):** bring special properties and
decrease the melting temperature and viscosity of the melt

Figure 4.1 *Structures of crystalline silica, amorphous silica, and a typical soda–lime–silica, glass. The main components of a silica glass and their role are also included (See insert for color representation of the figure)*

(NBO) atoms. Usually for glass production amounts of monovalent oxides, such as Na$_2$O, are added to act as a flux, decreasing considerably the melting temperature of the former and making the glass manufacture cheaper. However, binary Na$_2$O–SiO$_2$ glasses are too reactive under atmospheric conditions. To stabilize the resultant glass oxides of divalent cations, such as CaO, are also added. For instance, window glass is usually based on the soda-lime–silica (Na$_2$O–CaO–SiO$_2$) system. If glasses resistant to temperature and chemicals are needed oxides of trivalent elements are also included. Thus, Pyrex® glass belongs to the Na$_2$O–Al$_2$O$_3$–B$_2$O$_3$–SiO$_2$ system. On the other hand, the most important glasses used in medicine belong to the Na$_2$O–CaO–P$_2$O$_5$–SiO$_2$ system where a small amount of P$_2$O$_5$ is included because the inorganic component of bones is calcium phosphate.

In contrast to crystalline substances, whose phases have well defined and constant stoichiometry, glasses can have a continuous grade of their components. They can be synthesized in a practically infinite range of compositions with minimum variations from one to the other. Consequently, they can be designed to adjust to a specific necessity. Moreover, the properties of a glass optimized for an application can be adjusted by doping. This term usually refers to the inclusion in the glass composition of small amounts of other oxides, typically 1 wt% or less, to achieve additional properties. For instance, glasses were historically doped to obtain specific colors, like iron to get a green glass or copper for blue glass. As we will see later, glasses for clinical uses are also doped with inorganic elements exhibiting a therapeutic action: antimicrobial, osteoinductor, neuroprotective, and so on.

4.1.2   Properties of Glasses

An important property of glasses is the glass transition. Thus, whereas crystalline solids melt at a well defined temperature, in glasses does there is not a sudden change from solid to liquid when the temperature increases. At the glass transition temperature (T_g) the solid glass becomes a viscous liquid glass. T_g is in fact an interval of temperatures that depends greatly on the thermal history of the material: melting temperature, to what temperature it was cooled, and if it was subsequently treated. In structural terms, the solid glass and the liquid glass are similar. Nevertheless, the atom mobility is completely different: fixed in the solid glass state and very mobile in the liquid glass state. In short, a glass is a solid without the structural regularity and the long-range order of crystalline materials, that is to say, it seems a liquid but behaves as a solid.

The theoretical strength of most commercial glasses is very high because they contain a mixture of strong covalent and ionic bonds. Nevertheless, most glasses exhibit fragile behavior where fracture happens by propagation of cracks that originate in a deteriorated surface. Glasses could be strengthened by inhibiting the growth of cracks by obstruction of the crack pathway by an interface, or forcing the crack to work more in compression than in tension. Eliminating the defects of the surface by the dissolution of a small part of the surface can increase greatly the glass toughness until new flaws are formed. The mechanical properties of a glass also depend on the morphology. For example, in fibers, the bending strength and the flexibility of a glass increase when the diameter diminishes.

Glasses use to be considered homogeneous materials, but they can be inhomogeneous at a certain level. Even silica-based glasses, usually considered very homogeneous, are inhomogeneous on a scale smaller than the wavelength of visible light. Thus, variations in the composition or density of different parts of a glass have sometimes been detected. Phase separation is a relatively common phenomenon in glasses. Two types of immiscibility can be observed: the formation of a droplet phase in a matrix and an intertwined microstructure of two compositions. In the first case, if the phase in the matrix crystallizes a glass ceramic is formed. This type of material has many applications in industry and also in medicine, as will be described later. Its main advantages are the reinforced mechanical properties and the low thermal expansion coefficient.

For centuries, the more exploited properties of glasses were their transparency, their isotropic nature, and their outstanding chemical durability in different environments. However, for applications as biomaterials, glass compositions able to react with the biological fluids are designed. Moreover, although true glasses are amorphous, there are different forms in which a glass can be microstructurally designed through a liquid–liquid separation or devitrification.

4.1.3   Structure of Glasses

The atomic structure of glasses influences all their properties, especially bioactivity and degradation rate. Glasses do not present long-range order, therefore it is not possible to determine their structure by X-ray diffraction (XRD) or neutron diffraction (nD). Silicate glasses can be described by analogy with crystalline silicates as arrangements of the coordination tetrahedra [SiO_4]. In a glass network it is possible to define rings of tetraedra of different sizes. Amorphous silicates can have many different sizes whereas in the crystalline substances usually there are rings of only one or two sizes.

On the other hand, when other components, such as network modifiers are incorporated, the glass network is interrupted and NBOs are created to compensate for this incorporation. Another way to describe the glass is by using the network connectivity (NC) parameter that is 4 in the case extreme of pure silica, and 2 for chains of tetrahedra. However, it must take into consideration that NC is an average value and local variations can be found.

Bioactive glasses contain a great amount of network modifiers. In addition the $[PO_4]$ tetrahedron is specially inclined to have four NBOs. In this case the orthophosphate units created play an important role in the bioactivity of the glasses. Bioactive glasses tend to be quite depolymerized (i.e., they have low NC). In this sense a value close to 2 is optimal for the bioactivity that, on average, means that the structure is formed by chains of tetrahedrons, which seems to be optimal to promote the hydroxyapatite (HA) formation, required as we will see later for the bioactive behavior. Atomistic computational simulations have played an important role in our understanding of glasses. Simulations show that the fragmentation is more complex than simply chains of tetrahedra, that most bioactive glasses contain numerous orthophosphate groups not connected to any tetrahedron, and that there are a significant number of orthosilicate groups. Nevertheless, not all the $[PO_4]$ are Q^0 (see Section 4.3.6.1 for an explanation of Q^n notation) which means that the glass must have some P—O—Si connections. This is a controversial subject. Indeed, this connection is rare but it has been found in other compounds.

4.1.4 Synthesis of Glasses

Most commercial glasses are obtained by mixing the reactants (oxides, carbonates, nitrates, or sulfates) and heating to a high temperature (for a silicate glass usually around $1400\,°C$) until a melt is formed. The melt is then poured and quickly cooled until a glass is created (see the top of Figure 4.2). When ultrahigh purity is needed, as in the case of bioactive glasses designed for medical devices, it is necessary to use platinum crucibles. An important difference between glasses comes from how they are conformed. One way is by pouring the melt into a mold that is usually made of preheated graphite. Powdered glasses can be obtained by pouring the melt into water, creating a frit that is subsequently ground. In addition, the glass can be obtained as fibers by drawing strands from the melt in a rotating drum. Due to the low thermal conductivity of glasses, during the cooling process stresses appear that can crack and destroy the pieces of glass. For that reason glasses are annealed at a temperature where they have enough kinetic energy to relax the stresses (slightly higher than T_g) and then cooled slowly so as not to introduce new stresses. The mechanical properties of glasses are their weaker point, but they can be improved.

Glasses allow enormous flexibility in terms of conformation, that is to say, they can be obtained in many different shapes. For instance, a conformation method can produce materials able to be used in biotechnology and medicine, like the glass microspheres. They have two key applications in these areas: to be injected in the blood torrent or to be used to stabilize biological molecules. Glasses can also be obtained as fibers with two kinds of applications: structural materials and optical materials. Fibers can be used to obtain polymer matrix composites reinforced with glass fibers. The glass increases the polymer

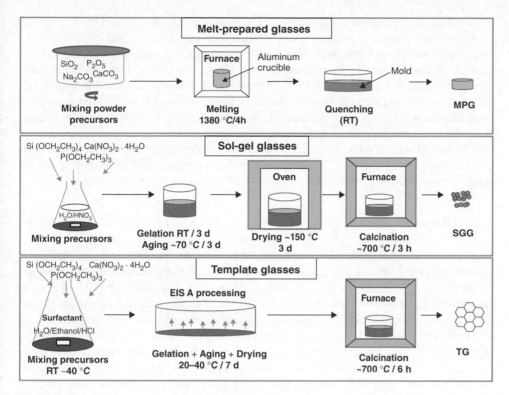

Figure 4.2 *Three methods to obtain bioactive glasses including standard synthesis conditions*

stiffness and the polymeric matrix contributes to avoiding fracture of the fibers. The methods to obtain fiber scaffolds will be described in Chapter 10.

High purity glasses and others that cannot be obtained by quenching of a melt due to the high temperatures used in this method can be obtained by other synthetic routes. One method is by vapor phase deposition starting from silicon tetrachloride, which can be boiled and reacted with oxygen to obtain silica. Another method to obtain high purity glasses is the sol–gel process. There are some glass compositions that cannot be obtained by melting which can be obtained by a sol–gel method, for example, those containing titania or binary compositions of SiO_2–CaO with silica content over 65 wt%. It should be noted that addition of Na_2O is not necessary in gel-derived glasses since it is a low temperature method and it is not necessary now to reduce the melt temperature. The sol–gel method starts with an organometallic compound that is hydrolyzed with water. The species formed then undergo a polycondensation reaction to form the bridge connections of the disordered glass network (See Chapter 5). With this method it is also possible to obtain multicomponent glasses by adding different oxide precursors to the silica network. This method is specially indicated for the preparation of high purity glasses. Other advantages include: (i) it is possible to obtain glasses with tailored porosity (from nanoscale on the microscale) and (ii) because it uses low temperatures, it is possible to incorporate in the glass composition molecules sensitive to temperature, such as polymers or drugs.

As will be see in Section 4.3.2, the surface of glasses plays a key role in the bioactivity. As one cannot study it at atomic scale resolution it is necessary to study it indirectly, for instance by modeling. Some interesting results are obtained: in pure silica glasses the surface tetrahedra are not Q^4. In multicomponent glasses the surface contains many small size rings, particularly two and three membered ones. Nevertheless, the fraction of two and three membered rings is lower in the bioactive glasses because the high number of modifiers allows a more relaxed structure. In summary, in spite of their complex chemistry, the structure of bioactive glasses and their surfaces can be greatly understood using basic concepts such as tetrahedral units and networks of coordination. The structural basis of bioactivity can be ascribed, at least superficially, to the structural characteristics of the defects and the distribution of calcium and sodium cations in the glass surface.

4.2   Glasses as Biomaterials

Bioactive glasses can be classified according to two criteria: the chemical composition of the network former and the synthesis method. With respect to the network former, bioactivity for silicate, phosphate, borate, and borosilicate glasses has been described. Until now the most investigated bioactive glasses contain a silica-like former although phosphate glasses are acquiring relevance, mainly as resorbable materials, and also it seems that glasses based on borate and borosilicate can play an important role in the future as biomaterials. Nevertheless, the most studied and used for clinical applications and approved by the regulatory organisations are the bioactive glasses based on silicate. The greater use of these glasses in implants is because they were the first well-known bioactive synthetic materials, and also because of the beneficial effect attributed to the soluble silicon that is released from bioactive silicate glasses in bone regeneration.

Regarding the synthesis method, the first bioactive glasses were obtained 40 years ago by the traditional method of quenching a melt. The studies to explain the bioactive behavior of silicate glasses demonstrated the bioactive response of the layer of silanol (Si−OH) groups formed on their surface when they came into contact with biological fluids. For that reason 20 years ago the synthesis of bioactive glasses by the sol–gel method was proposed (Figure 4.2). This method produces porous materials with high specific surface and porosity which leads to faster bioactive response kinetics than in classic bioactive melting glasses. Nevertheless, sol–gel glasses (SGGs) are scarcely used clinically because the improvement in bioactivity that they provide is not sufficient to justify the great cost required to bring the product to market. The sol–gel method is being investigated to obtain bioactive glasses in compositions or shapes difficult or impossible to obtain by melting, for example, for metallic substrate coating, or for obtaining fibers that can be used in scaffolds for tissue engineering or in biomedical composites.

Less than a decade ago, a new family of bioactive glasses was synthesized. Their synthesis mixes the sol–gel method with the principles of supramolecular chemistry by using amphiphilic molecules that act as structure directing agents. These glasses are obtained similarly to the mesoporous pure silica materials and display a mesostructure formed by an ordered porosity, in the mesoporous range (i.e., 2−10 nm in diameter), with a very narrow pore size distribution. These glasses are usually called template glasses (TGs), from the use of this type of molecules for their synthesis, and also mesoporous glasses, this being their more relevant structural feature. The bioactive response of TGs is still faster than that of

SGGs and, in addition, they display huge surface areas and pore volumes. Analagous to silica mesoporous materials, these are ideal materials to load with substances with biological activity, like growth factors, hormones, and peptides, that can penetrate into the mesopores and induce bone regeneration. For that reason, these materials conformed in scaffolds by methods that maintain the ordered mesoporosity, as can be done by rapid type prototyping, are promising candidates to apply in tissue engineering of bone. In Figure 4.2 are presented schematically and comparatively the three methods of synthesis of bioactive glasses.

4.2.1   First Bioactive Glasses (BGs): Melt-Prepared Glasses (MPGs)

The composition of glasses used in implants is quite different to that of glasses used in more conventional uses, such as windows, bottles, and so on. A very clear difference is the network former content, that is, the percentage of SiO_2, over 70 wt% for conventional glasses and around 45–50 wt% in highly reactive, and for this reason bioactive, glasses (see Figure 4.3). Also in this figure it is shown how to calculate the NC of a glass composition using the Stevels theory. The first bioactive glass, indeed the first artificial bioactive material, was a glass of the $Na_2O–CaO–P_2O_5–SiO_2$ system, reported by Hench *et al.* in 1971. This glass, later denoted Bioglass® 45S5, contains 45% SiO_2, 24.5% Na_2O, 24.5% CaO,

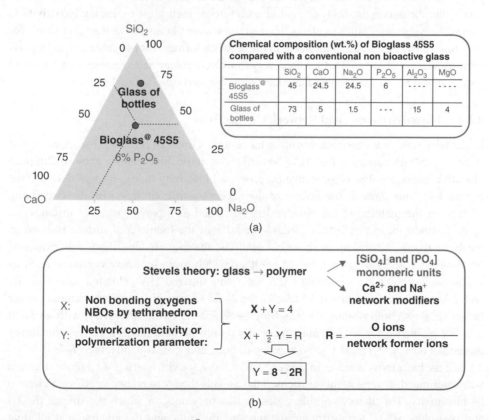

Figure 4.3   *Composition of Bioglass® 45S5 and of bottleglass. Bottom: calculation of the network connectivity for a glass following the Stevels theory*

6% P_2O_5, and 45% SiO_2 (in wt%) and was prepared by melting. This route involves mixing stoichiometric amounts of the precursors, that is, SiO_2, P_2O_5, Na_2CO_3, and $CaCO_3$, melting at temperatures close to 1400 °C, and then subsequent quenching in a graphite mold (blocks preparation) or in water (frit). When Bioglass® 45S5 was implanted in a rat femur it bonded strongly to bone, without the formation of a fibrous capsule characteristic of bioinert materials. This bone bonding ability, later described for other synthetic biomaterials, is usually called bioactivity, a valuable property to consider for the design of new biomaterials that must be in contact with bone.

4.2.2   Other Bioactive MPGs

Glasses with analogous compositions to Bioglass were investigated, looking for faster kinetics of bioactive response but, remarkably, the composition with the quickest bioactive response was the first one, that is, Bioglass® 45S5. Nevertheless, as a result of those studies regions with different *in vivo* behavior were defined in the phase diagram. Different silica contents, calcium oxide and lime were investigated, keeping the P_2O_5 content constant at 6 wt% (or 2.6 mol%). Greater phosphorus contents led to bone necrosis and with smaller phosphorus contents the reactivity of the glasses obtained was too low. It was possible to verify that the maximum SiO_2 content of a traditional melt glass to display bioactivity is 60 mol%. For greater silica contents, inert glasses were obtained. On the other hand, for very low silica percentages the content in the network former was not sufficient and a glass network was not obtained. Furthermore, very low CaO contents led to glasses that dissolved totally, that is to say, degradable materials were obtained (see Figure 4.4).

4.2.3   Bioactivity Index and Network Connectivity

Bioactivity was later described for other bioceramics, mainly certain glass ceramics and some crystalline ceramics like HA. Nevertheless, these bioceramics showed different bioactive behaviors. For comparative purposes, a bioactivity index, I_B, was defined (see Figure 4.4). This index is the inverse of the necessary time (in days) for direct bonding of 50% of the material to the bone and is expressed as a percentage. For instance, an I_B of 5 would mean that it took 20 days for 50% of the biomaterial surface to bond to the bone tissue. According to those calculations, the highest bioactivity was exhibited by Bioglass® 45S5 with an I_B of 12.5. Other widely used bioactive ceramics, such as apatite/wollastonite glass ceramic (A/W GC) and sintered HA exhibited values of 3.2 and 2.3, respectively. Figure 4.4 includes the I_B of several bioceramics, including other bioactive glass compositions, such as Bioglass® 55S4.3 (with 55% SiO_2) with an I_B of 3.2. For bioinert ceramics like alumina or silicon nitride (and melting glasses containing more than 60% SiO_2) there was no bonding to bone and, consequently, their I_B is 0.

Once the bioactivity window in the $Na_2O–CaO–SiO_2$ (with 6 wt% P_2O_5) phase diagram was determined, several authors proposed the Stevels theory to relate the NC of a glass to its bioactivity. This theory considers glasses like polymers, in which the silicate $[SiO_4]$ and phosphate $[PO_4]$ tetrahedra are the monomeric units, and the alkaline and alkaline earth ions, in their network modifiers role, interrupt the connectivity between tetrahedra, generating NBOs that do not connect two tetrahedra, in contrast to bonding oxygens (BOs)

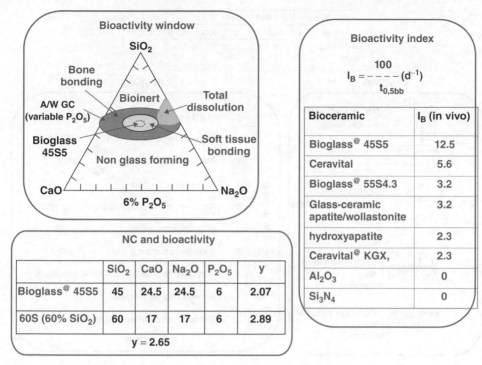

Figure 4.4 *Bioactivity window for melt glasses and two parameters that can be used to quantify the bioactive response of melt glasses: (i) by using the network connectivity (NC) and (ii) the* in vivo *response by the defined bioactivity index,* I_B

that behave as bridges between two silicon atoms in adjacent tetrahedra. Thus, increasing the silica content in a glass reduces the number of NBOs by tetrahedron (X).

In this theory arises a polymerization parameter, Y, that measures the NC (see Figure 4.3). So if

$$X + Y = 4 \text{ and}$$

$$X + 1/2\, Y = R, (\text{with } R = O \text{ ions/network formers ions})$$

$$Y = 8 - 2R$$

Hence, it was possible to calculate that for the glass with greater I_B, Bioglass 45S5, its polymerization parameter is 2.07, whereas for the inert glass with 60% of SiO_2 it is 2.89. It was established that the maximum limit for a melt glass to display bioactivity was $Y = 2.65$ (Figure 4.4).

4.2.4 Mechanism of Bioactivity

Bioactive glasses bond chemically to osseous tissues by the formation on their surface of a hydroxycarbonate apatite (HCA) layer. This is the widely known bioactivity mechanism proposed by Hench in 1991 for silica-based glasses and glass ceramics. It comprises 11 stages. The first five can also be reproduced under *in vitro* conditions (see Figure 4.5).

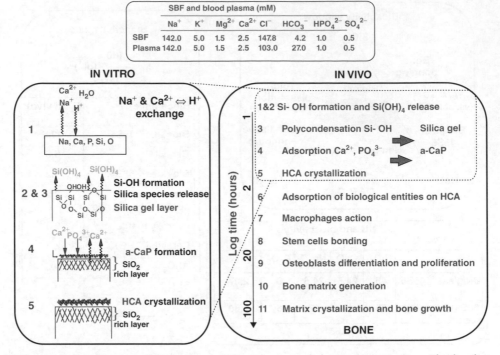

Figure 4.5 *Mechanism of bioactivity for silica glasses and glass ceramics* in vivo. *The five first stages can also be reproduced* in vitro *in solutions mimicking blood plasma, such as SBF (See insert for color representation of the figure)*

Despite this mechanism being initially formulated for $Na_2O-CaO-SiO_2-P_2O_5$ melt prepared glasses (MPGs) it is quite general and can be adapted to other silica-based bioceramics like SGGs. Thus, for MPGs and SGGs, the first three stages of this mechanism lead to the formation of a silica-gel layer, accompanied by an increase in the surface area of the material. The three stages are: (i) the formation of a silica-rich layer, (ii) the deposition of an amorphous calcium phosphate (aCaP) layer, and (iii) the crystallization on a HCA layer. In the six following stages cells and biological entities participate and so they can only be observed when the material is implanted.

4.3   Increasing the Bioactivity of Glasses: New Methods of Synthesis

4.3.1   Sol–Gel Glasses (SGGs)

Sol–gel glasses exhibiting *in vitro* bioactivity were first synthesized in 1991 by Li *et al*. The disordered glass network is obtained here by using a wet chemistry approach. The sol–gel synthesis of a bioactive glass is depicted schematically in Figure 4.2. This route starts with the preparation of a sol, obtained by mixing in an aqueous medium (generally including an acid or basic catalyst) different precursors, namely tetraethyl orthosilicate (TEOS), triethyl phosphate (TEP), and calcium nitrate tetrahydrate as sources of SiO_2, P_2O_5, and CaO, respectively. After mixing the precursors to obtain a sol, hydrolysis and polycondensation processes at room temperature yield a gel which is aged, dried, and stabilized to obtain a sol–gel glass. This synthesis method is time consuming, taking around 10 days. However, the maximum temperature needed, which is reached during the stabilization process, is seldom higher than 600 or 700 °C. A characteristic of gel-derived glasses is that they can be obtained with high purity and homogeneity in a great variety of porous structures with pore diameters ranging from nanometers to millimeters.

In SGGs it is not necessary to include components for decreasing the melting temperature, such as Na_2O as was added in the composition of MPGs. Thus, the $CaO-P_2O_5-SiO_2$ system was the most widely investigated for this type of glass. Moreover, P_2O_5-free SGGs exhibiting *in vitro* bioactivity were described. However, as will be explained later, phosphorus plays an important role in the bioactive response of these glasses. In general, the SGGs composition is simpler than that of MPGs. In Figure 4.6 the bioactivity windows of the most representative MPG and SSG systems are included. In addition, some authors added small amounts of extra oxides, such as MgO, ZnO, Ce_2O_3, or Ga_2O_3, to the ternary system, looking for specific added properties coming from the extra ions, including antimicrobial action, osteogenic character, or others.

On the other hand, the sol–gel method allows one to obtain glass compositions that are not possible to obtain by quenching of a melt. For instance, in SiO_2-CaO glasses the upper SiO_2 limit to avoid phase separation is 65 mol% whereas using the sol–gel method, glasses with SiO_2 contents up to 90 mol% can be obtained. The sol–gel route produces glasses with higher purity and homogeneity and with an expanded range of compositions and textural properties compared with MPGs, at notably lower synthesis temperatures. Sol–gel technology brings nanoporosity, leading to glasses with faster bioactivity kinetics. Moreover, this technology allows one to obtain materials in very different forms, including fibers, coatings, foams, and microparticles. However, the current presence of SSGs in the market is still scarce. This may be because these improvements are not enough to justify the costs associated with their marketing.

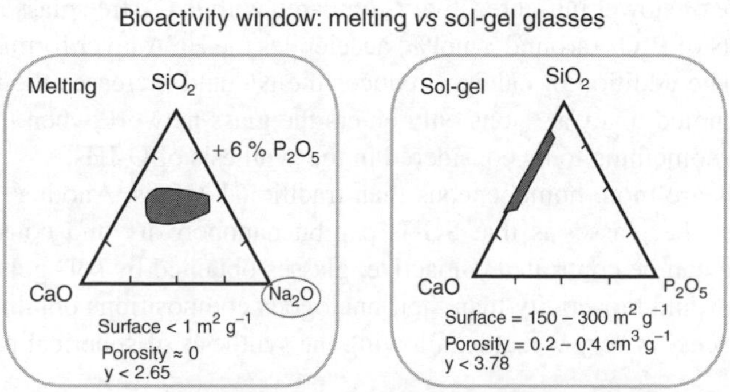

Bioactivity window: melting *vs* sol-gel glasses

Melting — SiO_2 — + 6 % P_2O_5 — CaO — Na_2O — Surface < 1 $m^2 g^{-1}$ — Porosity ≈ 0 — y < 2.65

Sol-gel — SiO_2 — CaO — P_2O_5 — Surface = 150 – 300 $m^2 g^{-1}$ — Porosity = 0.2 – 0.4 $cm^3 g^{-1}$ — y < 3.78

		SiO_2	CaO	Na_2O	P_2O_5	Y	Bioactivity
Connectivity coefficient and bioactivity in glasses	45S5-melt	45	24.5	24.5	6	2.1	+
	60s-melt	60	17	17	6	2.9	–
	58s-sol-gel	58	33	--	9	2.9	+
	68s-sol-gel	68	23	--	9	3.3	+
	91s-sol-gel	91	--	--	9	3.8	+

Figure 4.6 *A comparison of bioactivity windows of MPGs compared with SGGs. Differences between both families of glasses in terms of the network connectivity parameter are also included*

MPGs are dense materials and hence their specific surface and porosity are practically zero. Consequently they need a low connectivity between tetrahedra ($NC < 2.65$) to increase the reactivity by producing sufficient leaching of alkaline and alkaline earth ions to the medium to make possible the chemical reactions that yields the bioactive bond formation. Nevertheless, sol–gel glasses exhibit high specific surface area and porosity as intrinsic characteristics. For that reason, their reactivity in a physiological medium is favored. Thus, bioactivity for glasses with very high SiO_2 contents, even higher than 90%, has been described, as, for example, 91S sol–gel glass whose polymerization parameter, Y, is 3.8, very close to the maximum possible value of 4 for pure silica. Therefore, the bioactivity window for SGGs is significantly different from that of melt glasses, as observed in Figure 4.6. The presence of nanopores is behind this behavior. The silica network is interrupted by protons which behave as network modifiers. Moreover, gel-derived glasses contain a great amount of silanol (Si–OH) groups and for this reason their connectivity is actually minor compared with a melt glass of the same composition.

The low temperature used in SGGs allows the synthesis of porous scaffolds and the inclusion of drugs, growth factors, or polymers to obtain organic–inorganic hybrids (O-IHs). In addition, it is possible to make monolith or nanometric (or submicrometric) particles just by varying the catalyst (acid or base, respectively) and the TEOS to water molar ratio used during the sol preparation. Furthermore, this process allows one to obtain phosphate glasses and also amorphous titania, TiO_2. Up to now titania-containing glasses have only been applied to thin films synthesis and sol–gel phosphate glasses are too soluble and brittle to be used in implants.

The most investigated bioactive sol–gel glasses belong to the SiO_2–CaO–P_2O_5 or SiO_2–CaO systems. The role of the different glass components in textural properties (surface area and porosity) and bioactive behavior has been studied. It was found that SiO_2 and CaO were essential to obtain a bioactive response, but P_2O_5 was not essential. However, small amounts of P_2O_5 modified the mechanism of bioactivity. The final result was that, in spite of slower initial reactivity compared with P_2O_5-free glasses, the presence of small amounts of P_2O_5 (around 3 mol%) accelerates the HCA layer formation indicative of bioactivity. The addition of calcium reduces the NC and increases the nanopore size. It must also be noted that Ca^{2+} ions only enters the glass network when the temperature reaches 400 °C, something to be considered in the synthesis of O-IHs.

In brief, SSGs are more homogeneous than traditional MPGs. Another significant difference between the glasses is that SGGs exhibit nanoporosity and consequently their degradation rate can be controlled. Bioactive, glasses obtained by sol–gel exhibit greater degradation rates and bioactivity than their analogous compositions obtained by melting. The sol–gel process is very versatile, allowing the synthesis of spherical particles, fibers, or foams.

4.3.2   Composition, Texture, and Bioactivity of SSGs

The most relevant features of SGGs are their textural properties, that is, surface area and porosity. In Figure 4.7 are included the surface area, pore volume, and pore diameter, calculated from nitrogen adsorption measurements, for different SGGs all containing 4 wt% of P_2O_5. As observed, the surface area (S_{BET}: Brunauer, Emett, and Teller surface area) increases whereas the pore volume and pore diameter decrease with increasing amounts

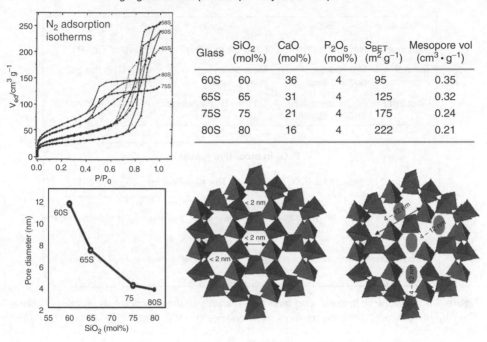

SoL-gel glasses: composition/porosity relationship

Glass	SiO$_2$ (mol%)	CaO (mol%)	P$_2$O$_5$ (mol%)	S$_{BET}$ (m^{2}g^{-1})	Mesopore vol (cm$^3\cdot$g^{-1})
60S	60	36	4	95	0.35
65S	65	31	4	125	0.32
75S	75	21	4	175	0.24
80S	80	16	4	222	0.21

Figure 4.7 *Sol–gel glasses: composition porosity relationship*

of silica in the glasses. As is observed, a low CaO content (related with a high SiO$_2$ content) increases micropore formation. This explains why the glass with lower CaO content is the one with higher surface area but lower pore volume. This effect is visualized in the schematic representations at the bottom of the figure.

On the other hand, Figure 4.8 collects the role of different glass components in the structure, textural properties, and bioactivity. The role of the network formers and network modifiers in the glass structure and in the textural properties has been described. Now, we will focus on the role of the essential glass components in the bioactivity. For MPGs, the role of each of oxide was determined. Thus, although in the glass surface there are abundant silanol groups (Si–OH), a consequence of the NBOs in the surface, their concentration increases considerably as a result of the leaching of Na$^+$ and Ca^{2+} ions to the liquid medium. As we previously saw, this high concentration of silanols plays an essential role in the bioactivity. In addition, as a consequence of the superficial degradation of glasses Si(OH)$_4$ is formed. The importance of the soluble silica is that it has high osteoinductor character, favoring the transformation of non-differentiated cells in osteoblasts.

Regarding the network modifiers, that is, Na$_2$O and CaO, they play an important role in the ionic interchange that takes place initially when the glass contacts with physiological fluids. In this way, an increase in the surface silanol groups takes place, as was mentioned. Moreover, as a result of the Ca^{2+} leaching, an increase in the calcium concentration takes place in solution which increases the medium supersaturation, which facilitates the formation of the HCA layer characteristic of bioactive materials.

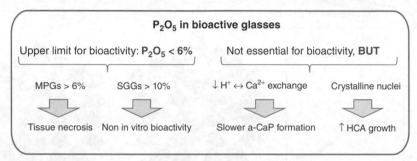

Figure 4.8   *Structural, textural, and bioactive features of the components of sol–gel glasses. The role of P_2O_5 and the limits for bioactive behavior of MPGs and SGGs are also included*

Of the basic components of bioactive glasses, the role of phosphorus was the more controversial and interesting. Indeed, it was verified that it is possible to obtain bioactive glasses without the presence of this element. Nevertheless the synthesis of bioactive glass was not possible without the presence of calcium. The role attributed to phosphorus is to favor the nucleation and the crystalline growth of the HCA phase, but is not an essential component in the glass composition since the nano-HCA layer can be formed from phosphate ions in the surrounding solution. In Figure 4.8 the effect and limits of P_2O_5 for bioactive behavior are compiled.

4.3.3   Biocompatibility of SGGs

Figure 4.9 displays two typical strategies to evaluate the biocompatibility of SGGs under *in vitro* and *in vivo* conditions. The *in vitro* study shows osteoblast cultures on the surface of three different bioactive glasses of the SiO_2–CaO–P_2O_5 system, and on the HCA layer formed on the glass surfaces after seven days soaked in simulated body fluid (SBF). As can be seen, the best osteoblast attachment is obtained for the HCA layer. In the *in vivo* study in rabbit femur the great bioactivity of the sol–gel glass can be seen, visualized by the direct apposition of new formed bone to the bioactive glass implant without the detection of fibrous tissue.

4.3.4   SGGs as Bioactivity Accelerators in Biphasic Materials

Due to the excellent bioactivity of the sol–gel glasses, their application field can be expanded by preparing mixed materials, where the glasses act as a complement to induce

Biocompatibility of sol-gel glasses

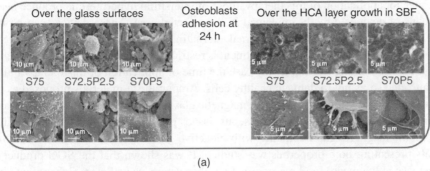

(a)

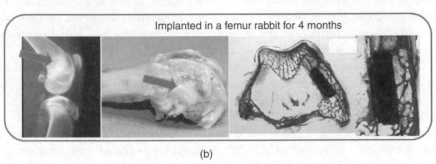

(b)

Figure 4.9 *(a) In vitro biocompatibility studies of $SiO_2-CaO-P_2O_5$ glasses, all with 25 mol% of CaO. (b) In vivo biocompatibility studies of a 80% $SiO_2-17\%$ CaO–3% P_2O_5 sol–gel glass*

or to increase bioactivity. Several mixed materials were investigated including systems for releasing therapeutic action substances to diminish the problems derived from the biomaterial implantation: sol–gel glass–polymer–drug, magnetic bioactive materials for the treatment of osseous tumors by hyperthermia, sol–gel glass-magnetic component, and HA-based materials with improved kinetics of the new apatite formation due to the presence of the glass, SGG–HA.

SGG–polymer–drug systems. The synthesis of these materials to obtain systems for controlled release of drugs, such as anti-inflammatory and antibiotics, was investigated. *In vitro* studies showed that the drug release depends on its solubility and the porosity and capability to absorb liquid of the piece matrix. If bioactive sol–gel glasses are used as the inorganic component in these systems, both bioactivity and a system of controlled release is achieved. In this way, the osseous integration is improved and the drug release is favored, due to the ionic interchange between the glass and medium. Gentamicin release from sol–gel glass implants was studied under *in vivo* conditions in New Zealand rabbits. Gentamicin levels found in the organs indicate that these implants are useful as drug release systems in bone during the 12 weeks of study.

SGG–magnetic glass ceramic. The combination of both materials may be an alternative for treating osseous tissues affected by tumors, since it will be possible to obtain bioactive and magnetic pieces, able to regenerate bone and apply a hyperthermia treatment. This treatment was discovered 70 years ago and is based on the higher sensitivity to high temperatures of cancer cells in comparison with healthy ones. Among the techniques for application of this therapy, those of interstitial hyperthermia produce heat by means of devices directly implanted in the affected area. These devices can be magnetic materials, which under an external alternating field generate heat due to hysteresis losses or parasite currents. Heat treatment is restricted to the desired volume, not affecting the surrounding tissues. In any case the time of application of this therapy must be controlled so as not to damage healthy cells. Among those materials proposed as thermoseeds in interstitial hyperthermia, magnetic glass ceramics are interesting alternatives for the substitution and filling of osseous tissues presenting tumors. The possibility to combine magnetic glass ceramics with bioactive glasses, synthesizing biphasic materials presenting both properties was studied. It was shown that the SGG produces an increase in the macro- and mesoporosity of the system, as well as a higher reactivity in SBFs, conferring high bioactivity to these biphasic materials. However, the presence of the SGG seems to partially inhibit the iron incorporation into the crystalline phases of the glass ceramic, which decreases the magnetic behavior of the material.

SGG–HA. As was previously pointed out, HA is the most similar (chemically and structurally) material to the mineral component of bone and for this reason presents an excellent biocompatibility with living tissues. However, as was pointed out, the HA bioactivity is lower than that of bioactive glasses. Biphasic SGG–HA materials were investigated and new apatite was found to exhibit faster kinetics of formation than pure HA. A biphasic mixture 30% CaO–70% SiO_2–HA was coated with a calcium phosphate layer after 12 h in SBF, whereas the surface of pure HA was not modified even after 45 days of immersion.

4.3.5 Template Glasses (TGs) Bioactive Glasses with Ordered Mesoporosity

TGs constitute a new generation of structurally unique materials exhibiting order on the mesoscopic-scale (2–50 nm) and disorder on the atomic-scale. These glasses are synthesized by the combination of supramolecular chemistry and sol–gel technologies (see Figure 4.2). Thus, materials with unique nanostructural characteristics and extremely high surface areas and pore volumes are obtained. Consequently, some TG compositions exhibit the fastest *in vitro* bioactive kinetics for an artificial material measured to date. Moreover, the ordered mesoporous arrangement allows the use of TGs in drug delivery systems for bone tissue disease treatment, an added value of these materials which increases its future applicability. The pore channels formed within the TGs are separated by amorphous silica walls and arranged periodically on lattices, like the atoms in ordinary crystals. Figure 4.10 represents an illustrative overview of the main differences between SGGs and TGs. These differences can be summarized basically on the meso-scale. TGs exhibit an ordered mesoporous arrangement in the range of 2–10 nm which is not present in SGs and MPGs. However, on the atomic scale, both glasses are similar, showing an amorphous glass network.

The ordered mesoporous arrangement produces surface areas and pore volumes significantly higher than glasses obtained by melt or sol–gel methods with analogous composition. This mesoporous arrangement presents narrow and controllable pore distribution on the mesoscopic scale which allows added value in their use as drug delivery systems. As can be observed in Figure 4.11, TGs show an improvement in textural properties over SGGs.

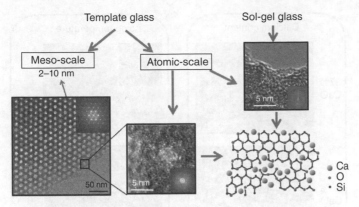

Figure 4.10 *Structural features at atomic and mesoscale of template glasses and sol–gel glasses*

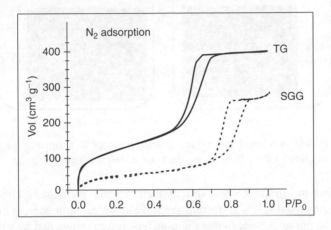

Figure 4.11 *N₂ adsorption isotherms of a SGG and a TG*

   TG synthesis is based on a combination of supramolecular chemistry and sol–gel technology because uses surfactants, as structure directing agents, and the glass network is formed by condensation of the inorganic precursors. However, this synthesis requires special strategies that lead to the presence of a multicomponent inorganic system that includes CaO. The success of the synthesis of TGs is based on the use of an evaporation induced self-assembly (EISA) process. The self-assembly starts with a homogeneous solution of glass precursors and surfactant prepared in ethanol/water with an initial concentration, $c_0 \ll$ cmc, the critical micelle concentration. The concentration of the system is progressively increasing by ethanol evaporation which leads to self-assembly of silica–surfactant micelles and further organization into a liquid crystalline mesophase (Figure 4.2). As can be observed, after a thermal treatment, the TG is obtained.

   This synthesis allows tailoring of the textural and structural features which offers numerous advantages in the control of the bioactive response of TGs. As with pure silica mesoporous materials, the formation of an ordered mesoporous arrangement is mainly guided by parameters such as surfactant nature and concentration, solvent, additives, pH, temperature,

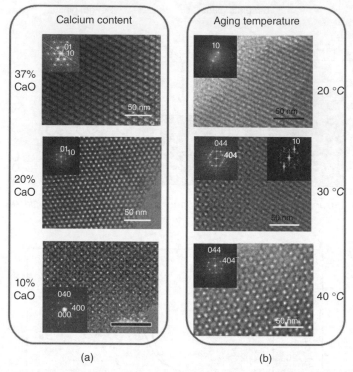

Figure 4.12   *TEM images and Fourier transform diagrams of TGs showing the mesostructural variation as a function of the calcium content (a) and aging temperature (b)*

and so on, which determine the final structural and textural parameters. However, in these multicomponent systems, it was demonstrated that both calcium content and aging temperature play an important role in these parameters. It was shown that by varying the CaO amount in $SiO_2-CaO-P_2O_5$ TGs, the porous structure could be controlled from 3D-cubic to 2D-hexagonal structures. Figure 4.12a shows transmission electron microscopy (TEM) images exhibiting a progressive evolution from a 2D-hexagonal structure (*p6mm*, plane group) to a 3D-bicontinuous cubic structure (*Ia-3d*, space group) with decreasing CaO content in the glass from 37 to 20 and 10%.

Moreover, the evaporation temperature during the EISA process appears to be an effective parameter to control the final mesostructure for a given ternary composition. The structure changes from a 3D-bicontinuous cubic structure to a 2D-hexagonal structure on decreasing the solvent evaporation temperature (Figure 4.12b). This could be explained taking into account that when the temperature increases, the hydrophilicity of the micelles decreases, allowing 3D cubic structures. The possibility to tailor the pore structure and textural properties at the nanometric level allows an exhaustive control in the final properties such as bioactivity/reactivity and capability to load and controlled release biologically active molecules. This possibility extends the range of applicability of these materials.

Concerning the *in vitro* bioactivity, TGs exhibit more bioactivity than conventional MPGs and SGGs due to their outstanding values of surface area and porosity. Figure 4.13 summarizes clearly such differences. TEM images confirm the presence of needle-like crystals of

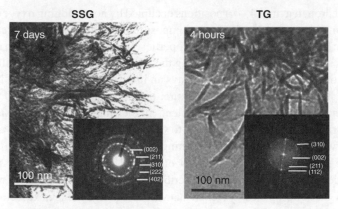

Figure 4.13 *TEM images and electron diffraction patterns corresponding to the material formed on the surface of glasses after soaking in SBF*

HA phase formed on the surface of a SGG and a TG with similar composition after soaking them in SBF. Both images exhibit identical settings, but the incubation times are markedly different. While for the SGG, the time is 7 days, for the TG it is only 4 hours. There is currently controversy regarding which are the factors that contribute to this great improvement in bioactivity. Although, in part, it is believed that the composition of the TG also plays a predominant role, similar to those found for melt and sol−gel glasses, the enhanced textural properties are recognized to be the dominating contribution to the superior *in vitro* bioactivity, which has been attributed mainly to the highly ordered arrangement of uniform sized mesopores. In the next section the structure at the atomic level of each glass family will be studied to establish which factors are dominant in the bioactive process.

4.3.6 Atomic Length Scale in BGs: How the Local Structure Affects Bioactivity

Glass bioactivity depends on both the glass dissolution rate and the capability to be coated by an apatite-like phase when in contact with physiological fluids. These parameters can be controlled by tailoring the composition and the features at the atomic length scale. Several studies using advanced characterization techniques, including high resolution transmission electron microscopy (HRTEM) and solid state nuclear magnetic resonance (NMR) have determined the complex amorphous structure at the atomic scale for all the types of glasses.

MPGs, SGGs, and TGs, containing phosphorus in their composition, exhibit two different domains: (i) an amorphous SiO_2−CaO phase and (ii) calcium phosphate clusters embedded in the amorphous phase. In this section the main aspects of these domains and their role in the bioactive response will be addressed.

4.3.6.1 The SiO_2−CaO Amorphous Component

The SiO_2−CaO−(Na_2O) amorphous component is mainly formed by an interconnected network of $[SiO_4]$ tetrahedra with network modifiers (such as Na^+ or Ca^{2+}) which break the Si−O−Si linkages. The structural building blocks of silicates are commonly described using the Q^n notation, where n denotes the number of bridging oxygen (BO) atoms at the

[SiO$_4$] tetrahedron, leaving $(4 - n)$ positions occupied by non-bridging oxygen (NBO) ions. Silica NC is mainly dictated by the glass composition and the synthesis process, which will determine the final bioactivity/reactivity properties. In general, an increase in network modifiers provokes a decrease in NC, increasing the NBO units in the three families of glasses.

Moreover, the glass synthesis method plays an important role in the silica NC. Bioactive MPG compositions require an open structure, built primarily by Q^2 and Q^3 units, which could explain why glasses with SiO$_2$ contents higher than 60% do not exhibit bioactivity. Such a structure provides high access to Ca and Si for aqueous attack, allowing ionic interchange with the protons of the medium which leads to silica gel layer formation and increases the surface area and the amount of Si–OH groups (not initially present in the MPGs surface). These features were described as limiting factors in the mechanism of bioactivity in glasses proposed by Hench that will be detailed in the next section. Moreover, for SGGs, sol–gel technology allows one to obtain materials with a lower interconnected silica network, which yields materials with inherent nanoporosity, surface area, and Si–OH groups. This could explain why in SGGs the bioactive kinetics is accelerated with respect to MPGs and why bioactivity for SGGs with silica contents as high as 90 mol% can be achieved. This situation is similar in TGs, where the ^{29}Si NMR studies showed analogous distribution of Q units, that is, Q^4, Q^3, and Q^2 units. Therefore, the silica environment cannot explain their accelerated bioactive behavior with respect to SGGs (Figure 4.13). This behavior can be explained at the nanoscopic length scale since the ordered mesoporous arrangement provokes outstanding textural properties which notably increase the accessibility of the surrounding medium to the silica environment, as previously reported.

4.3.6.2   The Local Structural Role of P: Calcium Phosphate Clusters

In general, it was reported that phosphorus additions to a bioactive glass led to a faster HCA formation in *in vitro* bioactive assays than with their CaO–SiO$_2$ and pure silica counterparts. Thus, the presence of phosphorus in (NaO$_2$)–CaO–SiO$_2$–P$_2$O$_5$ glasses triggers questions regarding its structural role, primarily its relationship with the Ca and Si environments. NMR and HRTEM studies have reported that the joint presence of calcium and phosphorus in a glass composition leads to the formation of calcium phosphate clusters, which are embedded in the SiO$_2$–CaO amorphous network. A schematic illustration of the P structural role in bioactive glasses is shown in Figure 4.14. HRTEM studies, in MPGs and SGGs showed the presence of small domains of a crystalline phase, composed of Ca and P, inside the amorphous SiO$_2$–CaO network. These Ca-P clusters may act as nucleation sites that increase the crystallization kinetics of the HCA layer.

In the case of TGs, initial studies based on HRTEM coupled with energy dispersive X-ray spectroscopy (EDS) showed a homogeneous element distribution over tens of nanometers. However, a deep study by NMR clarified the presence of domains of amorphous calcium phosphate (aCP) homogeneously distributed at the pore walls, which cannot be observed by HRTEM (Figure 4.14). The schematic picture corresponding to TGs shows the distribution of aCP clusters in the mesoporous arrangement. This allows an inherent high accessibility of the surrounding medium to these clusters, which act, as in other glass families, as nucleation sites of HCA. Such high accessibility is not present for SGGs as can be seen in the corresponding schematic illustration of these glasses included

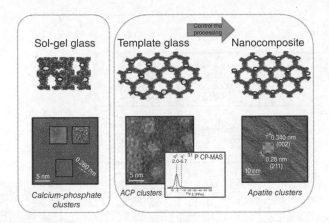

Figure 4.14 *Nanostructural characteristics of SGGs, TGs, and nanocomposite systems*

in Figure 4.14. The high accessibility of the nuclei sites could explain the accelerated bioactivity exhibited by these glasses.

Recently, the possibility to synthesize nanocomposite systems formed by nanocrystalline apatite particles embedded in the amorphous SiO_2–CaO matrix which exhibit improved properties with respect to TGs has been reported. This was possible thanks to an exhaustive control of the crystallization rate of aCP clusters present in TGs to nanocrystalline apatite particles (~ 12 nm in size). Such control was possible with certain variations in the TG synthesis conditions, like the use of surfactants with largerer polar heads and decreasing the initial micellar/precursor concentration (c_0). These nanocomposite materials exhibit a similar bioactive response to TGs but with an added value concerning the improvement in the cell proliferation response due to the presence of these nanocrystalline clusters instead of the aCP clusters in TGs.

4.3.7 New Reformulation of Hench's Mechanism for TGs

The original Hench mechanism of bioactivity described in Figure 4.5 was proposed for bioactive melt glasses and glass ceramics. This mechanism was adapted for SGGs in terms of the quicker speed of the stages 1–3 because in these glasses the silanol-rich layer was already present in the glass surface before the *in vitro* studies. This explained the quicker bioactive response of SGGs. However, in TGs, these steps are notably accelerated, and may even be partially avoided. This is one reason for the extremely enhanced bioactivity observed from TGs, although the CaP clusters formed in the P_2O_5-containing glasses also play a significant role in the accelerated formation of HCA.

Furthermore, some TG compositions with a 2D-hexagonal mesoporous arrangement and a high CaO content of 37 mol% exhibited a biomimetic mechanism with a sequential transition aCP → octacalcium phosphate (OCP) → HCA maturation, similar to the *in vivo* biomineralization process. To date, the accepted mechanism in other bioactive glasses involves the crystallization of an apatite-like phase directly from aCP. This biomimetic bone mineralization can be observed by TEM (Figure 4.15). The previously mentioned OCP formation as an intermediate step to the HCA crystallization is mainly promoted by a local

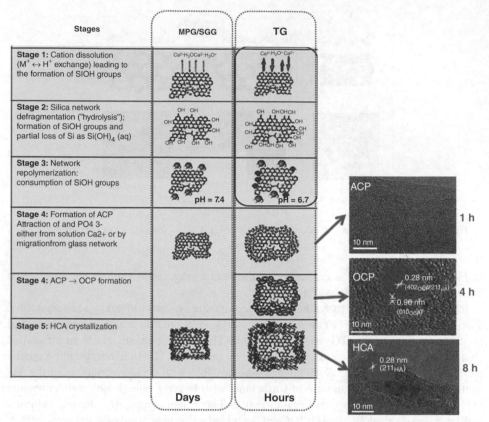

Figure 4.15 *Bioactivity mechanism of melt and sol–gel glasses compared with the extremely bioactive template glasses*

decrease in pH values below 7, being the consequence of their high surface area together with the large calcium content in this TG. This biomimetic behavior suggests bioactive reformulation of Hench's mechanism for this new glass family as shown in Figure 4.15.

4.3.8   Including Therapeutic Inorganic Ions in the Glass Composition

As was indicated, in recent years small additions of other components have been added to the basic glass composition looking for extra biological properties in the materials besides bioactivity. Glasses with fluorine were synthesized by the partial substitution of CaO by CaF_2. It was verified that this substitution did not significantly affect the bonding behavior to bone. Indeed it only produced a reduction in the glass dissolution rate in the glasses that were totally soluble, as a consequence of the low Ca content. Furthermore, substitutions of MgO by CaO or K_2O by Na_2O exerted litle influence on the bioactive bond.

On the other hand, trivalent ion oxides, like Al_2O_3 and B_2O_3, when included in the glass composition produced a modification of the processing parameters and the ratess of dissolution of the surface. Specifically, Al_2O_3, unlike B_2O_3, can inhibit bone formation. The maximum alumina content for which a glass continues to be bioactive depends on the glass composition but usually it varies between 1 and 1.5 mol%. Other elements of the periodic table were tried to improve the structural characteristics and bioactivity of glasses without obtaining significant improvements, hence bioactive melt glasses belong generally to the $SiO_2 - Na_2O(K_2O) - CaO\ (MgO) - Al_2O_3 - B_2O_3$ system.

At a later stage, other ions were added to glass compositions searching for a specific therapeutic activity, including Ag^+ (antibacterial), Sr^{2+} (antiosteoporotic: inhibit osteoclast activity), and many others such as Zn^{2+}, Ga^{3+}, Ce^{3+}, or Cu^{2+}. In all cases the bioactive response decreased although the glasses had important extra properties such as osteogenic or antimicrobial character.

4.4   Strengthening and Adding New Capabilities to Bioactive Glasses

4.4.1   Glass Ceramics (GCs)

The main advantages of devitrified glasses (that is glasses where a crystalline phase is grown into an amorphous glass) are generally observed in the mechanical properties, where the crack propagation can be stopped by the presence of a glass/crystal interface and in thermal expansion where the presence of a crystalline phase can help to control the net thermal expansion of a glass. Important features and application areas of glass ceramics as implants are shown in Figure 4.16.

Glass ceramics were discovered in 1959. They can be considered a combination of a glass and a ceramic. Glasses have different structures and microstructures visible under a microscope. In the glass microstructure a phenomenon called phase separation can occur that produces inhomogeneous glasses. On the other hand, ceramics, containing a maximum 1% in volume of glassy phase, are strong and tough and sometimes are required for other characteristics, like magnetic properties or biocompatibility.

Glass ceramics include both components which produce a material with improved properties with respect to each individual component. For the GCs synthesis, first a glass is obtained, denoted the base or parent glass, which is heated at high temperature to get a partial crystallization (Figure 4.16). From the interaction between glass and ceramic components, a material with extraordinary mechanical properties is obtained. A GC can be defined as a material formed by one or more vitreous phases and one or more crystalline phases with the crystals formed in the base glass. It is remarkable to recognize the possibility to control the formation and growth of crystals in a glass. The controlled process starts with the formation of nuclei (nucleation) that produces very small crystals in specific parts of the base glass. Numerous GCs with very different and specific characteristics can be obtained.

There are two methods to obtain GCs. The first involves the nucleation and crystallization processes within the base glass (internal or bulk processes). In the second, crystallization is controlled on the glass surface. In the internal nucleation, it is necessary to form nucleation sites by the presence of compounds added to the base glass. These agents accelerate the process in the first stages. The condition of the glass and the size of the crystals are preserved by quenching the material. GCs that practically do not expand when heated can be obtained. GCs with this behavior are obtained in the $SiO_2 - Al_2O_3 - Li_2O$ system with numerous technological applications. As nucleation agents TiO_2 or ZrO_2 are often used. When the crystals are not allowed to grow to more than 200 nm in size the GCs are transparent. That is, they seem to be glasses but with highly improved mechanical properties and very low thermal expansion coefficients.

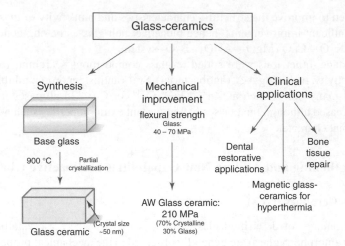

Figure 4.16   *Synthesis, flexural strength, and clinical applications of glass ceramics*

4.4.1.1   High Strength Moldable GCs for Dental Restoration

When the controlled surface crystallization method was used to obtain GCs the resultant material was very strong. However, it was considerably weakened if the surface was damaged. For this reason, controlled crystallization within the glass method is used. GCs are not as strong as many sintered ceramics, like ZrO_2, but they are strong enough and can be molded into the desired shape. A mixture of substances based on the SiO_2-Li_2O system was investigated, including metals or P_2O_5-like nucleating agents. Lithium phosphate phases were formed with the desired crystalline phase of lithium disilicate ($Li_2Si_2O_5$). When these crystals are heated they form an interlocking microstructure that reinforces the material that exhibits a flexural strength in the range 400–700 MPa. The great concentration of crystals (50–60% in volume) is responsible for obtaining materials much stronger than other GCs. However, the material contains some glass that softens if it is warmed up to 920 °C which is why it can be molded in different shapes. In brief, the material combines the high strength advantages of the ceramics with the moldability of glasses, both important characteristics of biomaterials used to repair teeth.

   Lithium disilicate GCs are commonly used to manufacture crowns and dental bridges which do not need to be reinforced with a metal. A very strong GC is used to make the base. Another GC containing apatite crystals is applied on their base and shaped to look like the original tooth. These materials not only exhibit remarkable mechanical properties but also amazing optical properties that gave them an appearance very similar to natural tooth. Tens of millions of units of these products were used after their discovery in 1991. Some commercial names are IPS e.max® Press or IPS e.max® Ceram.

4.4.1.2   Moldable and Machinable GCs

GCs containing Leucite ($KAlSi_2O_6$) crystals were obtained by controlled crystallization of the glass surface. Materials of this type can be shaped by cutting or molding. A commercial product based on these glass ceramics is IPS Empress®. More recent lithium disilicate GCs able to be shaped by molding or cutting have been produced (product IPS e.max® CAD).

Processing of Apatite/Wollastonite Glass Ceramic

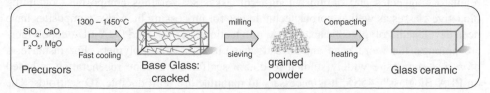

Some bioactive glass-ceramics and relevant features

Name	Inventor(s) year	Phases	Flexural Strength (MPa)	Applications
Ceravital®	Brömer and Pfeil 1973	Apatite Glass	100 – 150	Replacement of ossicular chain in the middle ear
Cerabone® A/W	Kokubo 1986	Apatite Wollastonite Glass	220	Iliac crest, vertebrae intervertebral discs
Ilmaplant®	Berger et al 1989	Apatite Wollastonite Glass	170	maxillofacial implants
Bioverit® I	Holland 1983	Apatite Flogopite Glass	100 – 160	Orthopedic surgery (spacers) Head and neck surgery
Bioverit® II		Mica Cordierite Glass		Stomatology (tooth root and veneer laminates)

Figure 4.17 *Processing of apatite/wollastonite glass ceramic. Some commercial glass ceramics used for bone regeneration and their applications and relevant properties are also included*

In Figure 4.17 shows a schematic of the synthesis of A/W GC, as well another glass ceramics designed for bone tissue repair.

As glasses, glass ceramics have also been used to obtain coatings. Some strategies used to obtain them are enameling, glazing, plasma spraying, radio frequency magnetosputtering deposition, or pulsed laser deposition. Each system has advantages and disadvantages in terms of homogeneity, layer thickness, facility of application, and tailored properties. Bioceramic coatings for implant devices will be described in Chapter 9.

4.4.2 Composites Containing Bioactive Glasses

Another strategy to improve the mechanical properties of bioactive glasses was to obtain bioactive glass–polymer composites where the glass particles or fibers behave as a reinforcing phase. It must be noted that the outstanding properties of bone in tension, compression, and with cyclical loads, that is, fatigue resistance, arise because this living tissue is a natural composite of collagen and a ceramic, in concrete nano-crystalline HCA. Collagen confers high flexural and tensile strengths to bone, whereas the ceramic component brings stiffness and compressive strength.

Many biodegradable, natural, and synthetic polymers were studied to obtain composites with bioactive glasses. Natural polymers, such as collagen, gelatin, and many others, were investigated. However, biomedical companies rather prefer to use synthetic polymers because they have greater control of the polymer properties, satisfy the requirements of regulatory agencies and diminish long-term risks. Scaffoldings composites containing bioactive glasses as well as porous technologies for processing bioactive composites have been developed. Polyglycolic acid, PGA, polylactic acid, PLA, and PGA-PLA copolymers are amongst the more used and investigated biodegradable polymers. For instance, a bioactive composite widely studied to obtain scaffolds for bone regeneration is based on PLA-Bioglass® 45S5. It is necessary to manufacture reproducible 3D scaffolds. The main factor affecting the mechanical properties and structural integrity of scaffolds is their porosity, that encompasses volume, pore size orientation, and pore interconnectivity. Thus, the development of fabrication technologies able to produce scaffolds with tailored porosity is essential.

4.4.3   Sol–Gel Organic–Inorganic Hybrids (O-IHs)

Some clinical applications require properties that cannot be achieved with the bioactive glasses and glass ceramics that are hard and rigid and difficult to be functionalized with specific functional groups. For instance, in bone repair applications it is very important to gain flexibility and toughness without losing bioactivity. O-IHs are synthesized by combining organic and inorganic components with interactions at the molecular scale that must be indistinguishable at submicrometric level. This is a key difference between O-IHs and composites because a composite is a combination of two or more distinguishable components. These materials, developed in the early 1980s, have received several names such as Ormosils (Organic Modified Silicates), Ceramers (Ceramic polymers), or O-IHs. Another name sometimes used is nanocomposites, from the nanometric or submicrometric scale of the interactions between the components. However, some authors consider this nomenclature a mistake because it can bring some misunderstanding. To date no medical device employing O-IHs has been approved by the regulatory bodies. However, their outstanding properties indicate that they are materials for the future having enormous potential in a variety of biomedical applications.

The mixtures at the atomic level of molecular O-IHs are advantageous compared with composites where (i) the polymeric component very often masks the bioactivity of glass, (ii) the degradation rates of polymer and ceramic are different, and (iii) it is very difficult to get a good interface. In hybrid materials, the properties can be greatly controlled by the selection of the components and the mixing ratio. Up to know most of the investigated O-IHs for clinical applications have been silica-based materials. Indeed they could be considered a modification of sol–gel glasses where, during the sol preparation, a polymeric precursor such as polydimethyl siloxane, PDMS, is added, to produce $CaO-SiO_2-PDMS$, hybrids. There are two methods to obtain covalent bonds between silica and polymer. The first uses a siloxane to introduce silica into a polymer. The second uses a siloxane coupling agent to bridge between a silicate network and the polymer, for instance 3-glycidoxypropyl trimethoxysilane (GPTMS). Thus a series of gelatin-siloxane hybrids have been prepared from gelatin that was grafted with GPTMS. Because of the importance of obtaining bioactivity, very often calcium was added. However, calcium it is not incorporated into the silica network until 400 °C, whereas O-IHs are usually heated at 100 or 150 °C to preserve the integrity of the polymeric component. Consequently, calcium usually appears precipitated on the surface or chelated by the carboxylic groups of the polymers.

The development of O-IHs has led to the production of new materials with an outstanding combination of properties, including toughness, controlled degradation, or chemical functionality. Nevertheless, the synthesis of hybrids is a complex process and to optimize the material is difficult with so many variables to be investigated. The potential is enormous but its development will take time. Since they are new products, transferring to clinical products is going to require time and will be expensive. However, O-IHs could be the answer to mimic the hierarchic structure and properties of living tissues and to produce materials that fulfill the surgeon's requirements.

4.5   Non-silicate Glasses

There are some inorganic glasses not based on silicate that are receiving growing interest in clinical applications. In this section the more studied systems, that is, phosphate glasses and borate (and borosilicate) glasses will be described. Phosphate glasses are soluble in a safe way in the organism. Glasses based on borate networks are soluble and have shown a great potential in wound regeneration, as in the treatment of ulcers in diabetics. Other glasses, such as fluoride- and chalcogenide (arsenic, antimony, tellurium, selenium, and sulfur)-based glasses are finding applications in medicine, some of them associated with their ability to conduct infrared laser light.

4.5.1   Phosphate Glasses

Phosphate glasses, based on P_2O_5 as the glass network former and CaO and Na_2O as network modifiers, can be completely dissolved in the human body, producing ionic species usually present in the organism. Furthermore, the solubility of these glasses can be controlled by modifying their composition. Consequently they can be used as resorbable implants. Therapeutic ions like Sr^{2+}, Zn^{2+}, or F^-, which are released during the degradation process, stimulate the bone growth, the wound healing, and avoid infection. All these capabilities make these glasses very versatile biomaterials for tissue regeneration. Phosphate glasses are synthesized at slightly lower temperatures than silicate glasses, that is, 800–1300 °C. They were also obtained by the sol–gel process, but the resultant materials are fragile and too soluble for most biomedical applications. Phosphate glasses are usually divided into several categories depending on the P_2O_5 content: ultraphosphates, when it is higher than 50 mol%, polyphosphates when is lower than 50 mol%, and metaphosphates when it is equal to 50 mol%. In addition, we can have the invert glasses when the P_2O_5 content is less than 33.3 mol%. Their properties depend not only on the P_2O_5 content but also on the size and charge of the network modifiers.

A safe non-toxic dissolution is the main reason why the phosphate glasses are of interest as biomaterials. In ultraphosphates the rupture of the P–O–P connection is necessary and this happens simply with atmospheric humidity. In polyphosphates this breakage is not necessary because they dissolve by solution of intact phosphate chains which happens in aqueous solution but at a slower rate. In general, the solubility decreases with decreasing P_2O_5 content. When these glasses are dissolved they produce an acid medium. However, when decreasing the P_2O_5 content the pH is close to neutral (7.0). Nevertheless, an acid pH produces a remarkable increase in the glass dissolution although at a slower rate. If in a P_2O_5–CaO–Na_2O glass the Na is replaced by Ca the cell attachment and proliferation increases due to the decrease in the glass solubility. Thus, very high Na_2O contents inhibit the cellular growth. On the other hand, the inclusion of TiO_2 is very effective in controlling the solubility of these glasses, which allows modulation of the adhesion and cellular proliferation. Replacing Ca in Ti-metaphosphate with Sr does not have the expected effects in terms of the reduction of pH. If Zn is included in P_2O_5–CaO–Na_2O glasses (from 1 to 20%) toxic effects are obtained over a specific Zn content which is very important to control in very degradable glasses. *In vitro* studies with cells produce distinct results when discs or the extracts containing the ions released by the glasses are investigated. This demonstrates the complexity and the importance of an appropriate programming of the investigations where living cells participate.

On the other hand, degradable phosphate glass fibers with good mechanical properties are of interest as materials for fractures fixation (composites, polymers reinforced with fibers, meshes, or woven fiber constructs) or soft tissue substitution. Thus, it is possible to reinforce degradable polymers with phosphate glass fibers that maintain the stability and mechanical properties during the last stages of the polymer degradation, maintaining an elastic modulus between 15 and 40 GPa. Therefore they can be used for repairing cortical bone fractures.

In conclusion, phosphate glasses are soluble in aqueous solutions and have similar composition to mineralized tissues. Therefore they are of interest as degradable materials for bone regeneration and controlled release of ions with therapeutic activity. The solubility of these glasses is closely related to their composition and structure and control of the degradation and liberation of ions is critical for their successful application. Composites based on degradable polymers such as PLA and phosphate glasses are a very promising way to modulate the mechanical properties to match them to those of the surrounding tissues.

4.5.2 Borate Glasses

All the bioactive glasses require calcium, normally over 10% in CaO, but after that the glass composition can be widely varied and can be tailored for a specific need. This makes them particularly useful for promoting bone regeneration. In borate glasses the network former is B_2O_3 and the composition can contain an arrangement of oxides of alkaline metals, like Li, Na, or K, alkaline-earth metals, such as Mg, Ca, Sr, or Ba, and transition metals, like Fe, Cu, Zn, Ag, or Au. As the glass degrades *in vivo*, these elements are released at a biologically acceptable rate. Because of their lower chemical durability, some borate bioactive glasses degrade faster and convert more completely to a HA-like material, when compared to typical bioactive silicate glasses such as Bioglass® 45S5, S53P4 or 13-93.

Unlike silicate glasses, borate glasses degrade directly from the HCA layer on the glass surface without the previous formation of a rich borate layer. This is because borate ions are very soluble in the body fluids, similarly to phosphate glasses. The products released from borate glasses go through the organism without problems and are excreted in the urine. The lack of a diffusion layer allows borate glasses to complete the reaction without a significant reduction in the kinetics of dissolution. To establish differences between silica and borate, glasses with intermediate compositions were synthesized, replacing 1/3, 2/3, and 3/3 of the SiO_2 content by B_2O_3. Only the borate glass totally dissolved reaching the theoretical weight loss of 58% whereas those containing silica stopped at 35 and 42%. Thus, the silica rich layer is acting as a conductor layer for the ion dissolution.

Borate bioactive glasses support cell proliferation and differentiation *in vitro* as well as tissue infiltration *in vivo*. Borate bioactive glasses have also been shown to serve as a substrate for drug release in the treatment of bone infection. An important concern associated with borate bioactive glasses was the toxicity of boron released to the solution as borate ions, $(BO_3)^{3-}$. It should be noted that silicate glasses react 1 or 2 orders of magnitude slower than those of borate. For this reason, in conventional "static" *in vitro* culture conditions, some borate glasses were observed to be toxic to cells because the pH can reach values as high as 9 or 10. However, the toxicity was diminished in "dynamic" culture conditions which means that it is necessary to investigate these glasses in the future by using bioreactors. Scaffolds of a borate bioactive glass, denoted as 13–93B3, having a composition obtained by replacing all the SiO_2 in 13–93 glass with B_2O_3 were obtained. Moreover, borate glass pellets implanted in rabbit tibiae produced boron concentrations in the blood far below the toxic level.

In orthopedic applications bioactive glasses have been used for bone regeneration. For this reason the emphasis is generally made in the possibility of forming HCA and osteoblasts type cells differentiate in osteocytes. However, the formation of soft tissues, such as blood vessels (angiogenesis) is also critical. For this reason, the administration of growth factors, such as vascular endothelial growth factor (VEGF) was widely studied to induce cellular growth in bone implants, but they are very expensive. On the other hand, it was reported that the human body uses Cu^{2+} ions to regulate the angiogenesis and these ions can be easily added to the bioactive glasses composition for a slow and controlled release. In addition, the release of these ions can be much more prolonged than a biochemical factor loaded in a bioactive glass.

Many people suffer from diabetes which is usually associated with vascular-deficient wounds that only cure when the vascular tissue is repaired. Bioactive borate glasses exhibited promising results in this application. Thus, small additions of Cu and Zn significantly increase the number of blood vessels adjacent to bioactive glasses. Moreover, in some bioactive glasses the addition of certain elements shifts the HCA formation to calcium carbonate, an HCA precursor that is more quickly dissolved by the osteoclasts than HCA. On the other hand, on doping the glasses with Cu, Sr, Zn, or Fe inhibition of the HCA formation was observed, depending on the size of the ion. For ions with radii smaller than 0.08 nm the inhibition of HCA formation was severe (Zn^{2+}, Cu^{2+}) or moderate (Fe^{2+}, Fe^{3+}, Mg^{2+}). This was explained by the addition of small ions to the HCA structure blocking the formation of other unit cells and stopping the HCA formation.

Borate glass technology is still in its early days. However, it exhibits very interesting results in bone growth applications and the treatment of diabetic wounds. In addition, the future of the borate glasses seems very promising in orthopedics, dental regeneration and in drug delivery systems.

4.6   Clinical Applications of Glasses

4.6.1   Bioactive Silica Glasses

Bioactive silica-based glasses are used for repairing bone defects in orthopedics, dentistry, including maxillofacial, ENT (ear, nose, and throat), cranial, and facial Surgeries. The first commercial product with Bioglass® 45S5 was Douek-Med™ (Mr. Ellis Douek Professor

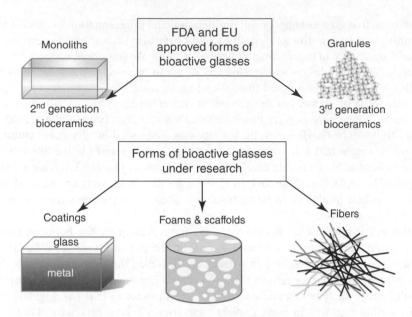

Figure 4.18   *Commercial forms of bioactive glasses*

of ENT surgery at Guy's Hospital, London, England) launched in 1985. It had the shape of cones and was used for the middle ear substitution. The device worked properly, but it was only available in two sizes and each clinical case had to be adapted which limited its commercial success. In 1988, a device also composed of Bioglass® 45S5 was launched for filling the voids caused by tooth extractions; it was called an Endosseous Ridge Maintenance Implant, ERMI® (University of Florida, Gainesville, FL). Both devices exhibit low solubility when implanted and are generally considered as second generation ceramics.

From then, all the commercial products with bioactive glasses have been in particulate form (Figures 4.18 and 4.19 and Table 4.1). In this form a great surface increase takes place with a subsequent huge increase in the ions leached from the glass to the medium. It was reported that soluble silicon when released stimulates genes to produce bone. For this reason, Bioglass® 45S5 in particulate form has an osteostimulatory effect and is considered as a third generation biomaterial. Most of the human clinical tests with Bioglass® 45S5 were made using PerioGlas (Novabone Products LLC), a particulate material for periodontal tissue repair formed by glass particles in the 90–710 µm range. Similar commercial products are Biogran® (Global Headquarters, FL) with a narrower particle size distribution in the 300–360 µm range, and BonAlive® (BoneAlive Biomaterials Ltd., Turku, Finland) based on the slightly different S53P4 glass formed by 1–2 mm granules when proposed for vertebral column applications. Surgeons generally mix the glass particles with blood from the defect so that the mixture acquires the consistency of putty that can be pushed within the bone defect. The blood also acts to introduce cells and natural growth factors in the defect which additionally promote living tissue regeneration. PerioGlas alone behaved as well as or better than an open debridement, an allograft or membrane and HA.

Other dental and maxillofacial clinical studies have shown that the bioactive glass particles promote remineralization at the site of the defect, including the repair of the orbital

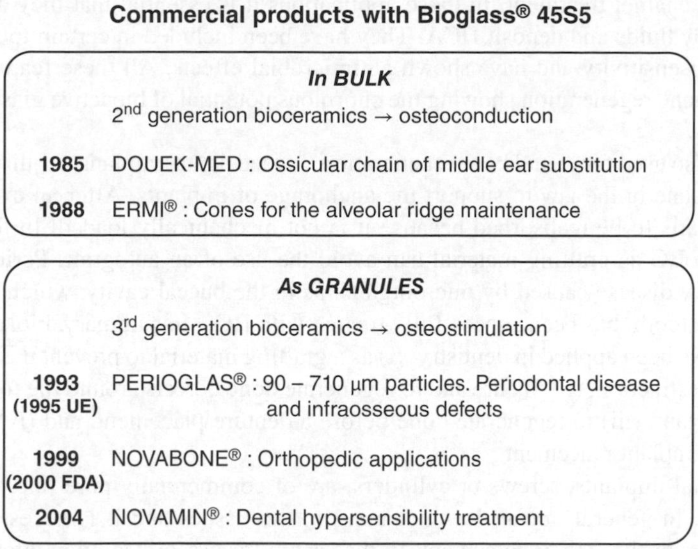

Commercial products with Bioglass® 45S5

> ### In BULK
>
> 2nd generation bioceramics → osteoconduction
>
> **1985**   DOUEK-MED : Ossicular chain of middle ear substitution
>
> **1988**   ERMI® : Cones for the alveolar ridge maintenance

> ### As GRANULES
>
> 3rd generation bioceramics → osteostimulation
>
> **1993**   PERIOGLAS® : 90 – 710 µm particles. Periodontal disease
> **(1995 UE)**   and infraosseous defects
>
> **1999**   NOVABONE® : Orthopedic applications
> **(2000 FDA)**
>
> **2004**   NOVAMIN® : Dental hypersensibility treatment

Figure 4.19   *Commercial products based on Bioglass® 45S5*

Table 4.1   *Composition (wt%) of some bioactive glasses*

Glass	Commercial product	SiO_2	Na_2O	CaO	P_2O_5	K_2O	MgO	B_2O_3	SrO
Bioglass® 45S5	PerioGlass	45.0	24.5	24.5	6.0	–	–	–	–
	NovaBone	"	"	"	"				
	Novamin	"	"	"	"				
	Biogran	"	"	"	"				
S53P4	BoneAlive	53.0	23.0	20.0	4.0	–	–	–	–
Sr10	StronBone	44.1	24.0	21.6	5.9	–	–	–	4.4
$P_{50}C_{35}N_{15}$	–	–	9.3	19.7	71.0	–	–	–	–
13-93B3	–	–	6.0	20.0	4.0	12.0	5.0	53.0	–
13-93B1	–	34.4	5.8	19.5	3.8	11.7	4.9	19.9	–

floor underneath the eye or the sinus floors. Moreover, NovaBone (Novabone Products LLC), a particulate material very similar to PerioGlas, was approved in 1999 for repair bone defects in orthopedics. Another glass compositions are StronBone, similar to Nova-Bone but containing strontium that is slowly released from the glass.

Dentistry uses biomaterials more often than any other area of medicine. The mouth is a specially complicated location because it contains hard and soft tissues and has plenty of microorganisms. Biomaterials must not have friendly surfaces for microorganism colonization, but they must be biocompatible and bioactive because they have to bond to hard and sometimes soft living tissues. Other features required are esthetic appearance, high mechanical resistance, rigidity and hardness for restorative dentistry, and the possibility to be shaped, and sterilized, on site for certain applications. BGs are able to support the regeneration of alveolar bone around the root of the tooth. Moreover they can be used for dentine and enamel treatment. In those applications it is essential that they dissolve partially in body fluids and deposit HCA. They have been included in certain toothpastes for dental hypersensitivity and have shown antimicrobial effects. All these features together induce the bone regeneration showing the enormous potential of bioactive glasses in dental applications.

BGs are also used for dental bone regeneration. Indeed dental implants require a sufficient amount of bone in the jaw to support the anchorage of implants. After an extraction, the jaw bone starts to be reabsorbed because it is not mechanically loaded. In this situation the use of a BG as grafting material can avoid the use of an autograft. Periodontitis, an inflammatory disease caused by microorganisms in the buccal cavity, which can yield to the loss of a tooth, has been successfully treated with BGs. In summary, bioactive glasses granules have been applied in dentistry: (i) as a grafting material to prevent the loss of bone after a root extraction, (ii) to cure and to regenerate bone defects around the tooth root and a metal implant, (iii) to regenerate bone before a denture placement, and (iv) to fill bone voids in an implant placement.

Most dental implants, screws or cylinders, are of commercially pure (c.p.) titanium or their alloys. In general, metal does not bond to the tissues and it is necessary to fix it mechanically or through bone cement. In the search for osseointegration attempts to coat metal with glasses were carried out. This approach has many limitations, including the glasses fragility, glass detachment from the substrate, glass solubilization and its crystallization during the processing. Thus, these attempts have been unsuccessful, as will be explained in Chapter 9.

Until now the wider clinical application of bioactive silica glasses has been for tooth hypersensitivity treatment, incorporated into a tooth paste commercialized as NovaMin® (GlaxoSmithKline LLC, Wilmington, NC). Here Bioglass® 45S5 particles adhere to dentine, blocking the dentinal tubules and relieving the pain. The small particles (20 μm), which cannot be detected by the user, exhibit a great surface area and the necessary solubility. Thus when brushing with NovaMin®, Bioglass particles are bonded to dentine. There, they release ions and increase the pH causing HCA precipitation that occludes the tubules until the permanent filling of the cavities takes place. Similar results were observed when patients were treated with a toothpaste containing S53P4 particles of 20 μm in size.

Another important property of BGs is their antimicrobial capability. It is well known that the colonization by certain microorganisms produces periodontitis and bone resorption. A high concentration of S53P4 grains of around 20 μm produces a toothpaste with antibacterial properties. This was attributed to the pH increase and to the fast increase in the osmotic pressure. Ag^+ ions with well known antimicrobial activity can be introduced into melt glasses, replacing the Na^+ ions. Numerous studies were made in Ag-doped melt and sol–gel bioactive glasses. The results are still controversial, for instance, glasses releasing Ag^+ ions too quickly were antimicrobial but also cytotoxic. Recent studies are including in the glass composition other ions with possible antibacterial activity, such as Zn^{2+} and Ga^{3+}. Moreover, studies are presented regarding the antibacterial capabilities of nanosized Bioglass.

Bioactive glasses can also be incorporated in biostable polymeric composites reinforced with BG fibers, and in composites with degradable polymers such as PLA, PGA, PLA-PGA copolymers as well as polycaprolactone (PCL). Bioactive glasses have been incorporated in composites, like Vitoss-BA (Orthovita Inc) composed of collagen, Bioglass, and tricalcium phosphate $(Ca_3(PO_4)_2)$ proposed to fill bone orifices, or Origen (Orthofix) developed for the same application composed of Bioglass, demineralized bone matrix (DBM), and gelatin. In addition, bioactive glasses were added to ionomers glass cements. They are powdered glasses of fluoraluminosilicate mixed with polymers soluble in water, generally acrylic. They can release F^- ions to the medium.

Patents have been also deposited in other areas proposing the use of bioactive glasses in products of personal care (Schott) or in formulations for reinforcing the hair (L'Oreal). Approaches of tissue engineering are still not used widely in clinic, except when the surgeon often mixes blood and bone marrow with the glass granules, hoping to activate some stem cells that already belong to the patient. To incorporate osteoprogenitors cells within scaffolds could be the best solution if healthy bone is needed in great defects. The production and uses of scaffolds for bone tissue regeneration will be treated in Chapters 10 and 11.

Silica bioactive glasses have applications in bone repair, but in terms of tissue regeneration, the maximum potential of these glasses has not been reached yet. A reason for this is the difficulty in obtaining amorphous bioactive scaffolds because they crystallize at the high temperatures required to obtain the scaffolds forming dense materials. In addition, they are also not suitable to obtain fibers or coatings of metallic implants. These facts could explain the moderate success of these bioactive glasses in bone regeneration applications if compared with other bioactive ceramics, such as calcium phosphates.

In addition, recent studies have shown the proangiogenic potential of bioactive glasses, which should provide benefits for their application for soft tissue repair. Indeed bioactive glasses have an exceptional set of properties, such as the ability to degrade at a controllable rate and to be converted into an apatite-like material, to bond to hard and soft tissues, and to release ions during the degradation process. These ions have a beneficial effect on osteogenesis (formation of bone) and angiogenesis (formation of blood vessels), which is critical for applications in tissue regeneration and the healing of soft tissue wounds. Recent attempts to stimulate angiogenesis have focused on the delivery of growth factors, such as vascular endothelial growth factor (VEGF) and basic fibroblast growth factor (bFGF), gene therapy, and cell-based therapy. Finally, very recent results indicate that bioactive silica glasses may also have a beneficial effect on chondrogenesis (formation of cartilage).

In summary, silica BGs are characterized by their capacity to bond to both soft and hard tissues. There has been much research but at the present they are mainly applied in a limited number of dental and orthopedic applications. At the moment BGs are used in grain form as bone fillers in the increase of the alveolar bridge. One of the main challenges to increase the use of bioactive glasses is the design of products and packages that can be manipulated and be used more easily by the clinicians. To use bioactive glasses like thermosetting components that cure with light offers a possibility of developing dental products which can be transformed *in situ* into bone filler, implants, or other reconstruction biomaterials. Bioactive glasses help the osseointegration of biostable polymeric implants and increase the capacity for regeneration of the bone tissue of biodegradable polymeric scaffoldings. Small bioactive glass particles in toothpastes have potential for the treatment of hypersensitivity and to increase the mineralization of enamel and dentine. The capacity of small particles of bioactive glasses to reduce the viability of oral microorganisms offers interesting possibilities to develop materials that can be used, for example, in fillings for tooth decay treatment. The speed of dissolution and the bioactivity of glasses can be adjusted with changes in the composition. This offers a possibility of developing new products for dental applications. For example, the slow dissolution of osteoconductive glass particles and fibers can be of interest for composites reinforced with fibers, or in composites with biodegradable polymers.

4.6.2   Inert Silica Glasses

Specific glass compositions of glasses have been used for tumor irradiation. This approach is based on two facts: the radiation damages the DNA of cells, and cancerous cells lose their ability to regenerate DNA. Brachytherapy with radioactive glass microspheres has been used for 25 years to irradiate inoperable malignant liver tumors. Compared with the external, the *in situ* irradiation is weaker, of shorter rank and more directly directed to the tumor, avoiding damage to the healthy adjacent tissues. Microspheres of glass are made radioactive by the incorporation in their composition of radioisotopes emitting beta, gamma, or alpha radiation, like rare earth oxides of Y, Sm, Ho, or Dy.

For this purpose bioinert glasses based on aluminosilicates and biodegradable glasses based on borate/borosilicate glasses have been used or investigated. Bioinert aluminosilicates microspheres of $10-30\,\mu m$ containing $^{90}Y_2O_3$ were obtained by flame spheroidization and successfully used for the treatment of hepatocellular carcinoma (cancer of the liver) with patients surviving for more than five years after the treatment. Moreover, biodegradable radioactive glass microspheres based on borate and borosilicate glasses are under investigation for radiation sinovectomy in the treatment of rheumatoid arthritis joints. Radioactive microspheres are also under investigation for other clinical applications, like the treatment of cancer of the kidney, brain, chest, prostate, and pancreas. Besides the therapy of radiation *in situ*, glass microspheres are investigated as vehicles for drug release. In Chapter 14, recent techniques for treating cancer, based on the use of nanometric sized particles, are presented. The small size of nanoparticles makes it possible for them to reach more organs and tissues in the human body because they are not detected by macrophages and concentrated in the liver, as happens with particles of micrometric size.

4.6.3   Phosphate Glasses

The current clinical uses of phosphate glasses are far fewer than those of silicate glasses. Their main interest comes from their high solubility (by several orders of magnitude) in a physiological medium that can be tailored by changing the glass composition. Phosphate glasses are investigated as porous fibers or to obtain scaffolds. Thus, an important research line is the synthesis of phosphate glass fibers to reinforce degradable polymers, such as PLA to obtain degradable composites that could be used in the fixation of bone fractures and hard tissue regeneration. Moreover, phosphate glasses could have interest in soft tissue applications, for instance ligament, muscle, or cartilage regeneration. Another suggested application of these glasses is for nerve regeneration and neural repair. A promising dental application of phosphate glasses is the synthesis of glasses releasing F^- ions which are giving promising results for controlling caries in children. It is also possible to use these glasses to release therapeutic ions, such as Sr^{2+}, for osteoporosis treatment, or Cu^{2+}, Zn^{2+}, or Ag^+ with antibacterial properties.

4.6.4 Borate Glasses

Bioactive borate glasses have been more recently investigated as biomaterials than silica glasses. Theoretically, borate glasses display an important lack because they cannot release silicon, a gene expression stimulator to regenerate bone. However, two well known bioactive glasses, Bioglass® 45S5 (silica based) and 13-93B3 (borate glass), were compared in a critical sized rat calvarial defect (4 mm). After 12 weeks of implantation, borate fibers appeared totally covered by bone whereas the silicate glass particles were not. In spite of the comparison it is not ideal because fibers and particles would behave in a very different way, the study was indicative of the interest of the research in bioactive borate glasses for bone regeneration applications. Furthermore, bioactive borate nanofibers have demonstrated their utility for the treatment of soft tissue wounds. For instance, in the treatment of diabetic's wounds materials are needed that promote hemostasis (ability to form blood clotting) and exhibit antimicrobial properties. Both characteristics are exhibited by borate bioactive glass nanofibers. Thus, one months treatment with borate glass nanofibers administered twice per week was able to cure a pressure ulcer with prevalence of over two years in a diabetic patient. Other future applications of calcium-containing borate glasses come from their ability to form in a few weeks hollow cavities in particles or fibers. After four weeks of *in vivo* study, the cavities were filled by soft tissue and blood vessels. Other clinical applications could come from the use of hollow HCA shells ($\sim 150\,\mu m$), obtained by the treatment of borate glasses with a phosphate solution, for the *in situ* release of drugs.

Recommended Reading

Books

Hench, L.L. (2013) *An Introduction to Bioceramics*, 2nd edn, Imperial College Press, London, U.K.

Hench, L.L., Jones, J.R. and Fenn, M. (eds) (2011) *New Materials and Technologies for Healthcare*, Imperial College Press, London, U.K.

Holand, W. and Beall, G.H. (2012) *Glass Ceramic Technology*, 2nd edn, John Wiley & Sons, Inc., Hoboken, NJ.

Jones, J.R. and Clare, A.G. (2012) *Bio-Glasses an Introduction*, John Wiley & Sons, Ltd, Chichester, U.K.

Planell, J.A., Best, S.M., Lacroix, D. and Merolli, A. (eds) (2009) *Bone Repair Biomaterials*, CRC Press, Boca Raton, FL.

Rimondi, L., Bianchi, C. and Verné, E. (eds) (2012) *Surface Tailoring of Inorganic Materials for Biomedical Applications*, Bentham Science Publishers.

Shelby, J.E. (2005) *Introduction to Glass Science and Technology*, Royal Society of Chemistry, Cambridge, U.K.

Vallet-Regi, M., Manzano, M. and Colilla, M. (2012) *Biomedical Applications of Mesoporous Ceramics: Drug Delivery, Smart Materials and Bone Tissue Engineering*, Taylor & Francis, CRC Press, Bosa Roca, USA.

Vallet-Regi, M. and Vila, M. (2010) *Advanced Bioceramics in Nanomedicine and Tissue Engineering*, Trans Tech Publication, Switzerland.

Ylänen H. (2011) *Bioactive Glasses, Materials, Properties And Applications*, Woodhead Publishing, Oxford.

Review Articles

Ahmed, I., Lewis, M., Olsen, I. and Knowles, J.C. (2004) Phosphate glasses for tissue engineering. Part 2. Processing and characterization of a ternary-based P_2O_5–CaO–Na_2O glass fiber system. *Biomaterials*, **25**, 501–507.

Gorustovich, A.A., Perio, C., Roether, J.A. and Boccaccini, A.R. (2010) Effect of bioactive glasses on angiogenesis: a review of in vitro and in vivo evidence. *Tissue Eng. B*, **16**, 199–207.

Hench, L.L. (1998) Bioceramics. *J. Am. Ceram. Soc*, **81**, 1705–1728.

Hench, L.L., Day, E.D., Holland, W. and Rheinberger, V.M. (2010) Glass and medicine. *Int. J. Appl. Glass Sci*, **1**, 104–1117.

Izquierdo-Barba, I., Salinas, A.J. and Vallet-Regi, M. (2013) Bioactive glasses: from macro to nano. *Int. J. Appl. Glass Sci*, **4**, 149–161.

Jones, J.R. (2013) Review of bioactive glass: from Hench to hybrids. *Acta Biomater*, **9**, 4457–4486.

Kaur G., Pandey O.P, Singh K, *et al.* (2013) A review of bioactive glasses: their structure, properties, fabrication, and apatite formation. *J. Biomed. Mater. Res. A*, doi: 10.1002/jbm.a.34690

Kokubo, T. (1991) Bioactive glass ceramics: properties and applications. *Biomaterials*, **12**, 155–163.

Pan, H.B., Zhao, X.L., Zhang, X. *et al.* (2010) Strontium borate glass: potential biomaterial for bone regeneration. *J. R. Soc. Interface*, **6**, 1025–1031.

Salinas, A.J., Esbrit, P. and Vallet-Regi, M. (2013) A tissue engineering approach based on the use of bioceramics for bone repair. *Biomater. Sci.*, **1**, 40–51.

Salinas, A.J. and Vallet-Regi, M. (2013) Bioactive ceramics: from bone grafts to tissue engineering. *RSC Adv.*, **3**, 11116–11131. doi: 10.1039/C3RA00166K

Rahaman, M.N., Day, D.E., Bal, B.S. *et al.* (2011) Bioactive glass in tissue engineering. *Acta Biomater*, **7**, 2355–2373.

Vallet-Regí, M., Izquierdo-Barba, I. and Colilla, M. (2012) Structure and functionalisation of mesoporous bioceramics for bone tissue regeneration and local drug delivery. *Philos. Trans. R. Soc. A*, **370**, 1400–1421.

5

Silica-based Ceramics: Mesoporous Silica

Montserrat Colilla
Departamento de Química Inorgánica y Bioinorgánica, Facultad de Farmacia,
Universidad Complutense de Madrid, CIBER en Bioingeniería, Biomateriales y
Nanomedicina (CIBER-BBN), Spain

5.1   Introduction

The unique properties of ordered mesoporous silicas (OMS) have made these matrices attractive for a wide range of applications, such as adsorption [1], catalysis [2, 3], sensing [4], separation [5], and, more recently, biomaterials science [6–9]. Among the potential biomedical applications of OMS, such as drug delivery and bone tissue regeneration, the possibility of being used as multifunctional platforms for smart drug delivery, gene therapy, imaging, hyperthermia-based therapy, biosensing, and theranostics has opened up promising expectations in nanomedicine [10–16].

The main features of OMS are summarized in Figure 5.1. These materials are structurally unique, exhibiting order on the mesoscopic scale (2–50 nm) and disorder on the atomic scale. The pore channels formed within the materials are separated by amorphous silica walls and arranged periodically on lattices. OMS display high surface areas and pore volumes, regular and tunable pore sizes and pore morphologies, stable mesostructures, controllable particle morphologies and sizes, non-toxic, and biocompatible behavior and the possibility to functionalize the silica walls with diverse organic groups [7, 11, 14].

When OMS are envisioned for use in biomedicine, their properties must be optimized in order to meet the requirements of the biomedical application for which they are intended. This chapter focuses on understanding the fundamentals of the synthetic approaches developed so far that permit modulatation of chemical composition, structure, pore size, surface area, and morphology and particle size of mesoporous materials. Since functionalization of OMS constitutes a paramount issue for their performance as drug delivery systems and

Bioceramics with Clinical Applications, First Edition. Edited by María Vallet-Regí.
© 2014 John Wiley & Sons, Ltd. Published 2014 by John Wiley & Sons, Ltd.

Characteristics of ordered mesoporous silicas

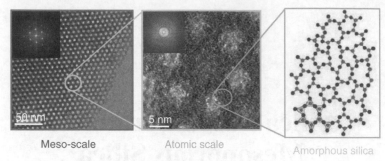

Meso-scale   Atomic scale

Amorphous silica

➧ Ordered mesopore arrangements.

➧ Narrow pore size distributions in the 2–50 nm range.

➧ High pore volumes ($\sim$1 cm^3 g^{-1}).

➧ High surface areas (500–1500 m^2 g^{-1}).

➧ High density of silanol groups (Si-OH): Functionalization.

Figure 5.1   *Main features of ordered mesoporous silicas (OMS)*

smart nanodevices, the main strategies used to link organic moieties to the surface of these matrices are also described.

5.2   Discovery of Ordered Mesoporous Silicas

To comprehend the discovery of OMS we have to move back to the first global energy crisis in the early 1970s, when alternatives to fuels derived from crude oil became necessary. In 1972, Mobil Corporation developed a process to convert a methanol feed into a hydrocarbon fraction containing aliphatic and aromatic compounds in the gasoline boiling range and with good gasoline properties. The zeolite ZSM-5 (Zeolite Secony Mobil #5) [17], was used as catalyst for this MTG (methanol-to-gasoline) process. However, the yield of this reaction was not high enough and the process was not commercialized. The main drawback was the deactivation of the catalyst due to coke formation from side reactions of reactants/products with limited diffusion within the small cavities of zeolites, which are microporous crystalline aluminosilicates.

This arena motivated researchers to find porous materials with larger pores than zeolites capable of improving their applicability as catalysts, adsorbents, host materials for nanoscale fabrication, and so on. However, the pore sizes available in zeolites increased only slightly, the highest values being those of VPI-5 in the late 1980s [18], with pore sizes around 1.2–1.3 nm.

It was in the 1990s when OMS with a regularly ordered pore system in the 2–10 nm range and pore walls essentially composed of amorphous silica were independently discovered by two research groups: the group at Waseda University in Japan [19, 20], headed by Kuroda; and the group working at Mobil Oil Corporation headed by C.T. Kresge [21–23]. The emergence of OMS represented a groundbreaking discovery that motivated many research activities in hundreds of laboratories around the world.

5.3   Synthesis of Ordered Mesoporous Silicas

Although Japanese scientists reported somewhat earlier the synthesis of OMS, the pathway described in the publication is rather specific and difficult to generalize. On the other hand, the publications of the Mobil Oil researchers described a pathway which is generalizable, and has been used for a wide variety of other materials in later years.

The synthesis route described by Kuroda and coworkers relies on the use of kanemite $(NaHSi_2O_5 \bullet H_2O)$, a sheet silicate composed of single-layered sheets of SiO_4 tetrahedra. Thus, it is possible to exchange Na^+ cations in the interlayer space of kanemite with organic cations, such as surfactants of the alkyltrimethylammonium type (Figure 5.2) [19]. The mechanism proposed for the formation of these mesoporous materials, which were called FSM-n (Folded Sheet Materials, where n indicates the number of C-atoms in the surfactant chain) consists in the intercalation of the surfactant between the kanemite sheets. Then, the subsequent bending of the sheets, followed by inter-sheet condensation led to the formation of a hexagonal array of pores, as depicted schematically in Figure 5.2. Certainly, one could think that at the high synthesis pH of these materials, the kanemite would be dissolved and reorganized to give the hexagonal mesostructure, driven by the surfactant molecules. However, later studies demonstrated that the properties of kanemite-derived OMS were noticeably different from those synthesized by direct templating [24].

The new family of OMS (M41S) reported by Mobil Oil researchers, the so-called MCM-n series (where MCM stands for Mobil Composition of Matter), exhibits different mesostructures. The most well-known representative materials are MCM-41 (two-dimensional (2D) hexagonal, space group $p6mm$), MCM-48 (three-dimensional (3D) cubic, space group $Ia\overline{3}d$), and MCM-50 (lamellar, space group $p2$) (Figure 5.3) [21–23].

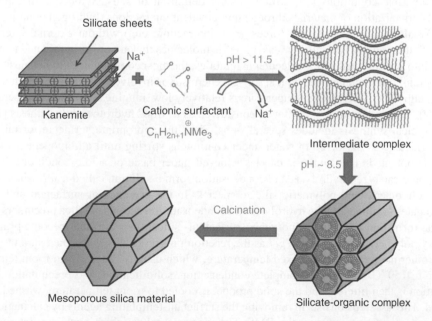

Figure 5.2   *Scheme of the formation mechanism of OMS derived from kanemite [19, 20]*

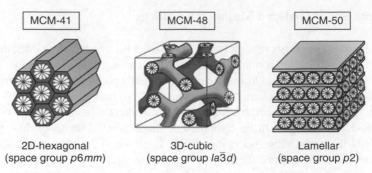

Figure 5.3 *Structures of M41S mesoporous materials: MCM-41, (2D-hexagonal, space group* p6mm*); MCM-48 (3D-cubic, space group* Ia3̄d*); and MCM-50 (lamellar, space group* p2*) [21–23]*

The use of supramolecular aggregates of ionic surfactants (long-chain alkyltrimethylammonium halides) as structure-directing agents (SDAs) of the inorganic mesostructure was revolutionary in the synthesis of OMS.

Throughout this chapter, we will focus on OMS that have been prepared by the supramolecular templating mechanism, where solubilized silica precursors are organized into mesostructured arrays under the influence of surfactant molecules.

5.3.1 Hydrothermal Synthesis

The most widely used approach to prepare mesoporous silica materials consists in combining the supramolecular self-assembly of surfactants with appropriate silica sources under hydrothermal conditions [25, 26]. A general definition of self-assembly is the spontaneous organization of materials through non-covalent interactions (hydrogen bonding, van der Waals forces, electrostatic forces, $\pi-\pi$ interaction, etc.) without external intervention. Self-assembly typically uses asymmetric molecules that are pre-programmed to organize into well-defined supramolecular assemblies, the most common used molecules being amphiphilic surfactants. The typical sol–gel chemistry is involved in the hydrothermal process, although the synthetic temperature is relatively low, ranging from room temperature to 150 °C and, therefore, it cannot be considered as a "true" hydrothermal synthesis [26].

A typical synthesis includes several steps. First, the appropriate surfactant or mixture of surfactants is dissolved in water under continuous stirring until a homogeneous solution is obtained. The synthesis can be achieved under basic or acidic conditions. Then, inorganic catalyzed hydrolysis and condensation form a solution called a *sol*, which contains oligomeric and polymeric silicate species. In the presence of surfactant and their aggregates (or micelles) the hydrolysis and condensation reactions of silica precursors lead to the formation of organic–inorganic (surfactant–silicate) species that become progressively polymerized and form a *gel* as the reaction progress. This *gel* is then submitted to hydrothermal treatment at a fixed temperature, which usually ranges from room temperature to 150 °C, to promote complete condensation, solidification, and precipitation. The solution is then filtered, and the solid product is cooled to room temperature, washed, and dried. The last step consists in removing the surfactant-templating agents by calcination or solvent extraction to produce the OMS. The experimental variables, such as silica source, type of surfactant, pH, temperature of reaction, and so on, used during the supramolecular self-assembly of OMS will drive the way the evolution of silica precursors and the self-assembly process occur and, consequently, they will determine the chemical, textural, and structural properties of the final OMS. In the following sections we will focus on understanding the different roles that the synthesis conditions play in the ultimate features of OMS. This knowledge is extremely important when aiming at designing mesoporous silicas for biomedical purposes, since the characteristics of these materials must be carefully tailored with regard to the clinical target.

5.3.1.1 Surfactants

Surfactants are amphiphilic molecules that consist of a hydrophilic, polar head group, and a long hydrophobic, nonpolar tail (Figure 5.4) [27]. The term "surfactant," which is an abbreviation of "surface active agent", has been coined from the high affinity of these molecules towards surfaces and interfaces. A fundamental characteristic of surfactants is their behavior in water. Once in contact with water, surfactant molecules self-organize to form aggregates called micelles. These micelles are formed because the hydrophobic tails of the surfactant tend to congregate to minimize contact with water, whereas their hydrophilic heads orientate to maximize the interaction with the aqueous medium. Micelles are generated at very low surfactant concentrations in water. The concentration at which micelles start to form is termed the critical micelle concentration (CMC), and is a significant parameter for a given surfactant. Usually a clear homogeneous solution of surfactant in water is required to get ordered mesostructures.

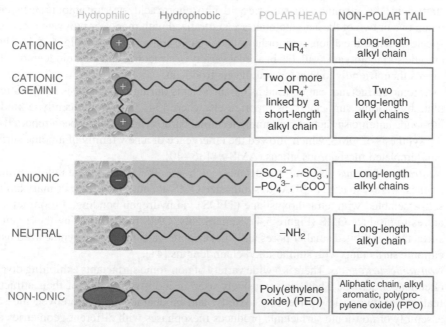

Figure 5.4 *Main types of surfactants used for the synthesis of OMS*

Surfactants employed in the synthesis of OMS can be classified, according to the chemical nature of the polar head group, into cationic, anionic, neutral, and non-ionic surfactants (Figure 5.4) [27].

1. *Cationic surfactants.* Among cationic surfactants, quaternary ammonium salts, having the general formula $C_nH_{2n+1}N(CH_3)_3Br$ ($n = 8$–22), are efficient for the synthesis of OMS. They are characterized by having quaternary ammonium polar head groups and long alkyl chains as nonpolar tails (Figure 5.4). Actually, cetyltrimethylammonium bromide ($C_{16}H_{33}N(CH_3)_3Br$, CTAB) was first used by Mobil Oil researchers as SDA to prepare the M41S family of OMS [21–23]. *Gemini* surfactants (Figure 5.4), a class of amphiphilic molecules incorporating two conventional surfactants connected by a spacer via covalent bonds (e.g., $[C_nH_{2n+1}(CH_3)_2N^+\text{-}(CH_2)_s\text{-}N^+(CH_3)_2C_mH_{2m+1}]\cdot 2Br^-$, also abbreviated as C_{n-s-m}) have also been used as SDAs for the synthesis of OMS [28–30].

   Tetra-head-group rigid bola-form surfactants, (denoted as $C_{n-m-m-n}$), have been also used as SDAs to synthesize OMS of the FDU-n (Fundan University) family [31]. They are a kind of multi-charged cationic surfactants, related to *gemini* surfactants, in which four hydrophilic head groups separated by different methylene chains are attached by one variable-length hydrophobic linker with a rigid biphenol group. Cationic fluorinated surfactants, such as 1,1,2,2-tetrahydroperfluorodecylpyridinium chloride and multi-head-group surfactants have also been reported as SDAs of different mesostructures [32, 33].

2. *Anionic surfactants.* Anionic salt surfactants, such as carboxylates, phosphates, phosphonates, sulfates, sulfonates, and so on, are widely produced and used in huge amounts in industry due to their high detergency and low cost. Various long-chain alkyl acids (fatty acid, *N*-acylamino acid, sulfuric acid, phosphoric acid, etc.) are anionic surfactants when one or two of the acid moieties are neutralized with mono- or polyvalent cationic species, such as metal ions, or inorganic and organic ammonium ions (Figure 5.4). These anionic surfactants are available in a great variety of structures and their biodegradability makes them friendly to humans and the environment.

   Anionic surfactants can be used as co-surfactants during the synthesis of OMS template by cationic surfactants, as reported elsewhere [34, 35]. More recently, a kind of lab-made anionic surfactant containing terminal carboxylic acids has been reported for the synthesis of OMS, which allowed the emergence of a new family of anionic surfactant templated mesoporous silicas (AMS-n) [36–40].

3. *Neutral surfactants.* Neutral surfactants, such as alkyl chains containing primary amines, were also used as SDAs to synthesize OMS at neutral pH [41]. Primary amines self-assembled with tetraethoxysilane (TEOS) via hydrogen bonding at room temperatures to produce OMS (Figure 5.4). Remarkably, OMS produced using these neutral surfactants exhibited smaller pores and thicker pore walls than those synthesized using cationic surfactants with similar alkyl chain lengths [41].

4. *Non-ionic surfactants.* There is a wide variety of non-ionic surfactants exhibiting diverse chemical structures. They are extensively used in industry because of their attractive features, including low price, low toxicity, and biodegradability. Moreover, the self-assembly of non-ionic surfactants produces mesophases with different geometries and pore arrangements, which make them very suitable to synthesize OMS with tunable textural and structural features.

Non-ionic surfactants alkyl chains containing poly(ethyleneoxide) (PEO) groups, $C_nH_{2n+1}(EO)_m$, with $n = 11-15$ and $m = 9-30$, have been used as the SDA of a family of OMS called MSU-n (Michigan State University) series [42].

However, among non-ionic surfactants, those that have received most interest are a class of triblock copolymers called Pluronics of the poly(ethyleneoxide)-poly(propyleneoxide)-poly(ethyleneoxide) type, usually referred to as EO-PEO-EO polymers (Figure 5.4). In a pioneering work [43], Pluronics of high molecular weights were used as SDAs in acidic solutions to produce the SBA-n (Santa Barbara Amorphous) series of mesoporous materials, which exhibit ordered mesostructures and tunable large pores. The most notable member of this family is SBA-15, which exhibits a 2D hexagonal structure. The pore sizes and the wall thicknesses of SBA-15 are bigger than those of hexagonally ordered MCM-41. Another remarkable feature of SBA-15 is that it contains microporous channels interconnecting adjacent mesopores [44–46]. Pluronics have also been reported for the synthesis of OMS of the KIT-n (Korean Institute of Technology) family [47, 48].

5.3.1.2   Silica Precursors

During the synthesis of OMS, the most common precursors for the formation of colloidal suspensions are silicon alkoxides, which consist of a silicon atom surrounded by several alkoxide groups as ligands. The most common silicon alkoxides are tetraalkoxysilanes $(Si(OR)_4)$, TEOS $(Si(OCH_2CH_3)_4)$ being the most suitable and efficient silica precursor for the synthesis of OMS. Alkoxysilanes undergo the main sol–gel reactions at the functional group level, that is, hydrolysis, condensation, and polymerization (Figure 5.5) with different reaction rates, depending on the pH of the solution [49, 50]. The hydrolysis of silica precursors and intermediate silicates can be divided into three steps: (i) first the oxygen atoms from water molecules initiate nucleophilic attack on silicon atoms of the silica precursors, (ii) then there is a transference of a proton from a water molecule to an alkoxide group of the silicon, and (iii) finally there is a release of an alcohol molecule (ROH). The same behavior is observed in organoalkoxysilanes, $R_xSi(OR)_{4-x}$, where $x = 1, 2,$ or 3. The rate of the hydrolysis reaction increases as the length of the alkoxide group of the silica precursors decreases. This fact can be explained because shorter alkoxide groups are better leaving groups due to their increased ability to donate electrons and also to their decreased steric hindrance for nucleophilic attack [49, 50].

Subsequently, a condensation reaction takes place to produce siloxane bonds (Si-O-Si) and either alcohol or water as by-product, depending on the reactants (Figure 5.5). Thus, when two partially hydrolyzed molecules link together (*water condensation*) they liberate a molecule of water. On the other hand, when a partially hydrolyzed silicon alkoxide, bearing a silanol group, condenses with a non-hydrolyzed silicon alkoxide (*alcohol condensation*) an alcohol molecule is released. Generally, condensation starts before hydrolysis is complete. Since water and alkoxysilanes are immiscible, a solvent compatible for both components, such as alcohol, is sometimes used as a homogenizing agent. Nevertheless, this process can proceed without added solvent, since alcohol produced as a by-product of the hydrolysis reaction is enough to homogenize the initially phase-separated system.

As the condensation reactions continue, dimers, and linear or cyclic silicate trimers and tetramers grow into large crosslinked silicon-containing polymer networks, during the so-called polymerization process (Figure 5.5).

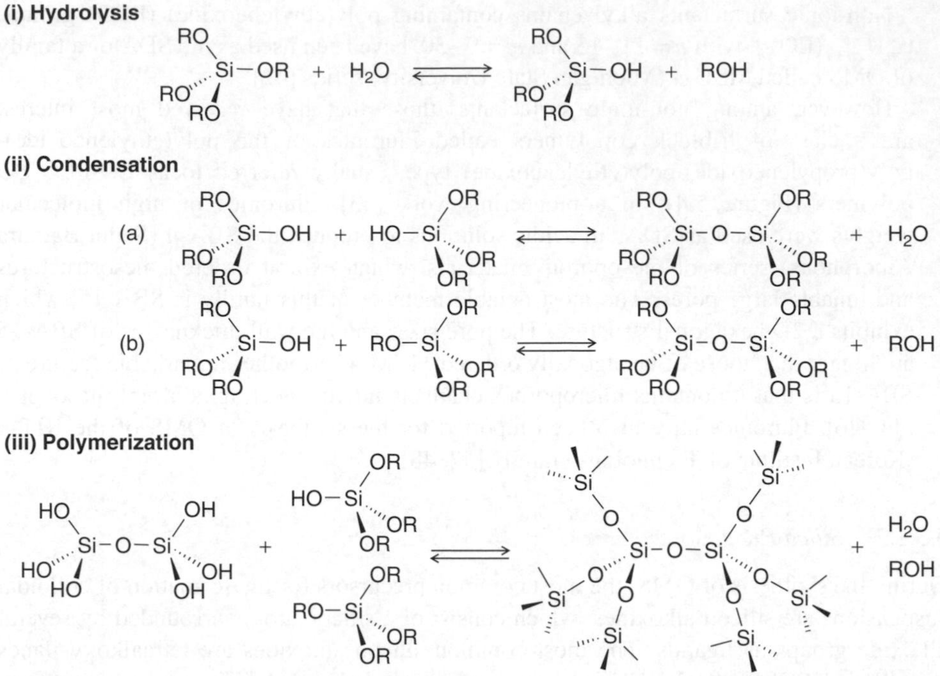

(i) Hydrolysis

(ii) Condensation

(iii) Polymerization

Figure 5.5 *Three main reaction steps of the sol–gel chemistry of alkoxysilanes in aqueous medium: (i) hydrolysis of an alkoxy group under building of silanol groups and the respective alcohol; (ii) condensation (a) of two silanol bearing species or (b) mixed condensation of a silanol group and an alkoxy group bearing species; and (iii) polymerization of silanol groups and alkoxy groups bearing species to give the silica network with a given crosslinking degree*

The rate of hydrolysis and condensation of silica precursors during the synthesis of OMS is highly dependent on the pH of the solution. In the case of silicon alkoxides, hydrolysis and condensation reactions typically proceed with the addition of either an acid or a base as catalyst, as will be discussed in the next sections.

5.3.1.3   Synthesis Temperature

One of the key factors to take into account during the synthesis of OMS is the temperature at which the reaction takes place. The reaction temperature is relatively low, in the -10 to $130\,^{\circ}\text{C}$ range and, actually, room temperature is very suitable to carry out the process. With the aim of setting the proper temperature there are two factors that must be considered, the critical micelle temperature (CMT) and the cloud-point (CP). The optimal synthesis temperature is normally greater than the CMT and smaller than the CP for a given surfactant. In the case of cationic surfactants, since they exhibit low CMT, the preparation of highly-ordered OMS is easily achieved at room temperature. On the other hand, anionic surfactants have higher CMT values and, therefore, when they are used as templates, the synthesis temperature is usually higher than room temperature. When non-ionic surfactants, such as block-copolymers, are used as templates, the synthesis temperature must be carefully chosen, since they become insoluble in water at the CP. For instance, during the synthesis of SBA-15 templated by Pluronic P123, the optimal synthetic temperature is 35–50 °C due to the CMT for the formation of micelles and the CP [43].

5.3.1.4 Synthesis Medium

As previously mentioned, the synthesis of OMS is usually carried out in an aqueous medium under acidic or basic conditions. Normally, neutral solutions are not appropriate to get ordered mesostructures because at pH 6–8.5 the polymerization and crosslinking rates of silicate species are so fast that the template cannot direct the formation of the mesostructure [41].

1. *Basic synthesis medium.* Under basic conditions (pH 9.5–12.5), the hydrolysis of silica precursors is accelerated because OH^- groups are excellent nucleophiles and efficiently attack the silicon atoms of the silica precursors. On the contrary, the condensation and polymerization rates decrease and even become reversible and, therefore, highly crosslinked silicates are unable to form. Although different silica precursors can be used during the synthesis of OMS under basic conditions (e.g., silica gels, sodium silicates, silica aerogels, TEOS, etc.), TEOS has been confirmed as the most suitable precursor in the laboratory [33]. NaOH, KOH, or NH_4OH are frequently used as basic catalysts. The most common OMS synthesized under basic conditions are MCM-41 and MCM-48 structures [22, 23].

2. *Acidic synthesis medium.* Huo *et al.* [33] reported for the first time the synthesis of OMS using CTAB as the surfactant, TEOS as the silica source, and HCl as the acidic catalyst. The resulting material was denoted SBA-3. Later, a series of OMS were synthesized using non-ionic surfactants under acidic conditions, such as SBA-15, SBA-16, and so on [43]. Under acidic conditions, the synthesis of OMS is accelerated by lowering the pH values of the solution, that is, a high acid concentration originates a fast precipitation rate. On the contrary, a low acid concentration leads to slow condensation rates of silica species. When strong acids such as HCl are used as catalyst, the ideal pH value during the synthesis of OMS is below 1. However, the acid concentration should not be excessively high to preserve the structural and textural properties of the resulting OMS. Contrary to what happened under basic conditions, in an acid medium the polymerization of the silicate species is irreversible, which can produce a synthesis failure once the gel is formed. Different silica sources, such as sodium metasilicate (Na_2SiO_3) or TEOS, are also suitable following this synthetic route, TEOS again being the most appropriate silica precursor.

5.3.1.5 Hydrothermal Treatment

Hydrothermal treatment is a widely used approach aimed at increasing the mesostructural order of OMS [43, 51, 52]. Although mesostructures have been already formed during the solution reaction, the hydrothermal treatment favors their reorganization, growth, and crystallization. The temperature during the hydrothermal treatment varies from 80 to 150 °C, the 95–100 °C range being the most widely used. Usually, this treatment is carried out under static conditions in appropriately sealed polypropylene or Teflon® bottles containing the mother liquid and using a high-pressure autoclave. The choice of the temperature and time of the hydrothermal treatment depend on the type of surfactant used as the SDA and also on the type of OMS to be synthesized. The hydrothermal treatment affects the textural properties, such as pore diameter, surface area, and so on, and may produce variations in the mesostructure of the resulting OMS, as will be discussed in Section 5.5.

5.3.1.6   Separation and Drying

As-synthesized mesostructured materials can be separated from the mother liquid by filtration, although centrifugation could sometimes be useful. Water is added during the washing step, followed by an alcohol such as ethanol. If the synthesis has been carried out under basic conditions, exhaustive washing with water until neutrality is achieved is essential to avoid residual NaOH destroying the mesostructure after surfactant removal by calcination. However, the washing step can be omitted when the synthesis is carried out under acid conditions, since HCl is volatile and can be fully removed together with the surfactant by calcination. Finally, the drying process for as-synthesized OMS is usually carried out at room temperature.

5.3.1.7   Surfactant Removal

Surfactant removal from as-synthesized mesostructured materials is essential to obtain the resulting OMS with appropriate porosity. The two main methods for surfactant removal are calcination and extraction with solvents.

1. *Calcination.* The most common method used for template removal is calcination because it is an easy and fast procedure and leads to complete elimination of surfactant. Organic surfactants are decomposed or oxidized under oxygen in an air atmosphere. However, the temperature programming rate should be low enough to avoid structural collapse provoked by local overheating. The most frequently used method is that reported by Kresge *et al.* [22], which consists of a two-step calcination, the first 1 hour under nitrogen to decompose surfactants and the second 5 hours in air or oxygen atmosphere to burn them out. Different modifications to this pioneer protocol have been performed, taking into account that the calcination process will affect the textural properties of the final OMS [53–55]. Thus, the higher the calcination temperatures the lower the surface areas, pore volumes, and surface silanol groups, and the higher the crosslinking degree and, therefore, the higher the hydrothermal stability of the OMS.

2. *Extraction with solvents.* Extraction is a mild and efficient method to eliminate surfactants from as-synthesized mesoporous materials without affecting the silica framework [44, 56–58]. For instance, ethanol or tetrahydrofuran (THF) can be used as organic solvents during the surfactant extraction. Usually a small amount of HCl is added during the extraction to improve the crosslinkage of frameworks and to diminish the effects on mesostructures [59]. Extraction with solvents offer some advantages compared to calcination, such as the possibility to obtain OMS with larger pore sizes in some cases, a higher amount of surface silanol groups and enhanced hydrophilic character. However, the main drawback of extraction is the fact that surfactants cannot be completely removed.

5.3.2 Evaporation-Induced Self-Assembly (EISA) Method

The evaporation-induced self-assembly (EISA) method combines classical sol–gel chemistry with surfactant self-assembly. This method is based on the formation of a hybrid mesostructured phase after solvent evaporation from dilute solutions containing inorganic precursors, the templating agent, and other additives [60]. The solution can be cast to synthesize monoliths, dip- or spin-coated to form thin films, or sprayed to form spherical-shaped microparticles or nanoparticles. EISA has also been used to prepare mesoporous non-silica oxides [61–64].

For the preparation of mesostructured silica films, TEOS dissolved in the organic solvent (usually ethanol, THF, or acetonitrile) is pre-hydrolyzed with a stoichiometric amount of water (catalyzed by acids such as HCl) at a temperature in the 25–70 °C range. The solvent evaporation permits the silica species to polymerize and condense around the surfactants. The polymerization rate gradually increases owing to the increasing acid concentration during the solvent evaporation; meanwhile templating by surfactant assembly takes place, leading to the formation of mineral-template hybrid mesophases. The hybrid mesophase is flexible as-synthesized, due to the incomplete inorganic polymerization. Aging of these soft hybrid mesophases under controlled humidity conditions and low temperature favors the upgrading of ordering [65]. Inorganic condensation can be increased in a subsequent step by heating or adding a condensation enhancer such as ammonia. The EISA method permits processing of a very flexible material in the shape of xerogels, monoliths [66–68], thin films [69–71], or aerosols [72, 73]. The last step to obtain OMS is the surfactant removal by calcination or solvent extraction. The EISA route is very interesting for the synthesis of mesoporous frameworks with different chemical compositions. It is especially useful for the synthesis of mesoporous bioactive glasses (also known as template glasses, TGs) in the SiO_2–CaO–P_2O_5 system [74, 75], as has been discussed in Chapter 4.

5.4 Mechanisms of Mesostructure Formation

Since the discovery of OMS in the earliest 1990s many research teams have focused on understanding the mechanism of formation of these materials. Albeit different mechanisms have been proposed for the formation of MCM-41 [76], all of them rely on the same principle: the surfactant molecules play a pivotal role by driving the inorganic mesostructure from soluble silica species. The main discrepancy among these methods is in regard to how the inorganic precursor interacts with the surfactant. Actually, the type of interaction between the surfactant and silica precursor governs the different synthesis pathways, formation models, and the resulting types of OMS.

Mobil Oil researchers proposed a "liquid crystal templating" (LCT) mechanism based on the resemblance between liquid crystalline surfactants assemblies (i.e., lyotropic phases) and M41S materials [21–24]. The common hallmarks were the dependence of mesostructure upon the hydrocarbon chain length of the surfactant (nonpolar tail), the surfactant concentration, and the presence of swelling agents [77]. In the case of MCM-41, a representative M41S material which exhibits hexagonally packed cylindrical mesopores, two main mechanistic pathways were hypothesized by Beck *et al.* (Figure 5.6) [23]:

1. **"True liquid crystal templating" (TLCT)**. This mechanism assumes that the silica precursors fill the water-rich spaces of the hydrophilic domains of a preformed hexagonal

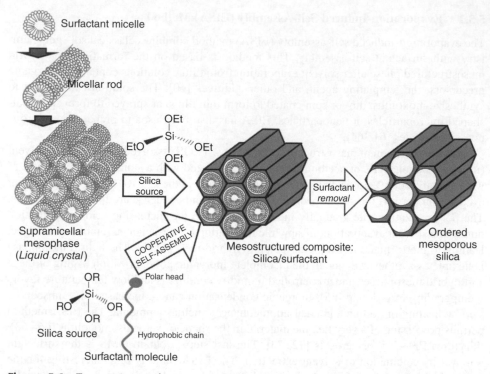

Figure 5.6 *Two main mechanisms proposed for the synthesis of OMS (2D-hexagonal mesostructure is displayed as illustrative example). The main difference takes place when adding the silica precursor (usually TEOS). Thus the "true liquid crystal template mechanism" (TLCT) involves the preformation of supramicellar mesophases before TEOS addition, whereas the "cooperative liquid crystal template mechanism" (CLCT) involves that the formation of supramicellar mesophases take place on the basis of a cooperative self-assembly of surfactant molecules and TEOS via organic–inorganic interaction [23]*

lyotropic liquid crystal phase and settle on the polar heads, placed at the external surface of the micelles.

2. **"Cooperative liquid crystal templating"** (CLCT). This mechanism considers that the silica precursor mediates, somehow, the ordering of the surfactants into the hexagonal arrangement.

In either case, the resulting organic–inorganic (surfactant–silica) mesostructured composite can be be held as a hexagonal array of surfactant micellar rods embedded in a silica matrix. Hence, the surfactant removal will produce open mesoporous cavities hexagonally arranged within the silica network of MCM-41.

It should be noted that the TLCT pathway was not fully accepted because the surfactant concentrations used for the synthesis of MCM-41 were well below the minimum concentration (CMC) required for hexagonal LC formation [77]. Moreover, ^{14}N NMR (nuclear magnetic resonance) studies have proved that the hexagonal LC phase is not formed during the synthesis of MCM-41 [78]. Nonetheless, under different synthesis conditions, the pioneer TLCT pathway has been validated [79]. On the other hand, the CLCT allows formation of the MCM-41 phase even at surfactant concentrations below the CMC when there is a cooperative self-assembly of the surfactant and the already added silica precursors.

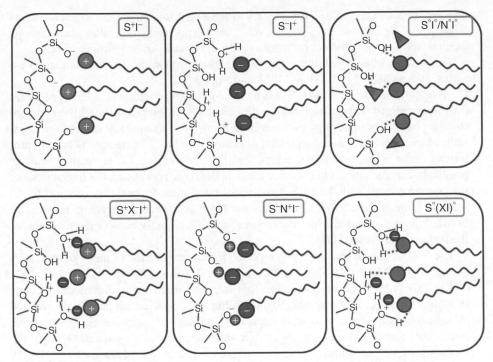

Figure 5.7 *Schematic representation of the interactions between the inorganic silica precursor and the head group of the surfactant, taking into account the pH of the synthesis. Electrostatic interactions: S^+I^-, $S^+X^-I^+$, S^-I^+, $S^-N^+I^-$; via hydrogen bond interactions: $S°I°/N°I°$, $S°(XI)°$*

The original approach has been expanded by a number of variations, using different surfactants as SDAs and variable synthesis conditions. Independently of the mechanistic approach, a pivotal condition for the formation of mesoporous materials is that there is an attractive interaction between the template and the silica precursor that ensures inclusion of the structure director without phase separation. This interaction between the polar head of the surfactant and the adequately ionized silica precursor can be controlled by carefully choosing the surfactant type and synthesis conditions. These interactions will be governed by the type of surfactant used (S) and synthesis conditions, which determine the reactivity of the inorganic precursor (I). The different interactions between the inorganic components and the head group of the surfactants, as schematically depicted in Figure 5.7, can be classified as follows [33, 41, 42, 80]:

S^+I^-: This is the type of electrostatic interaction that takes place when using cationic surfactants (S^+) under basic conditions, where silica species are present as anions (I^-).
This synthetic pathway has been followed to synthesize different mesoporous materials of the MCM-n series (e.g., MCM-41, MCM-48, and MCM-50) [23] SBA-n series (e.g., SBA-6 [81], SBA-2 [82], and SBA-8[83]) FDU-n series (e.g., FDU-2 [84], FDU-11 [31], and FDU-13 [31]), and so on.

$S^+X^-I^+$: This type of interaction takes place when using cationic surfactants (S^+) under acidic conditions (below the isoelectric point of silica, i.e., pH < 2), whereby the silica species are positively charged (I^+). To produce an interaction between equally positively charged surfactant and silica it is necessary to add a mediator anion (X^-, which is usually a halide ion, e.g., Cl^- or Br^-, SO_4^{2-}, NO_3^-, etc.). This interaction takes place through electrostatic Coulomb forcers or, more exactly, via double-layer hydrogen bonds.

This is the synthetic path that leads to different mesoporous materials of the SBA-n series, such as SBA-1 [33], SBA-2 [85], and SBA-3 [82].

S^-I^+: When an anionic surfactant (S^-) is used under acidic conditions, silica species are positively charged (I^+) and surfactant−silica electrostatic interaction would be expected to take place without needing any mediator. However, this strategy failed to produce ordered mesoporous materials, probably because the anionic surfactant is largely protonated under acidic conditions, which significantly weakens its interaction with the positively-charged silica. On the other hand, under basic conditions, the interactions of the counter cations with the surfactant and silicate ion under basic conditions are too weak to form suitable $S^-M^+I^-$ interactions. Thus, no efficient interaction takes place between the surfactants and the silica source, which is an indispensable requirement for the formation of ordered OMS.

The S^-I^+ approach has only been reported for the preparation of metal oxides, such as Mg, Al, Ga, Mn, Fe, Co, Ni, Zn, Pb, Zn, Sn, or Ti oxides [80, 86–88].

$S^-N^+I^-$: This is the type of interaction that happens when using anionic surfactants (S^-) under basic conditions, whereby again the silica species are negatively charged (I^-). In this case, contrary to what happened with cationic surfactants, the repulsive interactions that take place between anionic surfactants and silicate species hamper the formation of well-organized silica mesostructures. Hence, the addition of a positively charged mediator (N^+), usually an organoalkoxysilane containing amino groups (such as aminopropyltrimethoxy-silane, APS or N-trimethoxysilylpropyl-N,N,N-trimethylammonium chloride, TMAPS) is necessary to overcome the charge-matching effect and ensure the surfactant−silica interaction. This interaction also takes place via a double-layer hydrogen-bonding interaction. This synthetic pathway was proposed by Chen and coworkers to create a full family of mesoporous silica structures, the AMS-n series [36–40, 89].

$S°I°/N°I°$: This is the type of interaction that occurs via hydrogen bonds between neutral ($N°$) or non-ionic surfactants ($S°$) and uncharged silica species ($I°$). This synthetic pathway has been used to prepare different mesoporous materials, such as HMS and MSU disordered worm-like mesoporous silicas [41, 42, 79].

$S°(XI)°$: This is the double layer hydrogen-bonding interaction that occurs between non-ionic surfactants ($S°$) and X^-I^+ ion pairs, in an acidic medium (pH < 2), where I^+ stands for the positively charged silica species and X^- is an ion such as Cl^-, Br^-, I^-, SO_4^{2-}, NO_3^-, and so on. This synthetic route yielded OMS of the SBA-n series ($n = 11$, 12, 15, and 16) [43, 90], FDU-n ($n = 1$, 5, and 12) [91–93], and KIT-n ($n = 5$ and 6) [47, 48].

5.5 Tuning the Structural Properties of Mesoporous Silicas

When OMS are projected as ceramics for biomedical uses, their mesostructural characteristics must be tailored to meet the requirements of the targeted application. In this section we will tackle the different parameters that affect the micellar mesostructures and summarize the main synthetic strategies that permit tailoring of the pore diameter of OMS.

5.5.1   Micellar Mesostructure

The final mesostructures depend on the surfactant liquid-crystal phases (TLCT mechanism) or silica–surfactant liquid-crystal-like phases (CLCT mechanism). The factors that drive mesophases, which are displayed in liquid-crystal phase diagrams, include surfactant concentration and temperature, the packing parameter (g) or the hydrophilic/hydrophobic volume ratio (V_H/V_L) of the template molecule. The influence of these parameters on the mesophase is described here.

5.5.1.1   Critical Micelle Concentration

As defined in Section 5.3.1.1, the CMC is the concentration at which micelles start to form and is the most important single characteristic of a given surfactant. The CMC value depends on the chemical structure of the surfactant, and also on the temperature and presence or not of co-solutes [27]. A surfactant with low CMC value is a major condition to obtain highly-regular ordered mesostructures [94]. It should be highlighted that ordered mesostructures are always obtained if the CMC values of surfactants are between 0 and 20 mg l^{-1}. However, several synthetic approaches can be used to decrease the CMC values to yield ordered mesostructures when surfactants have CMC values in the 20–300 mg l^{-1} range. Surfactants with relatively high CMC values usually yield cubic mesostructures. However, the further increase in CMC hinders the preparation of ordered mesostructures [94].

5.5.1.2   Surfactant Packing Parameter

Micellar aggregates of surfactants organize according to different shapes yielding supramicellar aggregates (spherical or cylindrical micelles, lamellae, etc.). An accepted model to predict the resulting self-assembled structures when using ionic surfactants is that proposed by Israelachvili and coworkers [95]. This model considers that the geometric characteristics of the template drive their molecular packing that yields to supramolecular assemblies. Concretely, these geometrical considerations rely on the ratio of the polar head surface to the hydrophobic volume. The amphiphilic surfactant molecules are modeled as a conical fragment (the hydrophobic, nonpolar tail) joined to a spherical (hydrophilic, polar) head (Figure 5.8). The steric hindrance of the hydrophobic chain is defined by the ratio V/l, where V is the chain volume (plus any co-solvent between the chains) and l is the kinetic surfactant tail length. The contribution of the polar head is given by the effective surface area, a_0, at the aqueous-micelle surface. The value of the packing parameter $g = V/(a_0 l)$, relates the molecular structure of the surfactant to the architecture of the supramolecular aggregates. The limiting values of g can be calculated for any aggregate of known geometry by using the l value and an estimation of the aggregation number, that is, the number of molecules that constitute the micelle. The latter can be obtained from two relations: the aggregate surface to a_0 and the aggregate volume to V. An increase in g involves a decrease in the curvature of the micellar structure.

Figure 5.8 displays the different micellar structures compatible with a given g of a cationic surfactant [95]:

1. $g < 1/3$: Spherical micelles with an interior composed of hydrophobic hydrocarbon chains and an external surface composed of hydrophilic polar head groups facing water. The low surfactant g value permits the formation of spherical micelles with strongly positive spontaneous curvature. The hydrophobic core has a radius close to the length of the extended alkyl chain value.

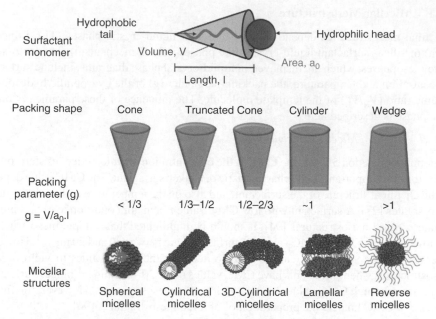

Figure 5.8   *Schematic representation of a surfactant molecule, where V is the hydrophobic group volume, l the kinetic surfactant hydrophobic tail length and a_0 the effective surface area of the hydrophilic head. The relationship between the packing parameter of cationic surfactant and the resulting micellar structures is also displayed (See insert for color representation of the figure)*

2. $1/3 < g < 1/2$: Cylindrical micelles with an interior composed of hydrophobic hydrocarbon chains and an external surface composed of polar head groups facing water. The cross-section of the hydrocarbon core is analogous to that of spherical micelles. The micellar length is highly variable so these micelles are polydisperse.
3. $1/2 < g < 2/3$: 3D cylindrical micelles.
4. $g \sim 1$: Surfactant bilayers build up lamellar micellar structures, where the hydrocarbon core has a thickness of about 80% of the length of two extended alkyl chains.
5. $g > 1$: Reversed or inverted micelles have a water core surrounded by the surfactant polar head groups. The alkyl chains together with a nonpolar solvent comprise the continuous medium. Like conventional micelles they can grow into cylinders.

Huo *et al.* [82, 85] were pioneers in considering the *g* parameter to explain the formation of different surfactant-templated oxide mesoporous structures. Hypothetically, the structure of the mesophase depends on the packing properties of the template molecules, that is, the value of *g*. The validity of this assumption has been demonstrated by carrying out complex studies that took into account diverse parameters, such as surfactant nature, pH, presence of co-solvents or co-surfactant, and their influence on the observed phase transitions, for fixed synthesis conditions. For silica mesoporous systems, the sequence of the mesostructure as a function of the *g* value of the cationic surfactant is cubic ($Pm\bar{3}n$) and 3D hexagonal ($P6_3/mmc$) with $g < 1/3$, 2D hexagonal ($p6mm$) with $1/3 < g < 1/2$, cubic ($Ia\bar{3}d$) with $1/2 < g < 2/3$, and lamellar with $g \sim 1$ (Table 5.1) [85, 96].

Table 5.1 *Relationship between the packing parameter* ($g = V/a_0 l$) *of cationic surfactants and mesostructure of the resulting OMS* [85, 96]

g	Micellar structure	Cationic surfactant	Ordered mesoporous silicas (OMS)
< 1/3	Spherical micelles	Single-chain surfactants with large head groups, for example, $C_nH_{2n+1}N(C_2H_5)_3X$ ($n = 12–18$); C_{m-s-1} ($m = 12–18$)	*Basic synthesis:* SBA-6 (cubic $Pm\bar{3}n$) SBA-7 (3D hexagonal $P6_3/mmc$) *Acidic synthesis:* SBA-1 (cubic $Pm\bar{3}n$) SBA-2 (3D hexagonal $P6_3/mmc$)
1/3–1/2	Cylindrical micelles	Single-chain surfactants with small head groups, for example, $C_nH_{2n+1}N(CH_3)_3X$ ($n = 8–18$)	*Basic synthesis:* MCM-41 (2D hexagonal, $p6mm$) *Acidic synthesis:* SBA-3 (2D hexagonal, $p6mm$)
1/2–2/3	3D cylindrical micelles	Single-chain surfactants with small head groups, for example, CTAB special surfactants with large hydrophobic polar head and double-chain surfactants with large head groups and flexible chains, for example, $C_{16}H_{33}(CH_3)_2N(CH_2)(C_6H_5)$, *gemini* C_{m-12-m}	*Basic synthesis:* MCM-48 (cubic $Ia\bar{3}d$)
1	Lamellar micelles	Double-chain surfactants with small head groups or rigid, immobile chains, for example, $C_nH_{2n+1}N(CH_3)X$ ($n = 20, 22$); *gemini* $C_{16-2-16}$	*Basic synthesis:* MCM-50 (lamellar structure) *Acidic synthesis:* SBA-4 (lamellar structure)

The packing parameter allows one to explain and predict the arrangement of surfactants that will drive the final mesostructure of the OMS. Nonetheless, g values depend not only on the inherent features of surfactants (such as shape, charge, or size) but also on the synthesis conditions, including pH, presence of organic additives, or the presence of inorganic salts. The main factors that affect packing of surfactants can be summarized as follows:

1. *Effective surface area of the hydrophilic head, a_0*: The greater the a_0 the smaller the g value, which promotes the formation of spherical mesostructures with high curvatures. For instance, cationic surfactants $C_nH_{2n+1}N(CH_3)_3Br$ ($n = 10–18$) usually lead to the formation of 2D hexagonal OMS (e.g., MCM-41). However, cationic surfactants

with larger hydrophilic heads, such as $C_nH_{2n+1}N(C_2H_5)_3Br$ usually induce 3D cage mesostructures (e.g., SBA-1). In the case of *gemini* surfactants (C_{m-s-m}), each hydrophilic head group is linked by a hydrocarbon chain, and a_0 can be tuned by changing the hydrocarbon length, that is, the s value. For example, when s gradually increases from 2 to 12 in the *gemini* surfactant $C_{16-2-16}$ the mesostructure evolves from lamellar to 2D hexagonal and to cubic bicontinuous under basic conditions.

2. *Hydrophobic tail*: In this case we must carefully consider the influence of the length (l) and volume (V) of the hydrophobic tail on the surfactant arrangement, although both parameters are closely related.

   Regarding surfactants with the same hydrophilic head group and having single tails, the length of the alkyl chain does not significantly affect the final mesostructure when the number of carbon atoms is lower than 20, since the V/l value, and consequently g, remain unaltered. Nevertheless, when the number of carbon atoms is higher than 20, the chains tend to curve, which leads to increased V and reduced l values, and hence increased g values. For example, lamellar mesostructures are usually obtained when using $C_nH_{2n+1}(CH_3)_3N^+$ ($n = 20$ and 22) surfactants.

   Surfactants exhibiting the same hydrophilic head group but different hydrophobic tails, that is, single or double, undergo different arrangements. The surfactant with double tails has a g value twice that of one with a single tail, which favors the formation of bilayers rather than spherical micelles. For instance, the *gemini* surfactant $C_{16-2-16}$ promotes the formation of lamellar mesostructures.

3. *Ionization degree of anionic surfactants*: The g values of anionic surfactants can change depending on their ionization degree, which can be tuned by the pH of the solution. For instance, the anionic surfactant acylglutamate, which is obtained by neutralizing *N*-lauroyl L-glutamic acid with a base, exhibits different micellar phases, lamellar, cubic, or hexagonal, the formation of the cubic phase being promoted by increasing the neutralization degree [97]. OMS exhibiting different mesostructures have been prepared by using the anionic surfactant *N*-myristoyl-L-glutamic acid (C14GluA) as template and TMAPS as co-SDA [39]. The packing of the micelle was controlled by simply adjusting the neutralization degree of the C14GluA surfactant. Different mesophases, ranging from tetragonal $P4_2/mnm$ (cage type, AMS-9), cubic $Fd\bar{3}m$ (cage type, AMS-8), to 2D hexagonal $p6mm$ (cylindrical, AMS-3), and an unknown mesophase (denoted as AMS-10) were obtained by decreasing the amount of NaOH that was added to the reaction system [39].

4. *Organic additives*: Surfactant solutions have a general trend to solubilize a certain amount of organic additives. The environment of solubilization of different additives in or around micelles can be correlated with their structural organizations and reciprocal interactions [98, 99]. Small organic molecules usually tend to localize near the micelle/water interface, whereas large molecules are absorbed in the core. Thus, micellar shapes and g values can vary, and consequently phase transformation, mesopores enlargement, or variations in morphologies of the ultimate OMS can occur.

5. *Inorganic salts*: The addition of inorganic salts to surfactant solutions reduces the electrostatic repulsion among the surfactant head groups, which results in a decrease in a_0 and therefore an increase in the g value. The effect of inorganic salts on the assembly of ionic surfactants depends on the radii of hydrated cations and anions. Less strongly hydrated ions have, in general, smaller ionic radii and bind more strongly on the surface of the surfactant micelles. Thus, small hydrated ions promote the formation of

2D hexagonal mesostructures rather than 3D cubic ones. When using cationic surfactants the anions' effect on micellization and arrangement follows the Hofmeister series: $SO_4^{2-}, HPO_4^{2-}, OH^-, F^-, HCOO^-, CH_3COO^-, Cl^-, Br^-, NO_3^-, I^-, SCN^-, ClO^-$. Che *et al.* [100] reported that under acidic conditions the *g* value of CTAB surfactant followed the order $SO_4^{2-} < Cl^- < Br^- < NO_3^-$, inducing the mesophase conversion from a high to a low curvature. Thus, well-ordered OMS with a variety of mesostructures were synthesized from the same silica source (TEOS) and surfactant (CTAB) but different inorganic acids. Thus, H_2SO_4, HCl, and HNO_3, led to the formation of cubic *Pm3̄n*, 2D hexagonal *p6mm*, and lamellar mesophases, respectively. However, the opposite effect has been reported for non-ionic surfactants of the EO-PO-EO type, SO_4^{2-} (HSO_4^-) $> NO_3^- > Br^- > Cl^-$ in acidic solution, producing the transformation from hexagonal *p6mm* to cubic *Ia3̄d*. This has been ascribed to the balance between the dehydration and the radii effects [101]. The reader can realize the complexity of this topic, for further discussion there are some reviews aimed at understanding the mechanisms of the anion effect on the formation of OMS [102, 103].

5.5.1.3   The Hydrophilic/Hydrophobic Volume Ratio

The hydrophilic/hydrophobic volume ratio (V_H/V_L) is a significant factor that directs the final mesophase when non-ionic surfactants are used as SDAs [104]. Usually, block-copolymers exhibiting high V_H/V_L ratios, such as F108 ($EO_{132}PO_{50}EO_{132}$), F98 ($EO_{123}PO_{47}EO_{123}$), F127 ($EO_{106}PO_{70}EO_{106}$), Brij700 ($C_{18}EO_{100}$), and so on, create mesophases with high curvatures and lead to cage-type cubic mesostructures. On the other hand, block copolymers with medium V_H/V_L ratios, such as P123 ($EO_{20}PO_{70}EO_{20}$) or B50-1500 ($BO_{10}EO_{16}$, BO = butylene oxide), usually direct the synthesis of mesostructures with medium curvatures (e.g., 2D hexagonal *p6mm* structures or 3D bicontinuous cubic *Ia3̄d*). In addition, the use of amphiphilic block copolymer blends as SDAs permits the design and control of the mesostructure phase and domain dimensions in a precise and easy fashion.

5.5.1.4   Surfactant Phase Diagrams

The degree of micellization, the shape of the micelles, and the aggregation of the micelles into liquid crystals depends on the surfactant concentration. Surfactant phase diagrams contain valuable information regarding all these aspects and allow one to establish the mesoporous silica structures in the case of the TLCT approach.

Figure 5.9 displays a schematic phase diagram for the cationic surfactant CTAB, chosen as an illustrative example, in water [105]. At very low surfactant concentration the surfactant is present as free molecules dissolved in solution and adsorbed at interfaces. At slightly higher concentration, the CMC1 is reached, where the individual surfactant molecules form small spherical aggregates, that is, micelles. At higher concentrations (CMC2) the amount of solvent available between the micelles decreases and hence micelles can come together to form elongated cylindrical micelles. At slightly higher concentrations, LC phases form. First, rod-like micelles aggregate to form hexagonal close-packed LC arrays. As the concentration increases, there is an evolution to cubic bicontinuous LC followed by lamellar LC phases. In some systems, at very high surfactant concentrations, inverse phases arise, where water is solubilized at the interior of the micelle and the head groups are oriented inwardly.

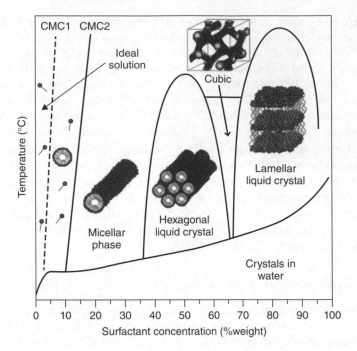

Figure 5.9   *Schematic phase diagram for CTAB surfactant in water*

5.5.2   Type of Mesoporous Structure

The mesoporous structure of OMS is a relevant factor for the potential biomedical application of these materials. For instance, as has been discussed in Chapter 4, mesopore arrangement highly affects the bioactive behavior of mesoporous bioactive glasses. Hence, 3D cubic mesostructures allow easier ionic exchange with the surrounding medium, that is, mass transport and diffusion processes, than 2D hexagonal mesostructures. In addition, the different mesostructure of OMS can also affect drug adsorption and release when these materials are used as drug delivery systems, as will be described in Chapter 12. When aiming at designing stimuli-responsive "smart" delivery devices, the possibility of achieving a perfect end-capping of mesopore entrances is of foremost relevance. In this sense, the mesoporous structure must be carefully chosen, as will be shown in Chapter 14.

To date many OMS with well-known structure have been designated, however, the number of distinctive mesostructures is much lower. Actually, since the structures of mesoporous materials are significantly limited by the liquid crystalline phase of surfactants, most OMS possess hexagonal or cubic structures with different symmetries, as displayed in Figure 5.10.

5.5.2.1   2D Mesostructures

Typical 2D-hexagonal OMS are MCM-41, FSM-6, SBA-15, SBA-3, and so on. These structures consist of hexagonally close packed cylindrical pore channels belonging to the *p6mm* space group. Among them, MCM-41 and SBA-15 are receiving growing attention for their potential applications in many fields, including catalysis, sensing, separation, biomaterials science, and so on.

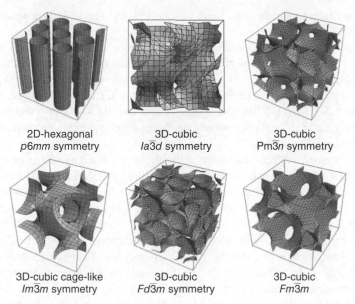

| 2D-hexagonal | 3D-cubic | 3D-cubic |
| *p6mm* symmetry | *Ia$\bar{3}$d* symmetry | *Pm$\bar{3}$n* symmetry |

| 3D-cubic cage-like | 3D-cubic | 3D-cubic |
| *Im$\bar{3}$m* symmetry | *Fd$\bar{3}$m* symmetry | *Fm$\bar{3}$m* |

Figure 5.10 *Reconstruction of mesoporous structures with different symmetries (See insert for color representation of the figure)*

MCM-41 is the simplest structure and can be prepared using a wide range of synthesis conditions, the most popular being that using CTAB as the SDA under basic conditions [21–23]. Pore channels of MCM-41 are usually approached as uniform cylinders of 1.5–3.5 nm in diameter. The pore wall thickness is around 1 nm and the surface area is usually higher than $1000 \, m^2 \, g^{-1}$. Micropores are not present in MCM-41.

The second main 2D structure is SBA-15, which is usually synthesized under acidic conditions using Pluronic block copolymers, P123 being the optimal template [43, 90]. SBA-15 exhibits uniform pore sizes in the 6.5–10 nm range, with pore wall thicknesses ranging from 3.1 to 4.1 nm, that is, much thicker than MCM-41, which result in higher thermal and hydrothermal stability. A remarkable feature of SBA-15 is the presence of a disordered micropore system in the silica walls interconnecting adjacent mesopores [45, 106] The structure of the micropores in the pore walls of SBA-15 remains unclear, but their presence arises from the partial occlusion of PEO chains of triblock copolymers into the silica matrix, which is a common effect for most PEO-containing surfactant-templated OMS [107]. The complementary microporosity of SBA-15 favors the diffusion of molecules, which is of foremost relevance in many fields, including catalysis, separation, or drug delivery.

5.5.2.2 3D Mesostructures

Many cubic mesostructures have been described, which can be divided into two main groups: bicontinuous cubic mesostructures and cage-type mesostructures.

1. *Bicontinuous cubic mesostructures.* MCM-48 is the most studied OMS exhibiting a bicontinuous cubic mesostructure. MCM-48 is defined by a so-called minimal surface, the gyroid [108], which divides the space into two enantiomeric separated 3D helical pore systems, forming a cubic bicontinuous structure (space group *Ia$\bar{3}$d*) (Figure 5.10). MCM-48 exhibits uniform mesopores of about 2 nm and disordered micropores of

0.5–0.8 nm interconnecting two main channels [109]. The wall thickness of MCM-48 is about 0.80 nm.

Large-pore OMS with bicontinuous cubic structure ($Ia\bar{3}d$) such as KIT-6 have been prepared using block copolymers as SDAs under acidic conditions and in the presence of some additives (e.g., butanol) [47, 110]. Additives strongly influence the V_H/V_L values of block copolymers and, therefore, the symmetry of the final mesophase, as described in Section 5.5.1.3. KIT-6 has pore diameter and wall thickness of about 8 and 3.5 nm, respectively, and exhibit complementary pores of about 1.7 nm interconnecting the two-group helical mesochannels.

2. *Cage-type mesostructures.* In addition to the above described mesostructures, exhibiting 1D and 3D uniform mesochannels, the vast majority of OMS have 3D cage-type pores (Figure 5.10). The ink-bottle model can explain such cage-type mesopores. We will briefly overview some of the most relevant OMS that exhibit mesostructures included within this category.

   SBA-6 exhibits a cage-type cubic $Pm\bar{3}n$ mesostructure, which can be described as a packing of spherical cages, distinguished by two types of cages with ordered arrangement [33, 81]. In the mesostructure of SBA-6, a B-cage is surrounded by 12 A-cages that are connected by mesopore openings of about 2 nm.

   SBA-16 exhibits a cage-type cubic $Im\bar{3}m$ mesostructure that can be described as the body-centered cubic symmetrical packing of spherical cages [81]. The pore size of SBA-16 is 5.4 nm, the size of the spherical cage being 8.5 nm and the minimum wall thickness 5.9 nm [111].

   FDU-2 presents a cage-type cubic $Fd\bar{3}m$ mesostructure consisting of an arrangement of spherical cages with the same space group as the diamond structure [84]. AMS-8 also displays face-centered cubic structure with $Fd\bar{3}m$ symmetry, which is composed of a bimodal arrangement of cages, 16 small (5.6 nm) and 8 large (7.6 nm) cages in the unit cell [37]. Large cages are interconnected through cage windows of 1.4–2.5 nm, whereas small cages and small–large interconnections are also present and occur via small windows of less than 0.5 nm.

   The FDU-1 mesostructure is a face-centered cubic $Fm\bar{3}m$ structure with 3D hexagonal intergrowth, the cubic phase being the main constituent [112]. Adjusting the hydrothermal treatment in the presence of inorganic salts, the pore cage diameters (from 9.5 to 14.5 nm) and the pore entrance diameters (from 4 to 8 nm) can be modulated. FDU-12 exhibits a cubic $Fm\bar{3}m$ mesostructure with a large cavity size of 10–12.3 nm without the intergrowth of the 3D hexagonal mesostructure [93]. A decrease in the synthesis temperature enlarges both the pore and the entrance sizes, the resulting structure being face-centered cubic close-packing of spherical cages, each connected to 12 nearest-neighboring cages [113]. KIT-5 is another OMS that presents a face-centered cubic $Fm\bar{3}m$ structure [48].

5.5.2.3 Lamellar and Disordered Mesostructures

OMS exhibiting lamellar structures are synthesized by SDAs with large g values or low V_H/V_L ratios, MCM-50 being the most popular one [21–23]. There are also other OMS, such as HMS [41], MSU [42], KIT-1 [114], TUD-1 [115], that exhibit disordered mesostructures consisting of a system of randomly packed short interconnected 1D pores [116].

Albeit disordered mesostructures cannot be defined by unit cells, symmetries, and space groups, their features, such as uniform pores, high surfaces areas, and easy surface modification may provide some opportunities in catalysis, adsorption, separation, and immobilization landscapes.

5.5.3   Mesopore Size

The pore size control is one of the most important subjects in the study of mesoporous materials. Mesopore size plays a key role in OMS when they are contemplated as ceramic matrices for drug delivery applications, as will be reviewed in Chapter 12. Pore size is the limiting factor for adsorption of molecules, since only molecules smaller than the mesopore diameter will be able to penetrate into the mesoporous cavities. In addition, drug diffusion out of the mesopores is strongly dependent on mesopore diameter.

In this section we will briefly summarize the main approaches developed so far aimed at tuning the pore size diameter in different OMS. For this purpose we will focus on the three main factors that permit variation of the pore size: surfactant characteristics, hydrothermal temperature and time, and the use of organic swelling agents.

5.5.3.1   Surfactant Characteristics

The pore size is highly dependent on the hydrophobic groups in surfactants. Increasing the alkyl chain length in cationic quaternary surfactants leads to larger pore diameters. Pioneer studies performed by Beck *et al.* [23] demonstrated that the pore diameter of MCM-41 could be enlarged from 1.8 to a maximum of up to 3.7 nm by increasing the surfactant chain length from C_8 to C_{18}. Later, the pore diameter of MCM-41 was adjusted from 1.6 to 4.2 nm by using a single or mixture of two surfactant(s) with alkyl chain lengths varying from C_8 to C_{22} [117]. The pore size of MCM-48 was also tuned in the 1.6–3.8 nm range by using (mixtures of) *gemini* surfactant C_{n-12-n} of different alkyl chain length ($n = 15$, 21) [118]. When using non-ionic surfactants such as triblock copolymers, different factors must be taken into account [94]: (i) an increase in the pore diameter can be achieved by increasing the molecular weight of the hydrophobic part of the surfactant; (ii) for different block architectures, the pore size of mesoporous materials templated by diblock copolymer is much larger than that templated by triblock copolymer. This fact can be explained by the bending of the triblock copolymer, which is needed to allow bond ends of the hydrophilic blocks to emerge from the micelle hydrophobic core; (iii) for different hydrophobic block compositions with the same hydrophobe molecular weight, templates with more hydrophobic moieties (e.g., PEO/PBO copolymers) favor larger pore diameters than copolymers with less hydrophobic parts (PEO/PPO); and (iv) for triblock copolymers with the same hydrophobic nature and repeating units, larger EO units favor smaller mesopore sizes.

5.5.3.2   Hydrothermal Treatment

An increase in the temperature and time of the hydrothermal treatment increases the pore diameter, surface area, and pore volume of the resulting OMS, leading to a decrease in the pore wall thickness [119]. For instance, the pore diameter of SBA-15 can be adjusted from 4.6 to 10 nm by increasing the hydrothermal temperature from 70 to 130 °C [120]. In addition, the pore size of SBA-15 can be tuned from 9.4 to 11.4 nm when changing the hydrothermal time from 6 hours to 4 days [120]. The same trend has been observed for OMS exhibiting cubic mesostructures, such as SBA-16 [110].

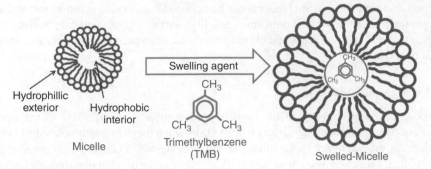

Figure 5.11 *Schematic representation of the performance of a hydrophobic molecule (trimethylbenzene, TMB) used as a micelle swelling agent to increase the pore diameter of mesophases*

5.5.3.3   Organic Swelling Agents

Surfactant solutions can solubilize certain hydrophobic organic additives. These organic molecules will locate in the inner region (core) of surfactant micelles, leading to swelled micelles, as depicted schematically in Figure 5.11 [27]. Common swelling agents used to expand the pore size of OMS are decane [121], 1,3,5-trimethylbenzene (TMB) [121, 122], decyldimethylamine [123], triisopropylbenzene [121], poly(propylene glycol) [124].

However, the amount of swelling agent used to enlarge the pore size has upper limits above which the mesostructural order would be lost or there could be changes in the mesophase. For instance, TMB amounts can enlarge pore sizes to 10 or 40 nm in basic CTAB or acidic triblock-copolymer surfactant systems, respectively. However, the resulting OMS would exhibit disordered pore arrangements. Actually, TMB can support the enlargement of the pore size of MCM-41 and SBA-15 to 6 and 13 nm, respectively, without losing the structural order [122]. The addition of higher amounts of TMB during the synthesis of SBA-15 produces a phase transition from hexagonal to a mesocellular foam phase, giving rise to the so-called mesocellular silica foams (MCFs) [122].

5.6   Structural Characterization of Mesoporous Silicas

There is not a single universal technique able to provide complete information about the mesoporous structure of OMS. For this reason, the structural characterization of ordered mesoporous materials is usually carried out combining the information derived from four main techniques: X-ray diffraction (XRD), electron microscopy (EM), nitrogen adsorption measurements, and NMR. In this section we will briefly describe the structural information that these techniques can provide concerning OMS.

OMS consist of amorphous silica walls, that is, they are not crystalline at the atomic level and, therefore, the XRD patterns of these materials show reflections only at low angles (0.2–7° 2θ range). XRD allows one to distinguish pore topologies and determine the average pore-to-pore distances in a periodical phase.

The position of an XRD peak in an XRD pattern is determined by the distance between atomic planes recognized by the Miller index (hkl), the so-called d-spacing (d_{hkl}), which can be calculated from the well-known basic Bragg law equation. To get information about the crystalline structure, a series of *d*-spacings must be calculated from the Bragg law. On the other hand, for a known unit cell, the d_{hkl} values can be calculated from unit-cell parameters. There are well-established relationships of crystal systems with *d*-spacing, indices of crystallographic planes (*hkl*) and lattice parameters.

Each mesoporous structure exhibits distinctive mesopore arrangements, giving rise to different XRD patterns. For instance, the mesostructure of SBA-15 consists of a honeycomb-like morphology, which is the result of hexagonal packing of unidimensional cylindrical pores. The XRD pattern of SBA-15 typically shows three to five reflections between $2\theta = 2°$ and $5°$. The reflections are produced because of the ordered hexagonal array of parallel silica tubes and can be indexed assuming a hexagonal unit cell as (10), (11), (20), (21), and (30). In some situations, the intensity of the XRD reflections might be small, which means a decreased domain size, but the material is still ordered. This can be observed in the XRD diffraction pattern of SBA-15 shown in Figure 5.12, where only one well-defined and another two broad peaks are observed.

EM has been revealed as the most powerful tool to determine the mesostructure of OMS [125, 126]. EM can provide information about mesostructure, periodicity, pore shape, and particle morphology of mesoporous materials. Actually, high-resolution transmission electron microscopy (HRTEM) can give detailed information about mesostructure, pore size, wall thickness, and so on which is difficult to obtain by using other techniques. TEM uses a high-energy electron beam to illuminate a sample, providing magnified images of the structure and composition information via energy-dispersive X-ray spectroscopy (EDX).

As mentioned above, XRD patterns of OMS in most cases show a few broad peaks, which make structure determination by XRD alone very difficult, even if the mesoporous material has 2D-hexagonal structure such as *p6mm*. In the particular case of SBA-15 (Figure 5.12), it is possible to obtain two TEM images with incidences of the electron beam parallel and perpendicular to the mesoporous channels, which gives us convincing evidence of *p6mm* symmetry and a one-dimensional channel system. Certainly, TEM images together with appropriate simulations allow the observation and discussion of (i) the 1D nature of the channels and (ii) their 2D hexagonal arrangement (*p6mm*), together with channel shape and wall thickness. However, these TEM images do not permit one to obtain information about the randomly arranged microporous interconnecting adjacent mesopores of SBA-15. For this purpose, the study of replicas such as a Pt nano-network structure using TEM is needed, as described elsewhere [127].

Regarding electron diffraction (ED) patterns for mesoporous materials, it follows the Bragg law. ED patterns are taken for single crystals and provide 2D structural information of materials, as shown in Figure 5.12 for SBA-15.

Scanning electron microscopy (SEM) is quite useful to determine the size and morphology of mesoporous crystals. When a small electron beam strikes the sample surface, back-scattered electrons and/or secondary electrons emitted from the material surface are collected and an image that reproduces the surface structure is produced [126, 128].

Adsorption analyses, normally of nitrogen molecules, have been employed to determine the surface area and the pore size distribution of ordered mesoporous materials. The adsorption isotherms can be classified into six types according to the IUPAC designation [129–132]. Typically, mesoporous materials exhibit type IV and V isotherms, characteristic of a multilayer adsorption followed by capillary condensation, a process associated

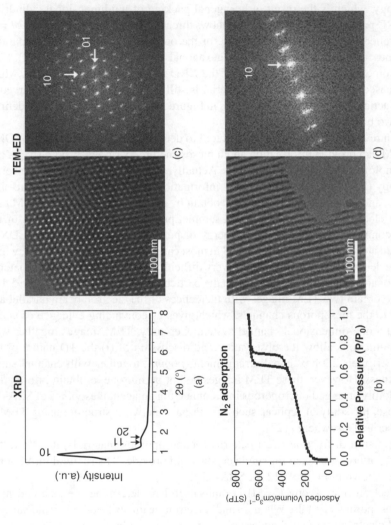

Figure 5.12 *Structural characterization of SBA-15 mesoporous material, chosen as a representative example, by different techniques: (a) XRD pattern at small angles; (b) nitrogen adsorption/desorption isotherms; (c) TEM image and ED pattern taken with the electron beam parallel to the mesoporous channels; and (d) TEM image and ED pattern taken with the electron beam perpendicular to the mesochannels*

with a sharp increase in the adsorption volume. In addition to isotherms, hysteresis loops can also provide relevant information regarding the pore structure and the type of pore connections. The hysteresis loop is attributed to thermodynamic factors and the effect of the pore system. Capillary condensation and evaporation shift toward higher and lower pressures, respectively, relative to the one at which gas and liquid coexist [130]. For example, the N_2 adsorption isotherm of MCM-41 shows a sharp capillary condensation step at a relative pressure (P/P_0) of 9.4 and no hysteresis between the adsorption and desorption branches is observed. On the contrary, SBA-15 exhibits relatively large pores interconnected by micropores. Hence, liquid nitrogen cannot desorb at the pressure corresponding to its pore size, because the micropore system has also been filled with liquid nitrogen, which produces a hysteresis loop (Figure 5.12). According to IUPAC, hysteresis loops can be classified into four types [132]. In the case of SBA-15, the H_1 loop at the capillary condensation region, with parallel and almost vertical adsorption and desorption branches, accounts for cylindrical mesopores and narrow pore size distributions (Figure 5.12).

The proper treatment of the data derived from N_2 adsorption–desorption isotherms using mathematical models that take into account geometrical considerations allows one to calculate the specific surface area, pore sizes (including mesopore and micropore diameters) and pore volume.

Since the OMS have amorphous silica frameworks, solid-state ^{29}Si-NMR provides valuable information about the coordination environments of silicon atoms, allowing determination of the crosslinking degree of the silica matrix [133, 134]. When different heteroatoms, such as phosphorus, aluminum, and so on, are incorporated within the silica network, ^{31}P and ^{27}Al NMR are also used to get information about the structure of the OMS. Actually, NMR has been demonstrated as a powerful tool to obtain structural information of different MBGs in the SiO_2–CaO–P_2O_5 system, as previously discussed in Chapter 4 [135, 136]. For OMS functionalized with organic moieties, ^{13}C-NMR can offer interesting information regarding the functionalization degree and the chemical nature of the organic functions.

5.7   Synthesis of Spherical Mesoporous Silica Nanoparticles

OMS can be synthesized with different morphologies, such as bulk [22, 43], fibers [36, 137, 138], rods [139, 140], films [69, 70], monoliths [66–68], spheres [141], and so on. Among the wide variety of existing morphologies, spherical mesoporous silica nanoparticles (MSNs), with particle sizes in the 50–300 nm range have found many potential applications in the biomedical field, especially in nanomedicine [14, 15, 142–144]. Bulk OMS exhibit irregular shape and particle size above several micrometers. In contrast, MSNs have uniform spherical shape and sizes on the nanometer scale, which offer several advantages, such as fast mass transport, effective adhesion to substrates, and good dispersity in solution.

The synthesis of MSNs for biomedical applications has to satisfy two conditions: (i) well-controlled nucleation and growth rate of the MSNs to produce uniform sizes in the 50–300 nm range, and (ii) absence of self-aggregation of the MSNs. There are two main approaches fulfilling these conditions that are widely used to produce MSNs: the aerosol-assisted synthesis and the modified Stöber method.

5.7.1   Aerosol-Assisted Synthesis

Aerosol or spray-drying methods are widely used in industrial processes for the massive and rapid preparation of MSNs [73, 142]. The process requires the use of an aerosol reactor device, as that depicted schematically in Figure 5.13. The method consists of atomizing a liquid solution into droplets, which are subsequently dispersed inside a gas stream forming an aerosol. Droplets can be generated using diverse atomizers, such as a powerful air flow, the use of ultrasonic nebulizers, and so on, whose operation relies on mechanically destabilizing the solution/atmosphere interphase. At this point, the first essential requirement for efficient aerosol processing is to use a stable initial solution, that is, the solution must be kinetically stable during the entire batch processing. To date, the most widely used method for aerosol-assisted synthesis of MSNs is the EISA process [60]. Therefore, the initial solution must contain the silica and/or other inorganic precursors, solvents, catalysts, and surfactants as templates.

The generated aerosol then flows through a tube to a heating zone at a temperature above the vaporization temperature of the volatile species (normally ethanol, THF, HCl, H_2O). This fast evaporation initiates the self-assembly process by concentrating gradually the system in inorganic precursors and surfactants. For given heating time and gradient, which depend on the heating chamber geometry, carrier gas flow rate, temperature, and pressure, the CMC is locally reached, which drives the micelles to form and self-assemble within

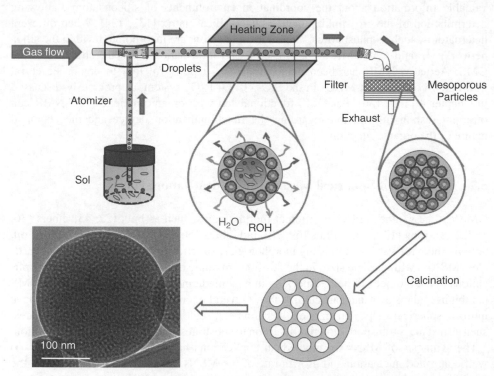

Figure 5.13 *Schematic representation of the synthesis of mesoporous silica nanoparticles (MSNs) by an aerosol-assisted process following the EISA process. TEM image of MSNs after calcination is also displayed*

the surrounding of the still flexible inorganic phase. The solvent evaporation during the drying step highly influences the final morphology of the spray-dried MSNs. The solidification of the droplet has to be slow enough to ensure the formation of a dense material, but fast enough to avoid coalescence and agglomeration. The whole process is under kinetic control, where the properties and structures developed at the different length scales are governed by the relative rates of each phenomenon [73]. Thus, the size and texture of the resulting MSNs depend on the relative rates of solvent evaporation and solvent formation. Usually, powders composed of solid particles are obtained, which can be retained, for instance in a filter, as displayed schematically in Figure 5.13. The last step of the synthesis is the surfactant removal, which leads to the final MSNs.

Synthesis of MSNs by a spray-drying technique combined with the EISA process offers many advantages derived from the high versatility of the procedure: (i) it is possible to use surfactants of different chemical nature (cationic, non-ionic), under both acidic and basic conditions; (ii) mesoporous mixed oxides nanoparticles, $(SiO_2-ZrO_2$ [145], SiO_2-CaO, P_2O_5 [146]) can be prepared following a "one-pot" route; (iii) the possibility of encapsulating maghemite nanoparticles of about 8 nm by a "one-pot" procedure from a colloidal $\gamma-Fe_2O_3$ suspension [147–149]; and (iv) the possibility of functionalization of the silica matrices with organosilanes following the "one-pot" procedure [147, 150]. All these possibilities make MSNs synthesized by spray-drying suitable for diverse biomedical applications (drug delivery, hyperthermia, imaging, etc.), which will be considered in Chapters 12–14.

5.7.2 Modified Stöber Method

The main shortcoming of MSNs synthesized by aerosol-assisted methods is their lack of uniformity in particle size and dispersity. This is why solution-based synthetic methods, which usually produce MSNs with uniform particle size and high dispersity, are preferred. Generally, solution synthesis of MSNs is carried out under basic and high dilution conditions in an alcohol–water mixture following the so-called modified Stöber method (Figure 5.14) [151]. The most widely used alcohols are methanol, ethanol, n-propanol, or n-butanol, while ammonia is used as catalyst for colloidal growth [142–156]. Under basic conditions, part of the silanol groups is deprotonated forming $Si-O^-$ species and, therefore, cationic surfactants are used to match the negatively charged silica surface.

This method creates monodisperse MSNs in sizes ranging from the micron to nanometer scale. It has been widely reported that the particle sizes can be tuned by controlling synthetic parameters, such as pH, reaction time and temperature [147–159], and by adding certain compounds such as organics and organosilanes [160–162]. These parameters are greatly associated with the hydrolysis and condensation of the silica sources (e.g., TEOS). Lu *et al.* [163] demonstrated that the particle sizes of MSNs gradually increased from 30 to 280 nm when the pH was progressively adjusted from 11 to 12.

Recently, Chiang *et al.* reported the control of MSNs particle sizes from 17 to 247 nm based on the Taguchi design method, by controlling the amount of TEOS, reaction time, and pH value [164]. They investigated the effect of the different parameters that influence particle size, resulting in the following sequence: pH (57%) > reaction time (29%) > amount of TEOS (13%). Thus, pH has the largest influence on the MSN size rather than the amount of silica source and the reaction time.

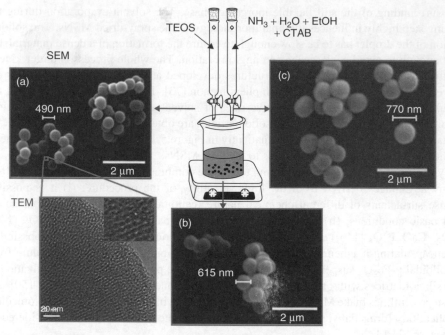

Figure 5.14 *Scheme of the synthesis of MSNs by the modified Stöber method, which requires controlling the pH value, surfactant, TEOS, and alcohol amounts and reaction time and temperature. SEM micrographs (a–c) of MSNs exhibiting different sizes synthesized by varying the TEOS amount and reaction temperature. TEM image of MSNs is also displayed. Reprinted from Chemical Engineering Journal, 137, 1, Manzano et al, Studies on MCM-41 mesoporous silica for drug delivery: Effect of particle morphology and amine functionalization, 30–37, 2013, with permission from Elsevier*

Finally, the particle sizes of MSNs can be tailored by the addition of certain compounds, such as organics [165, 166] and organosilanes [161, 167–169]. However, the addition of organosilanes into silica precursor solution has to be carefully considered, since it can affect the morphology of the synthesized MSNs.

5.8 Organic Functionalization of Ordered Mesoporous Silicas

The silica walls of OMS contain a high density of silanol groups on their surface, which can undergo organic modification by covalently grafting organic functionalities, giving rise to a full family of hybrid organic–inorganic mesoporous materials [170]. Organic functionalization allows modification of the chemical nature of the mesoporous silica matrix, incorporating organic groups containing hydrophobic chains, amine groups, sulfonic and carboxylic acids, and so on.

In fact, organic functionalization has made widespread the potential biomedical applications of OMS. Thus, functionalization has been found as the real milestone in the performance of OMS as drug delivery systems [11]. In addition, diverse moieties can be chemically grafted to the inner and outer surfaces of mesoporous silica, which permits their use as multifunctional platforms for the development of smart materials [13, 15]. The reader is encouraged to read Chapters 12–14 for a deeper study of this topic.

There are three main pathways to synthesize hybrid organic–inorganic OMS: (i) the later modification of the pore surface of a previously synthesized inorganic silica material ("grafting"). (ii) The simultaneous condensation of the corresponding silica and organosilica precursors ("co-condensation"), and (iii) the incorporation of organic groups as bridging components directly and specifically into the pore walls by the use of bis-silylated single-source organosilica precursors, giving rise to "periodic mesoporous organosilicas" (PMOs).

5.8.1   Post-synthesis Functionalization ("Grafting")

Grafting involves the later reaction of the silanol groups of OMS with organosilanes of the type $(RO)_3SiR'$ or less frequently chlorosilanes $ClSiR_3$ or silazanes $HN(SiR_3)_3$. During the post-synthesis functionalization or "grafting," the organic modification process is performed under anhydrous conditions once the free-surfactant OMS has already been obtained (Figure 5.15). Although the post-synthesis grafting leads to heterogeneous distributions of the functional groups, they result in well-defined structures controlled by the silica matrix. In addition, this method ensures that the organic groups are located in the outer surface of the mesopore walls, leading to higher functionalization degrees. Usually, post-synthesis functionalization is associated with a decrease in the porosity of OMS, whose magnitude depends on the size of the organic residue and the occupation degree. In addition, if the organosilanes react preferentially at the pore outlets, the diffusion of further molecules inside the mesoporous channels can be hampered, resulting in a heterogeneous distribution of the organic groups within the pores. If using extremely voluminous organic functions, they could totally block the pores.

Sometimes, the selective functionalization of specific regions of the mesoporous silica surface (external surface, pore surface, pore entrances) is needed to improve the performance of these materials as drug delivery systems [171]. To reach this goal, two different approaches relying on post-synthesis grafting have been developed. The first consists in starting from the free-surfactant OMS and a chemical agent exhibiting low-reactivity, which only produces the functionalization of the more accessible external surface. The subsequent treatment with more reactive agents leads to functionalization inside the pores [172]. The second approach consists in starting from the mesoporous matrix containing the surfactant inside the pores. After grafting a functional group at the external surface, the

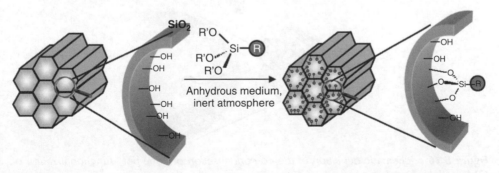

Figure 5.15 *Schematic depiction of the post-synthesis functionalization ("grafting") of OMS with organosilanes of the type $(RO)_3SiR'$, where R' is the organic functional group*

surfactant is removed, and the inner side of the pores further functionalized. However, to date, considering the wide variety of pore sizes and particle sizes of OMS, there is not a generally applicable method for their selective functionalization, which still remains a big challenge for the scientific community [171].

5.8.2   Co-condensation ("One-Pot" Synthesis)

The co-condensation method, also known as the one-pot route, involves the addition of the organic group, commonly an organosilane of the type $(RO)_3SiR'$, together with the silica source (usually TEOS) during the mesoporous synthesis, and all the functionalization process is performed in one step (Figure 5.16). When using the co-condensation method, the organic groups are linked to the outer and to the inner part of the silica walls. Consequently, the functionalization degree cannot exceed 40 mol% to avoid losing the mesoporous structure. It is important to highlight that surfactant removal must be carried out by solvent extractions since calcination would produce decomposition of the organic functions incorporated during the one-pot synthesis.

Being and coworkers have reported the site-selective functionalization of MSNs using a sequential co-condensation approach [173]. Using this strategy, functional groups were completely dispersed inside the channels, concentrated in parts of the mesopores, or exclusively placed on the external surface of colloidal MSNs, depending on the time of addition of the desired functionality. The authors state that, using this approach, several disadvantages associated with post-synthesis grafting can be avoided, such as the unwanted partial functionalization of the internal surface by diffusion of grafting reactants into the mesopores.

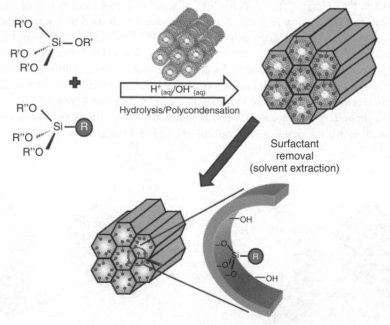

Figure 5.16 *Schematic depiction of the co-condensation or "one-pot" functionalization of OMS with organosilanes of the type $(RO)_3SiR'$, where R' is the organic functional group*

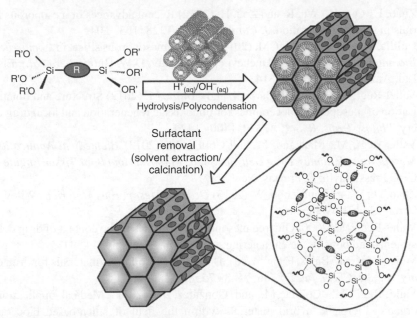

Figure 5.17 *Schematic representation of the synthesis route of periodic mesoporous organosilicas (PMOs) from bis-silylated organic bridging units of the type (R′O)₃Si-R-(OR′)₃, being R the organic bridge*

5.8.3 Periodic Mesoporous Organosilicas

The synthesis of PMOs relies on the use of bis-silylated organosilica precursors of the type $(R'O)_3Si\text{-}R\text{-}(OR')_3$ in the presence of surfactants as templates [174–176]. In PMOs, the organic bridges are integral components of the silica network. In fact, the organic units are incorporated in the 3D network structure of the silica matrix via two covalent bonds and, therefore, they are distributed completely homogeneously in the pore walls (Figure 5.17). These materials are promising candidates for a series of applications, including catalysis, sensors, drug delivery, and so on [177–180].

References

1. Sierra, I. and Pérez-Quintanilla, D. (2013) Heavy metal complexation on hybrid mesoporous silicas: an approach to analytical applications. *Chem. Soc. Rev.*, **42**, 3792–3807.
2. Corma, A. (1997) From microporous to mesoporous molecular sieve materials and their use in catalysis. *Chem. Rev.*, **97**, 2373–2419.
3. Díaz, U., Brunelab, D. and Corma, A. (2013) Catalysis using multifunctional organosiliceous hybrid materials. *Chem. Soc. Rev.*, **42**, 4083–4097.
4. Walcarius, A. (2013) Mesoporous materials and electrochemistry. *Chem. Soc. Rev.*, **42**, 4098–4140.

5. Zhao, L., Qin, H., Wu, R. and Zou, H. (2012) Recent advances of mesoporous materials in sample preparation. *J. Chromatogr. A*, **1228**, 193–204.
6. Colilla, M. and Vallet-Regí, M. (2011) Ordered mesoporous silica, in *Comprehensive Biomaterials*, vol. **4** (eds P. Ducheyne, K.E. Healy, D.W. Hutmacher *et al.*), Elsevier Ltd, Amsterdam, pp. 497–514.
7. Vallet-Regí, M., Izquierdo-Barba, I. and Colilla, M. (2012) Structure and functionalisation of mesoporous bioceramics for bone tissue regeneration and local drug delivery. *Philos. Trans. R. Soc. A*, **370**, 1400–1424.
8. Vallet-Regí, M., Manzano, M. and Colilla, M. (2013) *Biomedical Applications of Mesoporous Ceramics, Drug Delivery Smart Materials and Bone Tissue Engineering*, CRC Press, Taylor & Francis, New York.
9. Zhao, D., Wan, Y. and Zhou, W. (2013) *Ordered Mesoporous Materials*, Wiley-VCH Verlag GmbH, Weinheim.
10. Vallet-Regí, M. (2006) Ordered mesoporous materials in the context of drug delivery systems and bone tissue engineering. *Chem. Eur. J.*, **12**, 5934–5943.
11. Vallet-Regí, M., Balas, F. and Arcos, D. (2007) Mesoporous materials for drug delivery. *Angew. Chem. Int. Ed.*, **46**, 7548–7558.
12. Vallet-Regí, M., Colilla, M. and González, B. (2011) Medical applications of organic–inorganic hybrid materials within the field of silica-based bioceramics. *Chem. Soc. Rev.*, **40**, 596–607.
13. Li, Z., Barnes, J.C., Stoddart, J.F. and Zink, J.I. (2012) Mesoporous silica nanoparticles in biomedical applications. *Chem. Soc. Rev.*, **41**, 2590–2605.
14. Colilla, M., González, B. and Vallet-Regí, M. (2013) Mesoporous silica nanoparticles for the design of smart delivery nanodevices. *Biomater. Sci.*, **1**, 114–134.
15. Colilla, M., Vallet-Regí, M. (2013) Smart drug delivery from silica nanoparticles. In: *Smart Materials for Drug Delivery*, Vol. **2**, C. Álvarez-Lorenzo and A. Concheiro eds, Royal Society of Chemistry, pp. 63–82 .
16. Mamaeva, V., Sahlgren, C. and Lindén, M. (2013) Mesoporous silica nanoparticles in medicine – Recent advances. *Adv. Drug Deliver. Rev.*, **65**, 689–702.
17. Argauer, R.J. and Landolt, G.R. (1972) Crystalline zeolite ZSM-5 and method of preparing the same. US Patent 3,702,886.
18. Davis, M.E., Saldarriaga, C., Montes, C. *et al.* (1988) VPI-5: the first molecular sieve with pores larger than 10 angstroms. *Zeolites*, **8**, 362–366.
19. Yanagisawa, T., Shimizu, T., Kuroda, K. and Kato, C. (1990) The preparation of alkyltrimethylammonium-kanemite complexes and their conversion to microporous materials. *Bull. Chem. Soc. Jpn.*, **63**, 988–992.
20. Inagaki, S., Fukushima, Y. and Kuroda, K. (1993) Synthesis of highly ordered mesoporous materials from a layered polysilicate. *Chem. Commun.*, **8**, 680–682.
21. Beck, J.S. and Princeton, N.Y. (1991) Method for synthesizing mesoporous crystalline material. US Patent 5,057,296.
22. Kresge, C.T., Leonowicz, M.E., Roth, W.J. *et al.* (1992) Ordered mesoporous molecular sieves synthesized by a liquid-crystal template mechanism. *Nature*, **359**, 710–712.
23. Beck, J.S., Vartuli, J.C., Roth, W.J. *et al.* (1992) A new family of mesoporous molecular-sieves prepared with liquid-crystal templates. *J. Am. Chem. Soc.*, **114**, 10834–10843.

24. Vartuli, J.C., Kresge, C.T., Leonowicz, M.E. *et al.* (1994) Synthesis of mesoporous materials: liquid-crystal templating versus intercalation of layered silicates. *Chem. Mater.*, **6**, 2070–2077.
25. Soler-Illia, G.J.A.A., Sanchez, C., Lebeau, B. and Patarin, J. (2002) Chemical strategies to design textured materials: from microporous and mesoporous oxides to nanonetworks and hierarchical structures. *Chem. Rev.*, **102**, 4093–4138.
26. Wang, Y. and Zhao, D. (2007) On the controllable soft-templating approach to mesoporous silicates. *Chem. Rev.*, **107**, 2821–2860.
27. Holmberg, K., Jösson, B., Kronberg, B. and Lindman, B. (2003) *Surfactants and Polymers in Aqueous Solutions*, John Wiley & Sons, Ltd, Chichester.
28. Van Der Voort, P., Mathieu, M., Mees, F. and Vansant, E.F.J. (1998) Synthesis of high-quality MCM-48 and MCM-41 by means of the GEMINI surfactant method. *Phys. Chem. B*, **102**, 8847–8851.
29. Zana, R., Benrraou, M. and Rueff, R. (1991) Alkanediyl-.alpha.,.omega.-bis(dimethylalkylammonium bromide) surfactants. 1. Effect of the spacer chain length on the critical micelle concentration and micelle ionization degree. *Langmuir*, **7**, 1072–1075.
30. Xu, J., Han, S., Hou, W. *et al.* (2004) Synthesis of high-quality MCM-48 mesoporous silica using cationic gemini surfactant C12-2-12. *Colloids Surf., A*, **248**, 75–78.
31. Shen, S.D., Garcia-Bennett, A.E., Liu, Z. *et al.* (2005) Three-dimensional low symmetry mesoporous silica structures templated from tetra-headgroup rigid bolaform quaternary ammonium surfactant. *J. Am. Chem. Soc.*, **127**, 6780–6787.
32. Tan, B., Dozier, A., Lehmler, H.J. *et al.* (2004) Elongated silica nanoparticles with a mesh phase mesopore structure by fluorosurfactant templating. *Langmuir*, **20**, 6981–6984.
33. Huo, Q., Margolese, D.I., Ciesla, U. *et al.* (1994) Generalized synthesis of periodic surfactant/inorganic composite materials. *Nature*, **368**, 317–321.
34. Chen, F., Huang, L. and Li, Q. (1997) Synthesis of MCM-48 using mixed cationic-anionic surfactants as templates. *Chem. Mater.*, **9**, 2685–2686.
35. Lind, A., Spliethoff, B. and Lindén, M. (2003) Unusual, vesicle-like patterned, mesoscopically ordered silica. *Chem. Mater.*, **15**, 813–818.
36. Che, S., Liu, Z., Ohsuna, T. *et al.* (2004) Synthesis and characterization of chiral mesoporous silica. *Nature*, **429**, 281–284.
37. Garcia-Bennett, A.E., Miyasaka, T., Terasaki, O. and Che, S. (2004) Structural solution of mesocaged material AMS-8. *Chem. Mater.*, **16**, 3597–3605.
38. Garcia-Bennett, A.E., Kupferschmidt, N., Sakamoto, Y. *et al.* (2005) Synthesis of mesocage structures by kinetic control of self-assembly in anionic surfactants. *Angew. Chem. Int. Ed.*, **44**, 5317–5322.
39. Gao, C.B., Sakamoto, U., Sakamoto, K. *et al.* (2006) Synthesis and characterization of mesoporous silica AMS-10 with bicontinuous cubic Pn3m symmetry. *Angew. Chem. Int. Ed.*, **45**, 4295–4298.
40. Che, S., Garcia-Bennett, A.E., Sakamoto, K. *et al.* (2003) Mesoporous silica of novel structures with periodic modulations synthesized by anionic surfactant templating route. *Nat. Mater.*, **2**, 801–805.
41. Tanev, P.T. and Pinnavaia, T.J. (1995) A neutral templating route to mesoporous molecular sieves. *Science*, **267**, 865–867.

42. Bagshaw, S.A., Prouzet, E. and Pinnavaia, T.J. (1995) Templating of meso-porous molecular sieves by nonionic polyethylene oxide surfactants. *Science*, **269**, 1242–1244.

43. Zhao, D.Y., Feng, J.L., Huo, Q.S. *et al.* (1998) Triblock copolymer syntheses of meso-porous silica with periodic 50 to 300 angstrom pores. *Science*, **279**, 548–552.

44. Kruk, M., Jaroniec, M., Ko, C.H. and Ryoo, R. (2000) Characterization of the porous structure of SBA-15. *Chem. Mater.*, **12**, 1961–1968.

45. Lukens, W.W. Jr., Schmidt-Winkel, P., Zhao, D. *et al.* (1999) Evaluating pore sizes in mesoporous materials: a simplified standard adsorption method and a simplified Broekhoff – de Boer method. *Langmuir*, **15**, 5403–5409.

46. Miyazawa, K. and Inagaki, S. (2000) Control of the microporosity within the pore walls of ordered mesoporous silica SBA-15. *Chem. Commun.*, 2121–2122.

47. Kleitz, F., Choi, S.H. and Ryoo, R. (2003) Cubic Ia3d large mesoporous silica: syn-thesis and replication to platinum nanowires, carbon nanorods and carbon nanotubes. *Chem. Commun.*, 2136–2137.

48. Kleitz, F., Liu, D., Anilkumar, G.M. *et al.* (2003) Large cage face-centered-cubic Fm3m mesoporous silica: synthesis and structure. *J. Phys. Chem. B*, **107**, 14296–14300.

49. Brinker, C.J. (1988) Hydrolysis and condensation of silicates: effects on structure. *J. Non Cryst. Solids*, **100**, 31–50.

50. Brinker, C.J. and Scherer, G.W. (1990) *Sol-Gel Science: The Physics and Chemistry of Sol-Gel Processing*, Academic Press, London.

51. Linssen, T., Cassiers, K., Cool, P. and Vansant, E.F. (2003) Mesoporous templated silicates: an overview of their synthesis, catalytic activation and evaluation of the stability. *Adv. Colloid Interface Sci.*, **103**, 121–147.

52. Patarin, J., Lebeau, B. and Zana, R. (2002) Recent advances in the formation mech-anisms of organized mesoporous materials. *Curr. Opin. Colloid Interface Sci.*, **7**, 107–115.

53. Colilla, M., Balas, F., Manzano, M. and Vallet-Regí, M. (2007) Novel method to enlarge the surface area of SBA-15. *Chem. Mater.*, **19**, 3099–3101.

54. Zhang, F.Q., Yan, Y., Yang, H.F. *et al.* (2005) Understanding effect of wall structure on the hydrothermal stability of mesostructured silica SBA-15. *J. Phys. Chem. B*, **109**, 8723–8722.

55. Keene, M.T.J., Gougeon, R.D.M., Denoyel, R. *et al.* (1999) Calcination of the MCM-41 mesophase: mechanism of surfactant thermal degradation and evolution of the porosity. *J. Mater. Chem.*, **9**, 2843–2850.

56. Zhu, H.G., Jones, D.J., Zajac, J. *et al.* (2002) Synthesis of periodic large mesoporous organosilicas and functionalization by incorporation of ligands into the framework wall. *Chem. Mater.*, **14**, 4886–4894.

57. Burleigh, M.C., Markowitz, M.A., Spector, M.S. and Gaber, B.P. (2001) Amine-functionalized periodic mesoporous organosilicas. *Chem. Mater.*, **13**, 4760–4766.

58. Tanev, P.T. and Pinnavaia, T.J. (1996) Mesoporous silica molecular sieves prepared by ionic and neutral surfactant templating: a comparison of physical properties. *Chem. Mater.*, **8**, 2068–2079.

59. Inagaki, S., Sakamoto, Y., Fukushima, Y. and Terasaki, O. (1996) Pore wall of a meso-porous molecular sieve derived from kanemite. *Chem. Mater.*, **8**, 2089–2095.

60. Brinker, C.J., Lu, Y., Sellinger, A. and Fan, H. (1999) Evaporation-induced self-assembly: nanostructures made easy. *Adv. Mater.*, **11**, 579–585.

61. Yang, P.D., Zhao, D.Y., Margolese, D.I. *et al.* (1998) Generalized syntheses of large-pore mesoporous metal oxides with semicrystalline frameworks. *Nature*, **396**, 152–155.

62. Zhao, D.Y., Yang, P.D., Margolese, D.I. *et al.* (1998) Synthesis of continuous mesoporous silica thin films with three-dimensional accessible pore structures. *Chem. Commun.*, 2499–2500.

63. Yang, P.D., Deng, T., Zhao, D.Y. *et al.* (1998) Hierarchically ordered oxides. *Science*, **282**, 2244–2246.

64. Zhao, D.Y., Yang, P.D., Chmelka, B.F. and Stucky, G.D. (1999) Multiphase assembly of mesoporous – macroporous membranes. *Chem. Mater.*, **11**, 1174–1178.

65. Urade, V.N. and Hillhouse, H.W. (2005) Synthesis of thermally stable highly ordered nanoporous tin oxide thin films with a 3D face-centered orthorhombic nanostructure. *J. Phys. Chem. B*, **109**, 10538–10541.

66. Melosh, N.A., Davidson, P. and Chmelka, B.F. (2000) Monolithic mesophase silica with large ordering domains. *J. Am. Chem. Soc.*, **122**, 823–829.

67. Naik, S.P., Fan, W., Yokoi, T. and Okubo, T. (2006) Synthesis of a three-dimensional cubic mesoporous silica monolith employing an organic additive through an evaporation-induced self-assembly process. *Langmuir*, **22**, 6391–6397.

68. Feng, P., Bu, X., Stucky, G.D. and Pine, D.J. (2000) Monolithic mesoporous silica templated by microemulsion liquid crystals. *J. Am. Chem. Soc.*, **122**, 994–995.

69. Soler-Illia, G.J.A.A. and Innocenzi, P. (2006) Mesoporous hybrid thin films: the physics and chemistry beneath. *Chem. Eur. J.*, **12**, 4478–4494.

70. Sanchez, C., Boissiere, C., Grosso, D. *et al.* (2008) Design, synthesis, and properties of inorganic and hybrid thin films having periodically organized nanoporosity. *Chem. Mater.*, **20**, 682–737.

71. Edler, K.J. and Yang, B. (2013) Formation of mesostructured thin films at the air–liquid interface. *Chem. Soc. Rev.*, **42**, 3765–3776.

72. Rao, G.V.R., López, G.P., Bravo, J. *et al.* (2002) Monodisperse mesoporous silica microspheres formed by evaporation-induced self-assembly of surfactants templates in aerosols. *Adv. Mater.*, **14**, 1301–1304.

73. Boissiere, C., Grosso, D., Chaumonnot, A. *et al.* (2011) Aerosol route to functional nanostructured inorganic and hybrid porous materials. *Adv. Mater.*, **23**, 599–623.

74. López-Noriega, A., Arcos, D., Izquierdo-Barba, I. *et al.* (2006) Ordered mesoporous bioactive glasses for bone tissue regeneration. *Chem. Mater.*, **18**, 3137–3144.

75. Yan, X.X., Yu, C.Z., Zhou, X.F. *et al.* (2004) Highly ordered mesoporous bioactive glasses with superior in vitro bone-forming bioactivities. *Angew. Chem. Int. Ed.*, **43**, 5980–5984.

76. Ying, J.Y., Mehnert, C.P. and Wong, M.S. (1999) Synthesis and applications of supramolecular-templated mesoporous materials. *Angew. Chem. Int. Ed.*, **38**, 56–77.

77. Beck, J.S., Vartuli, J.C., Kennedy, G.J. *et al.* (1994) Molecular or supramolecular templating: defining the role of surfactant chemistry in the formation of microporous and mesoporous molecular sieves. *Chem. Mater.*, **6**, 1816–1821.

78. Chen, C.Y., Burkett, S.L., Li, H.X. and Davis, M.E. (1993) Studies on mesoporous materials II. Synthesis mechanism of MCM-41. *Microporous Mesoporous Mater.*, **2**, 27–34.

79. Attard, G.S., Glyde, J.C. and Göltner, C.G. (1995) Liquid-crystalline phases as templates for the synthesis of mesoporous silica. *Nature*, **378**, 366–368.

80. Huo, Q., Margolese, D.I., Ciesla, U. *et al.* (1994) Organization of organic molecules with inorganic molecular species into nanocomposite biphase arrays. *Chem. Mater.*, **6**, 1176–1191.

81. Sakamoto, Y., Kaneda, M., Terasaki, O. *et al.* (2000) Direct imaging of the pores and cages of three-dimensional mesoporous materials. *Nature*, **408**, 449–453.

82. Huo, Q., Margolese, D.I. and Stucky, G.D. (1996) Surfactant control of phases in the synthesis of mesoporous silica-based materials. *Chem. Mater.*, **8**, 1147–1160.

83. Zhao, D., Huo, Q., Feng, J. *et al.* (1999) Novel mesoporous silicates with two-dimensional mesostructure direction using rigid bolaform surfactants. *Chem. Mater.*, **11**, 2668–2672.

84. Shen, S., Li, Y., Zhang, Z. *et al.* (2002) A novel ordered cubic mesoporous silica templated with tri-head group quaternary ammonium surfactant. *Chem. Commun.*, 2212–2213.

85. Huo, Q., Leon, R., Petroff, P.M. and Stucky, G.D. (1995) Mesostructure design with Gemini surfactants: supercage formation in a three-dimensional hexagonal array. *Science*, **268**, 1324–1327.

86. Yada, M., Machida, M. and Kijima, T. (1996) Synthesis and deorganization of an aluminium–based dodecyl sulfate mesophase with a hexagonal structure. *Chem. Commun.*, 769–770.

87. Ulagappan, N. and Rao, C.N.R. (1996) Mesoporous phases based on SnO_2 and TiO_2. *Chem. Commun.*, 1685–1686.

88. Antonelli, D.M. and Ying, J.Y. (1995) Synthesis of hexagonally packed mesoporous TiO_2 by a modified sol–gel method. *Angew. Chem. Int. Ed.*, **34**, 2014–2017.

89. Han, L. and Che, S. (2013) Anionic surfactant templated mesoporous silicas (AMSs). *Chem. Soc. Rev.*, **42**, 3740–3752.

90. Zhao, D., Huo, Q., Feng, J. *et al.* (1998) Nonionic triblock and star diblock copolymer and oligomeric surfactant syntheses of highly ordered, hydrothermally stable, mesoporous silica structures. *J. Am. Chem. Soc.*, **120**, 6024–6036.

91. Yu, C., Yua, Y. and Zhao, D. (2000) Highly ordered large caged cubic mesoporous silica structures templated by triblock PEO–PBO–PEO copolymer. *Chem. Commun.*, 575–576.

92. Liu, X., Tian, B., Yu, C. *et al.* (2002) Room-temperature synthesis in acidic media of large-pore three-dimensional bicontinuous mesoporous silica with Ia3d symmetry. *Angew. Chem. Int. Ed.*, **41**, 3876–3878.

93. Fan, J., Yu, C., Gao, F. *et al.* (2003) Cubic mesoporous silica with large controllable entrance sizes and advanced adsorption properties. *Angew. Chem. Int. Ed.*, **42**, 3146–3150.

94. Yu, C., Fan, J., Tian, B. *et al.* (2003) Synthesis of mesoporous silica from commercial poly(ethylene oxide)/poly(butylene oxide) copolymers: toward the rational design of ordered mesoporous materials. *J. Phys. Chem. B*, **107**, 13368–13375.

95. Israelachvili, J.N., Mitchell, D.J. and Ninham, B.W. (1976) Theory of self-assembly of hydrocarbon amphiphiles into micelles and bilayers. *J. Chem. Soc., Faraday Trans. 2*, **72**, 1525–1568.

96. Landry, C.C., Tolbert, S.H., Gallis, K.W. *et al.* (2001) Phase transformations in mesostructured silica/surfactant composites. Mechanisms for change and applications to materials synthesis. *Chem. Mater.*, **13**, 1600–1608.

97. Kaneko, D., Olsson, U. and Sakamoto, K. (2002) Self-assembly in some N-lauroyl-L-glutamate/water systems. *Langmuir*, **18**, 4699–4703.

98. Nagarajan, R., Chaiko, M.A. and Ruckenstein, E. (1984) Locus of solubilization of benzene in surfactant micelles. *J. Phys. Chem.*, **88**, 2916–2922.

99. Zana, R. (1995) Aqueous surfactant-alcohol systems: a review. *Adv. Colloid Interface Sci.*, **57**, 1–64.

100. Che, S., Li, H., Lim, S. *et al.* (2005) Synthesis mechanism of cationic surfactant templating mesoporous silica under an acidic synthesis process. *Chem. Mater.*, **17**, 4103–4113.

101. Tang, J., Yu, C., Zhou, X. *et al.* (2004) The anion sequence in the phase transformation of mesostructures templated by non-ionic block copolymers. *Chem. Commun.*, 2240–2241.

102. Leontidis, E. (2002) Hofmeister anion effects on surfactant self-assembly and the formation of mesoporous solids. *Curr. Opin. Colloid Interface Sci.*, **7**, 81–91.

103. Wan, Y., Shia, Y. and Zhao, D. (2007) Designed synthesis of mesoporous solids via nonionic-surfactant-templating approach. *Chem. Commun.*, 897–926.

104. Kim, J.M., Sakamoto, Y., Hwang, Y.K. *et al.* (2002) Structural design of mesoporous silica by micelle-packing control using blends of amphiphilic block copolymers. *J. Phys. Chem. B*, **106**, 2552–2558.

105. Raman, N.K., Anderson, M.T. and Brinker, C.J. (1996) Template-based approaches to the preparation of amorphous, nanoporous silicas. *Chem. Mater.*, **8**, 1682–1701.

106. Joo, S.H., Ryoo, R., Kruk, M. and Jaroniec, M. (2002) Evidence for general nature of pore interconnectivity in 2-dimensional hexagonal mesoporous silicas prepared using block copolymer templates. *J. Phys. Chem. B*, **106**, 4640–4646.

107. Goltner-Spickermann, C. (2002) Non-ionic templating of silica: formation mechanism and structure. *Curr. Opin. Colloid Interface Sci.*, **7**, 173–178.

108. Monnier, A., Schüth, F., Huo, Q. *et al.* (1993) Cooperative formation of inorganic-organic interfaces in the synthesis of silicate mesostructures. *Science*, **231**, 1299–1303.

109. Solovyov, L.A., Belousov, O.V., Dinnebier, R.E. *et al.* (2005) X-ray diffraction structure analysis of MCM-48 mesoporous silica. *J. Phys. Chem. B*, **109**, 3233–3237.

110. Kim, T.W., Kleitz, F., Paul, B. and Ryoo, R. (2005) MCM-48-like large mesoporous silicas with tailored pore structure: facile synthesis domain in a ternary triblock copolymer-butanol-water system. *J. Am. Chem. Soc.*, **127**, 7601–7610.

111. Ravikovitch, P.I. and Neimark, A.V. (2002) Density functional theory of adsorption in spherical cavities and pore size characterization of templated nanoporous silicas with cubic and three-dimensional hexagonal structures. *Langmuir*, **18**, 1550–1560.

112. Matos, J.R., Kruk, M., Mercuri, L.P. *et al.* (2003) Ordered mesoporous silica with large cage-like pores: structural identification and pore connectivity design by controlling the synthesis temperature and time. *J. Am. Chem. Soc.*, **125**, 821–829.

113. Yu, T., Zhang, H., Yan, X. *et al.* (2006) Pore structures of ordered large cage-type mesoporous silica FDU-12s. *J. Phys. Chem. B*, **110**, 21467–21472.

114. Ryoo, R., Kim, J.M., Ko, C.H. and Shin, C.H. (1996) Disordered molecular sieve with branched mesoporous channel network. *J. Phys. Chem.*, **100**, 17718–17721.

115. Jansen, J.C., Shan, Z., Marchese, L. *et al.* (2001) A new templating method for three-dimensional mesopore networks. *Chem. Commun.*, 713–714.

116. Lee, J., Kim, J. and Hyeon, T. (2003) A facile synthesis of bimodal mesoporous silica and its replication for bimodal mesoporous carbon. *Chem. Commun.*, 1138–1139.

117. Jana, S.K., Mochizuki, A. and Namba, S. (2004) Progress in pore-size control of mesoporous MCM-41 molecular sieve using surfactant having different alkyl chain lengths and various organic auxiliary chemicals. *Catal. Surv. Asia*, **8**, 1–13.

118. Widenmeyer, M. and Anwander, R. (2002) Pore size control of highly ordered mesoporous silica MCM-48. *Chem. Mater.*, **14**, 1827–1831.

119. Kim, S.S., Karkamkar, A. and Pinnavaia, T.J. (2001) Synthesis and characterization of ordered, very large pore MSU-H silicas assembled from water-soluble silicates. *J. Phys. Chem. B*, **105**, 7663–7670.

120. Galarneau, A., Cambon, H., Di Renzo, F. *et al.* (2003) Microporosity and connections between pores in SBA-15 mesostructured silicas as a function of the temperature of synthesis. *New J. Chem.*, **27**, 73–79.

121. Jana, S.K., Nishida, R., Shindo, K. *et al.* (2004) Pore size control of mesoporous molecular sieves using different organic auxiliary chemicals. *Microporous Mesoporous Mater.*, **68**, 133–142.

122. Lettow, J.S., Han, Y.J., Schmidt-Winkel, P. *et al.* (2000) Hexagonal to mesocellular foam phase transition in polymer-templated mesoporous silicas. *Langmuir*, **16**, 8291–8295.

123. Sayari, A. (2000) Unprecedented expansion of the pore size and volume of periodic mesoporous silica. *Angew. Chem. Int. Ed.*, **39**, 2920–2922.

124. Park, B.–.G., Guo, W., Cuib, X. *et al.* (2003) Preparation and characterization of organo-modified SBA-15 by using polypropylene glycol as a swelling agent. *Microporous Mesoporous Mater.*, **66**, 229–238.

125. Thomas, J.M., Terasaki, O., Gai, P.L. *et al.* (2001) Structural elucidation of microporous and mesoporous catalysts and molecular sieves by high-resolution electron microscopy. *Acc. Chem. Res.*, **34**, 583–594.

126. Terasaki, O., Ohsuna, T., Liu, Z. *et al.* (2004) Structural study of meso-porous materials by electron microscopy. *Stud. Surf. Sci. Catal.*, **148**, 261–288.

127. Liu, Z., Terasaki, O., Ohsuna, T. *et al.* (2001) An HREM study of channel structures in mesoporous silica SBA-15 and platinum wires produced in the channels. *ChemPhysChem*, **4**, 229–231.

128. Kamiya, S., Tanaka, H., Che, S. *et al.* (2003) Electron microscopic study of structural evolutions of silica mesoporous crystals: crystal-growth and crystal-transformation from *p6mm* to *Pm-3n* with time. *Solid State Sci.*, **5**, 197–204.

129. Gregg, S.J. and Sing, K.S.W. (1982) *Adsorption, Surface Area and Porosity*, Academic Press, London.

130. Sing, K.S.W., Everett, D.H., Haul, R.A.W. *et al.* (1985) Reporting physisorption data for gas/solid systems with special reference to the determination of surface area and porosity. *Pure Appl. Chem.*, **57**, 603–619.

131. Jaroniec, M. and Kruk, M. (1999) Standard nitrogen adsorption data for characterization of nanoporous silicas. *Langmuir*, **15**, 5410–5413.

132. Kruk, M. and Jaroniec, M. (2001) Gas adsorption characterization of ordered organic – inorganic nanocomposite materials. *Chem. Mater.*, **13**, 3169–3183.

133. Steel, A., Carr, S. and Anderson, W. (1995) 29Si solid-state NMR study of mesoporous M41S materials. *Chem. Mater.*, **7**, 1829–1832.

134. García, A., Colilla, M., Izquierdo-Barba, I. and Vallet-Regí, M. (2009) Incorporation of phosphorus into mesostructured silicas: a novel approach to reduce the SiO_2 leaching in water. *Chem. Mater.*, **21**, 4135–4145.

135. Leonova, E., Izquierdo-Barba, I., Arcos, D. *et al.* (2008) Multinuclear solid-state NMR studies of ordered mesoporous bioactive glasses. *J. Phys. Chem. C*, **112**, 5552–5562.

136. Gunawidjaja, P.N., Mathew, R., Lo, A.Y.H. *et al.* (2012) Local structures of mesoporous bioactive glasses and their surface alterations in vitro: inferences from solid-state nuclear magnetic resonance. *Philos. Trans. R. Soc. A*, **370**, 1376–1399.

137. Yamauchi, Y., Suzukia, N. and Kimura, T. (2009) Formation of mesoporous oxide fibers in polycarbonate confined spaces. *Chem. Commun.*, 5689–5691.

138. Huo, Q.S., Zhao, D.Y., Feng, J.L. *et al.* (1997) Topological construction of mesoporous materials. *Adv. Mater.*, **9**, 974–978.

139. Yu, C.Z., Fan, J., Tian, B.Z. *et al.* (2002) High-yield synthesis of periodic mesoporous silica rods and their replication to mesoporous carbon rods. *Adv. Mater.*, **14**, 1742–1745.

140. Kosuge, K., Sato, T., Kikukawa, N. and Takemori, M. (2004) Morphological control of rod- and fiberlike SBA-15 type mesoporous silica using water-soluble sodium silicate. *Chem. Mater.*, **16**, 899–905.

141. Schacht, S., Huo, Q., Voigt-Martin, I.G. *et al.* (1996) Oil-water interface templating of mesoporous macroscale structures. *Science*, **273**, 768–771.

142. Vallet-Regí, M. (2012) Mesoporous silica nanoparticles: their projection in nanomedicine. *ISRN Mater. Sci.*, **2012**, 20Article ID 608548.

143. Wu, K.C.-W. and Yamauchi, Y. (2012) Controlling physical features of mesoporous silica nanoparticles (MSN) for emerging applications. *J. Mater. Chem.*, **22**, 1251–1256.

144. Knežvić, N.Ž., Ruiz-Hernández, E., Hennink, W.E. and Vallet-Regí, M. (2013) Magnetic mesoporous silica-based core/shell nanoparticles for biomedical applications. *RSC Adv.*, **3**, 9584–9593.

145. Colilla, M., Manzano, M., Izquierdo-Barba, I. *et al.* (2010) Advanced drug delivery vectors with tailored surface properties made of mesoporous binary oxides submicronic spheres. *Chem. Mater.*, **22**, 1821–1830.

146. Arcos, D., López-Noriega, A., Ruiz-Hernández, E. *et al.* (2009) Ordered mesoporous microspheres for bone grafting and drug delivery. *Chem. Mater.*, **21**, 1000–1009.

147. Julián-López, B., Boissiere, C., Chaneac, C. *et al.* (2007) Mesoporous maghemite–organosilica microspheres: a promising route towards multifunctional platforms for smart diagnosis and therapy. *J. Mater. Chem.*, **17**, 1563–1569.

148. Ruiz-Hernández, E., López-Noriega, A., Arcos, D. *et al.* (2007) Aerosol-assisted synthesis of magnetic mesoporous silica spheres for drug targeting. *Chem. Mater.*, **19**, 3455–3463.

149. Ruiz-Hernández, E., López-Noriega, A., Arcos, D. and Vallet-Regí, M. (2008) Mesoporous magnetic microspheres for drug targeting. *Solid State Sci.*, **10**, 421–426.
150. Liu, J.W., Jiang, X.M., Ashley, C. and Brinker, C.J. (2009) Electrostatically mediated liposome fusion and lipid exchange with a nanoparticle-supported bilayer for control of surface charge, drug containment, and delivery. *J. Am. Chem. Soc.*, **131**, 7567–7569.
151. Grün, M., Lauer, I. and Unger, K.K. (1997) The synthesis of micrometer- and submicrometer-size spheres of ordered mesoporous oxide MCM-41. *Adv. Mater.*, **9**, 254–257.
152. Buining, P.A., Liz-Marzán, L.M. and Philipse, A.P. (1996) A simple preparation of small, smooth silica spheres in a seed alcosol for Stöber synthesis. *J. Colloid Interface Sci.*, **179**, 318–321.
153. Bardosova, M. and Tredgold, R.H. (2002) Ordered layers of monodispersive colloids. *J. Mater. Chem.*, **12**, 2835–2842.
154. Okudera, H. and Hozumi, A. (2003) The formation and growth mechanisms of silica thin film and spherical particles through the Stöber process. *Thin Solid Films*, **434**, 62–68.
155. Pauwels, B., Tendeloo, G.V., Thoelen, C. *et al.* (2001) Structure determination of spherical MCM-41 particles. *Adv. Mater.*, **13**, 1317–1320.
156. Liu, S., Luc, L., Yang, Z. *et al.* (2006) Further investigations on the modified Stöber method for spherical MCM-41. *Mater. Chem. Phys.*, **97**, 203–206.
157. Qiao, Z.A., Zhang, L., Guo, M.Y. *et al.* (2009) Synthesis of mesoporous silica nanoparticles via controlled hydrolysis and condensation of silicon alkoxide. *Chem. Mater.*, **21**, 3823–3829.
158. Yano, K. and Fukushima, Y. (2003) Particle size control of mono-dispersed supermicroporous silica spheres. *J. Mater. Chem.*, **13**, 2577–2581.
159. Manzano, M., Aina, V., Areán, C.O. *et al.* (2008) Studies on MCM-41 mesoporous silica for drug delivery: Effect of particle morphology and amine functionalization. *Chem. Eng. J.*, **137**, 30–37.
160. Gao, C.B. and Che, S.A. (2010) Organically functionalized mesoporous silica by co-structure-directing route. *Adv. Funct. Mater.*, **20**, 2750–2768.
161. Huh, S., Wiench, J.W., Yoo, J.C. *et al.* (2003) Organic functionalization and morphology control of mesoporous silicas via a co-condensation synthesis method. *Chem. Mater.*, **15**, 4247–4256.
162. Huh, S., Wiench, J.W., Trewyn, B.G. *et al.* (2003) Tuning of particle morphology and pore properties in mesoporous silicas with multiple organic functional groups. *Chem. Commun.*, 2364–2365.
163. Lu, F., Wu, S.H., Hung, Y. and Mou, C.Y. (2009) Size effect on cell uptake in well-suspended, uniform mesoporous silica nanoparticles. *Small*, **5**, 1408–1413.
164. Chiang, Y., Lian, D., Leo, H.–.Y. *et al.* (2011) Controlling particle size and structural properties of mesoporous silica nanoparticles using the Taguchi method. *J. Phys. Chem. C*, **115**, 13158–13165.
165. Anderson, M.T., Martin, J.E., Odinek, J.G. *et al.* (1998) Surfactant-templated silica mesophases formed in water:cosolvent mixtures. *Chem. Mater.*, **10**, 311–321.
166. Gu, J., Fun, W., Shimojima, S. and Okubo, T. (2007) Organic–inorganic mesoporous nanocarriers integrated with biogenic ligands. *Small*, **3**, 1740–1744.

167. Sadasivan, S., Khushalani, D. and Mann, S. (2003) Synthesis and shape modification of organo-functionalised silica nanoparticles with ordered mesostructured interiors. *J. Mater. Chem.*, **13**, 1023–1029.

168. Wang, S.G., Wu, C.W., Chen, K. and Lin, V.S.Y. (2009) Fine-tuning mesochannel orientation of organically functionalized mesoporous silica nanoparticles. *Chem. Asian J.*, **4**, 658–661.

169. Radu, D.R., Lai, C.Y., Huang, J. *et al.* (2005) Fine-tuning the degree of organic functionalization of mesoporous silica nanosphere materials via an interfacially designed co-condensation method. *Chem. Commun.*, 1264–1266.

170. Hoffmann, F., Cornelius, M., Morell, J. and Froba, M. (2006) Silica-based mesoporous organic-inorganic hybrid materials. *Angew. Chem. Int. Ed.*, **45**, 3216–3251.

171. Brühwiler, D. (2010) Postsynthetic functionalization of mesoporous silica. *Nanoscale*, **2**, 887–892.

172. de Juan, F. and Ruiz-Hitzky, E. (2000) Selective functionalization of mesoporous silica. *Adv. Mater.*, **12**, 430–432.

173. Kecht, J., Schlossbauer, A. and Bein, T. (2008) Selective functionalization of the outer and inner surfaces in mesoporous silica nanoparticles. *Chem. Mater.*, **20**, 7207–7214.

174. Inagaki, S., Guan, S., Fukushima, Y. *et al.* (1999) Novel mesoporous materials with a uniform distribution of organic groups and inorganic oxide in their frameworks. *J. Am. Chem. Soc.*, **121**, 9611–9614.

175. Melde, B.J., Holland, B.T., Blanford, C.F. and Stein, A. (1999) Mesoporous sieves with unified hybrid inorganic/organic frameworks. *Chem. Mater.*, **11**, 3302–3308.

176. Asefa, T., MacLachlan, M.J., Coombs, N. and Ozin, G.A. (1999) Periodic mesoporous organosilicas with organic groups inside the channel walls. *Nature*, **402**, 867–871.

177. Hatton, B., Landskron, K., Whitnall, W. *et al.* (2005) Past, present, and future of periodic mesoporous organosilicas – The PMOs. *Acc. Chem. Res.*, **38**, 305–312.

178. Hoffmann, F. and Fröba, M. (2011) Vitalising porous inorganic silica networks with organic functions – PMOs and related hybrid materials. *Chem. Soc. Rev.*, **40**, 608–620.

179. Mizoshita, N., Tani, T. and Inagaki, S. (2011) Syntheses, properties and applications of periodic mesoporous organosilicas prepared from bridged organosilane precursors. *Chem. Soc. Rev.*, **40**, 789–800.

180. Wu, H.-Y., Shieh, F.-K., Kao, H.-M. *et al.* (2013) Synthesis, bifunctionalization, and remarkable adsorption performance of benzene-bridged periodic mesoporous organosilicas functionalized with high loadings of carboxylic acids. *Chem. Eur. J.*, **19**, 6358–6367.

6

Alumina, Zirconia, and Other Non-oxide Inert Bioceramics

Juan Peña López

Departamento de Química Inorgánica y Bioinorgánica, Facultad de Farmacia, Universidad Complutense de Madrid, Spain

Bioinert ceramics formed part of the first generation of bone substitutes early in the 1950s aiming to substitute without reaction with the living tissues [1]. The biomaterials of this first generation made use of available, off-the-shelf industrial materials whose initial application and development were not related to the medical field. They were selected more for their physical properties while chemically they were intended to be inert. Despite this inertness, all material implanted elicits a response from living tissue; in this case at least a fibrous tissue of variable thickness. The development of a fibrous capsule is increased due to the movement of the implant since there is nobiological or chemical bonding. Depending on the type of material implanted and, especially, the fixation required for an optimal mechanical performance, the formation of a thin fibrous layer does not alter a successful implantation [2].

However, the problem of an adequate biological and chemical fixation of a biomaterial has been largely overcome by the successive biomaterials generations [3, 4]. In fact the applications in which inert bioceramics are presently most clinical utilized: total hip replacement (THR) and total knee replacement (TKR) [5–9] do not require a good fixation but a hard surface with good wear behavior [10–17].

Besides their uses in orthopedics, bioceramics have been employed in additional applications that will be detailed in this chapter, although these are of much less clinical importance than their utilization in hip and knee reconstruction.

6.1 A Perspective on the Clinical Application of Alumina and Zirconia

The first application of alumina was its utilization as a replacement for the traditional heads of hip prostheses. This bioceramic (α-Al_2O_3) was later employed to fabricate acetabular cups, demonstrating adequate biocompatibility [9–11, 18, 19], corrosion resistance, and

Bioceramics with Clinical Applications, First Edition. Edited by María Vallet-Regí.
© 2014 John Wiley & Sons, Ltd. Published 2014 by John Wiley & Sons, Ltd.

wear rates. However, despite the improvement in conventional polyethylene cups in terms of wear debris generation, this type of material suffers from early failures due to its low fracture toughness [8, 9].

Alumina has become the reference material during the last 40 years as the component of THRs, both as ceramic heads and in ceramic on ceramic (CoC) bearings. Figure 6.1 shows the different possible combinations of the head and cup material, yielding, besides the already mentioned CoC bearings, MoM (metal on metal), MoUHMWPE (metal on ultra-high molecular weight polyethylene), and CoUHMWPE.

The increasing use of this bioceramic has been possible due to an improvement in the material's performance due to an enhancement of different aspects of the fabrication technology that will be discussed below. The enhancement in these materials performance has even led to a classification of alumina-based prostheses into different generations that end in one of the most utilized THA (total hip arthroplasty) bearings: Biolox Forte. The increase in the bending strength from 400 MPa in the first alumina of the 1970s to the 580 MPa Biolox Forte (CeraTec) and the decrease in the grain size from 4.5 to 1.8 μm, is a good example of the evolution of this material's performance, which has lowered the catastrophic *in vivo* failure from up to 13% to below 0.01% [12, 19, 20].

In the 1990s, zirconia was introduced as an alternative due to a substantially higher fracture toughness and strength, and extremely low wear [8, 21–24]. As a consequence of the tetragonal to monoclinic phase transformation toughening, which increases its crack propagation resistance, Y_2O_3-stabilized zirconia (yttria-tetragonal zirconia polycrystal, Y-TZP) exhibits the best mechanical properties of a single phase. However, due to this

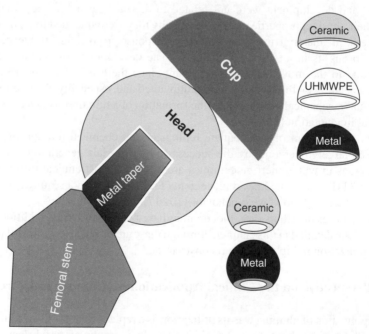

Figure 6.1 *Components of a total hip replacement indicating the different material head–cup combinations*

meta-stability, Y-TZP can suffer low temperature degradation in the presence of water. This process, also termed aging, is triggered by water molecules and results in surface roughening and micro-cracking which impact negatively on the wear performance. This was considered a minor problem until 2001, when a failure in the processing of some batches caused hundreds of failures. Although this was a limited and localized phenomenon, it constituted a shock for the commercial use of zirconia in orthopedics. The use of zirconia ceramics seems to have found an open field in dental restorations where, besides mechanical specifications, esthetic requirements (color, translucency) have a critical role [25–38].

Nowadays, besides the extensively used commercial alumina or zirconia-based products, there are new products, such as non-oxide ceramics (Si_3N_4, TiN, SiC), alumina- or zirconia-based ceramics nanostructured composites (Ce-TZP and Mg-PSZ–partially stabilized zirconia, micro-nano-alumina–zirconia composites, and zirconia toughened alumina, ZTA) [23, 39]. These different strategies are discussed at the end of this chapter. Table 6.1 summarizes the mechanical performance of some of the most utilized ceramics in THR and TKR.

6.1.1 Alumina

All the different phases of aluminum oxide transform into alpha-alumina (α-Al_2O_3) when heated above 1200 °C, in a close-packed hexagonal arrangement of oxygen ions. This material, due to the strong ionic and covalent chemical bonds between Al^{3+} and O^{2-}, is characterized by a high melting point, hardness, and chemical inertness [9, 15, 40].

The surface is characterized by a high wettability thanks to the chemisorbed layer of hydroxy groups that allows the bonding of molecules and proteins (Figure 6.2). This lubricating surface, together with the high hardness, that makes these components resistant to scratching, makes alumina-based components excellent candidates for arthroprostheses joints with low wear.

Table 6.1 *Mechanical properties of some of the most commonly employed materials in THA and TKA, compared to those of Co–Cr alloy and cortical bone*

		Bone (cortical)	CoCr	Al_2O_3	Mg-PSZ	Y-TZP	ZTA	Si_3N_4
Composition	units			Al_2O_3 99.9%	ZrO_2 – 8% MgO	ZrO_2 – 3% Y_2O_3	Al_2O_3 – 20% ZrO_2	
Density	g cm^{-3}	1.7–2.0	8.5	> 3.97	5.7	6.0	4.4	–
Strength	MPa	60–160	–	500	–	–	–	1000
Toughness (K_{IC})	MPa m$^{1/2}$	2–12	50–100	4–5	6–10	6–12	6–10	10
Vickers hardness	HV	–	350	1900	1200	1200	1700	2500
Thermal expansión conductivity	10^{-6} K^{-1}	–	14.0	8	7–10	11	8.5	–

Source: [8, 10, 12, 16, 19, 20].

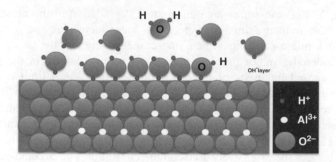

Figure 6.2   *Formation of a hydroxy layer on an alumina substrate*

Another characteristic derived from the strong chemical bonds is the low electric and thermal conductivity and the high melting point that hinders the shaping of alumina by casting, while the high hardness makes its machining complex and costly. The characteristic brittle fracture behavior, together with the moderate tensile and bending resistance, constitutes one of the main drawbacks of the mechanical performance of alumina.

The hydrothermal stability seems not to be a serious concern for alumina-based components, even in the case of BIOLOX®delta, due to the presence of 17% of Y-TZP. However, in the case of a composite where higher percentages of zirconium are used, the possible effect of aging must be considered.

Concerning its composition, medical grade alumina may contain up to a 0.5 wt% of sintering additives, namely magnesium and chromium oxides. The first is introduced to control the deleterious discontinuous grain growth, resulting in a denser ceramic, while CrO_3 is included to compensate the decrease in hardness due to the introduction of MgO. At the same time the precursors must be free from impurities, such as CaO, Na_2O, and SiO_2, that may be deleterious for the stability of medical grade alumina.

A proper selection of the raw materials used as precursors is one of the reasons for the improvement in the mechanical properties over the 40 years of clinical application. The considerable increase in the bending strength can be justified by the increase in density to values close to the theoretical one attributable to the reduction in the grain size.

Another factor that has contributed to the improvement of the mechanical properties is the introduction of hot isostatic pressing (HIP) that allows shaping at high pressures and temperatures, just below the sintering temperature, thus limiting the grain growth and yielding a high density. The fabrication procedure was also optimized by the employment of a laser for marking the components; otherwise the engraving methods previously applied induced the creations of defects that may have a deleterious effect on the final performance of the component.

Nevertheless, the most outstanding contribution in improving this performance is the systematic application of a proof test, a non-destructive check implanted in the fabrication process as a part of the final control on the finished components.

Recently, composites based on alumina and a metal niobium (Nb) [41] or tantalum (Ta) [42] have been prepared as an alternative femoral head material *in vivo*. This material combines the low wear of Al_2O_3 with the mechanical ductile behavior of the metal component.

The fabrication technology commonly employed in the preparation of an alumina ceramic component is summarized in Figure 6.3.

Besides the use of alumina in orthopedic components, it has been employed is many other applications in the medical field. Most of these applications, which are summarized in Table 6.2, are based on porous alumina or membranes, due to the additional benefits: large surface area, tunable pore size, and pore size distribution, low fabrication costs and so on [43].

Many of these applications require a modification/functionalization of the inert surface of alumina. Different types of molecules or ions linked to the substrate by different types of

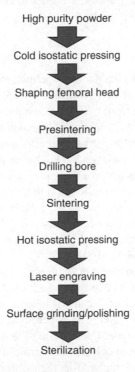

High purity powder

Cold isostatic pressing

Shaping femoral head

Presintering

Drilling bore

Sintering

Hot isostatic pressing

Laser engraving

Surface grinding/polishing

Sterilization

Figure 6.3   *Fabrication technology of a ceramic component*

Table 6.2   *Applications of Al_2O_3 in the biomedical field*

Applications	References
Implantable devices	
Systems to control and direct cell behavior	[44–48]
Drug delivery systems	[49–53]
Biosensors	[54–61]
Biotechnology	
Substrate for enzyme immobilization	[62–73]
Substrate for protein adsorption/purification	[74–79]
Environmental: removal of toxic components	[80–89]

bonds are required as a function of the final performance of the device. For example, when a better osseointegration is desired, different ions (Ca, Mg, Cr, Si, Mn, etc.) [90–92] and functional groups (-OH, -NH$_4$) [93] are integrated/linked by different mechanisms. Mixtures or coating with bioactive materials is also a considered strategy [94–98]. Silanization constitutes another example of how the surface can be functionalized with different groups in order to improve the required performance [43, 99]. A particular case that illustrates this chemical strategy is the enhancement of the adhesion between inert ceramic surfaces and the resins clinically applied to fix the pieces; this point will be discussed in detail below. On the other hand, self-assembled monolayer techniques have been employed to link different functional groups to the alumina substrate in order to control or elicit specific biological responses, or entrap proteins and peptides [99–101].

6.1.2   Zirconia

Three of the five polymorphs of zirconia (monoclinic, tetragonal, and cubic) have a role in the biomedical performance of this bioceramic. Pure zirconia cannot be used due to the extreme facility to transform from one phase to another, that is accompanied by shape and volume changes that may cause material degradation and cracking. This instability is triggered by either mechanical or chemical energy, and also by aging. The addition of stabilizing agents, such as MgO, CaO, or Y$_2$O$_3$, during the fabrication may contribute to limiting the phase transformations and, consequently, to improving the material mechanical performance. It must be taken into consideration that the mechanical behavior of zirconia components is of critical significance, not only in the biomaterials fields but also in other applications, such as solid electrolytes fuel cells and oxygen sensors based on the ionic conductivity generated by the presence of oxygen vacancies induced by the presence of trivalent cations [21–24, 35, 38, 102–104].

These oxygen vacancies seem to have a critical role in the water penetration inside the zirconia lattice as a consequence of its exposure to a humid environment [105]. Y-TZP, due to the presence of a great number of vacancies when compared to other zirconia ceramics, suffers a higher diffusion of water. The penetration of water radicals leads to lattice contraction that disrupts the tetragonal phase due to generation of tensile stresses in the surface grains. The tetragonal–monoclinic martensitic transformation affects part of the grains located at the surface of the implant that then suffer a volume increase, thus altering the neighboring grains and inducing micro-cracking that propagates through a nucleation and growth mechanism. Figure 6.4 represents the beneficial effects of this martensitic transformation when it occurs in the bulk and avoids the propagation of a crack, while at the surface level the presence of water leads to its degradation and the generation of different types of defects.

Unfortunately, some clinical reports seem to demonstrate that the degradation of 3Y-TZP does not only take place under extreme hydrothermal conditions but also under physiological conditions, thus restricting its long-term stability. This fact reinforced the need to find solutions to the aging problem when zirconia phases are considered alone or as part of a composite. In order to overcome this issue two strategies that will be detailed in the following sections have been adopted: minimizing the quantity of oxygen vacancies and/or decreasing the nucleation and growth kinetics by minimizing contact between grains [21, 33, 104].

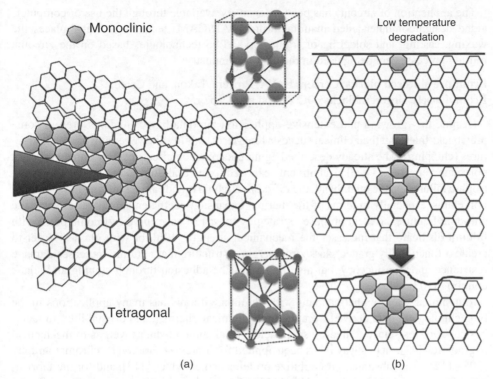

Figure 6.4 *Monoclinic to tetragonal martensitic transformation effect on the fracture tough-ness beneficial effect (a) and the undesirable low temperature degradation (b)*

In any case, there is reluctance to use any zirconia-based materials even though any of these materials can be considered "aging free", and the martensitic transformation kinetics can be tailored as a function of the different parameters that govern the microstructure (dopants, processing technology, raw components, etc.) to be sufficient low in order to avoid any degradation during the lifespan of the implant. In addition the mandatory application of aging accelerated tests (ISO standard for 3Y-TZP – ISO 13356) before launching any 3Y-TZP products allows assessment of their aging sensitivity. Although 3Y-TZP has almost disappeared from the orthopedic field, it still does have a strong potential due to its mechanical and esthetic properties, as will be discussed in the following section.

In recent years zirconia is being employed in different dentistry applications, such as implant and tooth-supported restorations, orthodontic brackets, endodontic post/dowels and abutments, and implant-borne crowns, as has been collected in the considerable number of reviews focusing on zirconia and dental applications [23, 30, 34–38]. Beside its excellent mechanical performance, zirconia has been proved to be biocompatible *in vivo* and *in vitro*, as well as osteoconductive. In addition, some studies have pointed out a certain bactericide capacity that ensures plaque reduction [38]. In esthetical terms zirconia com-pounds can be prepared with a high degree of translucency that, in addition, can be colored to match teeth.

The application of zirconia has been extensively available through the use of computer-aided design/computer-aided manufacturing (CAD/CAM) technology that replaces the waxing, casting, and soldering of frameworks. Two technologies based on the zirconia milling processes are being used in the dentistry industry:

From pre-sintered blocks: Cercon, LAVA, Procera, Etkon, and Cerec.
From densely sintered blocks: DC Zirkon and Denzir.

Despite this growing and extensive application zirconia-based implants present some drawbacks that limit their clinical success: long-term stability and veneering porcelain fractures (chipping). The already described aging phenomenon affects overall frameworks not porcelain veneered as well as abutments exposed to the oral fluids and may lead to catastrophic failure. On the other hand, the different zirconia brands are modifying the overlying porcelains, especially in terms of the thermal expansion coefficients, in order to solve the chipping fractures. Recent proposed strategies consist of covering the zirconia piece with lithium disilicate that increases the mechanical strength [106], or with glass in order to create a functionally graded glass/zirconia/glass structure that not only increases damage resistance and translucency, but also facilitates the adhesion through etching and silane coupling [107].

In a similar way to what happens with alumina, zirconia has many applications in the biotechnological field derived from the high chemical, thermal, and pH stability; in fact it has superior resistance to hydrolysis in acidic or alkaline media as well as to mechanical degradation [43]. In this way it has been applied for protein separation by chromatography [108–112], as a substrate to immobilize proteins and enzymes [113], and for the fabrication of enzyme-based bioreactors [114, 115]. Zirconia has also been employed in different devices related to environmental preservation: as a catalyst to detect or decompose toxic pollutants, bacteria, in metal ions removal from waste waters, and so on [116–118].

6.2   Novel Strategies Based on Alumina and Zirconia Ceramics

6.2.1   From Alumina Toughened Zirconia to Alumina Matrix Composite

ZTA is a composite that is formed by zirconia particles dispersed in an alumina matrix [9, 17, 39]. This composite combines the hardness of alumina with the improved strength and fracture toughness of the zirconia component. The zirconia particles are stabilized by inclusion in the dense and rigid alumina matrix, consequently, it is not necessary to dope with yttrium and no oxygen vacancies are created, thus reducing the diffusion of water into the zirconia lattice. In addition, for zirconia percentages below the percolation threshold, grains are isolated, and the propagation mechanism is hindered. Unfortunately, material degradation and strength decrease has been detected after aging in a Ringer solution.

A modified version of this composite, also known as alumina matrix composite (AMC), which includes SrO and Cr_2O_3 as additives, has been implanted and developed by Ceramtec under the trade name of Biolox delta®. The additions contribute to the creation of alumina grains with platelet-like morphology that strengthen the material by generating barriers to crack propagation. At the same time, a significant improvement in aging resistance (compared to Y-TZP) and excellent crack resistance have been observed [9, 17, 21, 39]. The first hip simulator test results indicate wear rates for AMC–AMC < ZTA–ZTA couplings that in any case are much lower when compared to alumina–alumina [12]. Due to its novelty there are not enough clinical results to reach any conclusion.

In order to improve the performance of this type of material, new types of composites are being developed such as [39, 119]:

- alumina toughened zirconia (ATZ) in which zirconia (3Y-TZP) is the major phase [7, 120]. These composites are aging sensitive, although this transformation is not so negative since roughness is not so strongly affected even after a long duration of aging.
- Nanocomposites processed from alumina and zirconia aiming at increasing the crack propagation threshold, tensile strength, and materials stability. Two strategies are being explored: alumina rich nanocomposites in which zirconia nanoparticles are evenly dispersed in micronic alumina grains – "micro-nano-composites" – and zirconia rich nanocomposites in which both phases are below 500 nm – "nano-nanocomposites" [21].
- Nano-structured ceramics that meet the translucency requirements for dental restoration can be prepared by means of new fast sintering techniques, such as spark plasma and microwave sintering, that ensure ultrafine, fully dense ceramic materials.

6.2.2   Introduction of Different Species in Zirconia

Ce-TZP (Ce-Y-TZP): This type of zirconia ceramics exhibits, when compared to Y-TZP, superior toughness (20 MPa) and reduced aging sensitivity, but lower strength (600 MPa vs. 1000 MPa). This lower mechanical performance, as well as the lack of translucency, is related to the difficulty in reducing the grain size of this type of compound (1.5–2 μm) as low as those observed for 3Y-TZP (0.5 μm). On the other hand, the better aging resistance can be explained considering the presence of tetravalent cerium cations and the absence of trivalent cations, thus decreasing the oxygen vacancies. However, the possible reduction of Ce^{4+} to Ce^{3+} after reducing atmosphere treatment may cause an increased aging sensitivity. These types of compounds are presently used in dental applications, although an enhancement of the densification through better sintering techniques is sought in order in improve the mechanical performance and translucency. Further alternatives lie in the combination of this compound with another oxide that influences its microstructure, such as alumina [121].

Divalent cations PSZ ceramics: Divalent cations such as Mg^{2+} and Ca^{2+} were among the first zirconia compounds tried. In fact Mg-PSZ ceramic femoral heads, which feature excellent toughness and smoothness, are being employed, to a minor extent, in the United States (Biopro, Huron, MI). In addition, their "a priori" excellent resistance to aging was fully demonstrated on retrieved prostheses that did not show any evidence of phase transformation *in vivo*. Despite the grain size distribution and mechanical properties, clearly inferior to well controlled Y-TZP, this aging immunity has renewed the interest in this material. Unfortunately, the resulting yellow to orange colors of Mg-PSZ precludes its possible application in the dental field [122].

Combination of trivalent and pentavalent ions: Y^{3+}/Nb^{5+}, Y^{3+}/Ta^{5+} the objective is to reduce the oxygen vacancies generated by the yttria doping through the introduction of both types of cations that allow minimization of total concentration of vacancies required for charge compensation.

6.2.3 Improvement of Surface Adhesion

With the growing application of zirconia ceramics in the field of prosthetic and implant dentistry, the problem of insufficient adhesion with intended synthetic surfaces or natural tissues has arisen [25]. Since traditional adhesive techniques based on silica-based ceramics do not work with zirconia (chemical resistance to acids used to etch and extremely aggressive abrasion methods that may create defects that may increase surface roughness) considerable research is being addressed toward surface modification that ensures an adhesive bonding:

- **Chemical bonding enhancement by silanization**. Silanes are employed in dental applications, such as restorative endodontic applications to increase the interaction between silica- or no-silica-containing-ceramic particles and the polymeric resins, thus achieving a tougher adhesion. However, the nonpolar and chemically stable surface of zirconia hinders its hydroxylation and, consequently, the interaction with silane coupling agents. Several groups are working with different silane coupling agents in order to enhance ZrO_2 luting [31, 32, 123–125].

  Besides the specific solution of zirconia-based composites, surface functionalization by means of a silane coupling agent is being developed for other hydroxylated surfaces, such as alumina, germanium, silicon, or glass [43, 99, 126, 127]. Thanks to the numerous silane compounds commercially available, a broad range of chemical functionality can be integrated onto these surfaces.
- **Deposition of a silica layer**. The tribochemical silanization by air abrading technology consists of effectuating a bombardment of the ceramic surface with alumina particles that have been coated with silica. The air abrading process with these particles results in the silanization of the surface with a certain degree of micromechanical retention, yielding an enhancement of the bond strength with resin cements. An alternative is Silicoating (marketed under the Silicoater-Technology or PyrosilPen-Technology) that consists of pyrolytically applying a silica coating onto the surface, followed by application of silane. This complex technique ensures an enhanced bond strength of resin cements to metals that do not degrade after thermocycling [27, 29, 99, 128–131]. Alternatively, new methods based on plasma spray [26] or molecular vapor deposition [132] are being employed to improve this deposition. Figure 6.5 shows schematically some of the strategies employed to functionalize the ceramic surface, thus increasing the adhesion to fixation resins.

Another strategy consists of enhancing the bone-bonding capacity by acidic or alkaline surface treatment [43]. The treatment with an acid such as H_3PO_4 is on occasions followed by an immersion in simulated body fluid (SBF) in order to induce the formation of an apatite layer [133, 134].

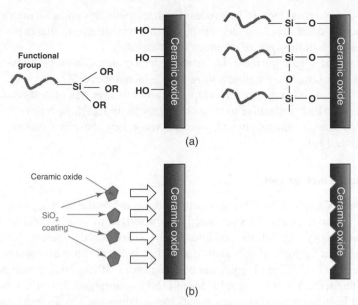

Figure 6.5 *Technologies employed to functionalize the surface of insert ceramic oxides: (a) silanization and (b) generation of a silica coating by air abrading*

6.3   Non-oxidized Ceramics

6.3.1   Silicon Nitride (Si_3N_4)

Silicon nitride has been extensively employed for different industrial applications (refractory elements, internal combustion engines, high temperature gas turbines, etc.) due to high hardness, thermal shock resistance, high strength, and fracture toughness [135–137]. Although it is one of the most studied ceramics and is commonly employed in high performance bearings, its application in the biomaterials field, specifically as an orthopedic material, is surprisingly recent [138–141]. Si_3N_4 is considered to be biocompatible since it does not result in adverse biological reactions, such as toxicity, mutagenicity, allergenicity, or carcinogenicity [142–144], in fact it has demonstrated the capacity to generate bone [145, 146]. Nevertheless, there is a certain concern about the oxidation of this ceramic which yields, as well as SiC, a several nanometers thick silicon oxide that may chip off and cause third-body wear. In addition it shows a curious feature: it is visible on plain radiographs as partially radiolucent material.

This material is presently employed in cervical spacers, spinal fusion devices, surgical screws and plates, bearings for spine disc surgery, and prosthetic hip and knee joints, most of them even with clinical results [147, 148]. Silicon nitride bearings for total joint applications are in development by Amedica Corporation (Salt Lake City, UT).

The considerable effort required to process (high temperatures under an inert atmosphere) and machine non-oxide ceramics seems to be one of the main reasons that explains the delay in the introduction of this type of ceramics in orthopedics.

The wear test of silicon nitride THA bearings demonstrates an excellent wear behavior, while strength and fracture toughness show a 50% increase over ZTA or AMC [138, 149]. Moreover, Si_3N_4, as other non-oxide ceramics, shows a lower risk of delayed failure since it is considered to be not sensitive to slow crack growth. In fact these types of materials are often referred to as *in situ* toughened, since no secondary phase is necessary to enhance crack resistance [39].

6.3.2   Silicon Carbide (SiC)

Silicon carbide has more than 200 polymorphs, depending on the stacking order of the double layer of Si and C atoms. This ceramic is characterized by its chemical inertness, high thermal conductivity, high elastic modulus, and low friction coefficient. Silicon carbide can be prepared in the form of thin films, mainly by chemical vapor deposition, but also by liquid phase epitaxy (LPE) and molecular beam epitaxy (MBE) yielding wafers with resistivities ranging from 1×10^5 to $1 \times 10^7 \, \Omega \, cm$ [150]. Amorphous silicon carbide (α-SiC), an electrical insulator with high wear resistance, is being used as an inert encapsulating coating for *in vivo* prostheses, specifically in commercial heart stents. On the other hand, SiC nanoparticles can be produced by different methods (laser ablation, electrochemical, low pressure microwave plasma, inductive coupled plasma, laser pyrolysis, etc.) [150]. In the bulk form, silicon carbide has been obtained by silicon infiltration into carbonaceous scaffolds derived from wood templates [151–153]. In such a bio-inspired way this material maintains the hierarchical organization of wood that somehow resembles that of bone. It has been demonstrated that this biomorphic SiC is capable of inducing the formation of an apatite layer when immersed in SBF. Silicon carbide and silicon nitride have demonstrated a suitable biocompatibility for use as medical implants and prostheses [144].

Due to its high resistance to corrosion in an aggressive medium such as body fluid, its wide band gap, and its mechanical properties SiC has been proposed as a candidate for smart implants and biosensors [150]. SiC, as a substrate, has been employed for coating neural probes as well as for the construction of myocardial biosensors [154]. Moreover, it is being employed as a hard coating for non-fouling coronary heart stents [155, 156]. In tissue engineering it has been used in hernia and breast reconstruction in the form of SiC nanowhiskers that optimize the mechanical strength and resistance to degradation [157].

Silicon carbide has increased strength and hardness when compared with alumina, but comparable fracture toughness values [135]. This last point, together with the uncertainty of the behavior of this ceramic, that is, formation of a SiO_2 layer on the surface that may suffer delamination [8], decreases its potential application in orthopedics.

References

1.   Vallet-Regi, M. (2010) Evolution of bioceramics within the field of biomaterials. *C.R. Chim.*, **13**, 174–185.
2.   Ratner, B.D., Hoffman, A.S., Schoen, F.J. and Lemons, J.E. (2012) *Biomaterials Science: An Introduction to Materials in Medicine*, Elsevier Science.

3. Navarro, M., Michiardi, A., Castano, O. and Planell, J.A. (2008) Biomaterials in orthopaedics. *J. R. Soc. Interface*, **5**, 1137–1158.
4. Burdick, J.A. and Mauck, R.L. (2011) *Biomaterials for Tissue Engineering Applications: A Review of the Past and Future Trends*, Springer, Vienna.
5. Heimke, G., Leyen, S. and Willmann, G. (2002) Knee arthroplasty: recently developed ceramics offer new solutions. *Biomaterials*, **23**, 1539–1551.
6. Zietz, C., Kluess, D., Bergschmidt, P. *et al.* (2011) Tribological aspects of ceramics in total hip and knee arthroplasty. *Semin. Arthroplasty*, **22**, 258–263.
7. Roualdes, O., Duclos, M.E., Gutknecht, D. *et al.* (2010) In vitro and in vivo evaluation of an alumina-zirconia composite for arthroplasty applications. *Biomaterials*, **31**, 2043–2054.
8. Rahaman, M.N., Yao, A., Bal, B.S. *et al.* (2007) Ceramics for prosthetic hip and knee joint replacement. *J. Am. Ceram. Soc.*, **90**, 1965–1988.
9. Piconi, C. (2011) Alumina, in *Comprehensive Biomaterials* (ed D. Paul), Chapter 1.105 Editor-in-Chief: , Elsevier, Oxford, pp. 73–94.
10. Gotman, I., Gutmanas, E.Y. and Hunter, G. (2011) Wear-resistant ceramic films and coatings, in *Comprehensive Biomaterials* (ed D. Paul), Chapter 1.108 Editor-in-Chief: , Elsevier, Oxford, pp. 127–155.
11. Yaszemski, M.J. (2003) *Biomaterials in Orthopedics*, Taylor & Francis.
12. Früh, H.J., Willmann, G. and Pfaff, H.G. (1997) Wear characteristics of ceramic-on-ceramic for hip endoprostheses. *Biomaterials*, **18**, 873–876.
13. Lappalainen, R. and Santavirta, S.S. (2005) Potential of coatings in total hip replacement. *Clin. Orthop. Relat. Res.*, **72–9**.
14. Jenabzadeh, A.-R., Pearce, S.J. and Walter, W.L. (2012) Total hip replacement: ceramic-on-ceramic. *Semin. Arthroplasty*, **23**, 232–240.
15. Willmann, G. (2001) Improving bearing surfaces of artificial joints. *Adv. Eng. Mater.*, **3**, 135–141.
16. Willmann, G. (2000) Ceramic femoral heads for total hip arthroplasty. *Adv. Eng. Mater.*, **2**, 114–122.
17. Sonntag, R., Reinders, J. and Kretzer, J.P. (2012) What's next? Alternative materials for articulation in total joint replacement. *Acta Biomater.*, **8**, 2434–2441.
18. Goodman, S.B., Barrena, E.G., Takagi, M. and Konttinen, Y.T. (2009) Biocompatibility of total joint replacements: A review. *J. Biomed. Mater. Res. Part A*, **90A**, 603–618.
19. Willmann, G. (1998) Ceramics for total hip replacement - what a surgeon should know. *Orthopedics*, **21**, 173–177.
20. Willmann, G. (2000) Ceramic femoral head retrieval data. *Clin. Orthop. Relat. Res.*, **22–8**.
21. Chevalier, J. and Gremillard, L. (2011) Zirconia as a biomaterial, in *Comprehensive Biomaterials* (ed D. Paul), Chapter 1.106 Editor-in-Chief: , Elsevier, Oxford, pp. 95–108.
22. Manicone, P.F., Iommetti, P.R. and Raffaelli, L. (2007) An overview of zirconia ceramics: basic properties and clinical applications. *J. Dent.*, **35**, 819–826.
23. Chevalier, J. (2006) What future for zirconia as a biomaterial? *Biomaterials*, **27**, 535–543.

24. Piconi, C. and Maccauro, G. (1999) Zirconia as a ceramic biomaterial. *Biomaterials*, **20**, 1–25.
25. Thompson, J.Y., Stoner, B.R., Piascik, J.R. and Smith, R. (2011) Adhesion/cementation to zirconia and other non-silicate ceramics: where are we now? *Dent. Mater.*, **27**, 71–82.
26. Derand, T., Molin, M. and Kvam, K. (2005) Bond strength of composite luting cement to zirconia ceramic surfaces. *Dent. Mater.*, **21**, 1158–1162.
27. Manso, A.P., Silva, N.R.F.A., Bonfante, E.A. *et al.* (2011) Cements and adhesives for all-ceramic restorations. *Dent. Clin. N. Am.*, **55**, 311–332.
28. Piascik, J.R., Wolter, S.D. and Stoner, B.R. (2011) Development of a novel surface modification for improved bonding to zirconia. *Dent. Mater.*, **27**, E99–E105.
29. Wolfart, M., Lehmann, F., Wolfart, S. and Kern, M. (2007) Durability of the resin bond strength to zirconia ceramic after using different surface conditioning methods. *Dent. Mater.*, **23**, 45–50.
30. Höland, W., Rheinberger, V., Apel, E. *et al.* (2009) Future perspectives of biomaterials for dental restoration. *J. Eur. Ceram. Soc.*, **29**, 1291–1297.
31. Aboushelib, M.N., Mirmohamadi, H., Matinlinna, J.P. *et al.* (2009) Innovations in bonding to zirconia-based materials. Part II: focusing on chemical interactions. *Dent. Mater.*, **25**, 989–993.
32. Aboushelib, M.N., Matinlinna, J.P., Salameh, Z. and Ounsi, H. (2008) Innovations in bonding to zirconia-based materials: Part I. *Dent. Mater.*, **24**, 1268–1272.
33. Kohorst, P., Borchers, L., Strempel, J. *et al.* (2012) Low-temperature degradation of different zirconia ceramics for dental applications. *Acta Biomater.*, **8**, 1213–1220.
34. Silva, N.R.F.A., Sailer, I., Zhang, Y. *et al.* (2010) Performance of zirconia for dental healthcare. *Materials*, **3**, 863–896.
35. Denry, I. and Kelly, J.R. (2008) State of the art of zirconia for dental applications. *Dent. Mater.*, **24**, 299–307.
36. Nakamura, K., Kanno, T., Milleding, P. and Ortengren, U. (2010) Zirconia as a dental implant abutment material: a systematic review. *Int. J. Prosthodont.*, **23**, 299–309.
37. Guess, P.C., Att, W. and Strub, J.R. (2012) Zirconia in fixed implant prosthodontics. *Clin. Implant Dent. Relat. Res.*, **14**, 633–645.
38. Hisbergues, M., Vendeville, S. and Vendeville, P. (2009) Zirconia: established facts and perspectives for a biomaterial in dental implantology. *J. Biomed. Mater. Res. Part B*, **88B**, 519–529.
39. Chevalier, J. and Gremillard, L. (2009) Ceramics for medical applications: a picture for the next 20 years. *J. Eur. Ceram. Soc.*, **29**, 1245–1255.
40. Kokubo, T. (2008) *Bioceramics and Their Clinical Applications*, Woodhead Publishing and Maney Publishing.
41. Rahaman, M.N., Huang, T., Bal, B.S. and Li, Y. (2010) In vitro testing of Al_2O_3–Nb composite for femoral head applications in total hip arthroplasty. *Acta Biomater.*, **6**, 708–714.
42. Huang, T.S., Rahaman, M.N. and Bal, B.S. (2009) Alumina–tantalum composite for femoral head applications in total hip arthroplasty. *Mater. Sci. Eng. C*, **29**, 1935–1941.

43. Treccani, L., Yvonne Klein, T., Meder, F. *et al.* (2013) Functionalized ceramics for biomedical, biotechnological and environmental applications. *Acta Biomater.*, **9**, 7115–7150.

44. Swan, E.E.L., Popat, K.C. and Desai, T.A. (2005) Peptide-immobilized nanoporous alumina membranes for enhanced osteoblast adhesion. *Biomaterials*, **26**, 1969–1976.

45. Lee, H.J., Kim, D.N., Park, S. *et al.* (2011) Micropatterning of a nanoporous alumina membrane with poly(ethylene glycol) hydrogel to create cellular micropatterns on nanotopographic substrates. *Acta Biomater.*, **7**, 1281–1289.

46. El-Said, W.A., Yea, C.-H., Jung, M. *et al.* (2010) Analysis of effect of nanoporous alumina substrate coated with polypyrrole nanowire on cell morphology based on AFM topography. *Ultramicroscopy*, **110**, 676–681.

47. Karlsson, M. and Tang, L. (2006) Surface morphology and adsorbed proteins affect phagocyte responses to nano-porous alumina. *J. Mater. Sci. Mater. Med.*, **17**, 1101–1111.

48. Yokoyama, Y., Tsukamoto, T., Kobayashi, T. and Iwaki, M. (2005) The changes of cell adhesion to collagen-coated Al_2O_3 by ion bombardment. *Surf. Coat. Technol.*, **196**, 298–302.

49. Liu, Z.-B., Zhang, Y., Yu, J.-J. *et al.* (2010) A microfluidic chip with poly(ethylene glycol) hydrogel microarray on nanoporous alumina membrane for cell patterning and drug testing. *Sens. Actuators B*, **143**, 776–783.

50. Wieneke, H., Dirsch, O., Sawitowski, T. *et al.* (2003) Synergistic effects of a novel nanoporous stent coating and tacrolimus on intima proliferation in rabbits. *Catheter. Cardiovasc. Interv.*, **60**, 399–407.

51. Das, S.K., Kapoor, S., Yamada, H. and Bhattacharyya, A.J. (2009) Effects of surface acidity and pore size of mesoporous alumina on degree of loading and controlled release of ibuprofen. *Microporous Mesoporous Mater.*, **118**, 267–272.

52. Kapoor, S., Hegde, R. and Bhattacharyya, A.J. (2009) Influence of surface chemistry of mesoporous alumina with wide pore distribution on controlled drug release. *J. Controlled Release*, **140**, 34–39.

53. Jansen, J., Treiner, C., Vaution, C. and Puisieux, F. (1994) Surface modification of alumina particles by nonionic surfactants – adsorption of steroids, barbiturates and pilocarpine. *Int. J. Pharm.*, **103**, 19–26.

54. Yin, T., Wei, W., Yang, L. *et al.* (2006) A novel capacitive immunosensor for transferrin detection based on ultrathin alumina sol–gel-derived films and gold nanoparticles. *Sens. Actuators B*, **117**, 286–294.

55. Tanvir, S., Pantigny, J., Morandat, S. and Pulvin, S. (2009) Development of immobilization technique for liver microsomes. *Colloids Surf., B*, **68**, 178–183.

56. Tanvir, S., Pantigny, J., Boulnois, P. and Pulvin, S. (2009) Covalent immobilization of recombinant human cytochrome CYP2E1 and glucose-6-phosphate dehydrogenase in alumina membrane for drug screening applications. *J. Membr. Sci.*, **329**, 85–90.

57. Tanvir, S., Morandat, S., Frederic, N. *et al.* (2009) Activity of immobilised rat hepatic microsomal CYP2E1 using alumina membrane as a support. *Nat. Biotechnol.*, **26**, 222–228.

58. Ekanayake, E.M.I.M., Preethichandra, D.M.G. and Kaneto, K. (2007) Polypyrrole nanotube array sensor for enhanced adsorption of glucose oxidase in glucose biosensors. *Biosens. Bioelectron.*, **23**, 107–113.

59.   Largueze, J.-B., Kirat, K.E. and Morandat, S. (2010) Preparation of an electrochem-
      ical biosensor based on lipid membranes in nanoporous alumina. *Colloids Surf., B*,
      79, 33–40.
60.   Yuan, J.H., Wang, K. and Xia, X.H. (2005) Highly ordered platinum-nanotubule
      arrays for amperometric glucose sensing. *Adv. Funct. Mater.*, **15**, 803–809.
61.   Darder, M., Aranda, P., Hernández-Vélez, M. *et al.* (2006) Encapsulation of enzymes
      in alumina membranes of controlled pore size. *Thin Solid Films*, **495**, 321–326.
62.   Hyndman, D., Lever, G., Burrell, R. and Flynn, T.G. (1992) Protein immobilization to
      alumina supports.1. Characterization of alumina-organophosphate ligand interactions
      and use in the attachment of papain. *Biotechnol. Bioeng.*, **40**, 1319–1327.
63.   Hyndman, D., Burrell, R., Lever, G. and Flynn, T.G. (1992) Protein immobilization to
      alumina supports. 2. Papain immobilization to alumina via organophosphate linkers.
      Biotechnol. Bioeng., **40**, 1328–1336.
64.   Ida, J., Matsuyama, T. and Yamamoto, H. (2000) Immobilization of glucoamylase
      on ceramic membrane surfaces modified with a new method of treatment utilizing
      SPCP-CVD. *Biochem. Eng. J.*, **5**, 179–184.
65.   Ida, J., Matsuyama, T. and Yamamoto, H. (2000) Surface modification of a ceramic
      membrane by the SPCP-CVD method suitable for enzyme immobilization. *J. Elec-
      trostat.*, **49**, 71–82.
66.   Yang, Z., Si, S. and Zhang, C. (2008) Study on the activity and stability of ure-
      ase immobilized onto nanoporous alumina membranes. *Microporous Mesoporous
      Mater.*, **111**, 359–366.
67.   Ticu, E.L., Vercaigne-Marko, D., Froidevaux, R. *et al.* (2005) Use of a protease-
      modified-alumina complex to design a continuous stirred tank reactor for producing
      bioactive hydrolysates. *Process Biochem.*, **40**, 2841–2848.
68.   Fadda, M.B., Dessi, M.R., Rinaldi, A. and Satta, G. (1989) Sandy alumina as substrate
      for economic and highly efficient immobilization of beta-glucosidase. *Biotechnol.
      Bioeng.*, **33**, 777–779.
69.   Dinnella, C., Stagni, A. and Lanzarini, G. (1997) Pectolytic enzymes co-
      immobilization on gamma-alumina spheres via organophosphate compounds.
      Process Biochem., **32**, 715–722.
70.   Itoh, T., Ishii, R., Hanaoka, T. *et al.* (2009) Encapsulation of catalase into nanochan-
      nels of an inorganic composite membrane. *J. Mol. Catal. B: Enzym.*, **57**, 183–187.
71.   Heilmann, A., Teuscher, N., Kiesow, A. *et al.* (2003) Nanoporous aluminum oxide
      as a novel support material for enzyme biosensors. *J. Nanosci. Nanotechnol.*, **3**,
      375–379.
72.   Pugniere, M., Juan, C.S., Colettipreviero, M.A. and Previero, A. (1988) Immobiliza-
      tion of enzymes on alumina by means of pyridoxal 5'-phosphate. *Biosci. Rep.*, **8**,
      263–269.
73.   Perazzini, R., Saladino, R., Guazzaroni, M. and Crestini, C. (2011) A novel and effi-
      cient oxidative functionalization of lignin by layer-by-layer immobilised Horseradish
      peroxidase. *Bioorg. Med. Chem.*, **19**, 440–447.
74.   Yeu, S., Lunn, J.D., Rangel, H.M. and Shantz, D.F. (2009) The effect of surface
      modifications on protein microfiltration properties of Anopore (TM) membranes. *J.
      Membr. Sci.*, **327**, 108–117.

75. Sun, L., Dai, J., Baker, G.L. and Bruening, M.L. (2006) High-capacity, protein-binding membranes based on polymer brushes grown in porous substrates. *Chem. Mater.*, **18**, 4033–4039.

76. Stoltenberg, R.M., Liu, C. and Bao, Z. (2011) Selective surface chemistry using alumina nanoparticles generated from block copolymers. *Langmuir*, **27**, 445–451.

77. Chang, C.S. and Suen, S.Y. (2006) Modification of porous alumina membranes with n-alkanoic acids and their application in protein adsorption. *J. Membr. Sci.*, **275**, 70–81.

78. Jain, P., Sun, L., Dai, J. *et al.* (2007) High-capacity purification of his-tagged proteins by affinity membranes containing functionalized polymer brushes. *Biomacromolecules*, **8**, 3102–3107.

79. Meder, F., Daberkow, T., Treccani, L. *et al.* (2012) Protein adsorption on colloidal alumina particles functionalized with amino, carboxyl, sulfonate and phosphate groups. *Acta Biomater.*, **8**, 1221–1229.

80. Javaid, A., Gonzalez, S.O., Simanek, E.E. and Ford, D.M. (2006) Nanocomposite membranes of chemisorbed and physisorbed molecules on porous alumina for environmentally important separations. *J. Membr. Sci.*, **275**, 255–260.

81. Paul, B., Martens, W.N. and Frost, R.L. (2011) Surface modification of alumina nanofibres for the selective adsorption of alachlor and imazaquin herbicides. *J. Colloid Interface Sci.*, **360**, 132–138.

82. Athanasekou, C.P., Romanos, G.E., Sapalidis, A.A. and Kanellopoulos, N.K. (2010) Ceramic-supported alginate adsorbent for the removal of heavy metal ions. *Adsorpt. Sci. Technol.*, **28**, 253–266.

83. Athanasekou, C.P., Papageorgiou, S.K., Kaelouri, V. *et al.* (2009) Development of hybrid alginate/ceramic membranes for Cd^{2+} removal. *Microporous Mesoporous Mater.*, **120**, 154–164.

84. Sabbani, S., Gallego-Perez, D., Nagy, A. *et al.* (2010) Synthesis of silver-zeolite films on micropatterned porous alumina and its application as an antimicrobial substrate. *Microporous Mesoporous Mater.*, **135**, 131–136.

85. Treccani, L., Maiwald, M., Zoellmer, V. *et al.* (2009) Antibacterial and abrasion-resistant alumina micropatterns. *Adv. Eng. Mater.*, **11**, B61–B66.

86. Hlavay, J. and Polyak, K. (2005) Determination of surface properties of iron hydroxide-coated alumina adsorbent prepared for removal of arsenic from drinking water. *J. Colloid Interface Sci.*, **284**, 71–77.

87. Boddu, V.M., Abburi, K., Talbott, J.L. and Smith, E.D. (2003) Removal of hexavalent chromium from wastewater using a new composite chitosan biosorbent. *Environ. Sci. Technol.*, **37**, 4449–4456.

88. Boddu, V.M., Abburi, K., Talbott, J.L. *et al.* (2008) Removal of arsenic(III) and arsenic(V) from aqueous medium using chitosan-coated biosorbent. *Water Res.*, **42**, 633–642.

89. Binh Thi Thanh, N., Peh, A.E.K., Chee, C.Y.L. *et al.* (2012) Electrochemical impedance spectroscopy characterization of nanoporous alumina dengue virus biosensor. *Bioelectrochemistry*, **88**, 15–21.

90. Howlett, C.R., Zreiqat, H., Odell, R. *et al.* (1994) The effect of magnesium-ion implantation into alumina upon the adhesion of human bone-derived cells. *J. Mater. Sci. Mater. Med.*, **5**, 715–722.

91. Feng, Q.L., Kim, T.N., Wu, J. *et al*. (1998) Antibacterial effects of Ag-HAp thin films on alumina substrates. *Thin Solid Films*, **335**, 214–219.

92. Pabbruwe, M.B., Standard, O.C., Sorrell, C.C. and Howlett, C.R. (2004) Bone formation within alumina tubes: effect of calcium, manganese, and chromium dopants. *Biomaterials*, **25**, 4901–4910.

93. Zhao, Q., Zhai, G.-J., Ng, D.H.L. *et al*. (1999) Surface modification of Al_2O_3 bioceramic by NH^+_2 ion implantation. *Biomaterials*, **20**, 595–599.

94. Kim, T.N., Feng, Q.L., Luo, Z.S. *et al*. (1998) Highly adhesive hydroxyapatite coatings on alumina substrates prepared by ion-beam assisted deposition. *Surf. Coat. Technol.*, **99**, 20–23.

95. Hamadouche, M., Meunier, A., Greenspan, D.C. *et al*. (2000) Bioactivity of sol–gel bioactive glass coated alumina implants. *J. Biomed. Mater. Res.*, **52**, 422–429.

96. Liu, J.L. and Miao, X.G. (2004) Sol–gel derived bioglass as a coating material for porous alumina scaffolds. *Ceram. Int.*, **30**, 1781–1785.

97. Verne, E., Bosetti, M., Brovarone, C.V. *et al*. (2002) Fluorapatite glass-ceramic coatings on alumina: structural, mechanical and biological characterisation. *Biomaterials*, **23**, 3395–3403.

98. Wang, Z., Xue, D., Chen, X. *et al*. (2005) Mechanical and biomedical properties of hydroxyapatite-based gradient coating on α-Al_2O_3 ceramic substrate. *J. Non-Cryst. Solids*, **351**, 1675–1681.

99. Schickle, K., Kaufmann, R., Campos, D.F.D. *et al*. (2012) Towards osseointegration of bioinert ceramics: introducing functional groups to alumina surface by tailored self assembled monolayer technique. *J. Eur. Ceram. Soc.*, **32**, 3063–3071.

100. Bertazzo, S., Zambuzzi, W.F., da Silva, H.A. *et al*. (2009) Bioactivation of alumina by surface modification: a possibility for improving the applicability of alumina in bone and oral repair. *Clin. Oral Implants Res.*, **20**, 288–293.

101. Bertazzo, S. and Rezwan, K. (2010) Control of alpha-alumina surface charge with carboxylic acids. *Langmuir*, **26**, 3364–3371.

102. Al-Amleh, B., Lyons, K. and Swain, M. (2010) Clinical trials in zirconia: a systematic review. *J. Oral Rehabil.*, **37**, 641–652.

103. Kelly, J.R. and Denry, I. (2008) Stabilized zirconia as a structural ceramic: an overview. *Dent. Mater.*, **24**, 289–298.

104. Lughi, V. and Sergo, V. (2010) Low temperature degradation -aging- of zirconia: a critical review of the relevant aspects in dentistry. *Dent. Mater.*, **26**, 807–820.

105. Nogiwa-Valdez, A.A., Rainforth, W.M., Zeng, P. and Ross, I.M. (2013) Deceleration of hydrothermal degradation of 3Y-TZP by alumina and lanthana co-doping. *Acta Biomater.*, **9**, 6226–6235.

106. Beuer, F., Schweiger, J., Eichberger, M. *et al*. (2009) High-strength CAD/CAM-fabricated veneering material sintered to zirconia copings – a new fabrication mode for all-ceramic restorations. *Dent. Mater.*, **25**, 121–128.

107. Zhang, Y. and Kim, J.-W. (2009) Graded structures for damage resistant and aesthetic all-ceramic restorations. *Dent. Mater.*, **25**, 781–790.

108. Lorenz, B., Marme, S., Muller, W.E.G. *et al*. (1994) Preparation and use of polyphosphate-modified zirconia for purification of nucleic-acids and proteins. *Anal. Biochem.*, **216**, 118–126.

109. Clausen, A.M., Subramanian, A. and Carr, P.W. (1999) Purification of monoclonal antibodies from cell culture supernatants using a modified zirconia based cation-exchange support. *J. Chromatogr. A*, **831**, 63–72.
110. Schafer, W.A., Carr, P.W., Funkenbusch, E.F. and Parson, K.A. (1991) Physical and chemical characterization of a porous phosphate-modified zirconia substrate. *J. Chromatogr.*, **587**, 137–147.
111. Wirth, H.J., Eriksson, K.O., Holt, P. *et al.* (1993) High-performance liquid-chromatography of amino-acids, peptides and proteins. 129. Ceramic-based particles as chemically stable chromatographic supports. *J. Chromatogr.*, **646**, 129–141.
112. Clausen, A.M. and Carr, P.W. (1998) Chromatographic characterization of phosphonate analog EDTA-modified zirconia support for biochromatographic applications. *Anal. Chem.*, **70**, 378–385.
113. Bellezza, F., Cipiciani, A. and Quotadamo, M.A. (2005) Immobilization of myoglobin on phosphate and phosphonate grafted-zirconia nanoparticles. *Langmuir*, **21**, 11099–11104.
114. Reshmi, R., Sanjay, G. and Sugunan, S. (2007) Immobilization of alpha-amylase on zirconia: a heterogeneous biocatalyst for starch hydrolysis. *Catal. Commun.*, **8**, 393–399.
115. Huckel, M., Wirth, H.J. and Hearn, M.T.W. (1996) Porous zirconia: a new support material for enzyme immobilization. *J. Biochem. Biophys. Methods*, **31**, 165–179.
116. Miller, T.M. and Grassian, V.H. (1995) Environmental catalysis – adsorption and decomposition of nitrous-oxide on zirconia. *J. Am. Chem. Soc.*, **117**, 10969–10975.
117. Lieberzeit, P.A. and Dickert, F.L. (2007) Sensor technology and its application in environmental analysis. *Anal. Bioanal. Chem.*, **387**, 237–247.
118. Pintar, A., Besson, M. and Gallezot, P. (2001) Catalytic wet air oxidation of Kraft bleaching plant effluents in the presence of titania and zirconia supported ruthenium. *Appl. Catal. Environ.*, **30**, 123–139.
119. Torrecillas, R., Moya, J.S., Díaz, L.A. *et al.* (2009) Nanotechnology in joint replacement. *Wiley Interdiscip. Rev. Nanomed. Nanobiotechnol.*, **1**, 540–552.
120. De Aza, A.H., Chevalier, J., Fantozzi, G. *et al.* (2003) Slow-crack-growth behavior of zirconia-toughened alumina ceramics processed by different methods. *J. Am. Ceram. Soc.*, **86**, 115–120.
121. Benzaid, R., Chevalier, J., Saddaoui, M. *et al.* (2008) Fracture toughness, strength and slow crack growth in a ceria stabilized zirconia-alumina nanocomposite for medical applications. *Biomaterials*, **29**, 3636–3641.
122. Roy, M.E., Whiteside, L.A., Katerberg, B.J. and Steiger, J.A. (2007) Phase transformation, roughness, and microhardness of artificially aged yttria- and magnesia-stabilized zirconia femoral heads. *J. Biomed. Mater. Res. A*, **83**, 1096–1102.
123. Matinlinna, J.P., Lassila, L.V.J. and Vallittu, P.K. (2006) The effect of a novel silane blend system on resin bond strength to silica-coated Ti substrate. *J. Dent.*, **34**, 436–443.
124. Matinlinna, J.P., Lassila, L.V.J. and Vallittu, P.K. (2006) The effect of three silane coupling agents and their blends with a cross-linker silane on bonding a bis-GMA resin to silicatized titanium (a novel silane system). *J. Dent.*, **34**, 740–746.

125. Yoshida, K., Tsuo, Y. and Atsuta, M. (2006) Bonding of dual-cured resin cement to zirconia ceramic using phosphate acid ester monomer and zirconate coupler. *J. Biomed. Mater. Res. Part B*, **77**, 28–33.

126. Ratner, B.D. and Hoffman, A.S. (2013) Physicochemical surface modification of materials used in medicine, in *Biomaterials Science*, 3rd, Chapter I.2.12 edn (eds B.D. Ratner, A.S. Hoffman, F.J. Schoen and J.E. Lemons), Academic Press, pp. 259–276.

127. Schickle, K., Korsten, A., Weber, M. *et al.* (2013) Towards osseointegration of bioinert ceramics: can biological agents be immobilized on alumina substrates using self-assembled monolayer technique? *J. Eur. Ceram. Soc.*, **33**, 2705–2713.

128. Janda, R., Roulet, J.F., Wulf, M. and Tiller, H.J. (2003) A new adhesive technology for all-ceramics. *Dent. Mater.*, **19**, 567–573.

129. Behr, M., Proff, P., Kolbeck, C. *et al.* (2011) The bond strength of the resin-to-zirconia interface using different bonding concepts. *J. Mech. Behav. Biomed. Mater.*, **4**, 2–8.

130. Kern, M. and Thompson, V.P. (1995) Bonding to glass infiltrated alumina ceramic: adhesive methods and their durability. *J. Prosthet. Dent.*, **73**, 240–249.

131. Özcan, M. and Vallittu, P.K. (2003) Effect of surface conditioning methods on the bond strength of luting cement to ceramics. *Dent. Mater.*, **19**, 725–731.

132. Piascik, J.R., Swift, E.J., Thompson, J.Y. *et al.* (2009) Surface modification for enhanced silanation of zirconia ceramics. *Dent. Mater.*, **25**, 1116–1121.

133. Faga, M.G., Vallee, A., Bellosi, A. *et al.* (2012) Chemical treatment on alumina-zirconia composites inducing apatite formation with maintained mechanical properties. *J. Eur. Ceram. Soc.*, **32**, 2113–2120.

134. Dehestani, M., Ilver, L. and Adolfsson, E. (2012) Enhancing the bioactivity of zirconia and zirconia composites by surface modification. *J. Biomed. Mater. Res. Part B*, **100**, 832–840.

135. Brook, R.J. (ed) (1991) *Concise Encyclopedia of Advanced Ceramic Materials*, Advances in Materials Science and Engineering, Pergamon, Oxford, p. ii.

136. Krstic, Z. and Krstic, V.D. (2012) Silicon nitride: the engineering material of the future. *J. Mater. Sci.*, **47**, 535–552.

137. Riley, F.L. (1991) Silicon nitride, in *Concise Encyclopedia of Advanced Ceramic Materials* (ed R.J. Brook), Pergamon, Oxford, pp. 434–437.

138. Bal, B.S., Khandkar, A., Lakshminarayanan, R. *et al.* (2009) Fabrication and testing of silicon nitride bearings in total hip arthroplasty winner of the 2007 "HAP" PAUL award. *J. Arthroplasty*, **24**, 110–116.

139. Bal, B.S. and Rahaman, M.N. (2012) Orthopedic applications of silicon nitride ceramics. *Acta Biomater.*, **8**, 2889–2898.

140. Mazzocchi, M., Gardini, D., Traverso, P.L. *et al.* (2008) On the possibility of silicon nitride as a ceramic for structural orthopaedic implants. Part II: chemical stability and wear resistance in body environment. *J. Mater. Sci. Mater. Med.*, **19**, 2889–2901.

141. Mazzocchi, M. and Bellosi, A. (2008) On the possibility of silicon nitride as a ceramic for structural orthopaedic implants. Part I: processing, microstructure, mechanical properties, cytotoxicity. *J. Mater. Sci. Mater. Med.*, **19**, 2881–2887.

142. Howlett, C.R., McCartney, E. and Ching, W. (1989) The effect of silicon-nitride ceramic on rabbit skeletal cells and tissue - an invitro and invivo investigation. *Clin. Orthop. Relat. Res.*, **244**, 293–304.

143. Neumann, A., Reske, T., Held, M. *et al.* (2004) Comparative investigation of the biocompatibility of various silicon nitride ceramic qualities in vitro. *J. Mater. Sci. Mater. Med.*, **15**, 1135–1140.

144. Cappi, B., Neuss, S., Salber, J. *et al.* (2010) Cytocompatibility of high strength non-oxide ceramics. *J. Biomed. Mater. Res. Part A*, **93**, 67–76.

145. Guedes e Silva, C.C., König, B., Carbonari, M.J. *et al.* (2008) Tissue response around silicon nitride implants in rabbits. *J. Biomed. Mater. Res. Part A*, **84**, 337–343.

146. Anderson, M.C. and Olsen, R. (2010) Bone ingrowth into porous silicon nitride. *J. Biomed. Mater. Res. Part A*, **92**, 1598–1605.

147. Neumann, A., Unkel, C., Werry, C. *et al.* (2006) Osteosynthesis in facial bones, Silicon nitride ceramic as material. *Hno*, **54**, 937–942.

148. Neumann, A., Unkel, C., Werry, C. *et al.* (2006) Prototype of a silicon nitride ceramic-based miniplate osteofixation system for the midface. *Otolaryngol. Head Neck Surg.*, **134**, 923–930.

149. Bal, B.S., Khandkar, A., Lakshminarayanan, R. *et al.* (2008) Testing of silicon nitride ceramic bearings for total hip arthroplasty. *J. Biomed. Mater. Res. Part B*, **87**, 447–454.

150. Oliveros, A., Guiseppi-Elie, A. and Saddow, S.E. (2013) Silicon carbide: a versatile material for biosensor applications. *Biomed. Microdevices*, **15**, 353–368.

151. González, P., Borrajo, J.P., Serra, J. *et al.* (2009) A new generation of bio-derived ceramic materials for medical applications. *J. Biomed. Mater. Res. Part A*, **88**, 807–813.

152. Liu, G., Dai, P., Wang, Y. *et al.* (2011) Fabrication of wood-like porous silicon carbide ceramics without templates. *J. Eur. Ceram. Soc.*, **31**, 847–854.

153. Will, J., Hoppe, A., Müller, F.A. *et al.* (2010) Bioactivation of biomorphous silicon carbide bone implants. *Acta Biomater.*, **6**, 4488–4494.

154. Hsu, J.-M., Tathireddy, P., Rieth, L. *et al.* (2007) Characterization of a-SiCx:H thin films as an encapsulation material for integrated silicon based neural interface devices. *Thin Solid Films*, **516**, 34–41.

155. Mani, G., Feldman, M.D., Patel, D. and Agrawal, C.M. (2007) Coronary stents: a materials perspective. *Biomaterials*, **28**, 1689–1710.

156. Saddow, S.E. (2012) *Silicon Carbide Materials for Biomedical Applications. Silicon Carbide Biotechnology*, Chapter 1, Elsevier, Oxford, pp. 1–15.

157. Deeken, C.R., Fox, D.B., Bachman, S.L. *et al.* (2011) Characterization of bio-nanocomposite scaffolds comprised of amine-functionalized gold nanoparticles and silicon carbide nanowires crosslinked to an acellular porcine tendon. *J. Biomed. Mater. Res. Part B*, **97B**, 334–344.

7

Carbon-based Materials in Biomedicine

Mercedes Vila
Dpt. Química Inorgánica y Bioinorgánica, Facultad de Farmacia, Universidad
Complutense de Madrid, CIBER de Bioingeniería, Biomateriales y Nanomedicina
(CIBER-BBN), Spain

7.1 Introduction

Carbon is unique regarding its structure–property relationship and, for the time being, this renders carbon as one of the most interesting topics in materials science and research.

The occurrence of various allotropes with very different properties and its ability to form stable substances with reaction partners with different electronegativity, results in various bonding possibilities which allow tailoring of materials response.

The broad range of applications of carbon materials in medicine is also the outcome of many preparation methods. Moreover, the structural versatility of the different allotropes also facilitates functionalization and other surface modification techniques to tailor/engineer the already outstanding properties, such as mechanical properties, high electrical conductivity, high thermal conductivity, optical absorbance, and so on.

7.2 Carbon Allotropes

When an element exists in more than one structural form with different molecular configurations, it is said that it exhibits allotropy [1, 2]. Twenty-five years ago, the world of carbon allotropes mostly focused on diamond and graphite-based materials, fibers, and compounds. However, it was significantly expanded with the discovery of the C60 buckminsterfullerene which led to the change in our way of thinking of materials at the nanoscale, with Kroto, Smalley, and Curl winning the Nobel Prize in Chemistry in 1996.

Bioceramics with Clinical Applications, First Edition. Edited by María Vallet-Regí.
© 2014 John Wiley & Sons, Ltd. Published 2014 by John Wiley & Sons, Ltd.

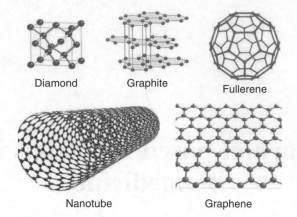

Figure 7.1   *Carbon arrangement of different carbon allotropes*

Following this, the discoveries of carbon nanotubes (CNTs) by Ijima and the isolation of graphene by Geim and Novoselov, which led to the Nobel Prize in 2010, provided materials with huge potential in the new era of nanotechnology, see Figure 7.1. Moreover, the development of new processing technologies allowed the application of carbon processed on other structures, such as amorphous carbon (a:C) materials in the form of bulk and thin films, which also displayed some of the typical properties of diamond, benefiting from both structures.

This new research era, as was expected, opened a new door to applications in biotechnology, profiting from the existing knowledge of the older traditional carbon structures that even now have an important market in the biomedical area [3].

7.2.1   Pyrolytic Carbon

Pyrolytic carbon (also called Pyrolite © carbon or Pyrocarbon) is a synthetic form of carbon which belongs to the family of turbostratic carbons, having a similar structure to graphite. In graphite, carbon atoms are covalently bonded in hexagonal arrays, and these arrays (graphene layers) are stacked and held together by weak interlayer binding. Pyrolytic carbon and other turbostratic carbons differ in that the layers are disordered, resulting in distortions within layers which form microscopic randomly-oriented zones.

Because of its similar elastic modulus to cortical bone and it thromboresistant character, it was proposed in 1996 as a coating for orthopedic implants, being commercialized by Tornier and Ascension orthopedics. Moreover, studies demonstrated the potential for biological fixation of pyrolytic carbon implants for prosthetic replacement of metacarpophalangeal joints [4].

Nevertheless, its most important market in the biomedical area started with its incorporation into heart valve design. Back in the 1960s artificial heart valves were constructed from plastics and metals which suffered from wear or failures because of blood clotting. It was Bokros and Gott who decided to apply a material intended for industrial applications, pyrolytic carbon, in new tests as a blood compatible candidate for coating cardiovascular components [5]. After all this time, pyrocarbon is still the most used material in more than 25 different valve designs in around 18 million patients per year [6].

7.2.2 Carbon Fibers

Although carbon fibers (CFs), also called graphite fibers, or carbon graphite (depending on the tensile modulus), date from the late 1800s, it was only in the 1960s that their commercial production started, mostly directed toward the aerospace industry to produce lightweight and resistant materials.

These fibers are made of carbon atoms bonded together onto microscopic crystals more or less aligned parallel to the long axis of the fiber (structural order only in the basal plane), making the fiber extremely strong for its size. The physical and chemical properties of the fibers are determined by their structure, and this is particularly important in the case of their application in biomedicine. The size of coherent crystallites in the carbon fiber lattice, corresponding to the crystallite thickness and height, is related to the fiber resistance to oxidation, and in a biological environment may influence the interaction between the fiber and the environment [7].

Carbon fibers are twisted together to form a tow, which may be used by itself or also reinforced with other materials. When the carbon fibers are reinforced in a carbon matrix, they are known as carbon fiber reinforced carbons (CFRCs) or carbon–carbon composites. Carbon fibers reinforced in a lightweight matrix, such as epoxy resin, polyester resin, or polyamide, are called carbon fiber reinforced polymers (CFRP).

Because of their light weight, flexibility, high strength, and high tensile modulus, several commercial products incorporated them in the 1980s as implant reinforcement fillers and, in spite of earlier failures, the materials have been adapted in the last decades to the specific conditions of biomedicine.

They were initially used as replacements for ligaments and tendons [8, 9] as they allowed the regeneration of tissues, but there was controversy over the mechanisms of disintegration into carbon particles which could be retained. This fact also influenced their use as scaffolds for tissue proliferation [10]. There were many studies after that, arriving to the conclusion that it was mandatory to control the type of fiber (physical, structural, and chemical properties) which influenced in many different parameters their integration *in vivo*. Control of the synthesis process was needed to design each material for each specific application in which it was to be applied. For example, carbon fibers in the form of a braid for soft tissue repair, or carbon fibers in the form of unwoven fabric as scaffolds for bone defects.

They were proposed in successful applications such as polymer reinforcement for the replacement of damaged vertebral bodies on the thoracolumbar spine. They showed potential when compared with titanium or surgical grade stainless steel, demonstrating high versatility and outstanding biological and mechanical properties [11]. The CFRC was also investigated as a unidirectional composite with high strengths ($1200\,MN\,m^{-2}$) and modulus ($140\,GN\,m^{-2}$) with an interlaminar shear strength of $18\,MN\,m^{-2}$. If the layers of fibers are also laid in two directions, it gives the composite material more isotopic properties. The implantation of these materials led to firm fixations with new formed bone adjacent to the implant. They were proposed for endosseous dental implants where the greater strength could be advantageous [12].

7.2.3 Fullerenes

In 1985 their discovery as a new allotrope of carbon [13], in which the atoms are arranged in a closed shell with the structure of a truncated icosahedron (with 60 carbon atoms arranged

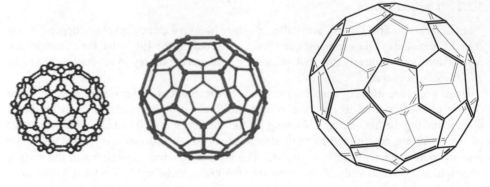

Figure 7.2   *Three different structure representations of fullerene, C60*

in a spherical structure in which every carbon atom forms a bond to three other adjacent atoms through sp^2 hybridization), see Figure 7.2, greatly expanded the number of known carbon allotropes, which until that time were limited to graphite, diamond, and a:C. They have been the subject of intense research, both for their unique chemistry and for their technological applications, especially in bio- and nanotechnology [14].

Initially, because of their lack of solubility in polar solvents, the research was focused on several functionalizations to make it useful for delivery into aqueous media [15].

Solubilization solutions were reported using cyclodextrins, calixarenes, phospholipids, liposomes, and polyvinylpyrrolidone, among others [16–20]. Hirsch and coworkers synthesized a dendrimeric fullerene derivative with a high level of solubility in water and, since then, various approaches have been used to induce hydrophilicity (e.g., –OH, -COOH, -NH$_2$) [21]. Nowadays, many fullerene-based compounds have been synthesized, and presented as potentially useful in anticancer or antimicrobial therapies, enzyme inhibition, controlled drug delivery, and diagnostic imaging [22].

They were found to be competitive inhibitors of recombinant affinity-purified HIV-1 protease. It was suggested by several researchers, that they could be applied successfully as antiviral activators. Studies started to check the antiviral activity in cells chronically infected and after incubation with a bis(monosuccinimide) derivative of *p,p'*-bis(2-aminoethyl)diphenyl-C60 compound [23]. It was demonstrated that it was active against human immunodeficiency virus type 1 (HIV-1) and HIV-2 [14]. This compound was viewed as a lead in the discovery of additional water-soluble fullerenes with greater virucidal and antiprotease activities. Years later, other approaches were designed and proved to inhibit different viruses, for example, fullerene aminoacid or cationic derivatives with human cytomegalovirus and hepatitis C replication, or water-insoluble fullerenes against enveloped viruses, such as semlimki forest or vesicular stomatitis viruses [24].

Because fullerenes absorb in the UV and visible region of the spectrum, they started to be applied for inducing photodynamic damage to biological systems and as photo-induced enzyme inhibition agents. C$_{60}$ is converted into its long-lived triplet excited state upon light irradiation and this triplet state, and its subsequent energy transfer to molecular oxygen, readily converts $^3O_2$ to $^1O_2$, a highly reactive singlet, as well as other reactive oxygen species (ROS) which are involved in both cellular signaling and cell damage [25]. This property of quenching or generating ROS could lead to their use as anticancer/antimicrobial agents but the potential toxicity that this implies could also compromise the biocompatibility of these systems.

In the presence of light they are also able to produce DNA and RNA cleavage, and it is believed there are two pathways for this to be possible [26]. The process mainly acts at guanine sites by the generation of singlet oxygen, or by energy transfer from the triplet state of fullerene to bases, which could be responsible for the oxidation of guanosines. Promising approaches were proposed by bearing an oligonucleotide chain able to interact with a single filament (or double chain). Moreover, new hybrid derivatives were synthesized by the introduction of a trimethoxyindole, which increased the selectivity and stabilization of the triple helix [27].

As drug or gene delivery carriers, they were tested linked with phosphonates designed to target osteoporotic bone tissue, or with paclitaxel for anticancer purposes [28, 29]. It was also proved that DNA-functionalized fullerenes were able to enter cells, increasing the lifetime of DNA in endosomes. In addition, the liophilicity of the sphere could be used to interact with the active site of various enzymes or could induce intercalability into biological membranes, destabilizing them and thus having potential as antibacterial agents [30–32].

Its capability of carrying a metal atom within the interior of the molecular cage (metallofullerenes/endofullerenes) would allow the isolation of reactive atoms from their environment.

Nevertheless, even now the extremely low water solubility of C60 and the need for functionalization of the fullerene core with various functional groups or molecules (which can change the photophysical properties and the ability to function as a "free radical sponge" and quench various free radicals), has retarded the advances in applied biomedicine research on these materials. Moreover, the drastic small size of these molecules (0.7 nm), has also limited their use in biomedicine, as fullerene solutions form aggregates presenting different properties than the original molecule, which has increased the complexity of controlling its behavior in biological systems.

7.2.4 Carbon Nanotubes

In some sense, CNTs were following in the footsteps of what was a unique discovery in carbon science, the C60 fullerene molecule. Although the fullerenes stimulated and motivated the large scientific community, their applications still remain undeveloped to date. Furthermore, since their discovery in 1991 by Ijima, CNTs have already compiled a remarkable list of superlatives [33]. CNTs were recognized as the ultimate carbon fiber with the highest strength of any material known and the highest thermal conductivity. They also were shown to possess outstanding field emission properties and, because of these properties, CNTs were initially the new focal point of scientists interested in their unique physical properties. With time, interest began to focus on their chemical properties and later they revolutionized biomedical research as they were considered as the new nanoparticles with a very interesting aspect ratio, incomparable with the existing ones [34].

Their name is derived from their long, hollow structure with the walls formed by one-atom-thick sheets of carbon, called graphene, see Figure 7.3. Basically, they are hexagonal networks of carbon atoms, a few nanometers in diameter and up to micron scale in length, formed by a rolled up graphite layer, this is what is called a single wall carbon nanotube

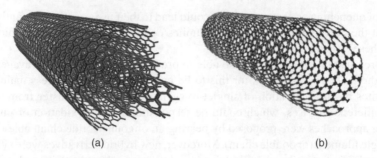

(a)   (b)

Figure 7.3   *Single wall (a) and multiwall (b) carbon nanotubes*

(SWCNT). If there are several graphene layers rolled up concentrically, they are called multiwall carbon nanotubes (MWCNTs). The chemical bonding between carbon atoms is of sp^2 type, and provides them with their unique strength.

Due to the chemical inertness of the graphitic walls, functionalization is often required in any application of these materials, maintaining the properties of the tubes themselves and the activity of the species attached to them. Functionalization could be done by covalent or non-covalent bonding and almost always begins with oxidation purification processes, which also remove any possible metal catalysts left from the synthesis [35]. These processes are also the first preparation step in the activation of chemically active sites for covalent bonding. Non-covalent functionalization involves van der Waals, $\pi - \pi$, and hydrophobic interactions of functional groups, and it has been suggested that although it preserves their aromatic structure and the electronic characteristics, it should not be used for drug delivery applications. On the other hand, enabling covalent attachment of functional groups to their surface provides a suitable platform for the attachment of almost any substance in a stable manner, for example, non-fouling polymers or organic chains for improving their solubility in water. Nevertheless, this kind of surface modification produces wall defects and affects imaging properties adversely.

In relation to their biocompatibility, the use of CNTs has been under discussion as at the beginning of this nanotechnology boom it was believed that they were toxic. However, recent studies on CNT surface functionalization have demonstrated that their biocompatibility is assured by a proper surface treatment and size, and that they can be excreted via the biliary pathway without causing any significant side effects [36]. Consequently, they have attracted great interest in nanomedicine.

They have recently been applied as scaffolds for tissue engineering, as a unique material or coated with biomolecules for nerve cell growth, and it has been reported that carbon nanofibers increased osteoblast functions and proliferation [37, 38]. They join mechanical strength with chemical stability and biological inertness and in some cases it has been demonstrated that they add some degree of control over the orientation of attached cells, also enhancing osteoblasts functions without the need for immobilization of bioactive molecules on their surface [39]. Moreover, CNTs were in the past used as additives to certain biomaterials to produce electrically conducting composites to respond to external stimuli to accelerate the osteointegration process [40, 41]. Electrical stimulation has been explored since the discovery of the presence of electrical potentials in mechanically loaded bones. CNTs and their 3D electrical conducting network have been previously used in combination with other ceramics and polymers in biomedicine as reinforcement for tendons and ligaments [42].

The use of CNTs as a reinforcement material for bioactive ceramics has also been widely studied as it has been demonstrated that collagen–CNT and polymer–CNT composites could be used as scaffolds for tissue engineering [43, 44]. It is also important to point out the property of the nanotubes for improving blood compatibility [45]. It has been reported that the platelet activation provoked by multiwall CNT–polymer composites is reduced compared to the polymer itself. All the above bioapplications of CNTs show the lack of toxicity in these cases, which explains the growing interest in the biomedical research community in the last few years.

The incorporation of CNTs in biosensing technology has made significant progress as their unique one-dimensional structure allows signals to be transported in a confined space, making them extraordinarily sensitive to chemical or electrical changes. These signals could be amplified by binding specific receptor species which could undergo selective interactions with the analyte [46].

Their ability to be loaded and functionalized with several substances and their nanometer size, which could allow them to permeate into cellular membranes, has been proved in different systems, demonstrating that they can deliver drug to neurons and cardiomyocytes, for example, and control the drug unloading mechanism by external factors such as changes in pH or by optical stimulation with near-infrared (NIR) wavelengths [47, 48].

Their functionalization with biologically active molecules has also permitted their application as gene transfection vectors or as hyperthermia agents, for cancer treatments, by using their ability to absorb in the NIR therapeutical window [49, 50]. This therapy is based on the energy transfer produced during the irradiation of a material generating vibrational energy, thus generating heat sufficient for cell destruction (above 40 °C).

Hyperthermia treatment thus has the advantage of being less risky to the body, with less side effects and the possibility of repeating treatment, as compared to surgery, chemotherapy, and radiation therapy.

7.2.5 Graphene

Among carbon-based materials, following close and taking over from CNTs, graphene represents one of the most promising "nanoparticles" of the last few years. A great effort has been addressed to the detailed investigation and optimization of its physical properties but the chemistry of graphene is still a rather unexplored field of research. Graphene is a two-dimensional atomic layer of sp^2 carbon and has a wide range of unique properties [51]. Thermal conductivity, mechanical stiffness, fracture strength, extraordinary electronic transport, and so on, and all of those confined in a nanosized sheet able of trespassing cell membranes and filter organs … Nevertheless, to take advantage of these features, the availability of graphene with a well-defined and controllable surface and interface properties, produced on a sufficient scale, is of critical importance. Robust methods for producing chemically functionalized graphene in large quantities are still lacking. The most common method for the covalent functionalization of graphene employs graphene oxide (GO), which is prepared by treating graphite particles with strong acids, which offers a new class of solution-dispersible polyaromatic platform for performing chemistry (see Figure 7.4) [52, 53] .

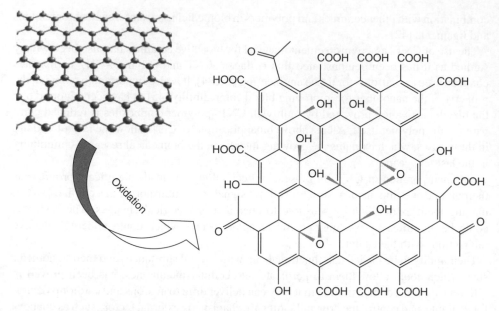

Figure 7.4 *Structure of graphene sheet and graphene oxide*

GO has its basal planes and edges covalently decorated with oxygen-containing functional groups so that it contains a mixture of sp^2- and sp^3-hybridized carbon atoms. Furthermore, manipulation of the size, shape, and relative fraction of the sp^2-hybridized domains of GO by reduction chemistry provides opportunities for tailoring its properties [54]. For example, as-synthesized GO is insulating but controlled de-oxidation leads to an electrically and optically active material that is transparent and conducting.

The defective nature of GO does not allow the observation of two-dimensional condensed matter effects. However, precisely because of its heterogeneous chemical and electronic structures, it can interact with a wide range of inorganic and organic species in non-covalent, covalent, and ionic manners so that functional hybrids can be synthesized.

Moreover, in contrast to pure graphene, GO is fluorescent over a broad range of wavelengths, owing to its heterogeneous electronic structure. The use of tunable fluorescence has already been demonstrated in biological applications for sensing and drug delivery [55]. Another interesting feature is the strong NIR optical absorption ability of GO-related materials that can be explored for use in *in vivo* photothermal therapy of cancer cells [56].

Besides all the properties of GO described above, its low cost, large production scale, and easy-processing, when compared, for example, with CNTs, makes this nanomaterial a promising nanoparticle for medical application where large scale production is needed.

Among the syntheses used to process graphene, such as epitaxial growth or graphite oxide reduction, acid exfoliation of graphite is probably the principal way of synthesis when large production is needed. Now, exfoliation of graphite by the Hummers method [57] is the only way of producing stable suspensions of quasi-two-dimensional carbon sheets which are a strategic starting point for large-scale synthesis.

In contrast to pristine graphite, the graphene-derived sheets in GO are heavily oxygenated, bearing hydroxy and epoxide functional groups on their basal planes, in addition to carbonyl and carboxyl groups located at the sheet edges. The presence of these functional groups makes GO sheets strongly hydrophilic and they disperse in water. The quality of graphene is lower than that from other techniques, but more important for these applications, is the intermediate step needed for further modification with desired functional species.

Graphene scientific research in the biomedical area has an advantage and is developing at extreme speed because it is adapting all the previous know-how from the long and exhaustive research on CNTs. All the surface chemistry is transferable from one system to the other and both materials in this particular area behave in a very similar way, with the advantage for graphene of its unique and smaller two-dimensional shape and aspect ratio.

Current challenges include controlling the chemical functionalization of graphene with functional units to achieve both a good processability in various media and a fine tuning of various physico-chemical properties of the 2D architecture, paving the way towards its technological application in chemical and biochemical fields, for example, producing anchoring points for additional surface groups, such as attachment of biomolecules (peptides, DNA, growth factors, etc.), and polyethyleneglycol (PEG) coatings for prolonging blood circulation half-life and avoiding agglomeration, and so on [58].

The optical properties of GO are also largely unexplored and could facilitate biological and medical research such as imaging and, therefore, diagnosis. Apart from the typical fluorescent labeling that can be performed on its surface and tracked easily, graphene can be brightly luminescent by several routes: its oxidation, non-equilibrium Dirac fermions, band gap manipulation, or mild oxygen plasma treatment, and so on [55]. For biomedical applications, its natural photoluminescence because of the functional groups on the surface, although it exists, is difficult to track inside the cells.

Long and branched PEG-coated CNTs and graphene have proved to exhibit prolonged blood circulation half-life, as is generally desired for nano-vehicles as it allows nanomaterials to repeatedly pass through tumor vascularization, benefiting tumor uptake via the enhanced permeability and retention effect (EPR) of cancerous tumors. Graphene, because of its 2D and smaller size (10–100 nm for this application), shows a better performance and distinctive behavior compared to CNTs [56].

It was found also that both materials have a strong optical absorption in the NIR region and are promising photothermal agents for *in vivo* tumor destruction. This therapy, as has been explained above, was developed for CNTs and, in the last few years, it has been adopted by the "graphene science".

PEG-GO, for example, has been proved to be internalized by the cells by endocytose and then they can be exposed to an NIR laser emitting in the "therapeutical window" where it is a non-invasive, harmless, and skin penetrating irradiation. Biological systems mostly lack chromophores that absorb in the NIR region. *In vitro* and *in vivo* experiments have been successfully carried out by Dai [59], Vila [60], Liu [61], and Markovic [56].

For the first time, PEGylated nanoscale graphene oxide (NGO) has been applied as a nanocarrier to load anticancer drugs via non-covalent physisorption and its cellular uptake has been studied, with satisfactory results [62]. The loading and release of doxorubicin (DOX) hydrochloride, for example, has been investigated as an anticancer drug. The loading ratio (weight ratio of loaded drug to carriers) of GO reached 200% more than with other nanocarriers, such as nanoparticles, that usually have a loading ratio lower than 100% [63]. It has also been recently reported that NGO functionalized with sulfonic acid groups followed by covalent binding of folic acid (FA) molecules, allows specific targeting of human breast cancer cells. Furthermore, controlled loading of two cancer drugs such as DOX and camptothecin (CPT) via $\pi-\pi$ stacking and hydrophobic interactions has been investigated [64].

GO has also been proposed as a biosensor, to be used in analytical techniques for sensitive and selective detection of proteins, for example, for the detection of cancer biomarkers [65, 66].

The GO bonding with different cationic polymers, such as polyethyleneimine (PEI), has also proved to be efficient in performing gene transfection and has shown advantages when compared with other nanoparticles. They have successfully used graphene as a non-toxic nano-vehicle for efficient gene transfection and the efficiency is even improved when subjected to irradiation in the NIR region [67].

As a completely different approach to nanomedicine, but also intended for the biomedicine area, one possible route to harnessing materials would be the incorporation of graphene sheets into a composite material. Small percentages of previously modified GO improve polymer elongation at break, leading to a tougher material performance and tribological behavior [68].

Some medical applications require hydrophobic materials without a cell-adhesive surface, such as devices in contact with human blood (e.g., artificial heart valves) [69] or joint prostheses in the friction area, while others need a cell-adhesive surface to assure complete tissue integration of the implanted material in the human body. Graphene may be another possibility as a biocompatible coating if it performs better than the other carbon layers, such as nano-diamond coatings or diamond-like carbon (DLC). For this purpose, physical techniques of graphene deposition, such as epitaxial growth, should be adapted to be competitive in cost and capable of deposition over large complex surfaces. It could also be useful as a coating of medical tools. All this research area is still in the early stages of development.

7.2.6   Diamond and Amorphous Carbon

Almost all the substitution prostheses and implants are made of metallic parts and no other materials have proved to perform better than them. Nevertheless, many adverse host responses to the implantation of medical metal implants are related to surface properties. The surfaces of medical implants and devices may be engineered to prevent the release of toxic elements and to induce a dynamic biomaterial–tissue response. Corrosion, fatigue, friction, and wear are some of the surface characteristics that are considered when designing biosensor and drug delivery devices, and more specifically, joint prostheses.

Although metal implant surfaces tend to be passivated in order to minimize the adverse host response, bioinert coatings has been widely applied as barriers in metal alloy implants to prevent the release of cobalt, chromium, nickel, aluminum, or vanadium which exhibit allergic, carcinogenic, and general toxic interactions, such as metallosis, with human tissues. Moreover, these coatings must have sufficient fatigue, wear, and adhesion properties to resist the large, cyclical stresses that are encountered. Carbon-based structures are the most used bioinert coatings as they can act as barrier diffusion layers and/or auto-lubricant coatings [70].

Diamond, where the carbon atoms are arranged in a variation of the face-centered cubic crystal structure, giving a tetrahedral unit where each C-atom is sp³-hybridized, has superlative physical features which mostly originate from this strong covalent bonding. It has the highest hardness and thermal conductivity of any bulk material. More specifically, the nanodiamond is the most interesting for these applications. The name nanodiamond is used for a variety of diamond at the nanoscale (the length scale of the crystals is approximately 1–100 nm) [71]. It has a rounded shape and an active surface keeping a diamond-like hardness. Moreover, its strength is compared to Teflon. In addition, its wear resistance (highly resistant to steel corrosion), and angstrom finishes of polished surfaces similar to the physical characteristics of rubber, has led to its use as auto-lubricant coatings for substituting oils [72]. Methods of nanodiamond synthesis are diverse, involving methods such as a gas phase nucleation at ambient to high pressure.

Studies on this material report the tribological and biological response of their application mostly as a coating of the friction part of joint prostheses (for example, the femur head working against the acetabular cup of the hip, or the femoral head working against the tibia component in the knee) where wear resistance and no cellular adhesion are demanded [73, 74]. It plays the role of a barrier protective coating and its low coefficient of friction is particularly useful in reducing the amount of wear debris generated during functioning. Any possible residues formed during the life-cycle of the implant were found to be harmless, unlike for most polymers and metals, and do not have adverse reactions from human monocytes or leukocytes. It also has high resistance to bacterial colonization when compared to medical steel or titanium. Besides total hip or knee replacement, it has been proposed as a coating for retinal microchips to restore retinal degeneration [75]. The studies on its hemocompatibility proved that it presents minimal plasma protein adhesion and no thrombogenic properties, so it has been used as a coating for heart valves, catheters, or stent coatings [76].

For the same purposes, other forms of a-C coatings with simple preparation methods have been proposed, as was the case for DLC or tetrahedral carbon (ta-C). a-C films are composed of sp² and sp³ hybridizations, and it is the relationship between the concentration of these which establishes their nomenclature and properties, see Figure 7.5 [77].

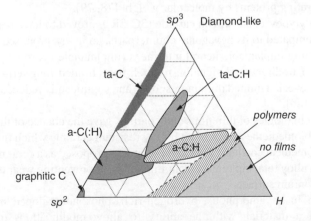

Figure 7.5 *Ternary phase diagram of amorphous carbon bonding*

Deposition methods were developed to produce a-Cs with increasing degrees of sp^3 bonding. Coatings with high sp^3 concentrations have properties similar to diamond and are called DLC and ta-C and, in the case where they have a small amount of hydrogen coming from the deposition techniques hydrogenated amorphous carbon (a-C:H). All a-C films have low friction coefficients and tribologically they perform differently depending on their surface characteristics, which confer on them different characteristics, such as hydrophobicity which controls their cell attachment interactions. These characteristics directly influence their tribological performance under different friction media, such as moderate/high humidity environments or aqueous lubricated sliding, such as found in biomedical applications [70].

7.3   Carbon Compounds

7.3.1   Silicon Carbide

Silicon carbide (SiC) is peculiar as it exists in a vast family of similar crystalline structures (polytypes), a phenomenon that is called polymorphism. It is formed in different structures which are variations of the same chemical formula, that are identical in two dimensions and differ in the third, as layers stacked in a sequence.

It is composed of tetrahedra of carbon and silicon atoms with strong covalent bonds in the crystal lattice which produces a very hard and strong material. Because of its superior chemical inertness, low density, high hardness, and elastic modulus, SiC has been widely used as an excellent abrasive over the last century. Nevertheless, in the last few decades with the high development of new materials processing techniques, this material has been processed into a high quality technical grade ceramic with also very good mechanical properties.

These features, together with low weight, no toxicity, and biocompatibility, determine SiC as a very promising material for implantable parts for biomedical purposes. It is also worth noting the transparency of wide band gap semiconductors for visible light, which could be used for *in situ* observation of living cell growth.

Porous SiC has been studied as a membrane for implantable biosensors. Free-standing porous SiC membranes were tested to be permeable to proteins and small molecules, and capable of separating proteins by molecular weight [78, 79].

One of the polytypes, cubic silicon carbide (3C-SiC), proved to have better *in vitro* biocompatibility compared to its hexagonal counterparts, so it was proposed for substituting the already existing implantable neural interfaces (implantable neuronal prosthetics offer the possibility of bi-directional signaling) which have limited long-term reliability. The 3C-SiC surface revealed limited immunoresponse and significantly reduced microglia compared to silicon [80].

The arising technologies of thin films and coatings gave the chance of the application of this material to biomedicine as a bioinert highly resistant ceramic which fulfilled the needs of protective biomedical coatings. For example, it was proposed as a ceramic coating material for titanium alloy total hip replacement implants, for preventing wear debris formation from the soft titanium surface.

SiC thin film in an amorphous, hydrogen-rich, phosphorus-doped modification (a-SiC:H) is such a material, with the ability to allow modifications to its electronic structure to be adapted to the electronic structure of proteins, transforming it into a passive "non-activating" biomaterial coating. It was found that electron transfer processes between protein macromolecules and the implant's surface directly result in an activation of proteins. When a-SiC:H coated metals are compared with the surface of metallic stents or uncoated stainless steel, proteins, platelets, and cells showed less significant adhesion and activation on the a-SiC:H surface than on metals, and the endothelization process was improved. Moreover, coating the a-SiC:H films with PEG significantly improved the antifouling characteristics for extended periods of time [81].

In a very recent study, amorphous silicon carbide (a-SiC) films, deposited by plasma-enhanced chemical vapor deposition (PECVD), were evaluated as insulating coatings for implantable microelectrodes [82].

7.3.2 Boron Carbide

As another of the hardest materials known, following cubic boron nitride and diamond, boron carbide B_4C, is an extremely hard boron–carbon ceramic material produced synthetically, commonly used in abrasive and wear-resistant products as well as in hard lightweight composite materials. It is characterized by a unique combination of properties that make it a material of choice for a wide range of engineering applications.

The primary structural units of boron carbide (BC) are the 12-atom icosahedra located at the vertices of a rhombohedral lattice of trigonal symmetry. The presence of icosahedra (essentially two pentagonal pyramids bonded together) within the BC structure is a consequence of elemental boron's ability to form caged structures of a variety of sizes [83].

First developed in the 1950s after its proposal in the 1930s by Locher, boron neutron capture therapy (developed for treating brain cancer) involves administering a boron-containing compound and then irradiating tumors in which the boron compound has become concentrated, see Figure 7.6 [84]. Nowadays, BC nanoparticles, nanodots, or nanopowder are spherical high surface area particles considered for application in biomedical and bioscience areas, and more specifically as a promising boron agent for this therapy, as it should be noted that at this time no single boron delivery agent fulfills all of the required criteria. With the development of new chemical synthesis techniques and increased knowledge of the biological and biochemical requirements needed for an effective agent and their modes of delivery, a number of promising new boron agents is emerging.

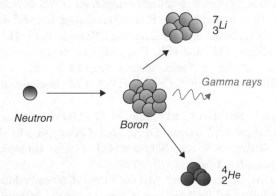

Figure 7.6 *Nuclear fission reaction in boron neutron capture therapy*

7.3.3   Tungsten Carbide

There are two forms of tungsten carbide (WC), a hexagonal and a cubic high-temperature form. The hexagonal form consists of close packed layers of metal atoms with layers lying directly over one another, with carbon atoms filling half the interstices, giving a regular trigonal prismatic structure.

Tungsten carbide is designed for wear resistant surfacing of parts operating under conditions of intense wear. It is used for the surfacing of drill bits, blast furnaces, mining tools, orthopedic surgical instruments, and dental burs.

References

1. Heimann, R.B., Evsyukov, S.E. and Koga, Y. (1994) Carbon allotropes: a suggested classification scheme based on valence orbital hybridization. *Carbon*, **35**, 289–299.
2. Yasuda, E., Inagaki, M., Kaneko, K. *et al.* (2003) *Carbon Alloys. New Concepts to Develop Carbon Science and Technology*, Elsevier, Amsterdam.
3. Krueger, A. (2010) *Carbon Materials and Nanotechnology*, Wiley-VCH Verlag GmbH, Weinheim.
4. Bravo, C.J., Rizzo, M., Hormel, K.B. and Beckenbaugh, R.D. (2007) Pyrolytic carbon proximal interphalangeal joint arthroplasty: results with minimum two-year follow-up evaluation. *J. Hand Surg.*, **32**, 1–11.
5. Brokos, J.C. (1977) Carbon biomedical devices. *Carbon*, **15**, 355–371.
6. Zilla, P., Brink, J., Human, P. and Bezuidenhout, D. (2008) Prosthetic heart valves: catering for the few. *Biomaterials*, **29**, 385–406.
7. Blazewicz, M., Blazewicz, S., Chlopek, J. and Staszkow, E. (1991) Structure and properties of carbon materials for biomedical applications, in *Ceramics in Substitutive and Reconstructive Surgery* (ed P. Vincenzini), Elsevier, Amsterdam, 189–195.
8. Miller, J.H. (1984) Comparison of the structure of neotendons induced by implantation of carbon polyester fibres. *J. Bone Joint Surg.*, **66**, 131–139.
9. Jenkins, D.H.R. (1985) Ligament induction by filamentous carbon fibre. *Clin. Orthop.*, **197**, 86–90.
10. Becker, H.P., Rosenbaum, D., Zeithammel, G. *et al.* (1996) Tenodesis versus carbon fiber repair of ankle ligaments: a clinical comparison. *Clin. Orthop.*, **325**, 194–202.
11. Ciappetta, P., Boriani, S. and Fava, G.P. (1997) A carbon fiber reinforced polymer cage for vertebral body replacement: technical note. *Neurosurgery*, **41**, 1203–1206.
12. Adams, D., Williams, D.F. and Hill, J. (1978) Carbon fiber-reinforced carbon as a potential implant material. *J. Biomed. Mater. Res.*, **12**, 35–42.
13. Kroto, H., Heath, J.R., ÓBrien, S.C. *et al.* (1985) C60: buckminsterfullerene. *Nature*, **318**, 162–163.
14. Bosi, S., Da Ros, T., Spalluto, G. and Prato, M. (2003) Fullerene derivatives: an attractive tool for biological applications. *Eur. J. Med. Chem.*, **38**, 913–923.
15. Jensen, A.W., Wilson, S.R. and Schuster, D.I. (1996) Biological applications of fullerenes. *Bioorg. Med. Chem.*, **4**, 767–779.
16. Andersson, T., Nilsson, K., Sundahl, M. *et al.* (1992) C60 embedded in γ-cyclodextrin: a water-soluble fullerene. *J. Chem. Soc., Chem. Commun.*, 604–606.

17. Atwood, J.L., Koutsantonis, G.A. and Raston, C.L. (1994) Purification of C60 and C70 by selective complexation with calixarenes. *Nature*, **368**, 229–231.
18. Garaud, J.L., Janot, J.M., Miquel, G. and Seta, P. (1994) Photoinduced electron transfer properties of porous polymer membranes doped with the fullerene C60 associated with phospholipids J. *Membr. Sci.*, **91**, 259–264.
19. Bensasson, R.V., Bienvenue, E., Dellinger, M. *et al.* (1994) C60 in model biological systems. A visible-UV absorption study of solvent-dependent parameters and solute aggregation. *J. Phys. Chem.*, **98**, 3492–3500.
20. Yamakoshi, Y.N., Yagami, T., Fukuhara, K. *et al.* (1994) Solubilization of fullerenes into water with polyvinylpyrrolidone applicable to biological tests. *J. Chem. Soc., Chem. Commun.*, 517–518.
21. Schlundt, S., Kuzmanich, G., Spänig, F. *et al.* (2009) Dendritic porphyrin-fullerene conjugates: efficient light-harvesting and charge-transfer events. *Chemistry*, **16** (15), 12223–12233.
22. Bakry, R., Vallant, R.M., Najam-ul-Haq, M. *et al.* (2007) Medicinal applications of fullerenes. *Int. J. Nanomedicine*, **2**, 639–649.
23. Schinazi, R.F., Sijbesma, R., Srdanov, G. *et al.* (1993) Synthesis and virucidal activity of a water-soluble, configurationally stable, derivatized C60 fullerene. *Antimicrob. Agents Chemother.*, **37**, 1707–1710.
24. Kaesermannm, F. and Kempf, C. (1997) Photodynamic inactivation of enveloped viruses by buckminsterfullerene. *Antiviral Res.*, **34**, 65–70.
25. Yamakoshi, Y., Umezawa, N., Ryu, A. *et al.* (2003) Active oxygen species generated from photoexcited fullerene (C60) as potential medicines: O-.bul.2 versus 1O2. *J. Am. Chem. Soc.*, **125**, 12803–12809.
26. Tokuyama, H., Yamago, S., Nakamura, E. *et al.* (1993) Photoinduced biochemical activity of fullerene carboxylic acid. *J. Am. Chem. Soc.*, **115**, 7918–7919.
27. Bergamin, M., Da Ros, T., Spalluto, G. *et al.* (2001) Synthesis of a hybrid fullerene-trimethoxyindole-oligonucleotide conjugate. *Chem. Commun.*, 17–18.
28. Zakharian, T.Y., Seryshev, A. and Sitharaman, B. (2005) A Fullerene-paclitaxel chemotherapeutic: synthesis, characterization, and study of biological activity in tissue culture. *J. Am. Chem. Soc.*, **127**, 12508–12509.
29. Gonzalez, K., Wilson, L., Wu, W. and Nancollas, G. (2002) Synthesis and In vitro characterization of a tissue-selective fullerene: vectoring C60(OH)16AMBP to mineralized bone bioorg. *Med. Chem.*, **10**, 1991–1997.
30. Nakamura, E., Isobe, H., Tomita, N. *et al.* (2000) Functionalized fullerene as an artificial vector for transfection. *Angew. Chem. Int. Ed.*, **39**, 4254–4257.
31. Da Ros, T., Prato, M., Novello, F. *et al.* (1996) Easy access to water-soluble fullerene derivatives via 1,3-dipolar cycloadditions of azomethine ylides to C60. *J. Org. Chem.*, **61**, 9070–9072.
32. Mashino, T., Okuda, K., Hirota, T. *et al.* (1999) Inhibition of E. coli growth by fullerene derivatives and inhibition mechanism. *Bioorg. Med. Chem. Lett.*, **9**, 2959–2962.
33. Iijima, S. (1991) Helical microtubules of graphitic carbon. *Nature*, **354**, 56–58.
34. Endo, M., Strano, M.S. and Ajayan, P.M. (2008) *Potential Applications of Carbon Nanotubes. Carbon Nanotubes*, Topics in Applied Physics, vol. **111**, Springer-Verlag, Berlin, Heidelberg, pp. 13–62.

35. Madani, S.Y., Naderi, N., Dissanayake, O. *et al.* (2011) A new era of cancer treatment: carbon nanotubes as drug delivery tools. *Int. J. Nanomed.*, **6**, 2963–2979.
36. Fischer, C., Rider, A.E., Han, Z.J. *et al.* (2012) Applications and Nanotoxicity of carbon nanotubes and graphene in biomedicine. *J. Nanomater.*, **2012**, 1–19, Article ID 315185.
37. Correa-Duarte, M., Wagner, N., Rojas-Chapana, J. *et al.* (2004) Fabrication and biocompatibility of carbon nanotube-based 3D networks as scaffolds for cell seeding and growth. *Nano Lett.*, **4**, 2233–2236.
38. Webster, T.J., Waid, M.C., McKenzie, J.L. *et al.* (2004) Nano-biotechnology: carbon nanofibers as improved neural and orthopaedic implants. *Nanotechnology*, **15**, 48–54.
39. Mattson, M.P., Haddon, R.C. and Rao, A.M. (2000) Molecular functionalization of carbon nanotubes and use as substrates for neuronal growth. *J. Mol. Neurosci.*, **14**, 175–182.
40. Supronowicz, P.R., Ajayan, P.M., Ullmann, K.R. *et al.* (2002) Novel current-conducting composite substrates for exposing osteoblast to alternating current stimulation. *J. Biomed. Mater. Res.*, **59**, 499–506.
41. Vila, M., Cicuendez, M., Sanchez-Marcos, J. *et al.* (2013) Electrical stimuli to increase cell proliferation on carbon nanotubes/ mesoporous silica composites for drug delivery. *J. Biomed. Mater. Res. Part A*, **101**, 213–21.
42. Deeken, C.R., Cozad, M.J., Bachman, S.L. *et al.* (2011) Characterization of bio-nanocomposite scaffolds comprised of amine-functionalized single-walled carbon nanotubes crosslinked to an acellular porcine tendon. *J. Biomed. Mater. Res. Part A*, **96**, 584–594.
43. Hirata, E., Uo, M., Takita, H. *et al.* (2011) Multiwalled carbon nanotube-coating of 3Dcollagen scaffolds for bone tissue engineering. *Carbon*, **49**, 3284–3291.
44. Shokrgozar, M.A., Mottaghitalab, F., Mottaghitalab, V. and Farokhi, M. (2011) Fabrication of porous chitosan/poly(vinyl alcohol) reinforced single-walled carbon nanotube composites for neural tissue engineering. *J. Biomed. Nanotechnol.*, **7**, 276–284.
45. Zhao, M.L., Li, D.J., Yuan, L. *et al.* (2011) Differences in cytocompatibility and hemocompatibility between carbon nanotubes and nitrogen-doped carbon nanotubes. *Carbon*, **49**, 3125–3133.
46. Yang, W., Thordarson, P., Gooding, J.J. *et al.* (2007) Carbon nanotubes for biological and biomedical applications. *Nanotechnology*, **18**, 412001.
47. Feazell, R.P., Nakayama-Ratchford, N. and Lippard, S.J. (2007) Soluble single-walled carbon nanotubes as longboat delivery systems for platinum (IV) anticancer drug design. *J. Am. Chem. Soc.*, **129**, 8438–8439.
48. Hampel, S., Kunze, D. and Haase, D. (2008) Carbon nanotubes filled with a chemotherapeutic agent: a nanocarrier mediates inhibition of tumor cell growth. *Nanomedicine*, **3**, 1105–1113.
49. Podesta, J.E., Al-Jamal, K.T. and Herrero, M.A. (2009) Antitumor activity and prolonged survival by carbon nanotube-mediated therapeutic siRNA silencing in a human lung xenograft model. *Small*, **5**, 1176–1185.
50. Burlaka, A., Lukin, S., Prylutska, S. *et al.* (2010) Hyperthermic effect of multi-walled carbon nanotubes stimulated with near infrared irradiation for anticancer therapy: in vitro studies. *Exp. Oncol.*, **32**, 48–50.

51. Zhu, Y., Murali, S., Cai, W. *et al.* (2010) Graphene and graphene oxide: synthesis, properties, and applications. *Adv. Mater.*, **22**, 3906.
52. Dreyer, D.R., Park, S., Bielawski, C.W. and Ruoff, R.S. (2010) The chemistry of graphene oxide. *Chem. Soc. Rev.*, **39**, 228.
53. Loh, K.P., Bao, Q., Ang, P.K. and Yang, J. (2010) The chemistry of graphene. *J. Mater. Chem.*, **20**, 2277.
54. Pei, S. and Cheng, H.M. (2012) The reduction of graphene oxide. *Carbon*, **50**, 3210–3228.
55. Chien, C.T., Li, S.S., Lai, W.J. *et al.* (2012) Tunable photoluminescence from graphene oxide. *Angew. Chem. Int. Ed.*, **51**, 6662.
56. Markovic, Z.M., Harhaji-Trajkovic, L.M., Todorovic-Markovic, B.M. *et al.* (2011) In vitro comparison of the photothermal anticancer activity of graphene nanoparticles and carbon nanotubes. *Biomaterials*, **32**, 1121.
57. Hummers, W.S. and Offeman, R.E. (1958) Preparation of graphitic oxide. *J. Am. Chem. Soc.*, **80**, 1339.
58. Yang, K., Feng, L., Shi, X. and Liu, Z. (2013) Nanographene in biomedicine: theranostic applications. *Chem. Soc. Rev.*, **42**, 530–547.
59. Robinson, J.T., Tabakman, S.M., Liang, Y. *et al.* (2011) Ultrasmall reduced graphene oxide with high near-infrared absorbance for photothermal therapy. *J. Am. Chem. Soc.*, **133**, 6825–6831.
60. Vila, M., Portoles, M.T., Marques, P.A.A.P. *et al.* (2012) Cell uptake survey of pegylated nanographene oxide. *Nanotechnology*, **23**, 465103.
61. Yang, K., Zhang, S., Zhang, G. *et al.* (2010) Graphene in mice: ultrahigh in vivo tumor uptake and efficient photothermal therapy. *Nano Lett.*, **10**, 3318–3323.
62. Liu, Z., Robinson, J.T., Sun, X. and Dai, H. (2008) PEGylated nanographene oxide for delivery of water-insoluble cancer drugs. *J. Am. Chem. Soc.*, **130**, 10876–77.
63. Yang, X., Zhang, X., Liu, Z. *et al.* (2008) High-efficiency loading and controlled release of doxorubicin hydrochloride on graphene oxide. *J. Phys. Chem. C*, **112**, 17554–17558.
64. Zhang, L., Xia, J., Zhao, Q. *et al.* (2010) Functional graphene oxide as a nanocarrier for controlled loading and targeted delivery of mixed anticancer drugs. *Small*, **6**, 537–544.
65. Li, H., Wei, Q., He, J. *et al.* (2011) Electrochemical immunosensors for cancer biomarker with signal amplification based on ferrocene functionalized iron oxide nanoparticles. *Biosens. Bioelectron.*, **26**, 3590–3595.
66. Tang, Z., Wu, H., Cort, J.R. *et al.* (2010) Constraint of DNA on functionalized graphene improves its biostability and specificity. *Small*, **6**, 1205–1209.
67. Feng, L.Z., Zhang, S. and Liu, Z. (2011) Graphene based gene transfection. *Nanoscale*, **3**, 1252–1257.
68. Gonçalves, G., Marques, P.A.A.P., Barros-Timmons, A. *et al.* (2010) Graphene oxide modified with PMMA via ATRP as a reinforcement filler. *J. Mater. Chem.*, **20**, 9927–9934.
69. Podila, R., Moore, T., Alexis, F. and Rao, A.M. (2013) Graphene coatings for enhanced hemo-compatibility of nitinol stents. *RSC Adv.*, **3**, 1660–1665.
70. Roy, R.K. and Lee, K. (2007) Biomedical applications of diamond-like carbon coatings: A review. *J. Biomed. Mater. Res. Part B*, **83**, 72–84.

71. Williams, O.A. (2011) Nanocrystalline diamond. *Diamond Relat. Mater.*, **20**, 621–640.
72. Papo, M.J., Catledge, S.A., Vohra, Y.K. and Machado, C. (2004) Mechanical wear behavior of nanocrystalline and multilayer diamond coatings on temporomandibular joint implants. *J. Mater. Sci. Mater. Med.*, **15**, 773–777.
73. Thomas, V., Halloran, B.A., Ambalavanan, N. *et al.* (2012) In vitro studies on the effect of particle size on macrophage responses to nanodiamond wear debris. *Acta Biomater.*, **8**, 1939–1947.
74. Yang, L., Sheldon, B.W. and Webster, T.J. (2009) Orthopedic nano diamond coatings: control of surface properties and their impact on osteoblast adhesion and proliferation. *J. Biomed. Mater. Res. Part A*, **91**, 548–56.
75. Xiao, X., Wang, J., Liu, C. *et al.* (2006) In vitro and in vivo evaluation of ultrananocrystalline diamond for coating of implantable retinal microchips. *J. Biomed. Mater. Res. Part B*, **77**, 273–281.
76. Narayan, R.J., Wei, W., Jin, C. *et al.* (2006) Microstructural and biological properties of nanocrystalline diamond coatings. *Diamond Relat. Mater.*, **15**, 1935–1940.
77. Robertson, J. (2002) Diamond-like amorphous carbon. *Mater. Sci. Eng., R*, **37**, 129–181.
78. Godignon, P. (2005) SiC materials and technologies for sensors development. *Mater. Sci. Forum*, **483–485**, 1009.
79. Rosembloom, A.J., Nie, S., Ke, Y. *et al.* (2004) Porous silicon carbide as a membrane for implantable biosensors. *Mater. Sci. Forum*, **457–460**, 1463.
80. Frewin, C.L., Locke, C., Saddow, S.E., and Weeber, E.J. (2011) Annual International Conference of the IEEE Engineering in Medicine and Biology Society (EMBC), pp. 2957–2960.
81. Zorman, C., Eldridge, A., Du, J. *et al.* (2011) Silicon carbide and related materials. *Mater. Sci. Forum*, **717–720**, 537–540.
82. Yakimova, R., Petoral, R.M. Jr.,, Yazdi, G.R. *et al.* (2007) Surface functionalization and biomedical applications based on SiC. *J. Phys. D: Appl. Phys.*, **40**, 6435–6442.
83. Domnich, V., Reynaud, S., Haber, R.A. and Chhowalla, M. (2011) Boron carbide: structure, properties and stability under stress. *J. Am. Ceram. Soc.*, **94**, 3605–3628.
84. Barth, R.F., Coderre, J.A., Vicente, G.H. and Blue, T.E. (2005) Boron neutron capture therapy of cancer: current status and future prospects. *Clin. Cancer Res.*, **11**, 3987–4002.

Part III
Material Shaping

8

Cements

Oscar Castaño[1,2,3] and Josep A. Planell[1,2,4]

[1]Biomaterials for Regenerative Therapies,
Institute for Bioengineering of Catalonia (IBEC), Spain
[2]CIBER de Bioingeniería, Biomateriales y Nanomedicina (CIBER-BBN), Spain
[3]Department of Materials Science and Metallurgy,
Polytechnic University of Catalonia (UPC), Spain
[4]Open University of Catalonia (UOC), Spain

Abbreviations

AC	alternating current
ACP	amorphous calcium phosphate
ALP	alkaline phosphatase
ASTM	American Society for Testing and Materials
BET	Brunauer–Emmett–Teller theory
BJH	Barrett–Joyner–Halenda theory
BSE	back-scattered electrons
BSP	bone sialoprotein
CaP	calcium phosphate
CDHA	calcium deficient hydroxyapatite
Col Iα2	pro-collagen Iα2
CPC	calcium phosphate cement
DCPD	dicalcium phosphate dihydrate – brushite
DLS	dynamic light scattering
DNA	deoxyribonucleic acid
ECM	extracellular matrix

Bioceramics with Clinical Applications, First Edition. Edited by María Vallet-Regí.
© 2014 John Wiley & Sons, Ltd. Published 2014 by John Wiley & Sons, Ltd.

EDS or EDAX	energy-dispersive X-ray spectroscopy
EPMA	electron probe microanalysis
FDA	Food and Drug Administration
FE-SEM	field-emission scanning electron microscopy
FWHM	full width at half maximum
HA	hydroxyapatite
hUCMSCs	human umbilical cord mesenchymal stem cells
ICDD	International Centre for Diffraction Data
ICP	inductively coupled plasma
IUPAC	International Union of Pure and Applied Chemistry
JCPDS	Joint Committee for Powder Diffraction Standards
LDH	lactate dehydrogenase
LPR	liquid-to-powder ratio
MI	mercury intrusion
mRNA	messenger ribonucleic acid
MS	mass spectrometry
MSCs	mesenchymal stem cells
MWCNT	multi-walled carbon nanotubes
OC	osteocalcin
OCP	octa-calcium phosphate
OES	optic emission spectroscopy
ON	osteonectin
OP	osteopontin
PDF	powder diffraction files
ppm	parts per million or $mg\,l^{-1}$
ppt	parts per trillion or $ng\,l^{-1}$
RT-PCR	real-time polymerase chain reaction
SBF	simulated body fluid
SEM	scanning electron microscopy
TCP	tricalcium phosphate:$Ca_3(PO_4)_2$
TTCP	tetracalcium phosphate
XPS or ESCA	X-ray photoelectron spectroscopy
XRD	X-ray diffraction

Glossary

Angiogenesis	The creation of new blood vessels from pre-existing ones.
Bioactivity	Ability to trigger a specific biological response. In bone tissue engineering, a material is usually considered bioactive when it is able to selectively precipitate bone-like hydroxyapatite onto its surface after immersion in a serum-like solution. However, there is a large controversy among scientists because use of this process leads to false positive and false negative results [1].
Biodegradability	Ability to be dissolved and transformed into new compounds by hydrolysis, oxidation, enzymes, bacteria, or osteoclasts [2].

Chemotaxis	Ability to recruit a particular type of cells by chemical signals. Special attention should be paid to the promotion of the trafficking and infiltration of MSCs to the damaged site [3].
Osteoblasts	Mononuclear cells originated from the differentiation of mesenchymal progenitor cells, responsible for the formation, and mineralization of new bone by the regulation of the local calcium and phosphate concentrations to trigger hydroxyapatite mineralization, through a previously synthesized extra cellular matrix formed by collagen and several glycoproteins. The process involves an auto-mineralization of the membrane to be trapped into the new formed bone, becoming then an osteocyte [4].
Osteoclasts	Multinucleated cells responsible for the degradation and resorption of the bone through the local acidification by an enzymatic reaction.
Osteoconduction	The ability to guide bone growth as a template facilitating attachment, proliferation, migration, and osteoblast differentiation to form new bone tissue [5].
Osteogenesis	The ability to create new bone tissue.
Osteoinduction	The ideal osteoinductive material should be capable of inducing ectopic bone formation when implanted into extra-skeletal locations; in other words, it should be osteogenic in soft tissues [6].
Vasculogenesis	The creation of new blood vessels by the new formation of endothelial cells from mesoderm cell precursors [7].

8.1   Introduction

8.1.1   Brief History

Calcium phosphate cements (CPCs) were meant to produce hydroxyapatite (HA), which is the calcium phosphate that usually results when the cements are mixed with or immersed in aqueous media. The name apatite was first used to describe a mineral that sometimes appeared to be a transparent green stone, or asparagus, by the father of German geology, Abraham Gottlob Werner, in 1786 [8, 9]. At the same time, its relationship with the mineral phase of bone was first described by the French chemist Joseph-Louis Proust [10], and then by the German chemist Martin Heinrich Klaproth, who was the first to chemically analyze apatite,[1] obtaining a composition of 53.75 lime (CaO) and 46.25 phosphoric acid P_2O_5 [11]. Other sources establish that in 1769 two Swedish chemists, Johan Gottlieb Gahn and Carl Wilhelm Scheele, showed that phosphoric acid could be obtained from bone ashes [12]. Nowadays, apatites are compounds that can be described with the chemical formula $Ca_{10}(PO_4)_6X_2$, where X can be an hydroxy (OH^-), or a halogen anion, like fluoride (F^-, fluorapatite) or chloride (Cl^-, chlorapatite), or a bigger anion like carbonate (CO_3^{2-}) or sulfate (SO_4^{2-}) [13]. In fact, pure apatites – those lacking even a small amount of halogen – have not been found in nature.

---

[1] He analyzed the "variety" of asparagus-stone. Depending on the chemical content, particularly OH^-, Cl^-, F^-, or CO_3^{2-}, the macroscopic features were different.

During the nineteenth century, surgeons were using a forerunner of current CPCs when dealing with fractures or to fill cracks. It was commonly known as "plaster of Paris", which is one of the most ancient construction materials. It is a hydrolyzed calcium sulfate ($CaSO_4 \cdot 2H_2O$) that forms a paste when mixed with water, and takes about 45 minutes to set. A Romanian surgeon, Themistocles Gluck, invented the first endoprosthesis made with ivory[2] powder embedded in a matrix of plaster of Paris, which was used in the first total wrist replacement performed in Berlin in 1890 [14]. However, the results were not good, probably due to the poor selection of patients, as well as a very fast hydrolysis in the medium, before new bone could be formed [15].

Tri-calcium phosphates ($Ca_3(PO_4)_2$) henceforth TCPs, have been known since the early nineteenth century [16], but they were used as bone grafts only after the First World War. According to Dorozhkin [16], in 1832, the difference between α-TCP and β-TCP had already been established. The war promoted studies in bone regeneration to raise the osteogenetic activity of bone in non-consolidated fractures and pseudarthrosis. Albee and Morrison, in a US Army hospital, tried, among other different substances, a suspension of TCP in water in the tibia of rabbits[3] at the point of fracture [17]. Osteogenesis was triggered by TCP combined with the natural healing process, and no toxicity was observed.

In 1923 [18], Eden tried to repair a bone defect in rabbits by filling the damaged area with TCP and calcium carbonate ($CaCO_3$). Although the material was difficult to resorb, he obtained the best results with a mix of gelatin[4] and a solution of calcium. Some years later, it was suggested that there was not enough evidence showing that TCP improved osteogenesis more effectively than powdery HA [19–22]. During this decade, the idea that calcium and phosphates (and some of their derivatives) in solution had a bioactive effect in bone as well as in cartilage was generally accepted [23, 24]. It is no wonder that scientists concluded that an organically derived compound of calcium phosphate in the diet enhanced the healing of fractures of the tibia better than an inorganic salt of calcium [25]. As a result, many efforts focused on increasing the content of the local calcium deposit, which was claimed to be the most important single factor to bring about fracture healing.

In 1934, Hadelman and Moore introduced several types of calcium phosphate particles into animal bone fractures to study osteogenesis in pseudarthrosis [19]. TCP was able to create a bridge connecting the two sides of a bone fracture; in other words, they discovered the importance of a bone guiding surface and, therefore, a pioneering scaffold. At the same time, researchers were beginning to recognize the importance of the calcium-to-phosphorus ratio, which should be close to 1.67 if it is to mimic the natural HA.[5] This issue was confirmed by Ray and Ward [20] in 1951, and later by Cutright and coworkers [26].

In 1950, a pioneer in CPCs, Kingery, published experimental data showing that cold-setting bonds were formed when CaO, H_3PO_4, HA, and water were mixed. But this seemed to be an isolated case, and not much attention was paid to such findings [27]. Nevertheless, it seemed that pioneers such as Kingery were laying down the foundations for future CPCs.

---

[2] Ivory is a white material derived from the teeth or tusks of several animals. It is basically composed of dentine, which is a HA with a deficient content of calcium and some low levels of carbonates.

[3] The reader can notice that no *in vitro* studies were performed. That is something that nowadays is discarded. In other words, *in vivo* studies usually are so expensive and controlled that they are designed after suitable *in vitro* ones.

[4] Gelatin is the result of denaturing the structure in a triple helix of collagen.

[5] Briefly, the ratio Ca/P determines how similar to HA the calcium phosphate will be. Bioceramics pioneers tried to mimic natural bone, which actually is insoluble at pH 7.4. Decades later, scientists noticed that the lower and more different the Ca/P ratio is, the higher the solubility in physiological fluid.

In the beginning of the 1970s, thanks to the development of better fabrication procedures, particularly furnaces, different calcium phosphates (CaP) and conformation methods could be tested. Biocompatibility, osteoconductivity, and biodegradation concepts were put forward. It was the age of the development of stable implants, made of SiO_2, Al_2O_3, or ZrO_2 due to their excellent mechanical properties, especially for hip joint and dental prostheses. Some of them were dense to exhibit good mechanical properties, but others were porous to improve bone colonization. Disappointing results led to the renewal of strategies and the evaluation of new approaches. From then on, different research groups focused their strategies towards the concept of bioactive grafts for bone regeneration; in other words, they were beginning to look for a material that degraded at the same rate as new bone was growing. The first option was quite obvious: HA. Roy and Linneham [28], for example, published a letter in *Nature* about the fabrication of a HA material derived from coral using a hydrothermal method. During this process, aragonite[6] from coral was replaced by HA, keeping its original microstructure and interesting macroporosity. There were other studies involving β-TCP, especially with animals, but results were not much better than those achieved with HA [29–31].

By the early 1980s, Brown and Chow had uncovered some new properties of TTCP (tetracalcium phosphate); when mixed with other CaP such as DCPD (dicalcium phosphate dihydrate) and an aqueous solution, a paste was formed that hardened in several minutes. This was the first formal CPC. Other sources assert that it was LeGeros *et al.* [32] who were the first to achieve it, but as Dorozhkin [33] remarked in his review, Kingery [27] cannot be forgotten in the sharing of this honor. Even after Kingery and before Brown, Chow, and Legeros, Monma and Kanazawa had succeeded in hydrating α-TCP to obtain CDHA (calcium deficient hydroxyapatite) [34], even though the reaction took a long time and did not offer any clinical application. Subsequently, much of the research into CPCs appeared increasingly in experimental and medical journals. The mechanism of the reactions of CPCs were detailed and enhanced by the use of different additives, and many of the compositions were modified. Blends with different calcium phosphates and different aqueous media were used to enhance biodegradation, self-setting, and osteogenesis in multiple applications. The discovery of self-setting CPCs was a major step forward, as it implied a great leap towards minimally invasive surgical techniques. CPCs meant that a bioactive paste could be directly injected into a fracture [35]. The golden age of CPCs was in the late 1990s and the beginning of the 21st century, when they were presented as promising bone substitutes and drug delivery systems [36].

8.1.2   Definition and Chemistry

Generally, cement is a substance that can act as a binder, either to itself and/or to other materials, and evolves from being in a liquid or viscous state to a solid one over a short period of time; in other words, it sets. Inorganic cements require a particularly aqueous reagent to be hydrolyzed and to set.

According to Driessens *et al.* [37] "a CPC is defined as a combination of a powder or a mixture of powders with water or an aqueous solution that sets upon mixing at room or body temperature due to the formation of at least one calcium phosphate, and that retains strength

---

[6] Corals are marine animals attached to the seabed that cover themselves with an exoskeleton of one of the allotropes of $CaCO_3$, aragonite. This process involves a mineralization by the polyps from calcium and carbonate ions dissolved in water.

upon soaking in water or Ringer's solution". The setting time is adjustable depending on the requirements of the application and, therefore, can be altered by a surgeon. Usually, the surgeon should have enough time to shape the paste into the defect before it sets completely. The set cement usually shows bioactive features.

Orthophosphate compounds are different from meta- (PO_3^-) and pyrophosphates ($P_2O_7^{4-}$) because they contain PO_4^{3-} anions. As only orthophosphates have been considered in this chapter, from now on they will be referred to as phosphates. Pyro- and meta-phosphates will be discarded because they can develop an extra-osseous calcification when hydrolyzed [38].

8.1.3   Description of the Different CaP Cements

CPCs constitute a large family of calcium phosphate compositions and combinations. Plots of solubility versus pH for the most important CaPs are shown in Figure 8.1. This chapter will focus on the most studied ones.

The main component of most important CPCs is TCP, but it has several polymorphs, and some are not stable at room temperature.

β-TCP or (whitlockite) $Ca_3(PO_4)_2$, stable at room temperature [39] but a minimum energy is needed for atomic diffusion and phase achievement, so normally a 600–700 °C thermal treatment is needed.

α-TCP, $Ca_3(PO_4)_2$, stable from around 1130 °C and above (for this reason, its fabrication requires high energy and a fast quench[7] even from temperatures around 1400 °C) [40].

γ-TCP, $Ca_3(PO_4)_2$, stable only at high pressure [41]. For this reason, it will not be considered in this chapter.

Apart from these, there are several other calcium phosphates that are normally mixed with α-TCP and/or β-TCP as additives (nuclei, hardening accelerants, binders, etc.). They are classified following their Ca/P ratio [39, 42] (see Table 8.1).

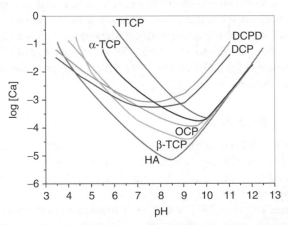

Figure 8.1   *Solubility phase diagrams of the concentration of calcium for CaP as a function of the pH (See insert for color representation of the figure)*

---

[7] Quenching is the fast cooling of a material, bypassing and avoiding low-temperature phase transformations.

Table 8.1   *Solubility product constants for some CaP at 25 and 37°C [33, 43]*

Name	Formula	Ca/P ratio	$-\text{Log } K_{sp}$ at 25 and 37°C	pH
Monocalcium phosphate monohydrate (MCPM)	$Ca(H_2PO_4)\cdot H_2O$	0.5	−1.14, very soluble	0.0–2.0
Monocalcium phosphate anhydrous (MCPA)	$Ca(H_2PO_4)$	0.5	−1.14, very soluble	0.0–2.0
Dicalcium phosphate dihydrate (DCPD), brushite	$CaHPO_4\cdot 2H_2O$	1	6.59–6.63	2.0–6.0
Dicalcium phosphate anhydrous (DCPA), monetite	$CaHPO_4$	1	6.90–7.02	2.0–6.0
Octacalcium phosphate (OCP)	$Ca_8(HPO_4)_2$ $(PO_4)_4\cdot 5H_2O$	1.33	96.6–95.9	5.5–7.0
Amorphous calcium phosphate (ACP)	–	1.2–2.2	–	5.0–12.0
Calcium-deficient hydroxyapatite (CDHA)	$Ca_9(HPO_4)$ $(PO_4)_5OH$	1.5–1.66	−85.1 (estimated)	6.5–9.5
Tricalcium phosphate α (α-TCP)	$\alpha\text{-}Ca_3(PO_4)_2$	1.5	25.5–25.5	Cannot be obtained from aqueous solution
Tricalcium phosphate β (β-TCP), whitlockite	$\beta\text{-}Ca_3(PO_4)_2$	1.5	28.9–29.5	Cannot be obtained from aqueous solution
Hydroxyapatite (HA)	$Ca_{10}(PO_4)_6(OH)_2$	1.67	58.4–58.6	9.5–12
Fluorapatite (FA or FAp)	$Ca_{10}(PO_4)_6F_2$	1.67	−120.0	7.0–12.0
Tetracalcium phosphate (TTCP)	$Ca_4(PO_4)_2O$	2.0	38–44 – 42.4	Cannot be obtained from aqueous solution

In addition, combinations or substitutions can be performed. Examples include several alkali metals such as Na^+ or K^+, or alkaline-earth metals such as Mg^{2+}, Sr^{2+}, or Ba^{2+}, or anions such as F^-, Cl^-, I^-, PO_4^{3-}, HPO_4^{2-}, $H_2PO_4^-$, SO_4^{2-}, HSO_4^-, CO_3^{2-}, HCO_3^-, or SiO_4^{2-}.

8.1.4   State of the Art

8.1.4.1   CaP Cements Fabrication Methods

There are several ways to fabricate CPCs. The most common method was for a long time the conventional ceramic process, which involved the mixing of different cement precursors (generally salts containing Ca, P, and the other different doping agents) and increasing the temperature to achieve two goals: enough energy for atom diffusion, and to reach the necessary thermodynamic stability of the phosphate crystal's desired structure. The way to achieve a powder was usually by milling using a planetary miller and sieving until the powder reached the desired particle size. Since, the size of the particles tended to be greater than 5–20 μm, the method could incorporate undesired wear particles from the milling recipient [35].

In the last decade, the trend has shifted towards nanotechnology to obtain nanometric particles in order to have a better specific surface area and, therefore, a better control of the kinetics of the reaction. Methods used to nanostructure CPCs described in the literature include the sol–gel method to prepare nanometric CaP particles, or flame spray synthesis, among others.

The sol–gel method basically consists of the gelation of a solution of several Ca and P organic salts from a controlled hydrolysis and condensation process, or its precipitation in basic media [44]. When this gel is mechanically stable, it is subjected to a thermal treatment to eliminate organics and to obtain the proper CaP phase. As this method derives from an ionic solution, the product obtained should exhibit the same range of size and distribution. However, temperature and thermodynamics also have an effect, and it is almost impossible to achieve particles under a few nanometers.

Flame spray synthesis is a technique that allows the fabrication of nanoparticles through the injection of an aerosol of a solution of different calcium and phosphate salt precursors, similar to the sol–gel ones, in a high temperature methane/oxygen flame [45, 46]. With this method, 20–50 nm size nanoparticles of amorphous calcium phosphates (ACPs) are obtained with a usual Ca/P rate of around 1.5.

8.1.4.2   Commercial CaP Cements

Even though CPCs were discovered in the early 1980s, they were not commercialized until a decade[8] later, initially for the treatment of maxillofacial damage [47] and later for skeletal fracture repair [48]. During the last 15 years, many CPCs have been commercially adapted for different applications, depending on several factors [35, 49] (see Table 8.2).

8.1.4.3   Recent Advances

The last decade involved a change of philosophy in the concept of cements. New strategies are taking advantage of the inherent properties of CPCs, trying to integrate them into more complex 3D architectures, such as scaffolds. Techniques such as rapid prototyping [51] or the combining of pre-set pieces of CPC [52] have been described.

Another strategy is the development of premixed CPCs in order to provide a single syringe to surgeons [53, 54]. Usually CPCs are powder mixed with a biocompatible organic fluid. The result is a paste that is hydrolyzed when in contact with the aqueous physiological fluid at the moment of the extrusion in a bone defect.

Finally, recent efforts have been made to adapt CPCs as delivery agents for drugs, ions, and bioactive molecules [36, 55, 56], acting as carriers.

However, these new formulations tend to increase the number of extra added components, especially polymers, in order to enhance and adapt the mechanical properties of the CPCs

---

[8] That this process is so long might surprise the reader, but in fact it is short-term compared with current average time ranges, which can easily reach 15 years.

Table 8.2 List of several commercial CPCs [33, 50]

Solid composition	Aqueous composition	End-product	Cement name	Company
ACP (50%) DCP (50%)	Buffered saline solution	CDHA	Biobone	Merck GmbH (winsix)
TTCP (73%) DCP (27%)	Saline solution	CDHA	BoneSource	Stryker
TTCP DCPD Trisodium citrate	H_2O Polyvinyl-pyrrolidone (PVP) Na-ortho-phosphate	CDHA	HydroSet®	
TTCP (49%) α-TCP (38%) Na glycero-phosphate	Solution of $Ca(OH)_2$ and H_3PO_4	CDHA	Cementek®	Teknimed LC
TTCP (49%) α-TCP (38%) Na-glycero-phosphate dimethylsiloxane	Solution of $Ca(OH)_2$ and H_3PO_4	CDHA	Cementek® LV	
α-TCP, CaCO3, MCPM (probably similar to old Callos™)	Aqueous sodium silicate	CDHA	SKaffold™	Skeletal Kinetics (US)
The same as SKaffold™ but with porogen to create macroporosity	Aqueous sodium silicate	CDHA	SKaffold ReNu™	
Probably similar to SKaffold™	–	CDHA	CAAP	
α-TCP (77%) $Mg_3(PO_4)_2$ (14%) $MgHPO_4$ (4.8%) $SrCO_3$ (3.6%)	Solution of $(NH_4)_2HPO_4$ 3.5 M	CDHA	KyphOs[Xii]	Medtronic Spine LLC
TTCP DCP	H_2O	CDHA	Rebone	Shanghai Rebone Biomaterials Co., Ltd (CHINA)
CaP (unknown composition)	H_2O	CDHA	Cem-Ostetic™	Berkeley Advanced Biomaterials (US)
α-TCP β-TCP ACP BCP HA	Phosphate buffered solution (PBS)	CDHA	MCPC	Biomatlante (FR)
α-TCP (61%) DCPA (26%) $CaCO_3$ (10%) HA (3%)	H_2O Na_2HPO_4	CDHA	Calcibon®	Biomet

(continued)

Table 8.2   *Continued*

Solid composition	Aqueous composition	End-product	Cement name	Company
Not found	Not found	CDHA	β-BSM®	ETEX
Not found	Not found		γ-BSM®	
Synthetic CaP, sodium carboxymethyl cellulose, $NaHCO_3$ Na_2CO_3	Phosphate buffered solution (PBS)		Carrigen	
CaP Hydroxypropyl methylyl cellulose (HPMC)	Sodium phosphate solution	CDHA	GRAFTYS® HBS	Graftys
CaP Organic phase (polysaccharide)			GRAFTYS® QUICK-SET	
DCPD (55%) TCP (45%)	Not available	Brushite	JectOS®	Kasios
Probably DCPD TCP ZrO_2 (Not available)			JectOS®+	
β-TCP (1.34 g) in solution 1	Solution 1: $Na_2H_2P_2O_7$ (0.025 g), H_2O Salts (0.05 M PBS, pH 7.4); Solution 2: MCPM (0.78 g), $CaSO_4 \cdot 2H_2O$ (0.39 g), H_2O, H_3PO_4 (0.05 M)	Brushite	VitalOS Cement®	Produites Dentaires SA
α-TCP (85%) $CaCO_3$ (12%) MCPM (3%)	H_2O Na_2HPO_4	CDHA	Norian SRS®	Synthes
No data	No data		Norian Drillable®	
β-TCP (73%) MCPM (21%) $MgHPO_4 \cdot 3H_2O$ (5%) $MgSO_4$ (< 1%) $Na_2H_2P_2O_7$ (< 1%)	H_2O Na hyaluronate (0.5%)	Brushite	chronOS Inject®	

Table 8.2 *Continued*

Solid composition	Aqueous composition	End-product	Cement name	Company
Similar to Noria SRS but with added reinforcing Bioresorbable fibers	Sodium hyaluronate based solution	CDHA	Norian Rein-forced Fast Set Putty®	
TCP TTCP	Sodium and calcium salts in water solution	CDHA	G-bone cement	Surgiwear Ltd
α-TCP CaCO$_3$ MCPM	Sodium silicate solution	CDHA	Callos®	Acumed

to cope with the new demands that tissue regeneration requires. Step by step, CPCs are becoming less and less like cement in order to be integrated into organic–inorganic hybrid biomaterials, whose features should still be evaluated.

8.1.4.4 View Point of the Clinicians

Feedback from clinicians is crucial, as there is an intrinsic knowledge shift between bioma-terials scientists and medical doctors. Usually, clinicians are more pragmatic, in the sense that they are looking for something efficient, cheap, and easy to handle and with minimum operator influence, which also offers good preliminary visual diagnostics and excellent clinical results. As a result, they will always compare CPCs with their gold standard and more used therapy: autologous bone.

To date, cements can compete in the handling aspect. In fact, several approaches have already been commercialized which minimize the influence of the operator in a surgical environment; these include Norian® Rotary Mixer or Skeletal Kinetics Flow Rotary Mixer, or premixed cements, which have been mentioned in Table 8.2, or VitalOS, a syringe in which two injectable pastes, the acidic calcium phosphate and the basic one, are mixed together in a mixer contained within the syringe at the time of implantation.

Resorption is also faster, especially in the new generation of brushite cements that avoid the inherent conversion to CDHA thanks to the addition of several additives such as citrates. The reason for this is that brushite resorption is faster than CDHA [57] and is usually preferred.

However, CPCs are brittle, with reasonable compressive strength but low tensile resis-tance, both dropping drastically with the essential requirement of porosity. In addition, they still have problems of wash-out integrity. Finally, CPCs are regulated under the 510(k) pre-market notification method of the FDA (Food and Drug Administration); in other words, they should demonstrate that as a new material they are substantially equivalent to another legally US marketed device and, therefore, an already approved one [58, 59]. Consequently, the low efficiency–cost ratio is one of the main drawbacks for commercialization.

All things considered, this means that clinicians are still skeptical about the advantages of CPCs, which are considered efficient biomaterials but they still lag behind autologous bone graft, which still remains the best choice for a large range of applications [60].

8.2   Calcium Phosphate Cements

8.2.1   Types

Generally, it is possible to differentiate between two types of cements: depending on the pH, they are transformed either into CDHA (pH > 6.5) or into brushite (DCPD pH < 6). There are other resulting compounds like OCP (octacalcium phosphate) or ACP, but they normally result in a transformation into CDHA [33] (see Table 8.3).

The CDHA obtained is generally poorly crystalline and nanosized, very similar to biological apatite. This low crystallinity is responsible for its higher solubility compared with HA, and therefore its exceptional resorption ability [61].

8.2.2   Mechanisms

The different reactions that take part in the cement self-setting process depend on many experimental factors – the composition of the cement, the stability of the different components, pH, liquid-to-powder ratio (LPR), and temperature, among others. The first step is lead by the fast solubility of the surface of the cement particles which, in the absence of stirring, creates a supersaturated region near the interface of the cement particle, and this induces the precipitation of CDHA or DCPD on the surface [62–65]. Intermediate compounds such as OCP or ACP can also be formed during this process. Usually, DCPD is then converted into CDHA, but this is normally undesirable in brushite cements due to the lower resorption rate of CDHA [66], as was commented before.

Precipitation of the CDHA or DCPD occurs through the formation of an entanglement of needles or platelets, which are responsible for the increase in the mechanical resistance of the resulting hardened cement [67].

Table 8.3   *Common CaP and reaction types for CPCs [33]*

Reaction scheme	Reaction type
β-**TCP** (or α-TCP or CDHA or HA) + **MCPM** (or H_3PO_4) + H_2O → **DCPD**	Acid–base
TTCP + DCPA → CDHA	Acid–base
ACP + H_2O → CDHA + nH_2O	Hydrolysis
β-**TCP** (or α-TCP) + H_2O → **CDHA**	Hydrolysis
TTCP + H_2O → CDHA + $Ca(OH)_2$	Hydrolysis

8.2.3   Relevant Experimental Variables

8.2.3.1   Composition of the Powder Phase

Usually, cement hardening is based on a classical acid–base reaction where one basic or near-basic calcium phosphate reacts with an acidic (or slightly acidic) calcium phosphate (see Table 8.3). There are different alternatives, such as including one of the calcium phosphates or a simpler phosphate dissolved in water. Then, many combinations can be performed.

There are also reactions based on the metastability of the CPC, such as α-TCP or ACP, which easily reacts in suitable conditions [68].

It is crucial that once the supersaturated solution has formed near the interface, there should be a nucleating surface that can act as a nucleation agent. In CDHA (Ca/P ~ 1.5) precipitation, for example, HA (Ca/P = 1.69) nuclei are previously formed to allow CDHA epitaxial growth [69]. This is the reason why some cements include nanoprecipitated HA in their formulations [68].

8.2.3.2   Liquid to Powder Ratio (LPR)

The LPR is calculated using Equation 8.1:

$$\frac{L}{P} = \frac{V}{W} \tag{8.1}$$

where V is the volume of the aqueous phase (usually deionized water) in milliliters, and W is the weight of the powder phase in grams. With this relation the level of fluidity of the cement is established. The setting time is modified because usually the more liquid the material, the longer the setting time. High L/P ratios dilute released ions and more time is needed to achieve the proper saturated state and, therefore, the desired hardness. However, smaller L/P ratios can promote a fast setting but also can inhibit the reaction process. In other words, CDHA or DCPD formed in the early stages can hinder the contact between the liquid phase and the rest of the cement, resulting in a slower diffusion controlled rate [70].

8.2.3.3   Composition of the Liquid Phase

Sometimes, the amount of liquid needed to fully react the cement powder is so high that the setting time becomes too long. There are several strategies to decrease the setting time, such as including additives in the liquid phase like acidic phosphates (H_3PO_4, NaH_2PO_4, or Na_2HPO_4) or sulfates (H_2SO_4 or $NaHSO_4$), which decrease the pH, increasing the solubility of the initial powder phase [71, 72].

There are other polymeric additives that can improve other aspects of the final cement microstructure, such as its porosity, slurry viscosity, setting time, mechanical features, resorption rate, and cell response. Examples of these additives are gelatin, hyaluronic acid, chitosan, alginate, collagen, and so on [73].

Citric acid, sodium citrate, chondroitin sulfate, or glycerol are common additives which can act as retardants in the solution, especially in brushite cements [74–77], in which the goal is to dissolve the brushite before CDHA formation (see Table 8.4).

Table 8.4   *Common additives used in CPCs*

Additive	Effect	References
Albumen	Increase porosity Decrease in compressive strength and Young's Modulus	[78, 79]
Alpha-hydroxy acids and salts	Reduction of water demand	[80]
Citric acid	Reduction of mixing time Modulation of setting time at low concentrations Increase mechanical resistance	[81, 82]
Cocarboxylase Arginine	Decrease CDHA crystal size Increase cell adhesion and proliferation	[83]
Collagen	Increase setting time Increase of adhesion and proliferation	[84, 85]
Gelatin	Decrease porosity Improve mechanical compression resistance Increase conversion rate to CDHA Increase adhesion and differentiation	[86, 87]
Glycerol	Increase setting time Increase injectability Decrease compression resistance	[54]
Mannitol	Increase porosity Decrease workability Decrease mechanical properties	[88]
$NaHCO_3$	Increase porosity. Pores $\sim 100\,\mu m$	[89]
PEO surfactants and PEO-sorbitan esters	Increase porosity Decrease in compressive strength and Young's Modulus	[90]
Poly/DL-lactic-co-glycolic acid) microparticles	Increase porosity	[91]
Sodium alginate	Decrease porosity Improve mechanical compression and tensile resistances Increase conversion rate to CDHA Increase adhesion and differentiation	[92]
Sodium chondroitin sulfate	Improve workability Increases setting time	[93]
Sodium dodecylsulfate (SDS) Hexadecyl-trimethyl-ammonium chloride (CTAC)	Increase porosity Decrease of compressive strength and Young Modulus Smaller crystals	[81, 94]
Sucrose fatty acid esters	Increase porosity Pores > 100 µm Decrease in compressive strength and Young's Modulus	[95]

8.2.3.4   Mixing Conditions

The mixing process is crucial to obtain reproducible results. It should be pointed out, that a surgeon should be able to mix the powder and liquid in the surgery in sterile conditions. Current commercial cements are sold at present as kits, in which tools are supplied already sterilized. Previously, surgeons had to mix the powder and liquid in a glass mortar; then, depending on the application, they had to fill the bone defect with a spatula or, alternatively, fill a syringe and inject the cement slurry into the proper location. In recent years, automated rotary mixers have proliferated, producing a reliable and reproducible cement paste with the proper viscosity [96]. Several brands offer the possibility to mix the powder and liquid directly in the syringe with a spatula, a method only applicable to low viscosity cements [93]. A typical scheme of the typical time points for Norian® cement can be observed in Figure 8.2.

8.2.3.5   Temperature of Synthesis

α-TCP, β-TCP, and TTCP are normally obtained by a thermal treatment of different calcium phosphates mixed in proper ratios. The temperature enables the formation of the desired phase by atomic diffusion, and provides also the thermodynamic conditions needed to obtain the material.

The temperature of the synthesis of the different calcium phosphates has a particular influence on particle size. Generally, the higher the temperature, the larger the particle. This growth is essentially promoted by atomic diffusion within a crystal network, which

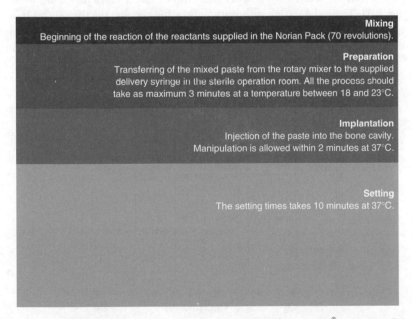

Mixing
Beginning of the reaction of the reactants supplied in the Norian Pack (70 revolutions).

Preparation
Transferring of the mixed paste from the rotary mixer to the supplied delivery syringe in the sterile operation room. All the process should take as maximum 3 minutes at a temperature between 18 and 23°C.

Implantation
Injection of the paste into the bone cavity.
Manipulation is allowed within 2 minutes at 37°C.

Setting
The setting times takes 10 minutes at 37°C.

Figure 8.2 *Typical scheme of the typical time points for Norian® cement. Data from http://www.synthes.com/MediaBin/International%20DATA/036.000.600.pdf*

occurs by either interstitial (Frenkel defects) or substitutional (vacancy-Schottky defects) mechanisms. The existence of point vacancies increases following an Arrhenius equation; in other words, the movement of atoms in the solid state increases with temperature. In fact, the diffusion coefficient D in Fick's laws obeys an Arrhenius law (see Equation 8.2) [97, 98]:

$$D = D_0 e^{-\frac{Q_d}{RT}} \qquad (8.2)$$

where D_0 is the pre-exponential factor or frequency factor independent of temperature, Q_d is the activation enthalpy required to initiate diffusion, T is the absolute temperature, and R is the universal gas constant.[9] This means basically that atomic diffusion depends exponentially on the temperature.

Temperature also influences the thermodynamically stable phase in the synthesis of a calcium phosphate. One example is the phase diagram of TCP (Figure 8.3), where it can be seen that a critical temperature ($\sim 1130\,^\circ$C) separates both polymorphs: α-TCP and β-TCP. Meanwhile, β-TCP is stable after reaching a temperature of around $700\,^\circ$C to form the crystalline phase [99], while α-TCP should be quenched from $\sim 1300\,^\circ$C after several hours.

8.2.3.6   Particle Size

Particles have a surface interface which makes contact with the aqueous liquid phase. The larger this surface, the more powder there is in contact with the liquid. The entire surface per mass, including internal particle porosity, is considered as specific surface area, and it is usually measured in $m^2\,g^{-1}$ [100]. A powder formed by small particles will involve a larger specific surface area than a powder made up of larger particles of the same weight (see Figure 8.4). Then, the setting reactivity will be higher, and the time to set reduced.

The bigger the particle, the more difficult it is for the aqueous phase to react with the bulk part of the particle. In this case, diffusion is the limiting step of the process, and usually

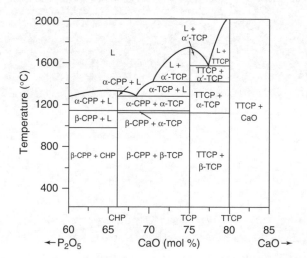

Figure 8.3   *Phase diagram of the system CaO–P_2O_5 [40]*

---

[9] $R = N_A k_B = 8.3145 \times 10^{-3}\ \text{kJ mol}^{-1}\ \text{K}^{-1}$ where N_A is the Avogadro number and k_B the Boltzmann constant.

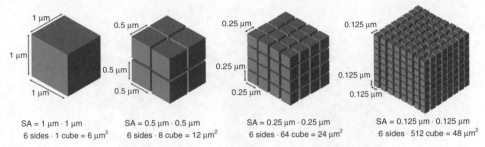

Figure 8.4 *Scheme showing the relationship between the particle size and specific surface area [101]*

provokes a slower final reaction and setting. But smaller particles also offer a higher number of nucleation sites, so as particle size influences reaction kinetics, the morphology of the resulting calcium phosphate is also particle size-dependent [102]. The higher degree of supersaturation attained in a solution with smaller particles results in the nucleation of much smaller crystals [103].

8.2.3.7 Heterogeneous Nucleation

The nucleation of the precipitates from the supersaturated area close to the cement interface is one of the main aspects that should be controlled in order to modulate the macroscale properties of the cement. The density of nuclei by unit of area of the interfacial substrate (I^h) can be expressed with an Arrhenius formula (see Equation 8.3) [104]:

$$I^h = K^h e^{-\frac{\Delta G_k^h}{RT}} \tag{8.3}$$

where K^h is the pre-exponential factor irrespective of the interface, ΔG_k^h is the activation energy for a heterogeneous nucleation, R is the universal gas constant, and T is the absolute temperature of the system, which for CPCs is usually 310.15 K.

Among other factors, ΔG_k^h depends on the angle θ between the crystal nuclei and the cement interface (see Equation 8.4):

$$\Delta G_k^h = \Delta G_k \frac{(2 + \cos\theta)(1 - \cos\theta)^2}{4} \tag{8.4}$$

where ΔG_k is the activation energy for a homogeneous nucleation. When $\theta \sim 0$, then $\Delta G_k^h \sim 0$, and acceleration of nucleation occurs. When θ is very high, nucleation is difficult to observe, so the best strategy to increase nucleation in CDHA cements is to add nanosized HA seeds into the mix, which act as crystallization nuclei [105].

8.2.4 Material Characterization

8.2.4.1 Setting and Cohesion

The setting time of CPCs begins at the moment the powder is mixed with the aqueous liquid phase and ends when the reaction is finished. This process should be slow enough to offer time for the surgeon to mold the paste and insert it in the damaged bone, while being fast

Figure 8.5 *Gillmore needles*

enough to provide the mechanical consistency to allow the cement to maintain the shape determined by the surgeon.

There are several methods for measuring the setting time of the cements. Two are standardized, such as the Vicat needle method (ASTM C191) [106] and the Gillmore needle method (ASTM C266-08e1, Figure 8.5) [107]. Both are based on the penetration of needles with a standardized weight to check if they leave a mark on the cement surface.

There are other non-destructive methods to assess the rate of progress of the reaction, such as AC (alternating current) impedance spectroscopy [108], pulse-echo ultrasound [109], and isothermal differential scanning calorimetry [110]. However, none of these are standardized as yet.

Usually, the average surgeon needs from 3 to 8 minutes to introduce the mix into the damage, depending on the application and its location [33] (see Figure 8.6). So, using the Gillmore needle method, two setting times can be defined based on a light and a heavier needle. The light needle defines the initial setting time (T_i), while the heavy one defines the final setting time (T_f), which determines the point after which no more modifications can be made in the set paste without causing cracking [111].

The cohesion time is defined as the minimum time needed to obtain a consistent cement paste in Ringer's solution[10] [112] at 37 °C. Normally this is evaluated by immersing a previously shaped cement paste in Ringer's solution at different time points [113]. In general, cohesion time should be more than 2 minutes but less than the initial setting time. Another

---

[10] One liter of an aqueous solution with 120 mM of Ca^{2+}, 109 mM of Cl^-, 4 mM of K^+, and 3 mM of Ca^{2+}.

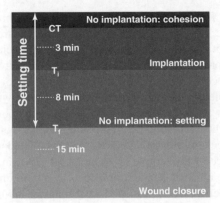

Figure 8.6 *Scheme of the relevant setting parameters for a typical CPC implantation, where CT is the cohesion time, T_i the initial setting time, and T_f the final setting time. Adapted with permission from [111]*

method was proposed by Bohner *et al*. [114], which was based on the immersion of a paste in solution at 37 °C and recording its weight in continuous mode.

Good cohesion avoids the release of cement particles into the medium, which can result in uncontrolled inflammation or a fatal thrombosis.

8.2.4.2 Chemical Composition and Microstructure

There are several techniques to assess chemical composition. Some of them are described here:

- **Inductively coupled plasma (ICP):** A destructive technique that can measure atomic composition with a detection limit of parts per million (ppm or $mg\,l^{-1}$) or parts per trillion (ppt or $ng\,l^{-1}$) depending on the sensor (OES: optical emission spectroscopy or MS: mass spectrometry). It is based on the ionization of the sample solution by a plume of an ICP of argon, and measures the intensity of the resulting electromagnetic radiation of the excited atoms by OES [115], or the atomic weight of the resultant fragments by MS [116], both with the help of previously prepared patterns. It can also be used to measure the concentration of the different elements released from the material to quantify its degradation.
- **Energy-dispersive X-ray spectroscopy (EDS or EDAX):** A non-destructive technique based on X-ray fluorescence, similar to techniques like X-ray photoelectron spectroscopy (XPS or ESCA), Auger electron spectroscopy (AES) or electron probe microanalysis (EPMA). It is based on bombarding the sample surface atoms with a high energy X-ray or high-energy beam of electrons. Valence electrons are excited and jump to higher energy levels, after which other external electrons fill the vacancy, emitting a secondary X-ray radiation that is collected by a detector [117]. Comparing the energy collected with a pattern related to atomic weight, a quantitative spectrum can be obtained. The accuracy of the results depends highly on the homogeneity, the porosity, and the roughness of the samples, owing to the loss of scatter radiation from the detector. Normally, these kind of devices are coupled inside a scanning electron microscope (SEM) or field-emission scanning electron microscope (FE-SEM).

- **SEM and FE-SEM:** Techniques that can produce images through electron bombardment on a conductive sample. Depending on how the electrons are reflected on the sample or emitted from it and how the detector collects these electrons, different images related to the surface properties are obtained. Secondary electron detectors are the most common detectors in conventional SEMs. The interactions of the electron beam with the near surface sample are monitored, producing an image with a large depth of field and displaying details at the nanometer scale. The images obtained offer information about the surface and the microstructure of the sample. On the other hand, back-scattered electrons (BSEs) images offer maps and cross sections related to the atomic number (Z) included on the surface of the sample up to about 50–200 μm in depth. Finally, a FE-SEM is a SEM whose electron beam is produced by a field emission gun [118] and is focused, with a smaller diameter than the usual tungsten filaments in usual SEMs as well as a higher intensity. This implies a higher magnification of about one order of magnitude, decreasing the noise and enhancing the resolution.

  In SEM and FE-SEM, the electron beam is so intense that the sample can be polarized and decomposed due to the high temperature that can be reached. Calcium phosphates usually have low conductivity, so electronic dissipation is usually poor. As a result, samples are coated with a sputtered conductive layer, such as gold or platinum, or graphite for EDS analysis.[11] Several microscopes make it possible to have a low or altered vacuum, or low voltages and smaller distances, but cannot reach such high magnifications, so the coating option is preferred.

8.2.4.3 Crystallinity and Homogeneity

Crystallinity is the level of order of the atoms or the molecules in a material. When this order is high, the compound is known as a crystalline material. When the level of order is 0, the compound is usually known as an amorphous or glassy material. When the level of order is low, the material is known as a semi-crystalline or, in the case of inorganic compounds, vitreous-ceramic material.

There are several ways to measure the crystallinity of a material. The most used one is X-ray diffraction (XRD), which uses a high penetrating X-ray beam [119]. The register of intensity and the angle of the diffracted beams can be related to the level of crystallinity, the crystal and glass ratio, the atomic and molecular distribution, the chemical and molecular composition, and the crystal size, if less than about 200 nm.

A diffractometer is a device that includes a goniometer, a source of X-rays and a detector. The most common is the Bragg–Brentano diffractometer, which is schematized in Figure 8.7. It only collects the diffracted beams that follow the Bragg's Law (see Equation 8.5):

$$2d \sin \theta = n\lambda \tag{8.5}$$

where d is the distance between planes with the same orientation, θ is the angle between the X-ray source and the detector, and the surface of the sample, n is an integer and λ is the wavelength of the radiation (the most typical radiation is from a copper source which is 1.5406 Å). The results usually are spectra collecting the distribution curves of the different

---

[11] Gold and platinum mask the signal of phosphorus in the EDS spectra.

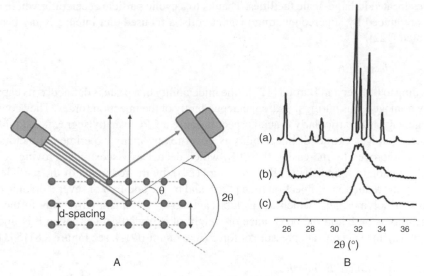

Figure 8.7 *(A) Scheme showing the interaction of the X-rays with a crystal. (B) Typical X-ray spectra of (a) bone mineral phase, (b) CDHA resulting from a hardened CPC, and (c) crystalline HA [120]*

crystalline planes included in the material. Usually, the narrower and taller these peaks, the higher the crystallinity (see Figure 8.7).

Eventually, the thickness of the peaks is also related to the size of the crystal domains or crystallites[12] through the Scherrer equation (see Equation 8.6), if the size is smaller than 100–200 nm [121],

$$\tau = \frac{K\lambda}{\beta \cos \theta} \tag{8.6}$$

where τ is the average crystallite size, K is the shape factor which is dimensionless and varies between 0.9 and 1 depending on the roundness of the crystal, and β is the full width at half maximum (FWHM) in radians of the peaks after subtracting the instrumental thickness component.[13]

The collected peaks can be compared with an XRD crystallographic database (such as the powder diffraction files (PDFs) from the International Centre for Diffraction Data (ICDD), the database of the Joint Committee for Powder Diffraction Standards (JCPDS), or others) to assess the different phases presented in the cement, or in the resultant CDHA or brushite.

Eventually, measurements of different areas of the cement can be carried out to check the homogeneity. These measurements can be combined with BSE maps.

In some exceptional cases, a high intensity X-ray beam with a very low wavelength is required. In this case, more accurate studies can be performed in synchrotron radiation

---

[12] Do not confuse crystal domain or crystallite with particle. A particle can involve only one crystallite or, in contrast, an agglomeration of crystallites, which is more common.

[13] To calculate $\beta_{instrument}$ an X-ray diffraction measurement of a pattern is required, such as the crystalline SiO_2 that is supplied with some equipment. However, for a better approach, it is recommended that a spectrum of lanthanum hexaboride (LaB_6) is measured .

sources located in large-scale facilities. Thanks to a cyclic particle accelerator where electrons are guided by superconducting magnet coils, a focused and intense X-ray beam is generated [122].

8.2.4.4 Injectability

According to Bohner and Baroud [123], "the injectability of a paste is defined as its capacity to stay homogeneous during injection, independently of the injection force". They consider homogeneity as a particularly crucial aspect because a CPC usually segregates liquid and powder when it is extruded. Injectability is governed by many experimental factors and it can be improved by increasing the LPR, which decreases viscosity, improving cement particle friction (a round shape allows rolling [124]), by the reduction of the particle size [123], by the reduction of injection time [125], and by the addition of several organic components or polymer gels, which act as lubricants [54].[14] A quantitative way to measure injectability is based on the percentage of weight of extruded cement when it is injected through a syringe at constant pressure or force (maximum 100 N, see Figure 8.8) [54, 125].

8.2.4.5 Mechanical Properties

Due to their brittleness and high porosity and, therefore, low mechanical resistance, CPCs can only be used in non-loading applications, because other applications usually involve high requirements of bending, torsion, tension, and compression. Their tensile resistance is very low, reaching a maximum of around 1–10 MPa [33], so compression tests are normally performed. Generally, a uniaxial mechanical testing machine is used, with two articulated compression plates at each end of the sample. Samples are usually cylindrical with a diameter of 6 mm, and their length varies between 3 and 12 mm. A minimum of 10 measurements is required due to the relative high dispersion found in the results, which is one of the main drawbacks of CPCs in clinical application [33].

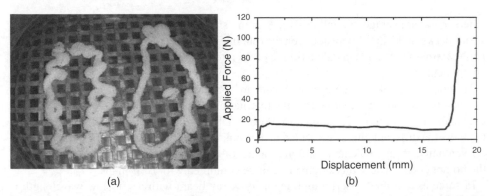

(a)   (b)

Figure 8.8   (a) Image of a cohesion test in Ringer's solution at 37°C. (b) Typical injectability test measuring the displacement of a syringe piston versus the applied force. Notice the dramatic increase of the resistance when compaction is taking place.

---

[14] The reader may note that the diameter of the needle is also a crucial aspect of injectability. However, minimum invasive surgery requires needles with diameters as small as possible (or the largest gauge possible).

The mechanical properties (particularly compressive strength) of a cement depend strongly on the porosity (see Equation 8.7) among other factors, such as the CDHA or brushite crystal size and morphology, the efficiency of the reaction, the shape of the particles, the different additives or defects included (unreacted particles that act as wedges, or the possibility of cracks formed by shrinkage, etc.).

$$S = S_0 \cdot e^{-bP} \tag{8.7}$$

where S is the strength, P the porosity, S_0 the strength for a dense sample, and b is an empiric constant [126].

The compression strength for most CDHA CPCs rarely exceeds 100 MPa, which is included in the trabecular bone range, whose strength is within the range of 1.5 and 45 MPa [127]. On the other hand, brushite cements do not reach strengths beyond 25 MPa [127]. Applications of CPCs usually involve direct contact with trabecular (spongy) rather than cortical (compact) bone.

8.2.4.6   Particle Size

The size of the particles directly influences the kinetics of the reaction and, as a result, all the other parameters and properties. Smaller particle sizes imply a higher specific surface area and thus more intensive contact with the fluid. As the diameter of the particles is smaller, the diffusion aspects are also minimized, giving precedence to the solubility constant of the cement in the liquid phase. The morphology of the resulting crystals is also affected, due to the different methods available to feed the resulting CDHA or brushite crystals. Consequently, it is normal to have needles or plates, depending on the initial particle size of the cement [103]. The setting time is usually lower when the particle size is decreased, and the mechanical compression strength resistance is usually increased because, among other factors, irregular macroporosity is decreased with a smaller particle size, favoring homogeneous micro- and nanoporosity [128].

Sintering and milling processes usually result in particle sizes between 1 and 100 μm. Other methods to nanostructure cement particles can be found in the literature, such as flame synthesis [46], the co-precipitation method [129, 130], or thermal decomposition [131].

One of the most usual way for measuring particle size is by assessing the size distribution by means of the laser diffraction technique (see Figure 8.9). Through several detectors located at different angles, scattered light can be collected and associated to different particle sizes [103] when a laser is applied to an ultrasonicated dispersion of the particles in ethanol.

Another similar method is dynamic light scattering (DLS), which differs in that the sensor is fixed, and the approach is focused on the Brownian movement of the particle in suspension. Smaller particles usually have smaller changes in their trajectory. The relation between the particle size and the particle motion is given by the Stokes–Einstein equation (see Equation 8.8):

$$D_h = \frac{k_B T}{3\pi \eta D_t} \tag{8.8}$$

where D_h is the particle diameter (also known as the hydrodynamic diameter), k_B is the Boltzmann constant, T the absolute temperature, η the dynamic viscosity (normally

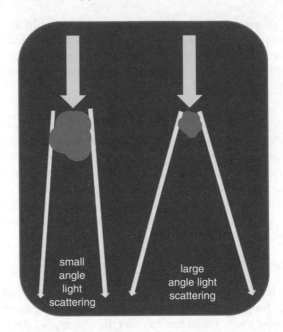

Figure 8.9 *Light scattering for small and large particles*

absolute ethanol $\eta_{20\,°C} = 1.2 \times 10^{-2}$ poise), and D_t is the translational diffusion coefficient, which is the result of the DLS analysis [132, 133].

An alternative method uses randomly collected SEM images. They are treated using an image processor focused on scientific and engineering applications such as ImageJ open-source image analysis software [134]. However, in this case, treatment of at least 25–100 images is required to obtain proper statistical reliability [135].

8.2.4.7   Surface, Porosity Analysis, and Density

As mentioned previously, surface assessment is crucial in order to understand the level of reactivity of cement and the kinetic behavior of a reaction. Porosity is closely related to the surface, because it can increase or decrease in size depending on the surface within the pore. Generally, for a constant amount of porosity percentage, a high amount of small pores will offer a larger specific surface area than a low amount of larger ones. However, as a condition for the proper bioresorption of implanted cements, it is well known that biomaterials for tissue regeneration should contain a certain level of interconnected macroporosity for efficient cell migration and proliferation. This macroporosity should offer diameters larger than 100 μm, big enough to allow a relatively free cell migration and favoring bone ingrowth and guiding [136–138]. On the other hand, micro- and nanoporosity are important to offer good permeability to fluids and allow efficient vascularization, which allows correct cell signaling, nutrient and oxygen transportation, and the elimination of residues. The cement should be colonized by cells and blood vessels, and the distance between them should not be higher than 100–200 μm [139].

The usual techniques to measure the specific surface and porosity consist in using poro-simeters based on the Brunauer–Emmett–Teller (BET) and the Barrett–Joyner–Halenda methods (BJH), which are useful approaches to understand how a gas is isothermically adsorbed and de-adsorbed [140], or porosimeters based on metallic mercury intrusion (MI) [140].

BET/BJH porosimetry allows the evaluation of not only the specific surface area, but also the pore size distribution. Some studies propose that through a specific surface, particle size can also be approached following Equation 8.9 if perfect, spherical particles or cubic particles, with a monodispersed size distribution, are considered.

$$\text{SSA} = \frac{B}{d \cdot \rho} \tag{8.9}$$

where SSA is the specific surface area, d is the particle diameter, ρ is the skeleton density of the cement, and B is a shape factor, which has a value of 6 for perfect spherical or cubic particles. For other shapes, B has other values [101].

For non-porous particles, the density ρ can be directly defined as the classical equation (see Equation 8.10):

$$\rho = \frac{W}{V} \tag{8.10}$$

where W is the weight and V the volume. However, when porous particles or volu-metric porous pieces are considered, a new term is defined: skeleton densities ρ_s (see Equation 8.11):

$$\rho_s = \frac{W}{V_s} \tag{8.11}$$

where V_s is the skeletal volume, the effective material density avoiding pores [141]. The skeletal density of powder cement can be easily measured by He pycnometry, which involves a device that can measure the 3D space inaccessible to a noble gas like helium, comparing two pressurized chambers [142, 143].

MI porosimetry is an approach to obtain the pore diameter distribution and the total poros-ity by introducing mercury into the sample at high pressure [89]. The pressure required and the resistance offered by the material allows the calculation of the pore diameter [144] according to the already developed Washburn equation (see Equation 8.12) [145]:

$$d_p = \frac{1470}{P_{Hg}} \tag{8.12}$$

where d_p is the pore diameter in μm and P_{Hg} is the mercury pressure. The higher P_{Hg}, the smaller the pore sizes found. The total porosity and skeletal volume can be obtained directly from the results through the Equation 8.13:

$$p = \frac{V_v}{V_t} \cdot 100 \tag{8.13}$$

where p is the percentage of total porosity, V_v is the volume of the void, which can be obtained directly from the MI curve, and V_t is the total volume of the sample [146].

As with particle size, porosity can also be approached through a scanning electron pho-tomicrography treatment, using ImageJ open-source image analysis software [147, 148].

8.2.5   Reaction Evolution of Cements

8.2.5.1   Kinetics

One of the most accepted models for describing cement reaction kinetics was proposed by Ginebra *et al.* [70]. They suggested that the degree and depth of reaction could be predicted by considering spherical cement particles, as is shown in Equation 8.14:

$$\alpha(\delta) = C_{2\delta} + \sum_{i=2\delta}^{d_{max}} \frac{\Delta C_i \alpha(d_i, \delta)}{100} \tag{8.14}$$

where $\alpha(\delta)$ is the degree of the reaction for a determined depth (δ), $C_{2\delta}$ is the volume percentage of the material with a particle size smaller than $d = 2\delta$, d_i is the diameter of the particle, and ΔC_i the volume percent of the material with particle size d_i.

The reaction rate for cements usually depends on two competing processes:

- **Solubility/hydrolysis:** the equilibrium between a solid and its ions in solution, that can be described by a characteristic solubility constant K_{sp} and offers information about the CPC degradation features, which also depend on other factors (temperature, pH, ionic strength, precipitation of a secondary phase, porosity, etc.). When diffusion processes are negligible, only K_{sp} governs the kinetics of the reaction, which usually happens in the early stages of the reaction [70]. The depth of the reacted particles can be described with the Equation 8.15:

$$\delta(t) = Kt \tag{8.15}$$

where δ is the depth of the hydrolyzed cement, K is a constant which depends on K_{sp}, and t is the time.
- **Diffusion:** the transport phenomenon that allows the liquid phase to reach the cement interface. It is favored by many factors, especially specific surface area, porosity, and hydrophilicity. It is common that when cement particles are too large, the outer part is dissolved. CDHA or brushite precipitates on the surface to create a shell, which masks the entering of more fluid to hydrolyze the rest of the particle, which remains unreacted (see Figure 8.10).When the shell is formed, the reaction is governed by the following formula, involving diffusion coefficients:

$$\frac{d\delta}{dt} = \frac{1}{\delta}k'D \tag{8.16}$$

where D is the diffusion coefficient of the liquid through the shell and k' is an empirical constant. Therefore :

$$\delta(t) = k't^{1/2} \tag{8.17}$$

Ginebra and coworkers [70] proposed a model which combines both rate limiting processes (Equation 8.18):

$$\delta(t) = KtI(t)_{t \leq t_0} + k't^{1/2}(t)_{t > t_0} \tag{8.18}$$

where $I(t)$ is the indicator function and t_0 is the time when diffusion becomes a limiting factor.

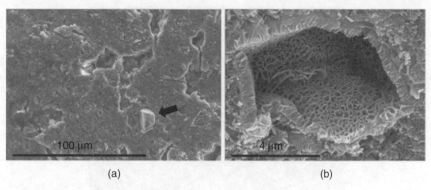

(a)   (b)

Figure 8.10 SEM micrographs of an unreacted particle, indicated by arrow, in (a) in a CPC hardened matrix, and the inner side of a masking shell after the loss of its unreacted particle (b)

8.2.5.2 Solubility and Ion Exchange in Aqueous Media

Hydrolysis and Precipitation. The solubility of the calcium phosphate is inherent to its composition and has its characteristic K_{sp}. However, the degree of the dissolved phosphate also depends on many variables as mentioned previously (pH, temperature, phosphate and calcium ions in the media, etc.). There are several calcium phosphates that can be transformed into more stable phases when they are hydrolyzed (see Table 8.1). Usually, at physiological temperature and at a pH which is slightly acid compared to basic media, CDHA is obtained, while in acid environments brushite is the stable phase (see Figure 8.1). Among all CPCs, it seems that the usual dominant mechanism at 37°C is the dissolution–precipitation, which involves hydrolysis of the CPC, saturation of the immediately aqueous environment close to the interface, and precipitation of CDHA or brushite (depending on the pH) onto this interface. Several authors [149–152] suggest that there are some reaction intermediates before CDHA is obtained, such as ACP or OCP through a topotactic reaction, especially observing the resulting crystals of CDHA, which are often obtained as pseudomorphs of typical OCP crystals (plate-like) instead of the needle-like crystals typical of HA (see Figure 8.11).

Degradation. There are three mechanisms of degradation of a CPC-based implant:

- **Dissolution:** Ionic crystal dissolution depends particularly on the K_{ps} of the resulting material (generally CDHA or brushite), but also other factors such as pH, temperature, or specific surface area. Brushite has a larger solubility than CDHA, however, it is spontaneously converted into CDHA in physiological media. The solubility of CDHA is quite low, but higher than pure HA.
- **Biological degradation:** Osteoclasts are multinucleated giant cells able to create a very acidic environment on the surface of the implant. The pH, which can be even lower than 4, is enough to increase the solubility of CDHA and HA. Usually, "dissolution pits" or etched crystals are evidence of an osteoclast-mediated degradation [153]. Interconnected macroporosity helps cells to migrate into the scaffold and to increase biodegradation, as well as providing the proper conditions for osteoclast adhesion, which is, for example, inhibited by basic pHs [154].

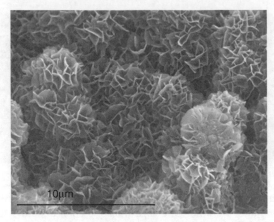

Figure 8.11   *A typical SEM micrograph of a hardened cement. The entanglement of the crystals derives from an increase in the compressive strength*

- **Erosion** is mainly due to the wear process that occurs on the interface between two materials under pressure and is connected to a relative movement or fatigue stress. It can be observed in the bulk material as well as at the interface of the implant. It usually leads to the release of debris and inflammation if released particles are too big to be phagocitized and eliminated.

8.2.6   Additives and Strategies to Enhance Properties

8.2.6.1   Nucleating Agents

As mentioned previously, the cement interface can be modified to favor or inhibit nucleation. Nucleation is critical in order to obtain a correct precipitation, and strategies to modify the nucleation rate will also modify the dissolution rate. On the one hand, the equilibrium is displaced through the elimination of one of the dissolved products, but on the other, precipitates are covering the rest of the cement surface. The most tested strategy is the epitaxy of CDHA introducing HA nuclei into the CPC formulation [105].

In contrast, nucleation can be inhibited by using, for example, Mg^{2+} ions, citrates, and pyrophosphates [155–158].

8.2.6.2   Accelerants/Retardants

A long list of additives has been studied to accelerate or inhibit reaction rates through the increase/decrease of the dissolution ratio or promoting/inhibiting CDHA or brushite nucleation. Accelerants are usually phosphates able to modify pH. They can be added in the same powder, such as MCPM (monocalcium phosphate monohydrate), or in the aqueous phase, such as acidic phosphates or sulfates (see Table 8.4). As was mentioned before, retardants such as citric acid can also be added [81].

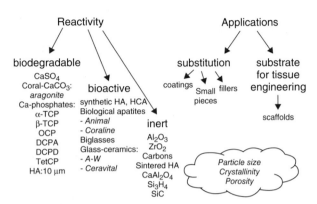

Figure 1.1 *Classification of bioceramics according to their reactivity. Particle size, crystallinity, and porosity are important factors to classify certain bioceramics, like apatites, in one group or the other. HA: hydroxyapatite, HCA: hydroxycarbonate apatite, A-W: apatite–wollastonite, TCP: tricalcium phosphate, OCP: octacalcium phosphate, DCPA: dicalcium phosphate anhydrous, DCPD: dicalcium phosphate dihydrate, TetCP: tetracalcium phosphate monoxide*

Bioceramics with Clinical Applications, First Edition. Edited by María Vallet-Regí.
© 2014 John Wiley & Sons, Ltd. Published 2014 by John Wiley & Sons, Ltd.

Chemical reactivity

➡ **Crystallinity**

➡ **Composition**

➡ **Particle size**

➡ **Defects**

➡ **Porosity**

➡ **Surface area**

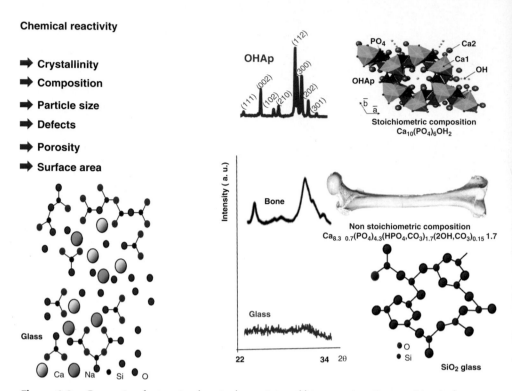

Figure 1.2 *Governing factors in chemical reactivity of bioceramics. Composition in between glasses (disordered) and crystals (ordered)*

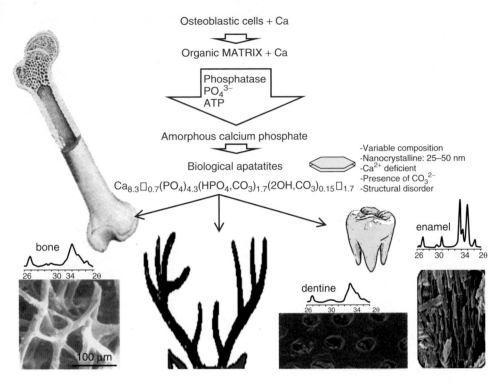

Osteoblastic cells + Ca

Organic MATRIX + Ca

Phosphatase
PO_4^{3-}
ATP

Amorphous calcium phosphate

Biological apatitites

$Ca_{8.3}\square_{0.7}(PO_4)_{4.3}(HPO_4,CO_3)_{1.7}(2OH,CO_3)_{0.15}\square_{1.7}$

-Variable composition
-Nanocrystalline: 25–50 nm
-Ca^{2+} deficient
-Presence of CO_3^{2-}
-Structural disorder

bone

26   30 34   2θ

100 μm

enamel

26   30   34   2θ

dentine

26   30   34   2θ

Figure 2.2 *Hard tissues in vertebrates are mainly formed by biological apatites*

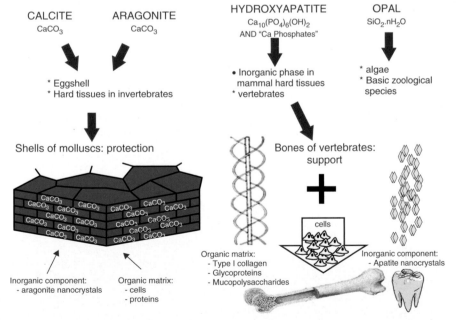

CALCITE
$CaCO_3$

ARAGONITE
$CaCO_3$

HYDROXYAPATITE
$Ca_{10}(PO_4)_6(OH)_2$
AND "Ca Phosphates"

OPAL
$SiO_2.nH_2O$

* Eggshell
* Hard tissues in invertebrates

• Inorganic phase in
  mammal hard tissues
* vertebrates

* algae
* Basic zoological
  species

Shells of molluscs: protection

Bones of vertebrates:
support

cells

$CaCO_3$

Inorganic component:
- aragonite nanocrystals

Organic matrix:
- cells
- proteins

Organic matrix:
- Type I collagen
- Glycoproteins
- Mucopolysaccharides

Inorganic component:
- Apatite nanocrystals

Figure 2.3 *The four most common inorganic phases in biomineralization processes. In all cases, they appear embedded in an organic matrix to form protection systems (molluscs, shells) or structural systems (vertebrates)*

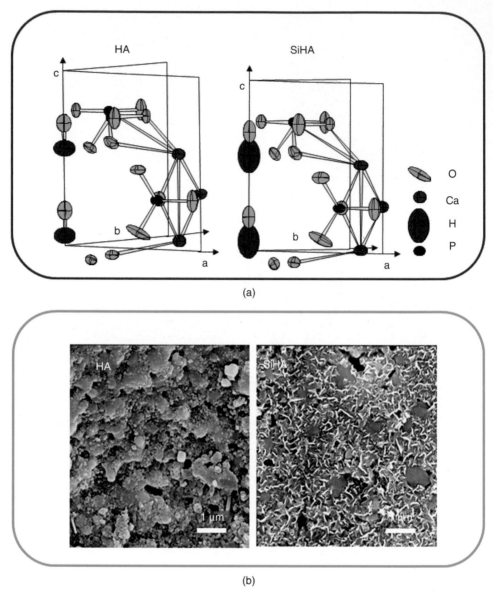

Figure 3.8 *(a) Anisotropic displacement of the atoms into HA and SiHA lattices. (b) Scanning electron micrographs of pure HA and SiHA after five weeks soaked in simulated body fluid*

$$Ca_{9.42}(PO_4)_{5.42}(HPO_4)_{0.58}(OH)_{1.42}$$

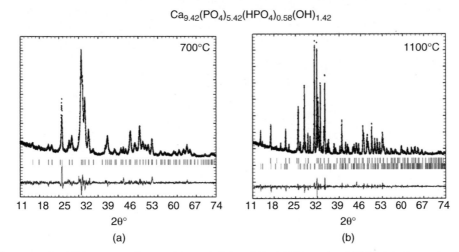

(a)   (b)

Figure 3.20   *XRD patterns for a calcium-deficient HA. (a) Treated at 700°C (single apatite-like phase) and (b) treated at 1100°C (biphasic HA/β-TCP mixture)*

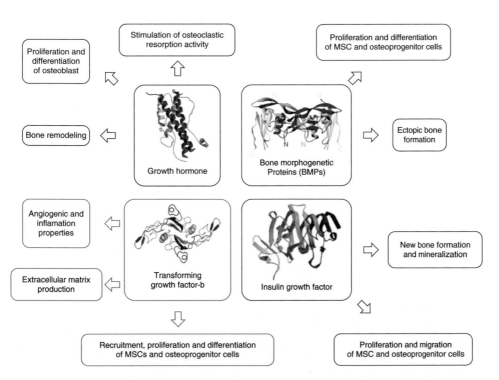

Figure 3.22   *Growth factors most used for bone regeneration and their biological activities*

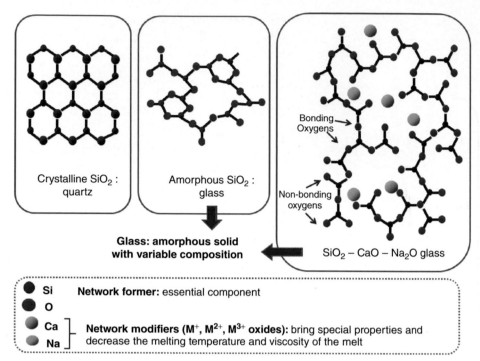

Crystalline SiO$_2$:
quartz

Amorphous SiO$_2$:
glass

Bonding
Oxygens

Non-bonding
oxygens

Glass: amorphous solid
with variable composition

SiO$_2$ – CaO – Na$_2$O glass

- Si   **Network former:** essential component
- O
- Ca ⎤
- Na ⎦ **Network modifiers (M$^+$, M^{2+}, M^{3+} oxides):** bring special properties and decrease the melting temperature and viscosity of the melt

Figure 4.1 *Structures of crystalline silica, amorphous silica, and a typical soda–lime–silica, glass. The main components of a silica glass and their role are also included*

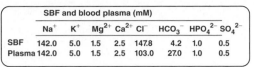

SBF and blood plasma (mM)								
	Na^+	K^+	Mg^{2+}	Ca^{2+}	Cl^-	HCO_3^-	HPO_4^{2-}	SO_4^{2-}
SBF	142.0	5.0	1.5	2.5	147.8	4.2	1.0	0.5
Plasma	142.0	5.0	1.5	2.5	103.0	27.0	1.0	0.5

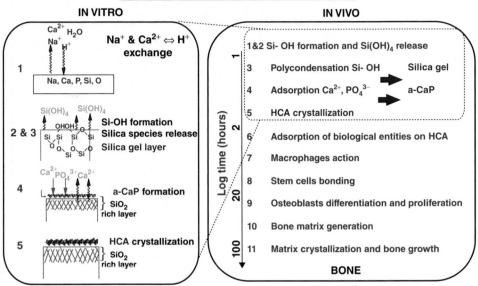

Figure 4.5 *Mechanism of bioactivity for silica glasses and glass ceramics* in vivo. *The five first stages can also be reproduced* in vitro *in solutions mimicking blood plasma, such as SBF*

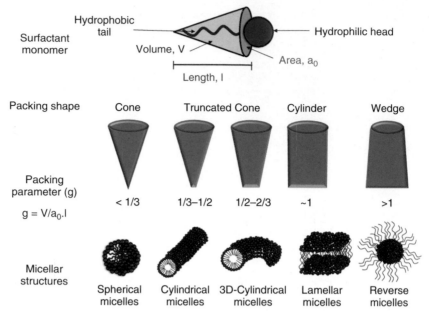

Figure 5.8 *Schematic representation of a surfactant molecule, where V is the hydrophobic group volume, l the kinetic surfactant hydrophobic tail length and a_0 the effective surface area of the hydrophilic head. The relationship between the packing parameter of cationic surfactant and the resulting micellar structures is also displayed*

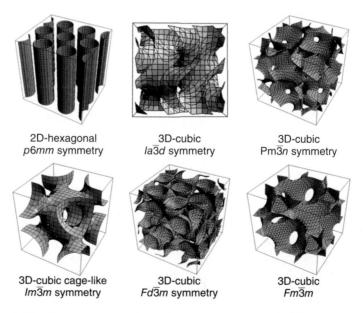

Figure 5.10 *Reconstruction of mesoporous structures with different symmetries*

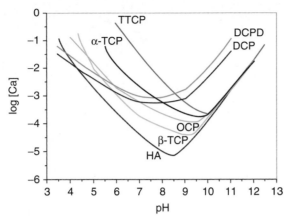

Figure 8.1 *Solubility phase diagrams of the concentration of calcium for CaP as a function of the pH*

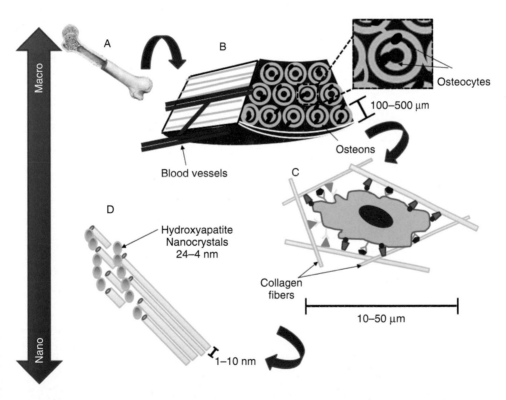

Figure 10.3 *Hierarchical organization of bone tissue at different length scales*

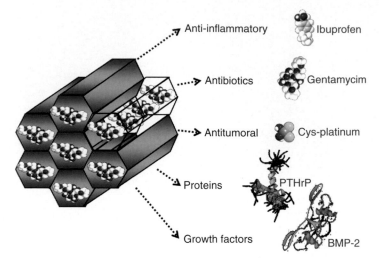

Figure 12.5 *Different drugs and biomolecules that have been loaded in ordered mesoporous silicas for drug delivery systems*

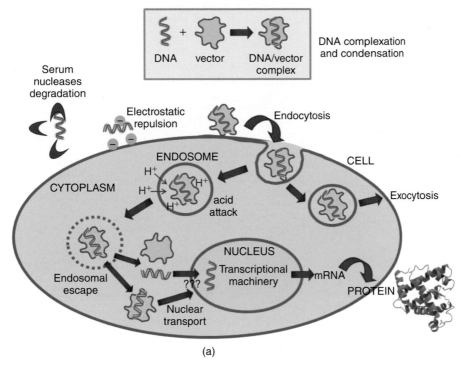

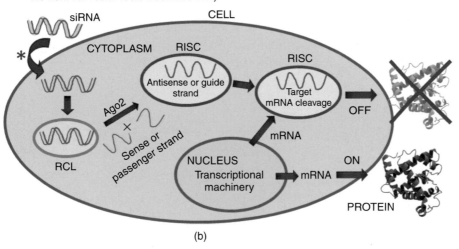

Figure 13.2 *(a) Cellular trafficking of nonviral vectors. (b) Simplified model of the RNAi mechanism in mammalian cells*

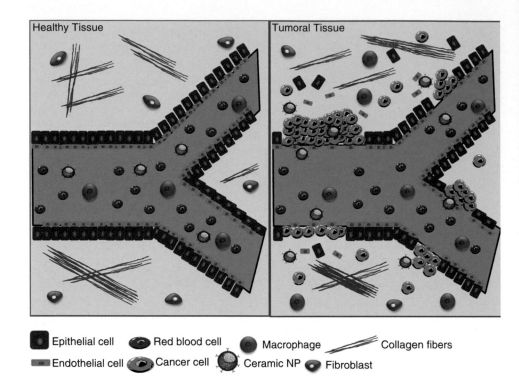

Figure 14.1 *Differences between healthy and tumoral tissues (EPR effect)*

8.2.6.3   Emulsifiers

Emulsifiers are usually added as porogens to induce interconnected macroporosity in the cements, minimizing the decrease in compression strength. An extensive list can be found in Table 8.4 and in the review of Dorozhkin [33].

They have the common feature that they can only be applied in pre-set cements. However, emulsifiers should not only be biocompatible, but also maintain porosity when injected.

8.2.6.4   Cation Doping

The doping of the cements with different cations and anions has introduced a great deal of versatility in CPC characteristics and results. In fact, natural apatite is not pure, and ions like F^- or Cl^- substituting OH^- can easily be found. Even slight differences can be found in the different HA located in the body. For example, it is quite common that $CO3^{2-}$ anions substitute $PO4^{3-}$ ones, or even OH^-.

For TCP, small ions, such as Mg^{2+}, Mn^{2+}, Cd^{2+}, Na^+, K^+, and Zn^{2+}, are used to stabilize β-TCP, while bigger divalent ions, such as Sr^{2+}, are more appropriate in the alpha phase stabilization [155, 159] (see Table 8.5).

The main ions that can be released by CPCs are calcium and phosphates. It has been suggested that calcium influences osteoblast proliferation [172], osteoclast regulation [173], and angiogenesis [174]. Meanwhile, phosphates seem to influence osteoblast apoptosis regulation [175], osteopontin (OP) production and regulation [176], and mineralization [177, 178].

8.2.6.5   Hybrid Biomaterials

Bone is a hierarchically ordered hybrid material formed by a mineral phase, which itself is mainly a result of poorly crystallized carbonated HA platelets, which can include some traces of other cations or anions; and an organic phase mainly formed by collagen type I,

Table 8.5   Typical doping elements used in CPCs

Doping element	Effect	References
F	Increase compression resistance	[160]
Fe	Magnetic properties	[161]
K and Na	Enhance antimicrobial properties Decrease setting reaction Decrease mechanical compression resistance	[54, 162, 163]
Mg	Decrease setting time	[160]
Sr	Modulation of setting reaction at low concentrations Increase degradation rate Enhance differentiation	[164–168]
Zn	Increase setting time Increase compression resistance Increase resorption	[160, 169–171]
Zn and Sr	Enhance osteoinduction	[170, 171]

other proteins, polysaccharides, and water [179]. The combination of these materials results in an elastic and hard material with a tensile mechanical strength comparable to mild steels [180–182].

A hybrid material is defined by IUPAC (International Union of Pure and Applied Chemistry) as "material composed of an intimate mixture of inorganic components, organic components, or both types of components. The components usually interpenetrate on scales of less than 1 μm" [183, 184].

The study of hybrid materials is a growing field of materials science, that has recently been the focus of attention of many research groups because they appear to be some of the most promising candidates for bone regeneration, owing to their excellent bioactivity and suitable mechanical properties. The combination of two or more components also promotes the synergy of their independent properties in one material.

Many combinations of CPCs and polymers have been assayed to improve different properties, especially mechanical tensile strength, but also to improve ion or drug release characteristics or porosity [73, 140, 185–188] (see Table 8.4).

8.2.6.6   Reinforcement Fibers

Fibers have offered a new method to enhance tensile mechanical properties of CPCs. But they also offer a way to interconnect pores, such as when biodegradable polymeric fibers are added [189, 190]. Temporal reinforcement is a good approach because it improves mechanical strength in the short-term, and after fiber dissolution, the void generated can be used by cells for migration and bone ingrowth [33, 191].

Another type of reinforcement that has been the focus of attention of several research groups is the use of multi-walled carbon nanotubes (MWCNTs), because of their extraordinary mechanical resistance. However, they are not biodegradable and controversy still exists about their use in parenteral implants [192–194].

8.2.7   Biological Characterization and Bioactive Behavior

Before proceeding with this section, several concepts should be clarified:

The process after adhesion in which progenitor cells progress from a primitive state to a fully functional osteoblast able to produce its own extracellular matrix (ECM) is divided into three stages: (i) proliferation, (ii) ECM development and maturation, and (iii) mineralization, with specific changes in their gene expression at each step [195].

8.2.7.1   Adhesion

Adhesion assays evaluate how the CPC enhances or facilitates the cell attachment onto proteins previously physisorbed on the surface of the CPC. Usually, cells are attached to the proteins through integrins, a pair of proteins directly connected to the cytoskeleton which can therefore promote signal transduction.

A qualitative assessment of the adhesion can be done through the visualization of the cell on the material by SEM or ESEM[15] (environmental scanning electron microscopy) after fixing, dehydration, and conductive coating (see Figure 8.12).

---

[15] ESEM allows measurement in a low vacuum and does not require a conductive sputtered layer, but maximum magnification is around 10-fold less.

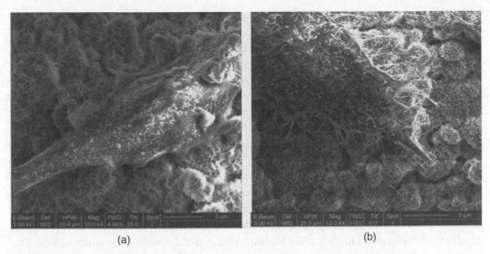

(a)   (b)

Figure 8.12   *Cell morphology as observed by HRSEM after 48 h of culture. (a) Corresponds to a coarse particle CPC and (b) to a fine particle CPC, with a more pronounced dorsal activity and more numerous philopodia. Reproduced with permission from [196]. © 2008, Mary Ann Liebert, Inc.*

It can also be studied using specific stains such as phalloidin,[16] which stains actin filaments of the cytoskeleton, allowing the behavior of the cell and the position of the focal adhesion points on the surface material[17] to be observed. Also, DAPI[18] can be added, a specific fluorophore to observe the nuclei. DAPI also allows the quantification of adhered cells. Time points usually go beyond six hours, and medium without serum is used to avoid the possible artifact effect of proteins.

8.2.7.2   *In vitro Cytotoxicity/Viability*

Commonly, *in vitro* cytotoxicity studies are carried out on osteoblast-like cell lines, using different cell staining methods to indirectly evaluate the number of viable/healthy cells present after having been cultured in a supernatant, which has previously been in contact with the CPC. One example is the use of one member of the family of tetrazolium salts that can be metabolized by most cell types, the MTT[19] assay, which measures the metabolic mitochondrial activity as an indication of the level of viable or healthy cells. These are able to enzymatically reduce MTT, obtaining an insoluble purple crystal of formazan, which is toxic for cells and eventually kills them [197]. Other similar reagents are XTT, MTS, WST-1.[20] In contrast, resazurin[21] is reduced to resorufin (pink, highly fluorescent) by oxidoreductases found mainly in the mitochondria of viable cells. As resorufin is excreted to the medium, this method allows the continuous monitoring of the

---

[16] Phalloidin is originally extracted from the toxic mushroom *amanita phalloides*.

[17] Focal points are the association of several integrins and need to be stained, for example, with Paxillin.

[18] DAPI 4′,6-diamidino-2-phenylindole.

[19] MTT 3-(4,5-dimethylthiazol-2-yl)-2,5-diphenyltetrazolium bromide.

[20] XTT (sodium 3′-[1-phenylamino)-carbonyl]-3,4-tetrazolium]-bis(4-methoxy-6-nitro)benzene-sulfonic acid hydrate; MTS 3-(4,5-dimethylthiazol-2-yl)-5-(3-carboxymethoxyphenyl)-2-(4-sulfophenyl)-2*H* tetrazolium; WST-1 2-(4-iodophenyl)-3-(4-nitrophenyl)-5-(2,4-disulfophenyl)-2*H*-tetrazolium.

[21] Resazurin 7-hydroxy-3*H*-phenoxazin-3-one 10-oxide, also known as alamarBlue®.

cytotoxicity of the material [170, 198, 199], which can be measured by either colorimetry or fluorimetry [200].

Another way to evaluate viable cells is the monitoring of the cell membrane integrity. For example, the live-cell protease can only be found active in live cells with a healthy membrane and becomes inactivated if the membrane loses its integrity. As live-cell protease cannot leave the cells, it should be monitored inside them [201]. Other options are the vital external dyes that are excluded from the cell if the cell membrane is healthy, but are found in the cell if there is any damage in the membrane, leading to a stained cytoplasm. Examples are trypan blue or propidium iodide [202, 203]. In contrast, the lactate dehydrogenase (LDH) is an internal catalyst enzyme present in many cells, that can be monitored in the extracellular medium if the cell membrane is compromised [204].

Live and dead assays involve a catalog of stains that are able to stain live and dead cells in green and red fluorescence, respectively. Calcein AM is an example that is cell membrane permeable and is fixed in the cytoplasma, yielding a green fluorescence. On the other hand, ethidium homodimer-1 labels nucleic acids of damaged-membrane cells with an intense red fluorescence [200]. Microscopy images can then be obtained and cells counted visually or using appropriate computer software such as ImageJ [134].

8.2.7.3   In vitro Bioactivity

Since the early 1990s, the most common way to test the bioactivity of CPCs and other bioactive materials has been by soaking the material at 37 °C in an acellular simulated body fluid (SBF) developed by Kokubo and coworkers [205, 206] with ion concentrations close to human blood plasma, in combination with tris-hydroxymethyl aminomethane (Tris) as buffer agent [206] (see Table 8.6). If materials are bioactive, then a layer of CDHA appears on the surface, as usually happens in CPCs.

SBF is not easy to prepare because of the spontaneous precipitation of some calcium phosphates (see Table 8.6 and Figure 8.13). A protocol was described by Cho *et al.* [207]. The hypothesis suggested by Kokubo *et al.* to validate this assay is that, as would happen *in vivo*, osteoblasts would be able to preferentially attach and proliferate onto this apatite surface to produce their own ECM of apatite and collagen. Then, the surrounded natural bone would bond to the newly formed ECM, forming a firm chemical bond [205]. Thus, the

Table 8.6   *Ion concentration of the different SBF compared with human plasma [205]*

Ion concentration (mM)	Human blood plasma	First SBF	Corrected SBF (c-SBF)	Revised SBF (r-SBF)	Newly improved SBF (n-SBF)
Ca^{2+}	2.5	2.5	2.5	2.5	2.5
Cl^-	103.0	148.8	147.8	103.0	103.0
HCO_3^{3-}	27.0	4.2	4.2	27.0	4.2
HPO_4^{2-}	1.0	1.0	1.0	1.0	1.0
K^+	5.0	5.0	5.0	5.0	5.0
Mg^{2+}	1.5	1.5	1.5	1.5	1.5
Na^+	142.0	142.0	142.0	142.0	142.0
SO_4^{2-}	0.5	0	0.5	0.5	0.5

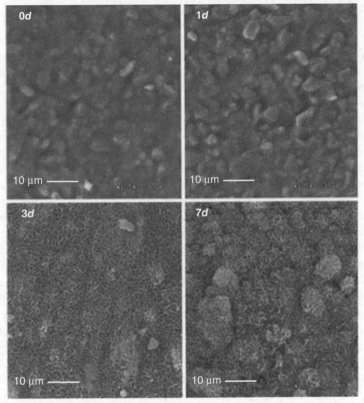

Figure 8.13 *Immersion tests of the CPC–alginate 3D porous scaffolds in simulated body fluid for different time points. SEM microstructure before 0 d and after immersion for 1d, 3d, and 7d. Tiny HA highly faceted nanocrystallites appeared at day 1, and grew over day 3 and 7. Reproduced with permission from [209]. Copyright © 2011, Elsevier*

ability of a CPC to develop a CDHA surface layer is an evidence of its chemical reactivity and can be associated to its potential ability to bond to bone *in vivo* [208].

8.2.7.4 Proliferation or Self-replication

The quantification of actively dividing cells is defined as cell proliferation evaluation. An approach can be the aforementioned dye methods used in viability but at different time points to observe a tendency, because the increase in cells number is usually more signifi-cant than cell growth.[22] The quantification can be done visually using an optical microscope in a hemocytometer, or by laser flow cytometry.

8.2.7.5 Differentiation

Stem cells and partially differentiated stem cells have the ability to proliferate, while becoming more specialized and forming new bone tissue (mineralization and osteogenesis)

---

[22] Cell division and cell growth are two processes given in cell proliferation.

in a proper environment with biochemical, topographical, and mechanical signals. In CPCs the use of osteosarcoma osteoblast-like cell lineages such as h-MG63,[23] h-SaOs-2, h-FOB, or r-MC3T3-E1 [210] is common, as they are available in unlimited number without requirements such as isolation or ethical approval, and yield more reproducible results. However, new requirements have gradually appeared in recent years in the bone regeneration field, particularly focused on the ability of bone biomaterials to be able to promote chemotaxis, osteogenesis, and osteoinduction. Translated into an *in vitro* assay, this means that a material should be capable of inducing differentiation of MSCs (mesenchymal stem cells) without the use of any osteogenic differentiation media [6]. Therefore, biomaterials should show a clear efficiency in more *in vivo*-extrapolating cells, such as primary cell lines, which have a better preclinical and clinical applicability [211]. In consequence, MSCs have gradually been used more often, or even co-cultures with other cell lineages, such as endothelial progenitor cells (EPCs), for angiogenesis.

To monitor the differentiation of osteoprogenitor cells there are several factors that are expressed and can be followed. For example, alkaline phosphatase[24] (ALP) is a metalloenzyme that is expressed in early stages of differentiation [212], where cells are partially differentiated to pre-osteoblasts, and it is soon observed on the cell surface and in matrix vesicles initiating mineralization [213]. Then, while other genes associated to the expression of osteocalcin (OC) and bone sialoprotein (BSP) are upregulated, ALP expression is decreased [214] meaning that fully differentiated osteoblasts are occurring. Other osteogenic markers can also be quantified, such as pro-collagen Iα2 (Col Iα2), OP, and osteonectin (ON). When pre-osteoblasts transform to fully differentiated osteoblasts, an elevated expression of BSP and OC occurs (see Figure 8.14). Usually, ALP and pro-collagen Iα2 are detected using their respective monoclonal antibodies.[25] In contrast, the rest can be detected using polyclonal antibodies [215]. Their respective associated mRNA (messenger ribonucleic acid) expressions can also be monitored by an *in situ* hybridization technique [216].

Real-time polymerase chain reaction (RT-PCR) is a quantitative technique that allows the determination and quantification of a targeted DNA (deoxyribonucleic acid) molecule, based on the general principle of polymerase chain reaction, and amplifying DNA more than a billion-fold. It requires a special thermocycler, including a camera that monitors the fluorescence originated by probes in each sample at frequent time points [217, 218].

Finally, several enzymes can also be quantified by staining, such as ALP with *p*-nitrophenyl phosphate as substrate in 2-amino-2-methyl-1-propanol alkaline buffer solution. Type I collagen produced by cells can also be monitored using a Sircol dye reagent [219].

8.2.7.6   *In vivo Bioresorption and Bone Remodeling*

Many models are used to test CPCs *in vivo*, however three of them are the more common ones: rabbit condyle [220] and rat calvaria [221] implantations, which are performed in a bony environment, and rat subcutaneous implantation, which is a method to test the osteoinduction capability of a material in a non-bony environment [6].

---

[23] h-human source and r-rat source.

[24] ALP phosphate-monoester phosphohydrolase. ALP is responsible for hydrolysis of monophosphate esters at a high pH (pH 8–10).

[25] Monoclonal or monovalent (specific) have affinity for the same antigen or for the same epitote.

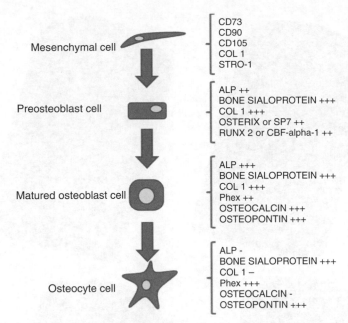

Figure 8.14 *List of common differentiation markers for MSCs to osteoblast and osteocyte* [6, 195]

The most common method to evaluate bioresorption and bone remodeling is by histological and histomorphometric procedures of euthanized animals at different time points after implantation. After the implant is extracted, it is washed, fixed, dehydrated, and embedded in a polymer. Then, several thin cross-sections are performed, using a microtome, to finally be stained with the proper dye in order to differentiate between the different tissues and cells with an optical microscope [222].

In recent years, micro-computed tomography [223] has increasingly been used because it allows the evaluation of the implant continuously without the need to kill the animal until the end of the study, thus minimizing the waste of animal lives.

8.3   Applications

8.3.1   Bone Defect Repair

8.3.1.1   Injectable Foams

In the last decade, the development of new systems to generate macropores in CPCs has involved the use of porogens and emulsifiers (see Figure 8.15). The possibility to obtain a manipulable calcium phosphate that maintains the porosity after foaming has become a difficult challenge. One successful strategy has been gas foaming using a porogen, allowing the inclusion of degradable alginate-fibrin microbeads, which encapsulated human umbilical cord mesenchymal stem cells (hUCMSCs) for a further release. Surprisingly,

Figure 8.15   *Typical μCT scans of the four different foamed blocks. Reproduced with permission from [52]. Copyright © 2011, Elsevier*

the mechanical properties were close to cancellous bone [224].Another approach was the use of a commercial porogen, such as Polysorbate 80 and bovine gelatine type B [225]. Typical current applications for CPCs are summarized in Table 8.7.

8.3.1.2   *3D Scaffolds*

3D scaffolds for tissue engineering are one of the most difficult challenges for CPCs researchers because of the hard-to-accomplish requirements of macroporosity and the decrease in mechanical properties associated with them (see Figure 8.16). Foamed and

Table 8.7   *Common applications of CPCs [33]*

Applications	References
Bone defects filler	[226–229]
Craniofacial/maxillofacial as bone filler	[230, 231]
Dental and orthopedic as link to natural bone	[232–235]
Direct pulp capping	[236, 237]
Drug delivery	[4, 238]
Scaffold for bone formation and provided histocompatible healing of periodontal tissues	[190, 239, 240]
Treatment of osteonecrotic lesions near articulating joints	[241, 242]
Vertebroplasty and kyphoplasty	[243–245]

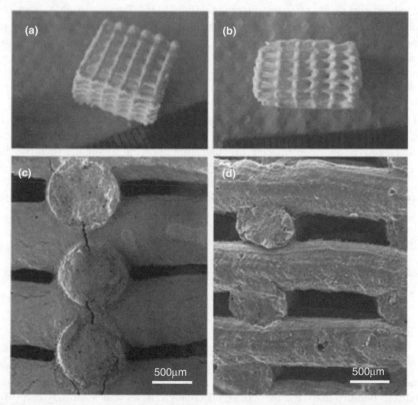

Figure 8.16   *Fabrication of a P-CPC scaffold by plotting of P-CPC and water-soluble polyvinylacetate (PVA) as supporting material. (a) Top view of the scaffold; (b) side view of the scaffold; (c,d) SEM images (side view) of P-CPC scaffolds plotted without (c) and with (d) PVA. Reproduced with permission from [188]. Copyright © 2012, John Wiley & Sons, Ltd.*

Table 8.8   *Methods for macroporosity introduction in CPCs [246]*

Macroporosity generation method	References
Addition of an acid solution to basic $NaHCO_3$ to liberate CO_2	[248]
Addition of soluble salts leaching	[249]
Freeze drying	[250]
Introduction of polymer microspheres	[251]
Introduction of polymer or organic compound for further leaching	[190]
Introduction of resorbable fibers	[252]
Porogen introduction	[52]
Rapid prototyping scaffolds	[253]
Sintering of previously set cement blocks	[246]
Template replication	[254]

porogened cements do not offer any interconnectivity [246]. Two approaches seem to solve the problem of interconnectivity and mechanical requirements: by casting a rapid proto-typed polymer resin [247], rapid prototyping of mixed organics-CPC [188] and reinforced CPCs with biodegradable polymers to create macropores [189] (see Table 8.8).

8.3.2   Drug Delivery Systems

One of the new applications that have appeared in recent years is the adaptation of CPCs as drug delivery systems, taking advantage of their excellent biocompatibility, their capacity to be set *in vivo* and their injectability, therefore allowing minimally invasive surgery [36] (see Table 8.9). To date, the main uses of CPCs as drug carriers can be divided into the following three groups: (i) low molecular weight drugs; (ii) high molecular weight biomolecules; and (iii) ions [238]. Hitherto, the release compounds assayed were antibiotics, anticancer, growth factors, and ions by three basic mechanisms:

- Controlled release by diffusion in the physiological media
- Release of the inner charged molecules after cement degradation
- Release of the inner charged molecules after formation of the apatite shell surrounding the CPC carrier.

8.4   Future Trends

CPCs have demonstrated fair efficiency for bone regeneration. Their numerous advantages still make them promising materials in the field. Further studies should be performed, especially to better understand their material characteristics and what they can offer for a specific application, considering that there is no universal biomaterial, and that the need for nanostructuration is fundamental. Cements have gradually been embraced in the wider field of composites by hybridizing their compositions in order that they may adapt to the new trends. Drug delivery and 3D macroporous scaffolds seem to be the two strategies with

Table 8.9 *Typical growth factors and drugs delivered by CPCs [4, 238]*

Growth factor	Function
Bone morphogenic protein (BMP)	Promote the recruitment of mesenchymal to the defect site and their differentiation into osteoblast avoiding osteoclast activity
Epidermal growth factor (EGF)	Regulate keratinocyte phenotype for wound healing
Fibroblast growth factor (FGF)	Promote osteoblast proliferation in order to increase angiogenesis
Insulin-like growth factor (IGF)	Promote proliferation of osteoblasts and osteoclasts
Placental growth factor (PGF)	Stimulate angiogenesis as well VEGF
Platelet-derived growth factor (PDGF)	Promote wound healing and tissue repair. Also stimulates osteoblast and osteoclasts proliferation as well as angiogenesis
Transforming growth factor-β (TGF-β)	Promote the proliferation and differentiation and ECM production of osteoprogenitor cells (partially differentiated cells), avoiding osteoclast progenitor cells
Vascular endothelial growth factor (VEGF)	Promote angiogenesis through the formation of capillary blood vessels
Drug	Function
Alendronate	Used for osteoporosis and several other bone diseases
Cephalexin	Antibiotic
Ceramide	Apoptosis promoter of certain (Type I PCD) cells
Ciprofloxacin	Antibiotic
Cis-platin	Anticancer agent
Doxycycline hyclate (DOXY-h)	Antibiotic
Gentamicin	Antibiotic
Human lactotransferrin (hLFI-II)	Antibiotic and fungicide
Ibuprofen	Non-steroidal anti-inflammatory and anti-pyretic drug
Methotrexate	Stop cell proliferation by inhibiting the metabolism of folic acid. It is used as anticancer agent, to treat autoimmune diseases, ectopic pregnancy, and for the induction of medical abortions
Tetracycline	Antibiotic
Vancomycin	Antibiotic

the most promising future. Accurate studies about interaction with the other components in the mixtures and how to optimize their relationships will be needed; however, there are many other competitors.

8.5 Conclusions

When CPCs were discovered two decades ago, expectation was high. Nowadays, CPCs seem to have lost much of their potential applicability behind other materials, which have demonstrated a better efficiency in specific niches. CPCs are still preferred in a restricted and relatively narrow field of bone applications. It seems that CPCs need to reinvent themselves to be able to meet the requirements of current regenerative medicine, even when this means that they may partially lose their identity as cement.

References

1. Bohner, M. and Lemaitre, J. (2009) Can bioactivity be tested in vitro with SBF solution? *Biomaterials*, **30**, 2175–2179.
2. Bandyopadhyay, A., Bernard, S., Xue, W. and Bose, S. (2006) Calcium phosphate-based resorbable ceramics: influence of MgO, ZnO, and SiO_2 dopants. *J. Am. Ceram. Soc.*, **89**, 2675–2688.
3. Sordi, V. (2009) Mesenchymal stem cell homing capacity. *Transplantation*, **87**, S42–S45.
4. Bose, S. and Tarafder, S. (2012) Calcium phosphate ceramic systems in growth factor and drug delivery for bone tissue engineering: A review. *Acta Biomater.*, **8**, 1401–1421.
5. Pietrzak, W.S. (2008) *Musculoskeletal Tissue Regeneration: Biological Materials and Methods*, Humana Press, A Part of Springer Science+Business Media, LLC.
6. Miron, R.J. and Zhang, Y.F. (2012) Osteoinduction: a review of old concepts with new standards. *J. Dent. Res.*, **91**, 736–744.
7. Risau, W. and Flamme, I. (1995) Vasculogenesis. *Annu. Rev. Cell Dev. Biol.*, **11**, 73–91.
8. Werner, A.G. (1786) Gerhard's Grundr., 281.
9. Werner, A.G. (1788) Geschichte, Charakteristik und kurze chemische Untersuchung des Apatits. *Bergmaen. J.*, **1**, 76.
10. Proust, J.-L. (1788) Lettre de M. Proust a M. d'Arcet sur un sel phosphorique calcaire naturel. *J. Phys.*, **32**, 241–247.
11. Klaproth, M.H. (1802) Beiträge zur chemischen Kenntnis der Mineralkörper. *Dritter Band, Rottmann Berlin*, **3**, 194–198.
12. Gottlieb-Gahn, J. http://www.britannica.com/EBchecked/topic/223488/Johan-Gottlieb-Gahn (accessed 09 November 2013).
13. Eisenberguer, S., Lehrman, A. and Turner, W.D. (1940) The basic calcium phosphates and related systems: some theoretical and practical aspects. *Chem. Rev.*, **26**, 257–296.
14. Gluck, T. (1890) *Autoplastik, Transplantation, Implantation vom Fremdkörpern: Vortrag, gehalten, in der Sitzung der Berliner Medicinischen Gesellschaft vom 23. April 1890*, Gedr. bei L. Schumacher.
15. Nilsson, M., Fernández, E., Sarda, S. *et al.* (2002) Characterization of a novel calcium phosphate/sulphate bone cement. *J. Biomed. Mater. Res.*, **61**, 600–607.
16. Dorozhkin, S.V. (2012) Calcium orthophosphates and human beings: a historical perspective from the 1770s until 1940. *Biomatter*, **2**, 53–70.
17. Albee, F. and Morrison, H. (1920) Studies in bone growth. *Annal. Surg.*, **71**, 32–38.
18. Eden, R.T. (1923) Untersuchungen über Vorgänge bei der Verknöcherung. *Klin. Wochenschr.*, **2**, 1798–1804.
19. Haldeman, K.O.and M. J. (1934) Influence of a local excess of calcium and phosphorus on the healing of fractures: an experimental study. *Arch. Surg.*, **29**:385-396.

20. Ray, R.D. and Ward, A.A. (1951) A preliminary report on studies of basic calcium phosphate in bone replacement. *Surg. Forum*, **2**, 429–434.

21. Jaffe, W.L. and Scott, D.F. (1996) Current concepts review – total hip arthroplasty with hydroxyapatite-coated prostheses. *J. Bone Joint Surg.*, **78**, 1918–1934.

22. Groves, E.W.H. (1927) Some contributions to the reconstructive surgery of the hip. *Br. J. Surg.*, **14**, 486–517.

23. Robinson, R. (1923) The possible significance of hexosephosphoric esters in ossification. *Biochem. J.*, **17** (2), 286–293.

24. Hunsberger, A. and Ferguson, L.K. (1932) Variations in phosphatase and inorganic phosphorus in serum during fracture repair. *Arch. Surg.*, **24**, 1052–1060.

25. Downs, G. (1937) Organic calcium in healing of fractures. *Am. J. Surg.*, **35**, 34–40.

26. Cutright, D.E., Bhaskar, S.N., Brady, J.M. *et al.* (1972) Reaction of bone to tricalcium phosphate ceramic pellets. *Oral Surg. Oral Med. Oral Pathol.*, **33**, 850–856.

27. Kingery, W.D. II, (1950) Cold-setting properties. *J. Am. Ceram. Soc.*, **33**, 242–246.

28. Roy, D.M. and Linnehan, S.K. (1974) Hydroxyapatite formed from coral skeletal carbonate by hydrothermal exchange. *Nature*, **247**, 220–222.

29. Bhaskar, S.N., Brady, J.M., Getter, L. *et al.* (1971) Biodegradable ceramic implants in bone: electron and light microscopic analysis. *Oral Surg. Oral Med. Oral Pathol.*, **32**, 336–346.

30. Getter, L., Bhaskar, S.N., Cutright, D.E. *et al.* (1972) Three biodegradable calcium phosphate slurry implants in bone. *J Oral Surg.*, **30** (4), 263–268.

31. Clarke, W.J., Driskell, T.D., Hassler, C.R. *et al.* (1973) Calcium phosphate resorbable ceramics, a potential alternative to bone grafting. IADR Program and Abstract, Vol. 52, Abstract No. 259.

32. Legeros, R.Z., Chohayeb, A. and Shulman, A. (1982) Apatitic calcium phosphates: possible dental restorative materials. *J. Dent. Res.*, **61**, 343.

33. Dorozhkin, S.V. (2011) Self-setting calcium orthophosphate formulations: cements, concretes, pastes and putties. *Int. J. Mater. Chem.*, **1**, 1–48.

34. Monma, H. and Kanazawa, T. (1976) The hydration of α-tricalcium phosphate. *Yogyo-Kyokai-Shi*, **84**, 209–213.

35. Bohner, M., Gbureck, U. and Barralet, J.E. (2005) Technological issues for the development of more efficient calcium phosphate bone cements: a critical assessment. *Biomaterials*, **26**, 6423–6429.

36. Ginebra, M.P., Traykova, T. and Planell, J.A. (2006) Calcium phosphate cements as bone drug delivery systems: a review. *J. Controlled Release*, **113**, 102–110.

37. Driessens, F.C.M., Planell, J.A. and Gil, F.J. (1995) Calcium phosphate bone cements, in *Encyclopedic Handbook of Biomaterials and Bioengineering* (eds D.L. Wise, D.J. Trantolo, D.E. Altobelli *et al.*), Marcel Dekker, Inc, New York.

38. Mandel, S. (1976) The structural basis of crystal-induced membranolysis. *Arthritis Rheum.*, **19**, 439–445.

39. Dorozhkin, S.V. (2010) Bioceramics of calcium orthophosphates. *Biomaterials*, **31**, 1465–1485.

40. Carrodeguas, R.G. and De Aza, S. (2011) alpha-Tricalcium phosphate: synthesis, properties and biomedical applications. *Acta Biomater.*, **7**, 3536–3546.

41. Zhai, S., Liu, X., Shieh, S. *et al.* (2009) Equation of state of tricalcium phosphate, gamma-Ca3(PO4)2, to lower mantle pressures. *Am. Mineral.*, **94**, 1388–1391.

42. Driessens, F. and Verbeeck, R. (1990) *Biominerals*, CRC Press, Boca Raton, FL.

43. Dorozhkin, S.V. and Epple, M. (2002) Biological and medical significance of calcium phosphates. *Angew. Chem. Int. Ed.*, **41**, 3130–3146.

44. Kalita, S.J., Bhardwaj, A. and Bhatt, H.A. (2007) Nanocrystalline calcium phosphate ceramics in biomedical engineering. *Mater. Sci. Eng., C*, **27**, 441–449.

45. Loher, S., Stark, W.J., Maciejewski, M. *et al.* (2004) Fluoro-apatite and calcium phosphate nanoparticles by flame synthesis. *Chem. Mater.*, **17**, 36–42.

46. Mohn, D., Doebelin, N., Tadier, S. *et al.* (2011) Reactivity of calcium phosphate nanoparticles prepared by flame spray synthesis as precursors for calcium phosphate cements. *J. Mater. Chem.*, **21**, 13963–13972.

47. Kamerer, D., Hirsch, B., Snyderman, C. *et al.* (1994) Hydroxyapatite cement: a new method for achieving watertight closure in transtemporal surgery. *Am. J. Otolaryngol.*, **15**, 47–49.

48. Constantz, B.R., Ison, I.C., Fulmer, M.T. *et al.* (1995) Skeletal repair by in situ formation of the mineral phase of bone. *Science*, **267**, 1796–1799.

49. Spies, C., Schnürer, S., Gotterbarm, T. and Breusch, S. (2010) Efficacy of Bone Source™ and Cementek™ in comparison with Endobon™ in critical size metaphyseal defects, using a minipig model. *J. Appl. Biomater. Biomech.*, **8**, 175–185.

50. Bohner, M. (2010) Design of ceramic-based cements and putties for bone graft substitution. *Eur. Cell. Mater.*, **20**, 1–12.

51. Gbureck, U., Hölzel, T., Klammert, U. *et al.* (2007) Resorbable dicalcium phosphate bone substitutes prepared by 3D powder printing. *Adv. Funct. Mater.*, **17**, 3940–3945.

52. Bohner, M., Van Lenthe, G.H., Grunenfelder, S. *et al.* (2005) Synthesis and characterization of porous beta-tricalcium phosphate blocks. *Biomaterials*, **26**, 6099–6105.

53. Carey, L.E., Xu, H.H.K., Simon, C.G. Jr., *et al.* (2005) Premixed rapid-setting calcium phosphate composites for bone repair. *Biomaterials*, **26**, 5002–5014.

54. Rajzer, I., Castaño, O., Engel, E. and Planell, J.A. (2010) Injectable and fast resorbable calcium phosphate cement for body-setting bone grafts. *J. Mater. Sci. - Mater. Med.*, **21**, 2049–2056.

55. Weir, M.D. and Xu, H.H.K. (2010) Osteoblastic induction on calcium phosphate cement–chitosan constructs for bone tissue engineering. *J. Biomed. Mater. Res. Part A*, **94**, 223–233.

56. Lopez-Heredia, M.A., Bernard Kamphuis, G.J., Thüne, P.C. *et al.* (2011) An injectable calcium phosphate cement for the local delivery of paclitaxel to bone. *Biomaterials*, **32**, 5411–5416.

57. Apelt, D., Theiss, F., El-Warrak, A.O. *et al.* (2004) In vivo behavior of three different injectable hydraulic calcium phosphate cements. *Biomaterials*, **25**, 1439–1451.

58. Food and Drug Administration, Medical Devices: Premarket Notification (510k), (2010).

59. Blokhuis, T.J. and Arts, J.J.C. (2011) Bioactive and osteoinductive bone graft substitutes: definitions, facts and myths. *Injury*, **42** (Suppl. 2), S26–S29.

60. Zimmermann, G. and Moghaddam, A. (2011) Allograft bone matrix versus synthetic bone graft substitutes. *Injury*, **42** (Suppl. 2), S16–S21.

61. Legeros, R.Z. (1988) Calcium phosphate materials in restorative dentistry: a review. *Adv. Dent. Res.*, **2**, 164–180.

62. Song, Y., Feng, Z. and Wang, T. (2007) In situ study on the curing process of calcium phosphate bone cement. *J. Mater. Sci. - Mater. Med.*, **18**, 1185–1193.

63. Lacout, J.L., Mejdoubi, E. and Hamad, M. (1996) Crystallization mechanisms of calcium phosphate cement for biological uses. *J. Mater. Sci. - Mater. Med.*, **7**, 371–374.

64. Fernández, E., Gil, F.J., Ginebra, M.P. *et al.* (1999) Calcium phosphate bone cements for clinical applications. Part I: solution chemistry. *J. Mater. Sci. - Mater. Med.*, **10**, 169–176.

65. Fernández, E., Gil, F.J., Ginebra, M.P. *et al.* (1999) Calcium phosphate bone cements for clinical applications. Part II: precipitate formation during setting reactions. *J. Mater. Sci. - Mater. Med.*, **10**, 177–183.

66. Giocondi, J.L., El-Dasher, B.S., Nancollas, G.H. and Orme, C.A. (2010) Molecular mechanisms of crystallization impacting calcium phosphate cements. *Philos. Trans. R. Soc. London, Ser. A*, **368**, 1937–1961.

67. Bohner, M. (2010) Resorbable biomaterials as bone graft substitutes. *Mater. Today*, **13**, 24–30.

68. Del Valle, S., Mino, N., Munoz, F. *et al.* (2007) In vivo evaluation of an injectable Macroporous Calcium Phosphate Cement. *J. Mater. Sci. - Mater. Med.*, **18**, 353–361.

69. Brown, P.W. and Fulmer, M. (1991) Kinetics of hydroxyapatite formation at low temperature. *J. Am. Ceram. Soc.*, **74**, 934–940.

70. Ginebra, M.-P., Fernández, E., Driessens, F.C.M. and Planell, J.A. (1999) Modeling of the hydrolysis of α-tricalcium phosphate. *J. Am. Ceram. Soc.*, **82**, 2808–2812.

71. Ginebra, M.P., Boltong, M.G., Fernández, E. *et al.* (1995) Effect of various additives and temperature on some properties of an apatitic calcium phosphate cement. *J. Mater. Sci. - Mater. Med.*, **6**, 612–616.

72. Bohner, M., Merkle, H.P., Landuyt, P.V. *et al.* (2000) Effect of several additives and their admixtures on the physico-chemical properties of a calcium phosphate cement. *J. Mater. Sci. - Mater. Med.*, **11**, 111–116.

73. Perez, R. A., Kim, H.-W., Ginebra, M.-P. (2012) Polymeric additives to enhance the functional properties of calcium phosphate cements. *J. Tissue Eng.*, **3** (1) doi: 10.1177/204173141239555.

74. Gbureck, U., Barralet, J.E., Spatz, K. *et al.* (2004) Ionic modification of calcium phosphate cement viscosity. Part I: hypodermic injection and strength improvement of apatite cement. *Biomaterials*, **25**, 2187–2195.

75. Barralet, J.E., Grover, L.M. and Gbureck, U. (2004) Ionic modification of calcium phosphate cement viscosity. Part II: hypodermic injection and strength improvement of brushite cement. *Biomaterials*, **25**, 2197–2203.

76. Tamimi-Mariño, F., Mastio, J., Rueda, C. *et al.* (2007) Increase of the final setting time of brushite cements by using chondroitin 4-sulfate and silica gel. *J. Mater. Sci. - Mater. Med.*, **18**, 1195–1201.

77. Mariño, F.T., Torres, J., Hamdan, M. *et al.* (2007) Advantages of using glycolic acid as a retardant in a brushite forming cement. *J. Biomed. Mater. Res. Part B*, **83**, 571–579.

78. Delgado, J.A., Harr, I., Almirall, A. *et al.* (2005) Injectability of a macroporous calcium phosphate cement. *Key Eng. Mater.*, **284–286**, 157–160.

79. Ginebra, M.P., Delgado, J.A., Harr, I. *et al.* (2007) Factors affecting the structure and properties of an injectable self-setting calcium phosphate foam. *J. Biomed. Mater. Res. Part A*, **80**, 351–361.

80. Barralet, J.E., Tremayne, M., Lilley, K.J. and Gbureck, U. (2005) Modification of calcium phosphate cement with alpha-hydroxy acids and their salts. *Chem. Mater.*, **17**, 1313–1319.

81. Sarda, S., Fernandez, E., Nilsson, M. *et al.* (2002) Kinetic study of citric acid influence on calcium phosphate bone cements as water-reducing agent. *J. Biomed. Mater. Res.*, **61**, 653–659.

82. Watanabe, M., Tanaka, M., Sakurai, M. and Maeda, M. (2006) Development of calcium phosphate cement. *J. Eur. Ceram. Soc.*, **26**, 549–552.

83. Vater, C., Lode, A., Bernhardt, A. *et al.* (2010) Modifications of a calcium phosphate cement with biomolecules – Influence on nanostructure, material, and biological properties. *J. Biomed. Mater. Res. Part A*, **95A**, 912–923.

84. Miyamoto, Y., Ishikawa, K., Takechi, M. *et al.* (1998) Basic properties of calcium phosphate cement containing atelocollagen in its liquid or powder phases. *Biomaterials*, **19**, 707–715.

85. Perez, R.A., Altankov, G., Jorge-Herrero, E. and Ginebra, M.P. (2012) Micro- and nanostructured hydroxyapatite–collagen microcarriers for bone tissue-engineering applications. *J. Tissue Eng. Regen. Med.*, **7**, 353–361.

86. Bigi, A., Bracci, B. and Panzavolta, S. (2004) Effect of added gelatin on the properties of calcium phosphate cement. *Biomaterials*, **25**, 2893–2899.

87. Ding, S.-J. and Shie, M.-Y. (2011) The significance of gelatin in calcium phosphate hybrid bone cement for attachment and differentiation of MG63 cells. *Adv. Eng. Mater.*, **13**, B246–B255.

88. Markovic, M., Takagi, S. and Chow, L.C. (2000) Formation of macropores in calcium phosphate cements through the use of mannitol crystals. *Key Eng. Mater.*, **192–195**, 773–776.

89. Del Real, R.P., Wolke, J.G., Vallet-Regi, M. and Jansen, J.A. (2002) A new method to produce macropores in calcium phosphate cements. *Biomaterials*, **23**, 3673–3680.

90. Ginebra, M.P., Planell, J.A., and Gil, F.J. (2005) Injectable, Self-Setting Calcium Phosphate Foam, EP 1787626B1, filed August 11, 2005 and issued October 16, 2013.

91. Simon, C.G., Khatri, C.A., Wight, S.A. and Wang, F.W. (2002) Preliminary report on the biocompatibility of a moldable, resorbable, composite bone graft consisting of calcium phosphate cement and poly(lactide-co-glycolide) microspheres. *J. Orthop. Res.*, **20**, 473–482.

92. Lee, G.-S., Park, J.-H., Won, J.-E. *et al.* (2011) Alginate combined calcium phosphate cements: mechanical properties and in vitro rat bone marrow stromal cell responses. *J. Mater. Sci. - Mater. Med.*, **22**, 1257–1268.

93. Bohner, M. (2000) Calcium orthophosphates in medicine: from ceramics to calcium phosphate cements. *Injury*, **31** (Suppl. 4), D37–D47.

94. Sarda, S., Fernández, E., Nilsson, M. and Planell, J.A. (2002) Influence of air-entraining agent on bone cement macroporosity. *Key Eng. Mater.*, **218-220**, 335–338.

95. Bercier, A., Gonçalves, S., Lignon, O. and Fitremann, J. (2010) Calcium phosphate bone cements including sugar surfactants: part one-porosity, setting times and compressive strength. *Materials*, **3**, 4695–4709.

96. Leung, K.S., Siu, W.S., Li, S.F. *et al.* (2006) An in vitro optimized injectable calcium phosphate cement for augmenting screw fixation in osteopenic goats. *J. Biomed. Mater. Res. Part B*, **78**, 153–160.

97. Dienes, G.J. (1950) Frequency factor and activation energy for the volume diffusion of metals. *J. Appl. Phys.*, **21**, 1189–1192.

98. Mehrer, H. (2007) *Diffusion in Solids: Fundamentals, Methods, Materials, Diffusion-Controlled Processes*, Springer, London.

99. Mirhadi, B., Mehdikhani, M. and Askari, N. (2011) Synthesis of nano-sized β-tricalcium phosphate via wet precipitation. *Process. Appl. Ceram.*, **5**, 193–198.

100. Ertl, G. (2010) *Reactions at Solid Surfaces*, John Wiley & Sons, Inc., Hoboken, NJ.

101. Hosokawa, M., Nogi, K., Naito, M. and Yokoyama, T. (2007) *Nanoparticle Technology Handbook*, Elsevier Science.

102. Tang, R., Orme, C.A. and Nancollas, G.H. (2003) A new understanding of demineralization: the dynamics of brushite dissolution. *J. Phys. Chem. B*, **107**, 10653–10657.

103. Ginebra, M.P., Driessens, F.C. and Planell, J.A. (2004) Effect of the particle size on the micro and nanostructural features of a calcium phosphate cement: a kinetic analysis. *Biomaterials*, **25**, 3453–3462.

104. Liu, C. and Shen, W.E.I. (1997) Effect of crystal seeding on the hydration of calcium phosphate cement. *J. Mater. Sci. - Mater. Med.*, **8**, 803–807.

105. Durucan, C. and Brown, P.W. (2002) Reactivity of alpha-tricalcium phosphate. *J. Mater. Sci.*, **37**, 963–969.

106. ASTM (2013) C191-08 Standard Test Methods for Time of Setting of Hydraulic Cement by Vicat Needle. Volume 04.01 Cement; Lime; Gypsum, American Society for Testing and Materials, West Conshohocken, PA.

107. ASTM (2013) C266-08e Standard Test Method for Time of Setting of Hydraulic-Cement Paste by Gillmore Needles. Volume 04.01 Cement; Lime; Gypsum, American Society for Testing and Materials, West Conshohocken, PA2013.

108. Liu, C., Huang, Y. and Zheng, H. (1999) Study of the hydration process of calcium phosphate cement by AC impedance spectroscopy. *J. Am. Ceram. Soc.*, **82**, 1052–1057.

109. Nilsson, M., Carlson, J., Fernandez, E. and Planell, J.A. (2002) Monitoring the setting of calcium-based bone cements using pulse-echo ultrasound. *J. Mater. Sci. - Mater. Med.*, **13**, 1135–1141.

110. Hofmann, M.P., Nazhat, S.N., Gbureck, U. and Barralet, J.E. (2006) Real-time monitoring of the setting reaction of brushite-forming cement using isothermal differential scanning calorimetry. *J. Biomed. Mater. Res. Part B*, **79**, 360–364.

111. Driessens, F.C., Planell, J.A., Boltong, M.G. *et al.* (1998) Osteotransductive bone cements. *Proc. Inst. Mech. Eng. H*, **212**, 427–435.

112. Marino, P.L. and Sutin, K.M. (2007) *The ICU Book*, Lippincott Williams & Wilkins.

113. Fernández, E., Boltong, M.G., Ginebra, M.P. *et al.* (1996) Development of a method to measure the period of swelling of calcium phosphate cements. *J. Mater. Sci. Lett.*, **15**, 1004–1005.

114. Bohner, M., Doebelin, N. and Baroud, G. (2006) Theoretical and experimental approach to test the cohesion of calcium phosphate pastes. *Eur. Cells Mater.*, **29**, 26–35.

115. Hill, S.J. (1999) *Inductively Coupled Plasma Spectrometry and its Applications*, Sheffield Academic Press.
116. Thomas, R. (2003) *Practical Guide to ICP-MS: A Tutorial for Beginners*, Taylor & Francis.
117. Goldstein, J. (2003) *Scanning Electron Microscopy and X-Ray Microanalysis*, Kluwer Academic/Plenum Publishers.
118. Egerton, R.F. (2005) *Physical Principles of Electron Microscopy: An Introduction to TEM, SEM, and AEM*, Springer.
119. West, A.R. (1985) *Solid State Chemistry and Its Applications*, John Wiley & Sons, Ltd.
120. Legeros, R.Z. (1993) Biodegradation and bioresorption of calcium phosphate ceramics. *Clin. Mater.*, **14**, 65–88.
121. Patterson, A.L. (1939) The Scherrer formula for X-ray particle size determination. *Phys. Rev.*, **56**, 978–982.
122. Hsu, H.-C., Tuan, W.-H. and Lee, H.-Y. (2009) In-situ observation on the transformation of calcium phosphate cement into hydroxyapatite. *Mater. Sci. Eng., C*, **29**, 950–954.
123. Bohner, M. and Baroud, G. (2005) Injectability of calcium phosphate pastes. *Biomaterials*, **26**, 1553–1563.
124. Ben-Nissan, B., Sher, D. and Walsh, W. (2003) Effects of spherical tetracalcium phosphate on injectability and basic properties of apatitic cement. *Key Eng. Mater.*, **240–242**, 369–372.
125. Khairoun, I., Boltong, M.G., Driessens, F.C.M. and Planell, J.A. (1998) Some factors controlling the injectability of calcium phosphate bone cements. *J. Mater. Sci. - Mater. Med.*, **9**, 425–428.
126. Ishikawa, K. and Asaoka, K. (1995) Estimation of ideal mechanical strength and critical porosity of calcium phosphate cement. *J. Biomed. Mater. Res.*, **29**, 1537–1543.
127. Ginebra, M.P. (2009) Cements as bone repair materials, in *Bone Repair Biomaterials* (eds J.A. Planell, S.M. Best and S. Lacroix), Woodhead Publishing Ltd, pp. 271–308.
128. Brunner, T.J., Bohner, M., Dora, C. *et al.* (2007) Comparison of amorphous TCP nanoparticles to micron-sized α-TCP as starting materials for calcium phosphate cements. *J. Biomed. Mater. Res. Part B*, **83**, 400–407.
129. Tao, J., Jiang, W., Zhai, H. *et al.* (2008) Structural components and anisotropic dissolution behaviors in one hexagonal single crystal of Î²-tricalcium phosphate. *Cryst. Growth Des.*, **8**, 2227–2234.
130. Lee, D.D., Rey, C., Aiolova, M., and Tofighi, A. (1995) Low temperature calcium phosphate apatite and a method of its manufacture. US 5783217 A, Etex Corporation, USA.
131. Mestres, G., Castaño, O., Navarro, M. *et al.* (2007) Micro and nanostructure evolution study of novel injectable calcium phosphate cements prepared by ceramic and sol-gel processes. *Tissue Eng.*, **13**, 1726.
132. Berne, B.J. and Pecora, R. (1976) *Dynamic Light Scattering: With Applications to Chemistry, Biology, and Physics*, Dover.
133. Maskos, M. and Stauber, R.H. (2011) Characterization of nanoparticles in biological environments, in *Comprehensive Biomaterials* (ed D. Paul), Chapter 3.319 Editor-in-Chief:, Elsevier, Oxford, pp. 329–339.

134. Rasband, W.S. (1997–2011) *ImageJ*, US National Institute of Health, Bethesda, MD.
135. Anmin, J., Kangmei, K. and Weitao, Y. (2007) An injectable cement: synthesis, physical properties and scaffold for bone repair. *J. Postgrad. Med.*, **53**, 34–38.
136. Cook, S.D., Thongpreda, N., Anderson, R.C. *et al.* (1987) Optimum pore size for bone cement fixation. *Clin. Orthop.*, **223**, 296–302.
137. Schliephake, H., Neukam, F.W. and Klosa, D. (1991) Influence of pore dimensions on bone ingrowth into porous hydroxyapatite blocks used as bone graft substitutes. A histometric study. *Int. J. Oral Maxillofac. Surg.*, **20**, 53–58.
138. Takagi, S. and Chow, L.C. (2001) Formation of macropores in calcium phosphate cement implants. *J. Mater. Sci. - Mater. Med.*, **12**, 135–139.
139. Jain, R.K., Au, P., Tam, J. *et al.* (2005) Engineering vascularized tissue. *Nat. Biotechnol.*, **23**, 821–823.
140. Espanol, M., Perez, R.A., Montufar, E.B. *et al.* (2009) Intrinsic porosity of calcium phosphate cements and its significance for drug delivery and tissue engineering applications. *Acta Biomater.*, **5**, 2752–2762.
141. Yang, W.C. (2003) *Handbook of Fluidization and Fluid-Particle Systems*, Taylor & Francis.
142. Lowell, S. and Shields, J.E. (1991) *Powder Surface Area and Porosity*, Springer.
143. Sangeeta, D. and Lagraff, J.R. (2005) *Inorganic Materials Chemistry: Desk Reference*, CRC Press.
144. Abell, A.B., Willis, K.L. and Lange, D.A. (1999) Mercury intrusion porosimetry and image analysis of cement-based materials. *J. Colloid Interface Sci.*, **211**, 39–44.
145. Pirard, R. and Pirard, J.P. (2000) Mercury porosimetry applied to precipitated silica, in *Studies in Surface Science and Catalysis* (eds G. Kreysa, J.P. Baselt and K.K. Unger), Elsevier, pp. 603–611.
146. Lowell, S. (2004) *Characterization of Porous Solids and Powders: Surface Area, Pore Size and Density*, Springer.
147. Uyanik, M.O., Nagas, E., Cubukcu, H.E. *et al.* (2010) Surface porosity of hand-mixed, syringe-mixed and encapsulated set endodontic sealers. *Oral Surg. Oral Med. Oral Pathol.*, **109**, e117–e122.
148. Hoey, D. and Taylor, D. (2009) Quantitative analysis of the effect of porosity on the fatigue strength of bone cement. *Acta Biomater.*, **5**, 719–726.
149. Kohn, M.J., Hanchar, J.M., Hoskin, P.W.O. *et al.* (2002) *Phosphates: Geochemical, Geobiological, and Materials Importance*, Mineralogical Society of America.
150. Chow, L.C. and Eanes, E.D. (2001) *Octacalcium Phosphate*, S Karger A.G., Basel.
151. Cheng, P.-T. (1987) Formation of octacalcium phosphate and subsequent transformation to hydroxyapatite at low supersaturation: a model for cartilage calcification. *Calcif. Tissue Int.*, **40**, 339–343.
152. Combes, C. and Rey, C. (2010) Amorphous calcium phosphates: synthesis, properties and uses in biomaterials. *Acta Biomater.*, **6**, 3362–3378.
153. Egli, R.J., Gruenenfelder, S., Doebelin, N. *et al.* (2010) Thermal treatments of calcium phosphate biomaterials to tune the physico-chemical properties and modify the in vitro osteoclast response. *Adv. Eng. Mater.*, **13**, B102–B107.
154. Kim, H.-M., Kim, Y.-S., Woo, K.-M. *et al.* (2001) Dissolution of poorly crystalline apatite crystals by osteoclasts determined on artificial thin-film apatite. *J. Biomed. Mater. Res.*, **56**, 250–256.

155. Boanini, E., Gazzano, M. and Bigi, A. (2010) Ionic substitutions in calcium phosphates synthesized at low temperature. *Acta Biomater.*, **6**, 1882–1894.

156. Lilley, K.J., Gbureck, U., Knowles, J.C. *et al.* (2005) Cement from magnesium substituted hydroxyapatite. *J. Mater. Sci. - Mater. Med.*, **16**, 455–460.

157. Grover, L.M., Gbureck, U., Wright, A.J. *et al.* (2006) Biologically mediated resorption of brushite cement in vitro. *Biomaterials*, **27**, 2178–2185.

158. Tang, R., Darragh, M., Orme, C.A. *et al.* (2005) Control of biomineralization dynamics by interfacial energies. *Angew. Chem. Int. Ed.*, **44**, 3698–3702.

159. Wang, L. and Nancollas, G.H. (2008) Calcium orthophosphates: crystallization and dissolution. *Chem. Rev.*, **108**, 4628–4669.

160. Julien, M., Khairoun, I., Legeros, R.Z. *et al.* (2007) Physico-chemical-mechanical and in vitro biological properties of calcium phosphate cements with doped amorphous calcium phosphates. *Biomaterials*, **28**, 956–965.

161. Fernández, E., Vlad, M.D., Hamcerencu, M. *et al.* (2005) Effect of iron on the setting properties of alpha-TCP bone cements. *J. Mater. Sci.*, **40**, 3677–3682.

162. Gbureck, U., Knappe, O., Grover, L.M. and Barralet, J.E. (2005) Antimicrobial potency of alkali ion substituted calcium phosphate cements. *Biomaterials*, **26**, 6880–6886.

163. Gbureck, U., Thull, R. and Barralet, J.E. (2005) Alkali ion substituted calcium phosphate cement formation from mechanically activated reactants. *J. Mater. Sci. - Mater. Med.*, **16**, 423–427.

164. Guo, D., Xu, K., Zhao, X. and Han, Y. (2005) Development of a strontium-containing hydroxyapatite bone cement. *Biomaterials*, **26**, 4073–4083.

165. Alkhraisat, M.H., Mariño, F.T., Rodrãguez, C.R. *et al.* (2008) Combined effect of strontium and pyrophosphate on the properties of brushite cements. *Acta Biomater.*, **4**, 664–670.

166. Hamdan Alkhraisat, M., Moseke, C., Blanco, L. *et al.* (2008) Strontium modified biocements with zero order release kinetics. *Biomaterials*, **29**, 4691–4697.

167. Pan, H.B., Li, Z.Y., Lam, W.M. *et al.* (2009) Solubility of strontium-substituted apatite by solid titration. *Acta Biomater.*, **5**, 1678–1685.

168. Capuccini, C., Torricelli, P., Boanini, E. *et al.* (2009) Interaction of Sr-doped hydroxyapatite nanocrystals with osteoclast and osteoblast-like cells. *J. Biomed. Mater. Res. Part A*, **89**, 594–600.

169. Li, X., Sogo, Y., Ito, A. *et al.* (2009) The optimum zinc content in set calcium phosphate cement for promoting bone formation in vivo. *Mater. Sci. Eng., C*, **29**, 969–975.

170. Pina, S., Vieira, S.I., Rego, P. *et al.* (2010) Biological responses of brushite-forming Zn- and ZnSr- substituted beta-tricalcium phosphate bone cements. *Eur. Cells. Mater.*, **20**, 162–177.

171. Pina, S., Vieira, S.I., Torres, P.M. *et al.* (2010) In vitro performance assessment of new brushite-forming Zn- and ZnSr-substituted beta-TCP bone cements. *J. Biomed. Mater. Res. Part B*, **94**, 414–420.

172. Kanatani, M., Sugimoto, T., Fukase, M. and Fujita, T. (1991) Effect of elevated extracellular calcium on the proliferation of osteoblastic MC3T3-E1 cells: its direct and indirect effects via monocytes. *Biochem. Biophys. Res. Commun.*, **181**, 1425–1430.

173. Zaidi, M., Datta, H.K., Patchell, A. *et al.* (1989) 'Calcium-activated' intracellular calcium elevation: a novel mechanism of osteoclast regulation. *Biochem. Biophys. Res. Commun.*, **163**, 1461–1465.

174. Aguirre, A., Gonzalez, A., Navarro, M. *et al.* (2012) Control of microenvironmental cues with a smart biomaterial composite promotes endothelial progenitor cell angiogenesis. *Eur. Cell. Mater.*, **24**, 90–106.

175. Meleti, Z., Shapiro, I.M. and Adams, C.S. (2000) Inorganic phosphate induces apoptosis of osteoblast-like cells in culture. *Bone*, **27**, 359–366.

176. Wu, X., Itoh, N., Taniguchi, T. *et al.* (2003) Requirement of calcium and phosphate ions in expression of sodium-dependent vitamin C transporter 2 and osteopontin in MC3T3-E1 osteoblastic cells. *Biochim. Biophys. Acta, Mol. Cell. Res.*, **1641**, 65–70.

177. Wang, D., Christensen, K., Chawla, K. *et al.* (1999) Isolation and characterization of MC3T3-E1 preosteoblast subclones with distinct in vitro and in vivo differentiation/mineralization potential. *J. Bone Miner. Res.*, **14**, 893–903.

178. Bohner, M., Galea, L. and Doebelin, N. (2012) Calcium phosphate bone graft substitutes: Failures and hopes. *J. Eur. Ceram. Soc.*, **32**, 2663–2671.

179. Fratzl, P., Gupta, H.S., Paschalis, E.P. and Roschger, P. (2004) Structure and mechanical quality of the collagen-mineral nano-composite in bone. *J. Mater. Chem.*, **14**, 2115–2123.

180. Reilly, D.T. and Burstein, A.H. (1974) The mechanical properties of cortical bone. *J. Bone Joint Surg.*, **56**, 1001–1022.

181. Wagoner Johnson, A.J. and Herschler, B.A. (2011) A review of the mechanical behavior of CaP and CaP/polymer composites for applications in bone replacement and repair. *Acta Biomater.*, **7**, 16–30.

182. Giesen, E.B.W., Ding, M., Dalstra, M. and Van Eijden, T.M.G.J. (2001) Mechanical properties of cancellous bone in the human mandibular condyle are anisotropic. *J. Biomech.*, **34**, 799–803.

183. Mcnaught, A.D. and Wilkinson, A. (1997) *UPAC Compendium of Chemical Terminology – the "Gold Book"*, 2nd edn, Oxford Blackwell Scientific Publications, Oxford.

184. Nic, M., Jirat, J., Kosata, B., and Jenkins, A. (2012) Online Gold Book, http://goldbook.iupac.org (accessed 11 November 2013).

185. Dorozhkin, S. (2009) Calcium orthophosphate-based biocomposites and hybrid biomaterials. *J. Mater. Sci.*, **44**, 2343–2387.

186. Verron, E., Khairoun, I., Guicheux, J. and Bouler, J.-M. (2010) Calcium phosphate biomaterials as bone drug delivery systems: a review. *Drug Discov. Today*, **15**, 547–552.

187. Arcos, D. and Vallet-Regí, M. (2013) Bioceramics for drug delivery. *Acta Mater.*, **61**, 890–911.

188. Lode, A., Meissner, K., Luo, Y. *et al.* (2012) Fabrication of porous scaffolds by three-dimensional plotting of a pasty calcium phosphate bone cement under mild conditions. *J. Tissue Eng. Regen. Med.* doi: 10.1002/term.1563

189. Xu, H.H.K. and Quinn, J.B. (2002) Calcium phosphate cement containing resorbable fibers for short-term reinforcement and macroporosity. *Biomaterials*, **23**, 193–202.

190. Xu, H.H.K., Quinn, J.B., Takagi, S. and Chow, L.C. (2004) Synergistic reinforcement of in situ hardening calcium phosphate composite scaffold for bone tissue engineering. *Biomaterials*, **25**, 1029–1037.

191. Canal, C. and Ginebra, M.P. (2011) Fibre-reinforced calcium phosphate cements: a review. *J. Mech. Behav. Biomed. Mater.*, **4**, 1658–1671.

192. Wang, X., Ye, J., Wang, Y. and Chen, L. (2007) Reinforcement of calcium phosphate cement by bio-mineralized carbon nanotube. *J. Am. Ceram. Soc.*, **90**, 962–964.

193. Ahn, G.-S., Seol, D.-W., Pyo, S.-G. and Lee, D.-H. (2011) Calcium phosphate cement-multi-walled carbon nanotube hybrid material (CPC-MWCNT Hybrid) enhances osteogenic differentiation. *J. Tissue. Eng. Regen. Med.*, **8**, 390–397.

194. Linkov, I. and Steevens, J. (2009) *Nanomaterials: Risks and Benefits*, Springer.

195. Aubin, J.E. (2008) Mesenchymal stem cells and osteoblast differentiation, in *Principles of Bone Biology*, 3rd, Chapter 4 edn (eds J.P. Bilezikian, L.G. Raisz and T.J. Martin), Academic Press, San Diego, CA, pp. 85–107.

196. Engel, E., Del Valle, S., Aparicio, C. *et al.* (2008) Discerning the role of topography and ion exchange in cell response of bioactive tissue engineering scaffolds. *Tissue Eng. Part A*, **14**, 1341–1351.

197. Khashaba, R.M., Lockwood, P.E., Lewis, J.B. *et al.* (2010) Cytotoxicity, calcium release, and pH changes generated by novel calcium phosphate cement formulations. *J. Biomed. Mater. Res. Part B*, **93B**, 297–303.

198. Borra, R.C., Lotufo, M.A., Gagioti, S.M. *et al.* (2009) A simple method to measure cell viability in proliferation and cytotoxicity assays. *Braz. Oral Res.*, **23**, 255–262.

199. Escobar, L., Rivera, A. and Aristizabal, F.A. (2010) Estudio comparativo de los métodos de resazurina y MTT en estudios de citotoxicidad en líneas tumorales humanas. *Vitae*, **17**, 67–74.

200. Stoddart, M.J. (2011) Cell viability assays: introduction, mammalian cell viability. *Methods Mol. Biol.*, **740**, 1–6.

201. Xia, Z., Grover, L.M., Huang, Y. *et al.* (2006) In vitro biodegradation of three brushite calcium phosphate cements by a macrophage cell-line. *Biomaterials*, **27**, 4557–4565.

202. Rossa, C., Marcantonio, E., Santos, L.A. *et al.* (2005) Cytotoxicity of two novel formulations of calcium phosphate cements: a comparative in vitro study. *Artif. Organs*, **29**, 114–121.

203. Pioletti, D.P., Takei, H., Lin, T. *et al.* (2000) The effects of calcium phosphate cement particles on osteoblast functions. *Biomaterials*, **21**, 1103–1114.

204. Hou, C.-H., Chen, C.-W., Hou, S.-M. *et al.* (2009) The fabrication and characterization of dicalcium phosphate dihydrate-modified magnetic nanoparticles and their performance in hyperthermia processes in vitro. *Biomaterials*, **30**, 4700–4707.

205. Kokubo, T. and Takadama, H. (2006) How useful is SBF in predicting in vivo bone bioactivity? *Biomaterials*, **27**, 2907–2915.

206. Kokubo, T., Kushitani, H., Sakka, S. *et al.* (1990) Solutions able to reproduce in vivo surface-structure changes in bioactive glass-ceramic A-W3. *J. Biomed. Mater. Res.*, **24**, 721–734.

207. Cho, S.-B., Nakanishi, K., Kokubo, T. *et al.* (1995) Dependence of apatite formation on silica gel on its structure: effect of heat treatment. *J. Am. Ceram. Soc.*, **78**, 1769–1774.

208. El-Ghannam, A. and Ducheyne, P. (2011) Bioactive ceramics, in *Comprehensive Biomaterials* (ed D. Paul), Chapter 1.109 Editor-in-Chief:, Elsevier, Oxford, pp. 157–179.

209. Lee, G.-S., Park, J.-H., Shin, U.S. and Kim, H.-W. (2011) Direct deposited porous scaffolds of calcium phosphate cement with alginate for drug delivery and bone tissue engineering. *Acta Biomater.*, **7**, 3178–3186.

210. Kartsogiannis, V. and Ng, K.W. (2004) Cell lines and primary cell cultures in the study of bone cell biology. *Mol. Cell. Endocrinol.*, **228**, 79–102.

211. Czekanska, E.M., Stoddart, M.J., Richards, R.G. and Hayes, J.S. (2012) Osteoblast; differentiation; primary cells; cell lines; in vitro cell models. *Eur. Cells Mater.*, **24**, 1–17.

212. Harris, H. (1990) The human alkaline phosphatases: what we know and what we don't know. *Clin. Chim. Acta*, **186**, 133–150.

213. Zur Nieden, N.I., Kempka, G. and Ahr, H.J. (2003) In vitro differentiation of embryonic stem cells into mineralized osteoblasts. *Differentiation*, **71**, 18–27.

214. Golub, E.E. and Boesze-Battaglia, K. (2007) The role of alkaline phosphatase in mineralization. *Curr. Opin. Orthop.*, **18**, 444–448.

215. Knabe, C., Driessens, F.C., Planell, J.A. *et al.* (2000) Evaluation of calcium phosphates and experimental calcium phosphate bone cements using osteogenic cultures. *J. Biomed. Mater. Res.*, **52**, 498–508.

216. Zreiqat, H., Markovic, B., Walsh, W.R. and Howlett, C.R. (1996) A novel technique for quantitative detection of mRNA expression in human bone derived cells cultured on biomaterials. *J. Biomed. Mater. Res.*, **33**, 217–223.

217. Reineke, J., Mo, Y., Wan, R. and Zhang, Q. (2012) Application of reverse transcription-PCR and real-time PCR in nanotoxicity research, in *Nanotoxicity*, Humana Press, pp. 99–112.

218. Valasek, M.A. and Repa, J.J. (2005) The power of real-time PCR. *Adv. Physiol. Ed.*, **29**, 151–159.

219. Oreffo, R.O.C., Driessens, F.C.M., Planell, J.A. and Triffitt, J.T. (1998) Growth and differentiation of human bone marrow osteoprogenitors on novel calcium phosphate cements. *Biomaterials*, **19**, 1845–1854.

220. Sanzana, E.S., Navarro, M., Macule, F. *et al.* (2008) Of the in vivo behavior of calcium phosphate cements and glasses as bone substitutes. *Acta Biomater*, **4**, 1924–1933.

221. Eliaz, N., Watering, F.C.J., Beucken, J.J.J.P. *et al.* (2012) *Biodegradation of Calcium Phosphate Cement Composites. Degradation of Implant Materials*, Springer, New York, pp. 139–172.

222. Theiss, F., Apelt, D., Brand, B. *et al.* (2005) Biocompatibility and resorption of a brushite calcium phosphate cement. *Biomaterials*, **26**, 4383–4394.

223. Chang-Chin, W., Chen-Chie, W., Dai-Hua, L. *et al.* (2012) Calcium phosphate cement delivering zoledronate decreases bone turnover rate and restores bone architecture in ovariectomized rats. *Biomed. Mater.*, **7**, 035009.

224. Chen, W., Zhou, H., Tang, M. *et al.* (2012) Gas-foaming calcium phosphate cement scaffold encapsulating human umbilical cord stem cells. *Tissue Eng. Part A*, **18**, 816–827.

225. Montufar, E.B., Traykova, T., Planell, J.A. and Ginebra, M.P. (2011) Comparison of a low molecular weight and a macromolecular surfactant as foaming agents for

injectable self setting hydroxyapatite foams: polysorbate 80 versus gelatine. *Mater. Sci. Eng., C*, **31** (7), 1498–1504.

226. Eriksson, F., Mattsson, P. and Larsson, S. (2002) The effect of augmentation with resorbable or conventional bone cement on the holding strength for femoral neck fracture devices. *J. Orthop. Trauma*, **16**, 302–310.

227. Müller, M. and Stangl, R. (2006) Norian-SRS-Augmentation bei Revision der Hüfttotalendoprothesen-Pfannenkomponente. *Der Unfallchirurg*, **109**, 335–338.

228. Panchbhavi, V.K. (2010) Synthetic bone grafting in foot and ankle surgery. *Foot Ankle Clin.*, **15**, 559–576.

229. Goff, T., Kanakaris, N.K. and Giannoudis, P.V. (2013) Use of bone graft substitutes in the management of tibial plateau fractures. *Injury*, **44** (Suppl. 1), S86–S94.

230. Friedman, C.D., Costantino, P.D., Takagi, S. and Chow, L.C. (1998) BoneSource™ hydroxyapatite cement: a novel biomaterial for craniofacial skeletal tissue engineering and reconstruction. *J. Biomed. Mater. Res.*, **43**, 428–432.

231. Aral, A., Yalçın, S., Karabuda, Z.C. *et al.* (2008) Injectable calcium phosphate cement as a graft material for maxillary sinus augmentation: an experimental pilot study. *Clin. Oral Implants Res.*, **19**, 612–617.

232. Stankewich, C.J., Swiontkowski, M.F., Tencer, A.F. *et al.* (1996) Augmentation of femoral neck fracture fixation with an injectable calcium-phosphate bone mineral cement. *J. Orthop. Res.*, **14**, 786–793.

233. Zhao, J., Liu, Y., Sun, W.-B. and Zhang, H. (2011) Amorphous calcium phosphate and its application in dentistry. *Chem. Cent. J.*, **5**, 40.

234. Sugawara, A., Fujikawa, K., Kusama, K. *et al.* (2002) Histopathologic reaction of a calcium phosphate cement for alveolar ridge augmentation. *J. Biomed. Mater. Res.*, **61**, 47–52.

235. Arisan, V., Anil, A., Wolke, J.G. and Özer, K. (2010) The effect of injectable calcium phosphate cement on bone anchorage of titanium implants: an experimental feasibility study in dogs. *Int. J. Oral Maxillofac. Surg.*, **39**, 463–468.

236. Zhang, W., Walboomers, X.F. and Jansen, J.A. (2008) The formation of tertiary dentin after pulp capping with a calcium phosphate cement, loaded with PLGA microparticles containing TGF-β1. *J. Biomed. Mater. Res. Part A*, **85A**, 439–444.

237. Lee, S.-K., Lee, S.-K., Lee, S.-I. *et al.* (2010) Effect of calcium phosphate cements on growth and odontoblastic differentiation in human dental pulp cells. *J. Endod.*, **36**, 1537–1542.

238. Ginebra, M.-P., Canal, C., Espanol, M. *et al.* (2012) Calcium phosphate cements as drug delivery materials. *Adv. Drug Delivery Rev.*, **64**, 1090–1110.

239. Hayashi, C., Kinoshita, A., Oda, S. *et al.* (2006) Injectable calcium phosphate bone cement provides favorable space and a scaffold for periodontal regeneration in dogs. *J. Periodontol.*, **77**, 940–946.

240. Varma, H., Rajesh, J., Nandakumar, K. and Komath, M. (2009) Calcium phosphate cement as a "barrier-graft" for the treatment of human periodontal intraosseous defects. *Indian. J. Dent. Res.*, **20**, 471–479.

241. Ng, V.Y., Granger, J.F. and Ellis, T.J. (2010) Calcium phosphate cement to prevent collapse in avascular necrosis of the femoral head. *Med. Hypotheses*, **74**, 725–726.

242. Civinini, R., Biase, P., Carulli, C. *et al.* (2012) The use of an injectable calcium sulphate/calcium phosphate bioceramic in the treatment of osteonecrosis of the femoral head. *Int. Orthop.*, **36**, 1583–1588.

243. Hisatome, T., Yasunaga, Y., Ikuta, Y. and Fujimoto, Y. (2002) Effects on articular cartilage of subchondral replacement with polymethylmethacrylate and calcium phosphate cement. *J. Biomed. Mater. Res.*, **59**, 490–498.

244. Hong, S.-J., Park, Y.-K., Kim, J.H. *et al.* (2006) The biomechanical evaluation of calcium phosphate cements for use in vertebroplasty. *J. Neurosurg.: Spine*, **4**, 154–159.

245. Khanna, A.J., Lee, S., Villarraga, M. *et al.* (2008) Biomechanical evaluation of kyphoplasty with calcium phosphate cement in a 2-functional spinal unit vertebral compression fracture model. *Spine J.*, **8**, 770–777.

246. Ginebra, M.P., Espanol, M., Montufar, E.B. *et al.* (2010) New processing approaches in calcium phosphate cements and their applications in regenerative medicine. *Acta Biomater.*, **6**, 2863–2873.

247. Guo, D., Xu, K. and Han, Y. (2009) The in situ synthesis of biphasic calcium phosphate scaffolds with controllable compositions, structures, and adjustable properties. *J. Biomed. Mater. Res. Part A*, **88**, 43–52.

248. Del Real, R.P., Ooms, E., Wolke, J.G. *et al.* (2003) In vivo bone response to porous calcium phosphate cement. *J. Biomed. Mater. Res. Part A*, **65**, 30–36.

249. Fernandez, E., Vlad, M.D., Gel, M.M. *et al.* (2005) Modulation of porosity in apatitic cements by the use of alpha-tricalcium phosphate-calcium sulphate dihydrate mixtures. *Biomaterials*, **26**, 3395–3404.

250. Qi, X., Ye, J. and Wang, Y. (2009) Alginate/poly (lactic-co-glycolic acid)/calcium phosphate cement scaffold with oriented pore structure for bone tissue engineering. *J. Biomed. Mater. Res. Part A*, **89**, 980–987.

251. Habraken, W.J.E.M., Wolke, J.G.C., Mikos, A.G. and Jansen, J.A. (2006) Injectable PLGA microsphere/calcium phosphate cements: physical properties and degradation characteristics. *J. Biomater. Sci., Polym. Ed.*, **17**, 1057–1074.

252. Zuo, Y., Yang, F., Wolke, J.G.C. *et al.* (2010) Incorporation of biodegradable electrospun fibers into calcium phosphate cement for bone regeneration. *Acta Biomater.*, **6**, 1238–1247.

253. Habibovic, P., Gbureck, U., Doillon, C.J. *et al.* (2008) Osteoconduction and osteoinduction of low-temperature 3D printed bioceramic implants. *Biomaterials*, **29**, 944–953.

254. Miao, X., Hu, Y., Liu, J. and Wong, A.P. (2004) Porous calcium phosphate ceramics prepared by coating polyurethane foams with calcium phosphate cements. *Mater. Lett.*, **58**, 397–402.

9

Bioceramic Coatings for Medical Implants

M. Victoria Cabañas
Departamento de Química Inorgánica y Bioinorgánica, Facultad de Farmacia,
Universidad Complutense de Madrid, Spain

9.1 Introduction

The chemical, physical, and biological properties of different bioceramics with clinical applications have been discussed throughout the different chapters. Each type of material used in implant devices exhibits specific advantages that make them particularly suitable for specific applications. However, it is difficult for a single material to fulfill all the required properties. This chapter is focused on the use of these bioceramics, in the form of coatings, aimed at extending their applications and achieving improvements in the performance and reliability of existing biomedical implants.

A wide range of bioceramics has been developed and produced but, despite their excellent biological performance, some of them are brittle and have poor fracture toughness, and they cannot be used in load bearing applications. Therefore, metals are generally used for implants where high strength is required. Metallic materials, such as stainless steel, cobalt-based alloys, and mostly titanium and its alloys, are widely used in biomedical devices and components, especially as hard tissue replacements, internal fixation devices, and in cardiovascular applications. The extended use of titanium and titanium-based alloys is due to their biocompatibility and excellent mechanical properties (tensile strength and fatigue strength) under *in vivo* conditions [1]. However, there are some problems associated with the use of these metallic implants: poor implant fixation due to the lack of osteoconductivity and osteoinductivity; and corrosion resistance and wear in biological environments leading to ions release and formation of wear debris. These drawbacks stem from some surface originated problems associated with the metallic implants. In this sense, the material

Bioceramics with Clinical Applications, First Edition. Edited by María Vallet-Regí.
© 2014 John Wiley & Sons, Ltd. Published 2014 by John Wiley & Sons, Ltd.

surface plays an important role in the response of the biological environment to the artificial devices because the biological reactions occur directly on the implant surface from the moment that it is placed into a body.

Therefore, sometimes it is necessary to modify the substrate surface when specific surface properties, different from those in the bulk, are required. For example, in order to accomplish biological integration, good bone formability is desired. In blood-contacting devices, such as artificial heart valves, blood compatibility is crucial. In other applications, good wear and corrosion resistance are also required.

In the following sections, the improvement of the above-mentioned features (bioactivity, blood compatibility, or wear and corrosion resistance) of existing clinical implants by using various surface modification technologies is discussed. However, the functionality of an implant depends not only on the surface modification but also on the ceramic used for this modification, that is, the chemical and physical features of said ceramic will determine the final application of the implant.

This chapter is divided into two parts dealing with bioactive ceramic coatings and bioinert ceramic coatings, respectively. The success of the first type depends on their ability to induce bone regeneration and bone ingrowth at the tissue/implant interface without an intermediate fibrous tissue layer in the fixation of joint prostheses and dental implants. For these applications the use of bioactive ceramics, such as calcium phosphates and bioglasses, is recommended. On the other hand, the characteristics of inert bioceramics, for example, their tribological properties, render them suitable as protective coatings when deposited onto metallic substrates, preventing the release of metal ions and reducing the release of wear debris in the articulation of artificial joints. The blood compatibility of carbon-based materials can minimize thrombus formation when used in cardiovascular devices, such as prosthetic heart valves or vascular stents.

Besides the surface modifications with a bioceramic mentioned above, new features can be introduced into an implant by incorporation of active biological molecules. Bioactive growth factors immobilized on the surface of orthopedic and dental implants are capable of inducing rapid cell functions, including cellular proliferation and differentiation activity, thus accelerating tissue regeneration. Also, the incorporation of anti-bacterial agents, such as antibiotics (vancomycin, gentamicin, tobramycin, etc.) or Ag-related agents into the coating, namely *antibacterial coatings*, is very interesting in order to prevent implant-associated infections due to bacterial adhesion on the implant surface, reducing the incidence of post-surgical infection. The behavior of ceramic coatings as drug delivery systems is dealt with in Chapter 12.

9.2   Methods to Modify the Surface of an Implant

Thin film is the general term used for coatings which modify and increase the functionality of a bulk surface or substrate. Different names can be used for such a film, coating, or layer, but they all fall into the broad subject area of surface engineering, which implies the modification of a substrate surface. The objective of surface modification is to improve the macroscopic surface properties for enhanced applications [2].

Thin films have distinct advantages over bulk materials. Since most processes used to deposit thin films are non-equilibrium in nature, the composition of thin films is not constrained by phase diagrams. Film properties, such as crystalline phase composition, microstructure, or surface morphology, depend on and can be modified by the deposition process; not all deposition technologies produce materials with the same properties. New advanced deposition technologies and hybrid processes are being used and developed to deposit advanced films unobtainable with conventional techniques.

As mentioned above, the biological response to biomaterials and devices and, therefore, the lifespan and performance of implants, is controlled largely by their surface chemistry and structure. The rationale for the surface modification of biomaterials is therefore straightforward: to preserve the key physical properties of the implant while only the outermost surface is modified to improve biointeraction. If such surface modification is properly effected, the bulk properties and functionality of the device will be unaffected, but the tissue-interface-related biocompatibility will be improved or changed.

Two chemical approaches can be considered to modify the surface of the material, although both approaches can also be used together (Figure 9.1):

1. **Deposited coatings**: a ceramic film, deposited onto the metal surface, which can be totally different from the substrate. That is, coating the surface with a new material without modifications to the substrate material.
2. **Conversion coatings or surface-modified layers**: chemical surface modification of a metallic substrate resulting in slight increase in thickness. In this case, the substrate material components are involved in the modified layer formation.

On the other hand, and although this issue will not be addressed here, it is necessary to mention that prior to the chemical surface modification, a mechanical surface modification is usually performed on the substrates, such as polishing, blasting, or grinding, in order to modify the surface roughness of the substrate and ensure good mechanical interlocking of coatings on metals, improving bonding strength and shear resistance. In this sense, substrate preparation is critical, and the resulting surface will influence the properties, structure, adhesion, and surface texture of the new deposited film.

9.2.1   Deposited Coatings

There are a very large and varied number of techniques to deposit coatings on a substrate. In fact, almost all types of materials can be deposited onto similar or dissimilar substrates. These techniques can be classified according to different features: as a function of the precursor state (gas, liquid, or solid); of the temperatures utilized for the deposition; or simply by their chemical and physical principles of operation. In this chapter, the technologies available for the deposition of bioceramic coatings have been classified as *physical techniques* and *chemical techniques* (Figure 9.2).

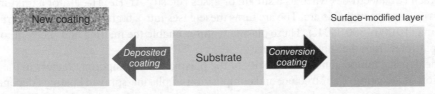

Figure 9.1   *Schematic picture showing two approaches to modifying the material surface*

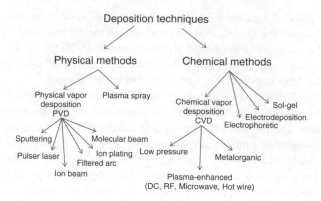

Figure 9.2   *Common deposition techniques to coat metallic implants with bioceramics*

Some of the more usual techniques for the deposition of biomedical coatings will be described briefly below. Each technique has its advantages and disadvantages. The choice of a given technique will depend on various selection criteria, specific application, and available resources.

9.2.1.1   Physical Techniques

During some surface modification processes, such as thermal spraying and physical vapor deposition (PVD), no chemical reactions are present. In this case, the formation of a layer onto the substrate is mainly attributed to the thermal, kinetic, and electrical energy.

Plasma Spraying.   Thermal plasma spraying is a process in which materials are thermally melted into liquid droplets and introduced energetically to the surface of the substrate on which the individual particles stick and condense [3, 4]. The film is formed by a continuous build-up of successive layers of liquid droplets, softened material domains, and hard particles. Thermal spraying requires a device that creates a high temperature flame or a plasma jet. There are many thermal spray processes: *plasma spraying*, also known as *powder plasma spray* or *atmospheric plasma spraying* (PS or APS); *vacuum plasma spraying* or *low-pressure plasma spraying* (VPS or LPPS); *suspension plasma spraying* (SPS); *liquid plasma spraying* (LPS); *high-velocity suspension flame spraying* (HVSFS); and *high velocity oxy-fuel* (HVOF) [5]. However, plasma spraying is nowadays the most popular commercial method to prepare biomedical coatings.

PS involves the spraying of molten material onto a substrate surface to provide a coating (Figure 9.3). It uses electrical energy as the source to create the plasma and can provide a very high temperature that is determined by the energy input. An electric arc is struck between two electrodes whereas a stream of gases (mostly Ar, He, H_2, N_2, or a mixture of them) passes through this arc. The arc turns these gases into a high speed, high temperature plasma ($10\,000-300\,000$ K). These plasma features enable the melting and acceleration of the particles from the supplied powder.

The density, temperature, and velocity of the plasma arc are important factors in the formation of a coating. The structure and composition of plasma-sprayed coatings depend on process parameters, such as atmosphere (plasma gas and flow), particle size of the powder, cooling rate, and so on. This method is usually considered to be a cold process, although the

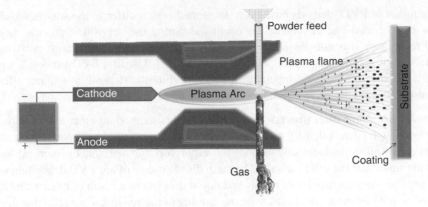

Figure 9.3 *Schematic diagram of the typical plasma spray set-up*

temperature of the sprayed particles may be rather high. Appropriate cooling techniques keep the substrate temperature below $100-150\,^\circ C$.

The advantages of plasma sprayed coatings include high deposition rates, ability to coat large areas, thick deposits, low capital cost, and low operating cost. However, it is not possible to coat homogeneously porous surfaces or complicated substrates, because it is a line-of-sight-process.

Physical Vapor Deposition (PVD). PVD is a general term used to describe a variety of vacuum techniques to deposit thin films by the condensation of a vaporized form of the solid material onto a surface [6]. The coating method involves physical processes. In a vacuum chamber, the solid source is evaporated or sputtered to form atoms, molecules, or ions that are subsequently transported to the substrate surface on which condensation take place leading to film growth (Figure 9.4).

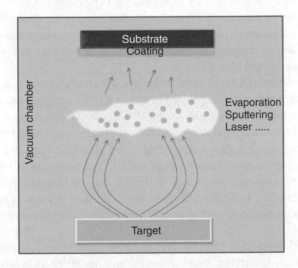

Figure 9.4 *A schematic diagram of the PVD process*

A number of PVD methods have been developed which differ in the way in which the vapor is generated and the conditions employed during the deposition process. In most PVD techniques it is not possible to uniformly coat non-planar substrates; only regions in the line-of sight of the vapor source are coated [7]. Usually, PVD processes can be carried out at low substrate temperatures, allowing thin film deposition without affecting bulk properties. Some PVD techniques are briefly described here:

Cathodic arc deposition process (Arc-PVD). In this method an electric arc, with continuous high current density and low voltage electricity is used to vaporize the cathode material while simultaneously ionizing the vapor, forming a plasma by microexplosions at the surface of the solid cathode. The main disadvantage of arc-PVD deposition is that it produces macroparticles of metals and liquid droplets as a result of intense and localized heating from the arc. To decrease the number of molten macroparticles that degrade the coating performance, *Filtered cathodic arc deposition (FCA)* processes were developed such that, before coating the substrate, all particles pass through a filtering electric field and focusing magnetic field.

Sputtering deposition. Sputtering is the process whereby atoms or molecules of a target material are ejected in a vacuum chamber by bombardment with high-energy ions and the ejected atoms must be able to move freely toward the substrate [8]. A DC glow discharge can be used to sputter conductive targets but radio frequency *(RF) sputtering* is preferred for insulating targets. A better control of the deposition process is obtained by using *ion beam sputtering*. A drawback inherent to all these sputter deposition techniques is the low deposition rate. The development of magnetron sputtering sources that provide relatively high deposition rates has increased the effectiveness of the sputtering process.

A *magnetron sputtering* system works on the principle of applying a specially shaped magnetic field to a diode sputtering target. The magnetic field allows an increase in the ionization efficiency and the ion current density. Then, magnetron sputtering is considered as a high-rate vacuum coating technique for depositing a wide range of materials. The substrate can be kept at a constant temperature during the deposition process (e.g., 100 °C) or be subjected to plasma irradiation without any additional external heating. Some of the main advantages of sputtering deposition include: uniform coating thickness, dense coating, and high purity films.

Pulser laser deposition (PLD) or laser ablation. This technique is based on the irradiation of a solid target by a focused pulsed laser beam resulting in a gaseous cloud, formed by electrons, atoms, molecules, and so on [9]. The plasma cloud expands, in vacuum or in a gaseous environment, and deposits onto a substrate as a film. As the energy source is located outside the chamber, it is possible to use ultrahigh vacuum as well as an inert or reactive gas atmosphere. One advantage of PLD technology is that the stoichiometry of the deposited coating is very close to that of the used target (stoichiometry transfer), a very interesting fact for the deposition of complex systems. This stoichiometric transfer is difficult to obtain with other PVD techniques, like evaporation or magnetron sputtering.

Ion beam assisted deposition (IBAD). This technique combines ion beam bombardment with simultaneous sputtering, evaporation, or another PVD technique in a high vacuum environment [10]. First, material sputtered/evaporated from the target deposits on a substrate forming a film. A second ion source, constituted by inert or reactive ions, is directly focused at the substrate to modify the properties of the growing film by some combination of bombardment or reaction within it. The concurrent ion bombardment differentiates IBAD from other film deposition techniques. It improves adhesion, and allows one to control morphology, density, crystallinity, and chemical composition of the deposited films when compared with similar films prepared by PVD without ion bombardment. It is a versatile technique, and a line-of-sight technique.

9.2.1.2 Chemical Techniques

Chemical Vapor Deposition. Chemical vapor deposition (CVD) is a process which involves heterogeneous chemical reactions of gaseous reactants on/near the surface of a heated substrate, thereby coating said substrate [11] (Figure 9.5). Usually, volatile by-products are produced, which are removed by gas flow through the reaction chamber.

Chemical reaction types basic to CVD include pyrolysis (thermal decomposition), oxidation, reduction, disproportionation, hydrolysis, and chemical transport. The temperature at which the coating is deposited is very important because it determines both the thermodynamics and the kinetics of the coating process.

The CVD processes may be carried out at *atmospheric pressure* (conventional CVD) or at *low pressure* (Low pressure chemical vapor deposition, LPCVD), both options require high temperatures to deposit the coat. The conventional CVD method uses thermal energy to activate the chemical reactions; these reactions can also be initiated using different energy sources, giving rise to other versions of CVD with higher deposition rates: *plasma enhanced chemical vapor deposition (PECVD)* or glow discharge CVD which use electron energy (plasma) as the activation method to enable low temperature deposition. There are three main processes based on PECVD: *microwave-plasma, hot filament,* and *plasma arc* depending on the source used to generate the plasma. *Photo-assisted chemical vapor deposition* (PACVD) is a process that relies on absorption of light to raise the substrate temperature and causes thermal decomposition of the precursor in the gas phase on the substrate surface. In *Flame assisted chemical vapor deposition (FACVD)* a flame source is used to initiate the chemical reaction. Other variants of CVD include *metal organic-chemical vapor deposition (MOCVD)* which uses metal organic compounds as precursors

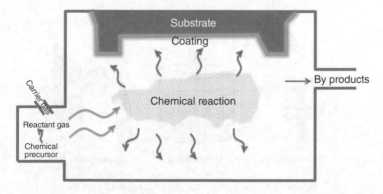

Figure 9.5 *Schematic illustration of the CVD process*

and *aerosol assisted chemical vapor deposition (AACVD)* which uses special precursor generation and delivery systems, unlike conventional CVD.

CVD is not a line-of-sight process. Therefore, it can be used to uniformly coat complex-shaped components and deposit films with good conformal coverage. CVD can provide highly pure materials with structural control at atomic or nanometer scale levels. The flexibility of using a wide range of chemical volatile precursors enables the deposition of a large number of materials. The use of sophisticated reactor and/or vacuum systems by CVD versions such as low pressure, plasma-assisted CVD, and PACVD tends to increase the cost of fabrication. However, other variants such as AACVD and FACVD that do not use sophisticated reactor and/or vacuum systems may provide alternatives for applications where production cost is an issue.

Sol–Gel Technique.   The sol–gel method is a wet-chemical coating technique. This technique has been described in previous chapters for powder and monoliths preparation. One important characteristic of this process is that before the gelation, the solution is ideal for the preparation of coatings [12, 13].

Sol–gel thin films are usually produced using *spin-coating, dip-coating, or spray-coating techniques* and inexpensive equipment is utilized (Figure 9.6).

In *spin-coating* a solution is dropped at the center of the spinning substrate and is spread by centrifugal force. The film is thus formed while a shearing force is applied between both interfaces of the drying solution layer. For complex-shaped substrates, the most commonly used sol–gel technique is *dip-coating*. In this process, the sample is dipped in the solution containing the precursors and then withdrawn at a constant rate, usually with the help of a motor. Gravitational draining and solvent evaporation, accompanied by further condensation reactions, result in the deposition of a solid film. In *spray-coating* the solution is pulverized onto the surface of the substrate using an aerosol generator or an atomizer.

Sol–gel is widely used to deposit ceramic coatings. Various parameters, such as the choice of precursor, concentration of precursor, solvent, pH, aging temperature, or aging time, determine the chemistry of the process and the quality/phase of the final coating.

The advantages of the sol–gel process are based on control of the chemical composition and microstructure of the coating, its ability to form a uniform coating over complex geometric shapes, and the use of simple equipment. A disadvantage of this

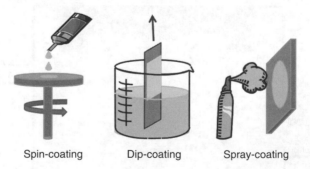

Spin-coating   Dip-coating   Spray-coating

Figure 9.6   *Different sol–gel coating techniques*

method is the difficulty in preparing thick films ($> 1 \, \mu m$), due to the formation of cracks. Sol–gel chemistry also offers new and interesting possibilities for the encapsulation of heat-sensitive and fragile biomolecules, enzymes, proteins, cells, and so on [14].

Electrophoretic Deposition (EDP).   This is based on the movement of electrically charged particles under an electric field applied to a stable colloid suspension, and the subsequent deposition onto a conductive substrate surface of opposite charge [15]. Two types can be distinguished, *cathodic* EDP and *anodic EDP*, depending on where the deposition occurs. The migration of the particles will be faster and more uniform when submicron and homogeneous particles form the stable colloid suspension. The liquid medium used to suspend the particles must have a dielectric constant that gives effective coating characteristics. Electrophoresis can be done at constant voltage or dynamic voltage, producing coatings with different characteristics. In general, a short time is required for coating (a few seconds to a few minutes).

This method represents an important technological alternative due to its simplicity, low cost, and chemical composition control of the coatings. In addition, it is a non-line of sight coating process, therefore substrates with wide areas and/or complex shapes can be easily coated. However, it is difficult to produce coatings without cracks and the bonding strength between the coating and the metal substrate is weak, hence a subsequent heat treatment must be carried out to sinter and bond the coating to the substrate.

9.2.2   Conversion Coatings

In this approach the surface modification is produced by a treatment that chemically transforms the outer layers of the metallic substrate surface. Unlike deposition techniques, the components of the substrate are involved in the layer formation. Usually, the produced layers are inorganic and show ceramic-like character. This surface modification can be found in the bibliography as conversion coatings [16], surface-modified layer [4], or ceramification [17].

The surface-modified layer may be formed by specific reactions between the substrate material and the environment. *Chemical methods*, such as chemical immersion, oxidation processes, passivation, and anodization of the metallic substrate lead to the formation of a ceramic layer on its surface.

On the other hand, *physical methods* such as ion implantation allow modification of the chemical composition at the near surface by penetration of energetic ions into the substrate surface via bombardment at low temperatures [10]. Diffusion of new elements, such as nitrogen, oxygen, or carbon, into the metal substrate at elevated temperatures is also used to produce surface-modified layers, obtaining a transformation of the surface from metal to ceramic. The advantage of the surface-modified layer is the graded composition and the absence of a clearly defined interface between the surface layer and substrate, making surface delamination less of a problem. Such conversion coatings represent an effective way to modify the surface of a material, or can be used as a pre-treatment to improve the adhesion of a final coating.

These methodologies are also used to improve the bioactivity and biocompatibility as well as the wear and corrosion resistance of the metallic implant devices.

9.3   Bioactive Ceramic Coatings

One of the main challenges in orthopedic surgery is to improve the clinical success of total hip arthroplasty/total knee arthroplasty (THA/TKA). To date, in the European Union and United States, more than 1 200 000 hip and knee arthroplasties are performed each year, with increasing tendency [18]. However, not only are joint replacement interventions, due fundamentally to diseases (osteoporosis or osteoarthritis) or trauma, on the rise; the revision of hip/knee implants due to implant failure has also increased.

In the clinical use, the endoprosthetic implant is usually fixed to the bone by using a polymeric cementing paste, polymethylmethacrylate, PMMA (cemented prostheses). Although this method ensures immediate fixation it poses some problems owing to the risk of necrotic damage of the bone cell by the heat liberated during *in vivo* polymerization or the generation of small particles that can migrate within the body (particle disease). On the other hand, this type of prosthesis is not suitable for younger active people, where more stable fixation and bone growth are needed. An alternative solution is to promote the adhesion by means of the so-called *biological* or *bioactive fixation*: cementless prostheses are utilized whose osseointegration is facilitated by coating the metallic part of the prosthesis with a bioactive ceramic. The concept of biological fixation of load-bearing implants using bioactive hydroxyapatite (HA) was proposed as an alternative to cemented fixation in the mid-1980s [19].

Nowadays, two approaches are considered in order to improve the osteoconductivity of titanium-based prostheses:

1. Coating the surface of the implant with a bioactive ceramic (deposited coatings).
2. Modifying the surface of the titanium to form a surface-modified layer, which is effective in inducing bone-like apatite formation.

For the first approach, as shown in Chapters 3 and 4, synthetic materials like HA-based calcium phosphate and bioactive glasses (BGs) are able to form bone-like apatite in their surface after implantation and can bond to the living bone throughout this apatite layer, without mediation of a fibrous connective tissue interface. This process is related to solution-mediated surface reactions, with time-dependent kinetics, that take place close to the surface of these bioactive ceramics. These bioactive ceramics differ in the resorption rate: HA resorbs very slowly compared to BGs. Hence, although HA and BGs are too brittle for use as bulk materials under loaded conditions, they are proposed as suitable coatings to provide stronger early fixation of cementless artificial prostheses. Bioactive ceramic coatings onto metallic implants serve different purposes:

- To promote a rapid fixation of the devices to the skeleton by enhancing bone ingrowth and stimulating osseous apposition to the implant surface.
- To act as surface protective barriers, minimizing or preventing metallic ion release and reducing the danger of corrosion.
- To serve as a carrier system for the slow and controlled release of biologically active substances, such as drugs or growth factors, at the implantation site.

9.3.1   Clinical Applications

Bioactive ceramic coatings have great potential for bone fixation applications in which the bioactive coating in direct contact with the bone can promote a faster attachment to bone tissue and provide a mechanically stable interface between implant and bone. The presence of the bioactive coating will improve fixation as well as the medical implant durability. Bioceramic coatings can be potentially applied in both orthopedic prostheses and dental implants (Figure 9.7), where metallic materials are generally used [8, 20].

Orthopedic prostheses include both reconstructive implants (hip, elbow, ankle, and knee implants) and fixation devices (bone plates, screws, or pins). The first are used to replace joints deteriorated by disease or injury. In artificial joints, such as a hip prosthesis, a bioceramic coating is usually applied to the metallic surface of the femoral stem (the entire stem or the proximal part only), and/or to the outer surface of the acetabular cup.

On the other hand, in a bone fracture, an early and complete restoration can be achieved by osteosynthesis. Metallic fixation, using bone screws and bone plates, is the common bone-fracture treatment method. For instance, screws are used to fix plates or other devices to the bones, whereas bone plates are used as bridging fixations. The presence of bioactive surfaces on these fixations can improve the osseointegration, decreasing the risk of infection and mechanical failure of fracture fixation.

The dental metallic implants are usually inserted into the jaw bones to replace single or multiple missing teeth, and are designed to help preserve bone structure. The commonly used endosseous implants are of root-form type, and bioactive ceramic coatings can be used to cover the surface of metal dental posts.

For these potential applications, and as discussed throughout this chapter, numerous studies have been performed by using ceramic coatings of HA, BGs, and BGCs (bioactive glass-ceramics). Various deposition methods as well as chemical modifications of the bioceramics have been developed and tested in order to obtain metallic implants with a good performance. Prior to any clinical trials, all new biomaterials must undergo *in vitro* and *in vivo* evaluation.

At present, HA coatings deposited mostly by PS have been used to coat metallic implants with commercial clinical performance, although other coating technologies have also found commercial applications. However, and in spite of their potential, there are no available commercial metallic implants coated with BGs yet.

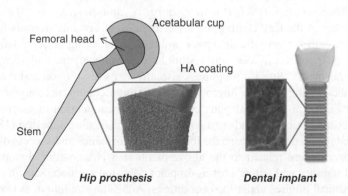

Figure 9.7   *A schematic picture showing a typical hip prosthesis and dental implant*

9.3.2 Calcium Phosphates-Based Coatings

The similarity between HA and the mineral phase of bone and tooth is suggested as the reason behind the early adoption of HA for coating biocompatible metallic implants. Plasma-sprayed HA coatings were first introduced to improve fixation between bone and implant [19]. The first reported clinical trials of HA coatings were with hip femoral stems, by Furlong and Osborn [21] and by Geesink [22] in the mid-1980s. After more than two decades of clinical use, it is generally accepted that the originally pursued benefits of HA coatings can be achieved: the combination of the metal mechanical characteristics with the osteoconductive properties of HA has proven suitable to stimulate an early integration and enhance fixation strength, but doubts still exist concerning the long-term stability of non-cemented orthopedic fixation.

The osseointegration process of a HA-coated implant is the result of the HA coating reaction with the environment. A schematic view of the different stages in this process was depicted in Figure 3.2. Among these consecutive events, dissolution of HA coating is a key step to induce precipitation of bone-like apatite on the implant surface and to enhance the bone regeneration process with the ability to even overcome a 1 mm gap between implant and bone.

On the other hand, the success of the implant depends on the stability of the ceramic coating which is controlled, fundamentally, by its chemical, physical, and mechanical properties. In this sense, the main *CaP coating characteristics* that determine its performance are collected below, although a careful balance of all of them is necessary: [5, 8, 20, 23]

- *The phase or chemical composition and crystallinity.* The presence of CaP phases in the coatings, which exhibit higher solubility in aqueous medium than HA (see Table 3.2), is desirable to accelerate bone formation. However, this faster dissolution decreases the stability and increases the potential loosening of the implant. Highly crystalline HA, on the other hand, dissolves slowly in physiological solutions, but this low dissolution rate can provide long-term stability of the implant in clinical use. Therefore, to prepare implants with predictable properties it is necessary to design and control the crystallinity and purity of the coatings. According to the International Organization for Standardization [24], the mass fraction of HA shall be 50% or greater and the maximum allowable level of other crystalline phases shall be 5%. The HA phase shall have a crystallinity value not less than 45%.

- *Surface characteristics (microstructure, porosity, and roughness).* The denser the microstructure of the CaP coating the lower the risk of bonding degradation during *in vivo* contact with the body fluids (fewer problems of cracking or degradation). Rough, textured, and porous surfaces could stimulate cell attachment and formation of an extra-cellular matrix. Surface roughness also affects its dissolution and the bone apposition on the coatings or bone ingrowth. An optimum coating porosity and roughness are important for in-growth of bone cells, but the accumulation of macropores at the substrate/coating interface leads to a weakening of the coating adhesion [18].

- *Mechanical properties* determine the long-term performance and success of an implant. These properties are related to the above-mentioned HA coating characteristics. An interfacial stability and strong coating-implant adhesion are necessary in order to prevent mechanical failures under load conditions. Adhesion strength of at least 35 MPa is desired [18]. Failure of the implant takes place when detachment of the coating occurs.

The different thermal expansion coefficients of HA and Ti/Ti alloys are relevant and may result in dangerous thermal residual stress. The residual stresses within the coating increase with thickness and may promote cracking at the substrate/coating interface. In a very thick coating, the outer layers tend to detach from the implant, resulting in poor adhesion, but a thin HA coating will be prematurely resorbed during the bone integration. Also, porous or crystalline coatings show poor adhesion with the substrate.

These coating characteristics depend on and can be modified by the deposition method utilized. There are plenty of methods in use nowadays to prepare HA coatings onto titanium (or titanium-based alloys). Some of the most commonly investigated and tested deposition methods are described below, but the basic principles of these deposition techniques were described in Section 9.2.

9.3.2.1 Plasma Spray Deposition

PS deposition [5, 19, 20, 25] is the most commercially used method for orthopedic devices, due to its high deposition rate and the ability to coat large areas. *In vivo* reports of plasma-sprayed HA-coated metallic implants have shown good bone-bonding ability and fast osseointegration, although it suffers from certain drawbacks, especially with regard to the long-term stability.

To prepare the coating, usually a crystalline HA with well-defined particle size and particle size distribution is used as powder feed. The powder is injected into a high temperature plasma flame; HA particles are partly melted and projected at high velocity onto the substrate surface to form the coating. Plasma sprayed coatings must be at least $50\,\mu m$ thick to completely cover the implant. The bonding of the coating to the metallic substrate is mainly due to mechanical interlocking, although chemical bonding cannot be totally excluded.

After plasma spraying has occurred, both the crystallinity and purity of the starting feedstock HA decrease due to the high temperature of the plasma and the rapid cooling rate of the molten fraction in contact with the metal. These phenomena provoke dehydration and decomposition of HA and the formation of an amorphous calcium phosphate (ACP). The dehydration produces the dehydroxylation of the HA, leading to oxyhydroxyapatite (OHA) and/or oxyapatite (OAP). The thermal decomposition of OHA/OAP results in different phases, such as α-tricalcium phosphate (α-TCP), tetracalcium phosphate (TTCP), and calcium oxide (CaO). These phases are heterogeneously distributed in the sprayed particle, formed by a solid core and a solidified shell. Rapid cooling upon impact yields an amorphous phase (ACP) in the outermost solidified shell, whereas the solid core is composed mostly of high temperature CaP phases [18]. Also, the rapid formation of the CaP coating produces cracks in the coating, increasing its brittleness.

Therefore, PS coatings are heterogeneous, constituted by crystalline and amorphous regions with inhomogeneous microstructure, producing rough areas within and on the coating surface [23]. As already mentioned, the ratio between ACP and HA phases determines the dissolution and biological behavior of HA coatings and, in turn, the long-term reliability of the implants. ACP should be present, in moderate quantities, on the surface coating to promote fast fixation to bone tissue and ensure good adhesion of the coating to the implant [26]. Thicker coatings, with weak adhesion to the substrate, HA particle release and delamination are also serious concerns for these PS coatings.

While plasma spraying is a well understood process, the control of variables is complicated and, depending on processing parameters, coatings with different features can be deposited. On the other hand, PS is not suitable for coating homogeneously porous substrates, surfaces of interest for bone ingrowth and implant fixation.

At present, this deposition technology is the object of widespread research in order to find more effective coatings, introducing dopants into the structure, such as Si [27, 28] or Ag [29], to improve the biological response or to provide antibacterial activity to the coatings, respectively. The use of modified thermal-spray technologies (see Section 9.2) such as *VPS, SPS, LPS, or HVOF* which allow one to obtain thinner coatings than those obtained by PS and a better control of microstructure or coatings with *good* adhesion are also being investigated [5].

Despite its drawbacks, nowadays, plasma-spraying is regarded as the most efficient and economical technique commercially available for coating implant devices. For example, Osprovit®, Osprovit®LC, and Osprovit®V (Eurocoating, S.p.a) are PS HA coatings, with different thickness, crystallinity, and resorption rates. These HA coatings are suitable for several types of devices, such as arthroplasty implants, trauma components or dental and spinal implants.

9.3.2.2   Alternative Deposition Techniques

Due to an increasing demand for better performance from HA coatings and in order to make a more adherent, thinner, and uniform bioactive coating on metals with high mechanical strength, alternative deposition techniques to PS have been explored, including PVD and CVD methods.

Magnetron sputtering allows the deposition of thin coatings ($< 4\,\mu m$) with strong adhesion and dense microstructure [8, 23, 30, 31]. Depending on the type of sputtering system and parameters used for the HA deposition, amorphous or crystalline calcium phosphate coatings can be obtained. If as-deposited coatings are amorphous, a subsequent heat treatment induces their crystallization; however, higher values of bone contact length were observed in as-deposited coatings implanted *in vivo* [32]. A nanocrystalline HA has been deposited using a right-angle RF magnetron sputtering approach, avoiding the need for *in situ* or *ex situ* annealing [33]. As in plasma-sprayed HA coatings, the coating composition may be quite different from that of the target material, depending on the deposition parameters. Although the surface morphology and surface roughness of the coatings are dependent on the synthesis conditions and post-deposition heat treatments, in general, the sputtered HA coatings are dense and continuous, and the bonding strength of the coatings exceeds 30 MPa. In addition to favorable mechanical properties and *in vitro* biocompatibility, sputtered-CaP coatings have proven useful in decreasing the amount of nickel released from a NiTi substrate [34] and have evidenced no adverse *in vivo* tissue response as well as a remarkable induction of bone growth at the interface [35, 36]. The magnetron sputtering technique has also been used to prepare other CaP coatings, such as Si-HA and Zn,Mg-doped HA [33, 37].

Another advantage of RF magnetron sputtering is that CaP-coated implants can be produced with almost the same surface roughness as uncoated titanium implants. Although sputtered coatings resolve some drawbacks of plasma-sprayed HA coatings, nowadays, this technology is not applied to coat commercial implants.

To deposit HA coatings by *pulsed laser deposition (PLD)*, a pulsed laser beam is focused onto a rotating HA target inside a vacuum chamber, with the substrate heated to a fixed temperature. Different lasers have been used to deposit CaP films [5]. The variation of the deposition parameters (wavelength and pulse duration of the laser beam, gas and water vapor pressure in the chamber, or substrate temperature) allows the preparation of amorphous or crystalline HA coatings with thickness from 0.04 to 10 μm [38, 39]. Usually, substrate temperatures in the range of 350–600 °C lead to the formation of highly crystalline and pure phase coatings with a composition close to that of the initial target [5]. Also, the processing parameters can provide HA coatings with different texture, showing different mechanical behavior: HA coatings with granular grains have higher resistance to delamination as compared to HA coatings with columnar morphology [40]. A careful control of the conditions allows one to obtain preferentially *c*-axis oriented HA films, which show better mechanical properties and a controlled dissolution behavior compared to randomly oriented HA coatings [41].

CaP coatings deposited by this technique show high adherence to the metal substrate so that no detachment from the substrate is observed. HA-PLD coatings with thickness of 2 μm and adhesion strength higher than 58 MPa were implanted *in vivo*, showing no sign of dissolution after six months and enhanced bone growth was generated in its surface [42]. Recently, PLD has been used to coat both the surface and the inner region of a titanium web scaffold with a thin crystalline HA layer [43].

Although thin, crystalline, and adherent coatings can be obtained by the PLD technique, the equipment is very expensive and may limit the scope of industrial applications for this methodology.

CaP coatings deposited on metallic substrates by **IBAD** have shown higher adhesion strengths compared to plasma-sprayed coatings [5, 10]. A typical deposition set-up includes an electron beam to vaporize a pure HA target, while an Ar ion beam is focused on the metal substrate to assist deposition [44]. The as-deposited coatings, with 700 nm thickness and bond strength ranging between 20 and 60 MPa, are amorphous and they crystallize after heat treatment in vacuum at 630 °C (1 h). Different phases, such as TCP or HA, are observed as a function of the ion-beam current used. Functional CaP coatings with graded crystallinity, which can promote better osseointegration, have been deposited by manipulating the substrate temperature during the IBAD [45]. The coatings show graded crystallinity with a nanocrystalline columnar structure near the substrate/coating interface and an amorphous structure at the top coating layer. The deposited coatings showed very high bonding strengths (> 85 MPa) without coating delamination. The high bond strength associated with the IBAD coatings is due to the presence of an atomic intermixed layer, as a consequence of the energetic ion bombardment which provokes a chemical bond at the coating/substrate interface.

IBAD has been used to prepare a commercially available dental implant coated with CaP used in clinical practice to replace lost teeth [46, 47]. After one year the implants were well maintained with osseointegration and no inflammation or bone loss was observed around the implants. The *in vivo* performance of IBAD coatings deposited onto metallic dental implants was comparable to plasma-sprayed HA coatings [48].

Besides PVD techniques, other deposition techniques based on vapor phase and non-linear processes, such as **metal organic CVD** [49], **aerosol CVD** [50], or **simultaneous**

vapor deposition [51] have also proven to be useful in the deposition of dense coatings of HA and other CaP phases.

Alternative methods based on solution processes, such as sol–gel deposition or the electrophoretic process have also been widely used for the preparation of bioactive coatings, mainly due to better control of the coating chemistry and structure, coating of complex pieces, simplicity, and low equipment cost. These methods enable the deposition of nanocrystalline apatite coatings, with crystal structures more similar to that of bone mineral if compared with high temperature depositions.

The **sol-gel method** allows one to obtain CaP coatings at a relatively low temperature. This technology has been investigated by several authors in order to optimize the deposition conditions of HA coatings [52, 53].

Bioactive apatite coatings were deposited onto Ti-based substrates by using the *dip-coating method* from aqueous precursor sols aged at different pH and temperature conditions, between 30 and 80 °C [54]. Annealing temperatures lower than 550 °C lead to homogeneous HA films with a smooth surface constituted by round-shaped HA crystals, smaller than 25 nm (Figure 9.8). The coating thickness is around 0.2 μm and it can be increased by dipping into the precursor sol several times. The formation of cracks/microcracks is observed when the thickness is higher than 1 μm or if the annealing temperature exceeds 700 °C (Figure 9.8). The efficiency of the process, which is interesting for an industrial application, is better when ethanol is added into the sol, allowing the coating of larger and complex surfaces more easily, with higher thickness and tensile strength adhesion higher than 20 MPa [55].

HA coatings have also been prepared by spin-coating with a sol prepared by mixing triethylphosphate and calcium nitrate, showing the biocompatibility of these HA coatings by culture of human osteosarcoma cell line [56].

Si-HA, F-HA, or Ag-antibacterial coatings showing the improved activity and functionality of cells compared to HA pure coatings have been investigated [57–59]. Also, a dip-coating method at room temperature has been developed to prepare chitosan/CaP composites, of great interest in orthopedic applications due to the outstanding properties of chitosan [60].

Adherent and homogeneous nanocrystalline HA films with mesopores between 4.5 and 10 nm have been deposited using non-ionic surfactant as porous former agent in the precursor solution [61]. These coatings are very promising candidates to be loaded with bioactive molecules confined inside the pores, which might help cell attachment to the surface.

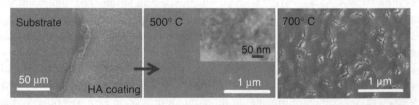

Figure 9.8 *Surface corresponding to HA coating deposited by sol–gel. Micrographs obtained by scanning electron microscopy (SEM) and transmission electron microscopy (TEM) in the insert*

Sol–gel deposition of nanosized CaP crystals (20–100 nm) has been used to coat a titanium alloy dental implant currently available for clinical use: Nanotite™ Implant (Biomet 3I Implant Innovations, Palm Beach Garden, FL, USA). The implant surface is described as being created by a particulate sol–gel deposition method using the discrete crystalline deposition of CaP particles with a surface coverage of approximately 50% [62, 63].

The *electrophoretic process (EDP)* is also a simple and inexpensive solution process. The method requires HA powders to be dispersed in a suitable solvent. In a typical experiment HA suspensions are prepared with HA powder in ethanol. The electrophoretic cell uses titanium as cathode and platinum as anode. Due to the simplicity of the process and the possibility of forming coatings with complex shapes, EDP has been widely used to deposit CaP coatings, including Si-HA, on different substrates for biomedical applications [15, 64]. Coating thicknesses can vary in a wide range, between 0.1 μm and more than 100 μm. However, the as-deposited coatings show poor adherence and post-heating is necessary to improve coating adhesion, although due to the uniformity and good sinterability of the coatings there is potential interest in this methodology.

In vivo studies performed with screw-type Ti6Al7Nb dental implants coated with HA deposited by EDP showed that these coatings lead to high integration between bone and implant by increasing the bioactivity of the implant, enhancing the mechanical properties of the screws and having a good osseointegration during the healing period compared to the uncoated ones [65].

9.3.2.3 Biomimetics and Other Solution-Based Processing Methods

Nanostructured materials offer much improved performances compared to their larger particle-sized counterparts due to their larger surface and unusual chemical synergistic effects [66]. In fact, bone is composed of HA nanocrystals in a collagen matrix. Therefore it is desirable to develop surfaces with nanoscale features to enhance the osseointegration of HA coatings. The most promising methods are those that produce CaP coatings at room temperature from aqueous solutions. On the other hand, almost all coating techniques described above generally involve high temperature in processing or post-processing so that any kind of drug or growth factors that improve the performance of the implants cannot be included. These molecules are introduced after the preformed coating process, mostly by adsorption, leading to a potential burst release effect of the loosely bonded drugs [67]. Therefore, instead of simple surface adsorption, co-precipitation of biological active molecules during the formation of the coating is gaining significant attention [68, 69].

The *biomimetic method*, inspired by the natural process of biomineralization, involves the heterogeneous nucleation and growth of bone-like crystals onto the substrate surface by immersing the implant into a solution at physiological conditions of temperature and pH. Usually metastable and supersaturated calcium phosphate solutions are used. In these conditions, a calcium phosphate layer is deposited on the surface of the substrate by consuming calcium and phosphate ions from the solution. The biomimetic CaP coating process was originally developed in the 1990s by Kokubo *et al.* [70], by using supersaturated solutions with ionic composition similar to that of human body plasma (simulated body fluid, SBF). By using the original procedure the nucleation and growth processes of apatite layer on substrate surfaces are very slow. Accelerated biomimetic strategies have been proposed such as:

- to modify the SBF solution employing 5/10 concentrated SBF solutions, introducing sodium hydrogen carbonate salts, or using Ca/P supersaturated calcifying solution [71, 72]
- To activate the titanium surface with active chemical groups (as will be described in Section 9.3.4).

The inorganic phase deposited from these supersaturated solution has chemical, morphological, and structural features similar to those of biological apatites. The structure of the crystals formed is more akin to that of bone mineral than are those of HA produced by plasma spraying or the other alternative methods mentioned above. Figure 9.9 shows the biomimetic coating deposited onto a HA sol–gel film soaked for seven days in SBF. The HA sol–gel coating reacts with the ions present in the SBF medium forming a new apatite coating on its surface. The biomimetic coating corresponds to a hydroxycarbonate apatite (HCA) of low crystallinity and consists of numerous needle-like crystallites with a Ca/P = 1.4 similar to that observed in other biological apatites [55].

Nanocrystalline HA coatings have a high specific surface area and they should be able to adsorb a sufficient amount of bone proteins and trigger osteoblast adhesion. Biomimetically deposited apatite coatings have shown to be the most favorable surface for proliferation and differentiation of osteoblast-like cells compared with other deposition techniques [73]. *In vivo* studies have shown that the bone is deposited directly onto the biomimetic apatite implants without any intervening soft tissue and early bone formation is promoted [74].

By tuning the conditions and composition of the solution in which the substrates are immersed different phases like ACP, nanocrystalline HA, Si(Sr)-HA, HCA, or OCP (octacalcium phosphate) have been deposited [74–76]. Due to the mild preparation conditions collagen/apatite or gelatin/apatite composite coatings were successfully deposited onto Ti and Ti-based alloy surfaces [77, 78].

The biomimetic process shows numerous advantages, including that it is simple, inexpensive, and applicable to porous substrates with complex surface geometries. However,

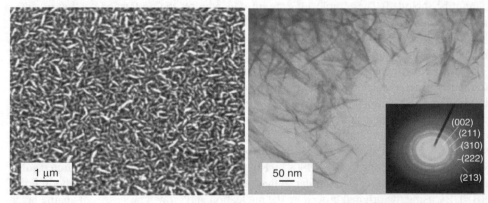

Figure 9.9 *Surface of HA coating deposited by a biomimetic method. SEM and TEM micrographs. Insert: electron diffraction pattern*

one of the most attractive features of this method is that biomimetic CaP coatings may incorporate drugs or bone-stimulant agents, such as growth factors, in order to enhance the bone-healing process in the peri-implant region. Although the use of these bioactive coatings as drug delivery systems will be considered in Chapter 12, it is noteworthy that these therapeutic agents incorporated by co-precipitation are gradually released over several days, enhancing their potential as controlled drug-delivery carriers. In this sense, the signal molecules–calcium phosphate co-precipitation on a substrate surface is a very promising process to achieve sustained regulation activity of the signal molecules by controlled and localized delivery of the signal molecules to implantation sites [68, 79].

PureFix™ Peri-Apatite™ is Stryker's Patented Technology for depositing biomimetic HA coatings on porous fixation surfaces of commercial orthopedic devices used for joint replacements. The HA coating is precipitated in an aqueous solution under conditions similar to those occurring in the body during bone formation and forms a coating, with an overall average thickness of 20 μm, which is highly crystalline and 100% HA [80, 81].

Another interesting solution-based approach to applying CaP coatings on metal implants is the ***electrochemical deposition or electrodeposition (ELD)***, a non-line of sight process, by which calcium phosphate coatings can be deposited onto a metal substrate in aqueous solutions by controlling the ion concentrations near the substrate.

The ELD process is based on the electrolytic decomposition of a metastable calcium phosphate aqueous solution [82]. A redox reaction produces supersaturation of OH^- ions near the electrode. This local effect induces heterogeneous nucleation on the metal substrate serving as electrode. For this, an anode and a cathode (titanium implant) are connected to a current generator. The application of a current density provokes the following electrochemical reaction in the cathode: $2H_2O + 2e^- \rightarrow H_2 + 2OH^-$. The increase in pH and concentration of phosphate ions in the interface electrolyte/substrate, can lead to the formation of an apatite phase on the implant surface.

The formation of H_2 gas hinders the deposition of the film, due to the formation of bubbles on the titanium surface, and leads to the formation of heterogeneous and deficient coatings. The use of H_2O_2, stirring, pulsed time electric fields, or pressure has been employed to avoid the negative effects of H_2 gas and to improve the adherence and coating uniformity [83–85].

As a function of the synthesis conditions, such as pH, temperature, and deposition time, current densities or composition of electrolyte, different CaP phases are deposited. Supersaturated electrolytes at pH 7.4 have been used for depositing HA coatings with a good uniformity and thickness [86]. If the ELD process is performed in high acidic solutions and the OH^- ions released are not sufficient to increase the pH, the DCPD and OCP deposit more easily. These CaP phases can act as HA precursors when they are soaked in SBF solution [87]. The surface morphology is also related to the deposition parameters: coatings deposited by using ultrasonic agitation contain CaP crystals with an acicular morphology (each crystal is composed of nanograins of much smaller size); coatings deposited without ultrasonic agitation are composed of globular deposits [88].

The introduction of amino acids into the electrolytic solution allows the deposition of HA coatings composed of nanocrystals, creating nanoscale characteristics at the coating surface. These electrodeposited coatings are denser and less porous than the coatings deposited without amino acids [89].

Some advantages of this deposition method are: the product can be obtained directly in a one-step process at low temperatures; rapid formation is possible (e.g., several micrometers per minute); and it can be applied to porous-type substrates. Owing to the mild conditions used during the ELD deposition, the incorporation of bioactive molecules into coatings is feasible.

Bonemaster® (Biomet Deutschland GmbH, Berlin, Germany) is an electrochemically deposited HA coating with 1/10th of the thickness of a plasma-sprayed HA coating and is able to reproduce the topography of the implant. These nanocrystalline HA coatings have shown favorable bone remodeling [90] and are commercially available for coat femoral stems or acetabular shells of orthopedic prostheses (Biomet Orthopedics).

The findings collected in this section show that currently there are many deposition techniques to prepare HA coatings on metallic implants, each of them with its own features. The absence of standardized technological procedures and the different equipment used for deposition of the coatings leads to the occurrence of CaP coatings with different composition, surface roughness, and adhesion to the substrate, which makes it difficult to compare results obtained by different authors/methods. It may be desirable to design surfaces with nanoscale features as a way to further enhance the osseointegration of HA coatings. There are sufficient experimental data available in the literature that prove the efficacy of CaP-based coatings for bone fixation purposes on uncemented prostheses, but the long-term reliability of these coated systems is not yet accepted. To date, the practical usefulness of these coatings has not been yet established.

9.3.3   Silica-based Coatings: Glass and Glass-Ceramics

BGs and BGCs are a group of ceramics used as alternative candidates to HA coatings on prosthetic implants.

As was mentioned in Chapter 4, bulk glasses in the $SiO_2-CaO-Na_2O-CaO$ system, such as Bioglass®, exhibit outstanding bioactivity and biocompatibility as well as osteoconductivity and osteoinductivity. In comparison with HA these silica-based BGs require less time for bone bonding, but the bioactive and biodegradable behavior depends on, and can be modified by, suitably adjusting both the composition and textural properties of the BGs. Another important characteristic of silica BGs is that their dissolution products have a stimulatory effect on cell genes aimed at tissue regeneration and angiogenesis [91].

Given these features, both BGs and BGCs-produced by controlled crystallization of glass-seem to be suitable ceramics for bone-contact applications, and worthy replacements for HA coatings. They could be used as coatings on non-bioactive metallic implants to improve the bone-implant adhesion, due to their bone bonding ability with no fibrous tissue formed between bone and implant. As was shown in Chapter 4, this process is related to the dissolution of the glass network, leading to the formation of a silica-rich gel layer and subsequent formation of a biologically active layer of HCA on their surface (Figure 4.5).

Many techniques are nowadays available to produce BG and BGC coatings on metallic substrates [92, 93]. Manufacturing of high-quality coatings on metal substrates is a complex issue, however, due to poor coating adhesion and/or glass degradation during the coating process, which leads to unsatisfactory mechanical properties. Great efforts have been devoted to optimizing the glass composition and processing parameters to deposit BG and BGC layers with high adhesive strength. Also, the high dissolution rates of these bioactive coatings are detrimental to the lifespan and mechanical stability of the prosthetic implant.

The enameling technique has been the conventional method to coat a metal substrate with a glass [94, 95]. In this technique, a suspension/slurry containing the glass powder is deposited onto a metallic substrate by painting, spraying, or dipping. (To ensure composition uniformity, the glass powders are often prepared through a melting and quenching process.) After drying, the glass is glazed by an adequate thermal treatment, usually temperatures exceeding 600 °C, forming a glass film fused onto the metallic implant. The firing atmosphere, time, and temperature of the process must be controlled in order to avoid the reaction of the glass with the metal substrate, implant degradation, or crystallization of the glass. The enameling technique is simple, inexpensive, and can be used to coat complex shapes of substrates.

Other deposition techniques, similar to those described before for calcium phosphate coatings, have been studied and used for the deposition of BG coatings, including *plasma spraying* [96, 97], *RF magnetron sputtering* [98, 99] *PLD* [100, 101], *laser cladding* [102], *sol–gel* [103, 104], or *EPD* [105, 106]. As already mentioned in the CaP coatings, the different techniques have also shown advantages and disadvantages for BGs deposition (see for instance references [93, 106]) and it is not easy to evaluate them in a comparative way. To date, coatings of BGs are not used clinically because the long-term stability of BG coatings is not yet satisfactory.

The main problems found with the coating of a metallic implant with BGs include:

- The development of thermal stresses at the interface, caused by the mismatch between the coefficient of thermal expansion (CTE) of the BG and the metal implant [95], which can cause cracking and delamination of the glass coating from the metallic substrate.
- Some silica-based BGs tend to crystallize during the coating deposition in cases where sintering is needed for a good coating. The crystallization may improve the mechanical properties of the coating, but it may impair the bioactivity [107]. In this regard, the sol–gel method allows fabrication of BGs in a larger compositional range than melt-derived ones.
- BGs deposited can react with the Ti-based alloys used in the fabrication of medical implants, and the glass/metal reactions that occur during firing are detrimental to adhesion and bioactivity [108].

Some solutions have been proposed to solve the above-mentioned problems. Composite BG/HA coatings are prepared in order to control and to improve both the solubility and bioactive behavior of these coatings while preserving a good adhesion to the substrate [109–111]. Also, modifications in glass composition, including substitutions in the SiO_2–CaO–Na_2O–P_2O_5 system of the CaO with MgO, Na_2O with K_2O [112], or the preparation of bioactive borate glass containing SiO_2, Al_2O_3, and P_2O_5 [113], have been developed to decrease the crystallization tendency of silicate-based glasses or to match the thermal expansion coefficient of the glass to that of the titanium alloy.

The introduction of an intermediate layer between the coating and the metallic implant, with an appropriate CTE, is another solution proposed to overcome the mismatch problem and improve adhesion [114, 115]. Also, the deposition of multiple layers with different compositions (functionally graded materials) can produce coatings with optimal properties. By controlling the gradient in the glass composition along the coating, good adhesion to the metal can be combined with rapid biofixation and long-term stability of the implant [116, 117].

The *in vitro* and *in vivo* behavior of BGs and BGCs coatings has been extensively studied [118–122], showing similar results to bulk BGs: SBF assays show the formation of an HCA layer on their surface and cell-seeding tests indicate optimal osteoblasts growth and proliferation on their surface. The *in vivo* studies show the potential of these BGs and BGCs coatings, that is, higher bone growth and osseointegration of BGs and BGCs-coated implants compared with uncoated metal was observed. On the other hand, a study performed by using BGs-coated and HA-coated titanium dental implants in human jaw bone showed that both were equally successful in achieving osseointegration and supporting final restorations after 12 months [123].

In spite of these good findings and their potential, BG coatings are not yet commercialized. Fabrication of coatings for biomedical implants involves a compromise between adhesion, mechanical stability, and bioactivity, but coatings that satisfy all these requirements are extremely difficult to develop.

9.3.4   Bioactive Ceramic Layer Formation on a Metallic Substrate

This section is focused on the chemical modification of the surface of metallic implants, Ti and Ti-based alloys, in order to induce a bone-like apatite formation on them, and then bond to living bone through this apatite layer. Different methods have been developed to obtain bioactive titanium but the most widespread and investigated was proposed by Kokubo *et al.* [124].

Some gels, such as silica and titania gels, form a bone-like apatite layer on their surfaces when placed in SBF [125]. This indicates that functional groups, such as Si-OH and Ti-OH, abundant on the gels, induce nucleation of the apatite. Therefore, it is expected that some metallic materials might be bioactive if their surfaces are modified with functional groups that enable effective apatite nucleation.

Based on these findings, and although titanium is generally chemically stable, since its surface is covered with a thin layer of passive titanium oxide, the following alkaline treatment was proposed to impart titanium bioactivity[124]: Titanium metal was introduced in 5 M NaOH aqueous solution at 60 °C (24 h). In these conditions a sodium titanate hydrogel layer, gradient structured, about 1 μm thick, is formed on the substrate surface (Figure 9.10). A detailed study [126] showed that the layer formed was nano-sized crystalline sodium hydrogen titanate ($Na_xH_{2-x}Ti_yO_{2y+1}$, with $0 < x < 2$ and $y = 2$, 3, or 4). This layer is not stable and the alkali-treated titanium is subjected to thermal treatment at 600 °C (1 h); then the gel layer was dehydrated and densified to form a nanocrystalline sodium titanate layer ($Na_xTi_2O_{2y+1}$; $y = 5$, 6, etc.) and TiO_2 (rutile) (Figure 9.10). The scratch resistance of the surface-modified layer formed by the NaOH treatment increased with the subsequent heat treatment, from 3 to 50 mN [127].

In order to study the bioactivity behavior of this treated Ti substrate, it was immersed in SBF. After immersion, the metal surface was constituted by dense and uniform bone-like apatite and there was no distinct boundary between the apatite–titanate composite and the metal substrate. Transmission electron microscopy and energy dispersive X-ray spectrometry confirmed that the apatite crystals so formed are nanometric in size, needle-like shaped, and exhibit a chemical composition similar to bone mineral, with a Ca/P atomic ratio of 1.65 after five days in SBF [128].

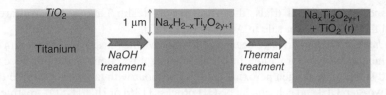

Figure 9.10   *Surface modification of Ti substrate due to NaOH and heat treatment*

The following mechanism was proposed for the apatite formation [129]: The Na^+ ions are released from the surface of the sodium titanate via exchange with H_3O^+ ions in the surrounding fluid, forming Ti-OH groups on the substrate surface while the solution pH increases. Then, the Ti-OH groups are negatively charged and react with the Ca^{2+} ions of the SBF, leading to calcium titanate. These Ca^{2+} ions accumulate on the surface and react with the phosphate ions, negatively charged, to form an ACP. The ACP is metastable and spontaneously transforms into stable crystalline bone-like apatite (Figure 9.11).

Similar results, in inducing the apatite-forming ability, have been found in other Ti-based alloys, such as Ti6Al4V and Ti6Al2NbTa, subject to the same alkaline and thermal treatment. Also, a uniform bioactive layer forms easily on the irregular surfaces of treated porous titanium implants [130].

In vitro studies by using cell culture experiments have shown enhancements in both differentiation of osteoblasts and bone nodule formation on this bone-like apatite layer formed in the treated Ti substrate in SBF [131, 132]. *In vivo* assays showed that treated Ti implanted in the living body forms an apatite layer on its surface and it bonds through this apatite to the bone. Studies performed implanting treated porous titanium into a rabbit's femur, show that the implant was penetrated with newly grown bone from the outer surface and was bonded with new bone [133].

Clinical trials of titanium cementless THA using a Ti6Al2Nb1Ta prosthesis subjected to the alkaline-heat treatment have been performed [134]. The clinical studies showed good osteoconduction without adverse effects. The hip joints were fully covered with newly grown bone and tightly fixed to the surrounding bone. These authors remarked the advantages of this method compared with the traditional PS HA coating, such as the simpler and inexpensive manufacturing process; it allows homogeneous deposition of bone-like apatite within a porous implant and the coating binds strongly to the substrate (solving the issue of delamination of the HA coating).

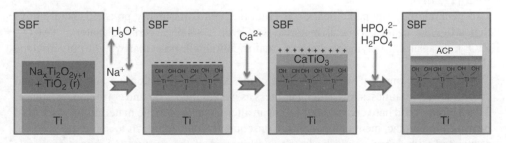

Figure 9.11   *Process of bone-like apatite formation on treated Ti metal in SBF*

After successful clinical trials, the Ministry of Health Labour and Welfare, Japan, approved this prosthesis for sale in August 2007 [127].

Other *chemical treatments* have been used to improve the bioactivity of titanium implants through inorganic layer formation, involving chemical reactions at the interface between metal and solution: hydrogen peroxide solution without/with tantalum chloride treatment [135], a two-step HCl and subsequent NaOH process [136] or HCl/H_2SO_4 treatment [137]. Other approaches based on incorporation of biological molecules which can be anchored to the substrate by biological functionalization of titanium have been performed to improve bioactivity, osteoconductivity, and osseointegration of the metal implants [138].

These chemical methods involve modifications of the metallic surface in order to include in it functional groups capable of inducing the nucleation of a biologically bone-like apatite layer onto the said surface, creating a direct chemical link between the layer and the substrate. They are considered enabling techniques from the point of view of surface modification of metallic biomaterials, due to their simplicity and flexibility.

Physical methods such as *ion implantation* of calcium and/or phosphorus have also been used to modify the chemical surface and to improve the biological properties of these metal substrates [5, 139, 140]. These two types of ions incorporated into the Ti surface are very important for HA nucleation and formation when the metal is placed in SBF, and also increased the corrosion resistance under both short-term and long-term exposures to test solutions. *In vitro* cell culture studies revealed improved bone cell adhesion, spreading, and growth on these ion-implanted titanium surfaces compared to non-implanted titanium. The main limitation of this technique is the expensive apparatus involved, and it is not appropriate for complex geometries.

9.4   Bioinert Ceramic Coatings

Total joint arthroplasty is a successful and accepted procedure for the treatment of diseases or trauma, although, as already discussed in the previous section, it presents some problems associated, mainly, with the osseointegration of the implants. This section is focused on additional limitations for the longevity of joint replacements: the articular wear caused by the friction between the articulating surface, that is, the debris which can lead to osteolysis and subsequent loosening. The load-bearing surfaces of THAs and TKAs are subjected to wear and it is acknowledged that the most serious consequences are those related to polyethylene (PE) debris. Improved lubrication reduces friction, thus reducing wear [17, 141].

In general, the orthopedic bearing is manufactured from highly polished metal alloys that articulate against a metal material (CoCr or Ti6Al4V), ceramic material (Al_2O_3, ZrO_2), or against a polymeric material, mainly ultrahigh molecular weight polyethylene (UHMWPE).

In Chapters 6 and 7 it was shown that oxide ceramics, like alumina and zirconia and carbon-based ceramics, show a high hardness as well as high wear and corrosion resistance. These bioinert and hard ceramics represent an alternative to metallic materials used for hard bearing joint replacement components. Consequently, one approach to solve the problem associated with debris would lie in the combination of the fracture toughness of metals with the wear performance of these bioinert ceramics. In addition, the ceramic coating on the metallic surface bearing acts as a protective layer, avoiding the release of metal ions and wear debris to the surrounding tissue. The aim of these hard coatings is to improve the tribological performance and lifespan of total joint replacements.

9.4.1 Titanium Nitride and Zirconia Coatings

This section is focused on hard coatings, containing transition metals, currently employed in the orthopedic industry, as well as on the most promising advances in this field. Titanium nitride deposited onto Ti6Al4V and CoCr alloys, two of the most frequently used components for THA and TKA, will be considered initially. Furthermore, the formation of a zirconia ceramic by surface modification of the metal implant is also detailed.

9.4.1.1   *Titanium Nitride Coatings*

The physical properties of titanium nitride (TiN) coatings such as high hardness, high scratch resistance, and low friction coefficient, together with the appropriate wettability with synovial fluids, have converted this material into an excellent alternative to enhance the wear properties of different metals used in arthroplasty [142, 143]. In fact, it is the hard ceramic coating that has the longest clinical application for total hip and knee replacements. TiN coatings in articulating surfaces of joint replacements, designed to articulate against UHMWPE or against other materials, are commercially available from Corin Medical and Endotec, Inc.

The most common methods to deposit TiN thin film are PVD, such as cathodic arc deposition or magnetron sputtering, and CVD [144, 145]. Modern PVD processes have been improved in order to avoid the weak adhesion between the substrate and the hard layer due to the absence of chemical reactions and diffusion phenomena in traditional low-temperature PVD. Hence, the medical device company Endotec, Inc. uses a proprietary PVD method to deposit UltraCoat®TiN coating on articulating surfaces of its hip, shoulder, and knee systems.

These golden colored TiN coatings show excellent *in vitro* and *in vivo* wear behavior, as demonstrated against PE and by an increased resistance to abrasive third-body wear compared with CoCr, ensuring a decrease in metal ions release from the substrate [146–148]. This point is more critical for CoCr components since one disadvantage of this compound is that Co, Cr, and small amounts of Ni can be released and they are known to cause allergic reactions. Adhesion strength of this TiN coating is higher on a hard CoCr than a softer Ti6Al4V substrate [149].

Despite the limited number of recovered joint replacements after surgical revision or *in vivo* data on wear performance of TiN coatings, some drawbacks related to the hardness of the TiN layer when compared to the substrate, especially with the softer titanium-based alloys, have been observed, mainly delamination and coating breakage. The sudden dissimilarities of hardness and elastic modulus across the coating/substrate interface causes a plastic deformation in the substrate material when a load is applied to the TiN coating. The TiN coating is unable to accommodate the deformation and fractures. Generated flakes, together with the possible presence of coating defects such as pinholes and embedded microparticles, formed during the deposition process and polishing, cause deterioration in the wear performance of these materials [143, 150].

As mentioned in Section 9.2, surface modification by ion implantation or nitrogen diffusion at high temperatures are alternative approaches employed to prepare wear-resistant bearings such as TiN. In this case, the surface-modified layer grows from the surface itself, thus ensuring an excellent conformity and strong adhesion and a more gradual transition that avoids a sudden mismatch [4, 10, 151, 152]. Wear resistance increases by orders of magnitude due to the surface hardening and the reduced friction that can be related to the formation of hard-phase carbide, nitride (TiN and Ti_2N) and oxide compounds. Moreover, nitrogen implantation also enhances the bone conductivity of titanium and increases the corrosion resistance, especially in low doses.

Merging the two above-mentioned approaches to the preparation of TiN coatings has been proposed, termed the duplex-process, to enhance the capacity of the deposited TiN coating to bear highly localized loads thanks to the underlying graded substrate that shows a progressive transition in hardness [153, 154]. For example, the combination of plasma nitriding of a titanium-based substrate with the deposition of a TiN layer produced by plasma-enhanced CVD provides a better adhesion to the substrate and an improvement in the wear performance of the substrate. Also, the deposition of TiN/Ti(C, N) multilayers and Ti(C, N) graded coatings has been performed in order to improve the adhesion and functional characteristics of these hard coatings [155].

Other modifications to TiN coatings have been considered by using chromium nitride (CrN) and chromium carbonitride (CrCN) coatings, which also showed a lower ion release from the CoCr alloy as well as a better cohesive strength and toughness, and a lower volume of wear debris compared to TiN when articulating against itself or conventional PE [156, 157].

9.4.1.2   Oxidized Zirconium

Zirconia is one of the hardest materials used to fabricate components that can withstand higher stresses (Chapter 6). The oxidation of a zirconium-based alloy, above 500 °C, yields an oxidized layer of ZrO_2, termed oxidized zirconium (OxZr) [17, 158].

This OxZr is not a ceramic surface coating, but a surface-modified layer, with better values of toughness and lifespan. This methodology sought to provide improvements in terms of resistance, roughening, and frictional behavior, as well as biocompatibility, thus avoiding the limitations of brittle monolithic ceramics. The transformation of the original metal surface into zirconia ceramic, also termed ceramification [17], has been in global commercial use for THA and THK under the name Oxinium^{TM} (Smith & Nephew Orthopaedics, Memphis, TN).

The oxidized layer of around 5 μm in thickness ensures the pursued tribological behavior of the surface while, at the same time, maintaining the mechanical performance given by the metallic substrate [159]. The surface of the zirconium-based prosthesis is transformed into a dense ceramic that is predominantly monoclinic ZrO_2, which does not suffer any type of phase transformation in an aqueous environment like that observed in yttria-stabilized zirconia components (Chapter 6). This well-known transformation from stabilized tetragonal to monoclinic zirconia implies a detrimental effect since it induces a reduction in the strength and a deterioration of the surface, thus lessening the wear performance.

A tuned oxidation procedure in terms of time and temperature produces a fine grained layer with uniform thickness and without internal flaws, exhibiting excellent integrity, due in part to grains that are columnar, staggered, nanostructural in size, and generally perpendicular to the surface [160].

The diffusion process allows the coating of complex-shaped components with a uniform layer that gradually turns into the underlying pure metal without any abrupt interphase, thus enhancing cohesion and adhesion of the oxide, even under shear and mechanical strain. *In vitro* simulator studies have shown that OxZr-bearing surfaces for TKA [161] dramatically decrease wear against PE compared with CoCrMo. This can be attributed to the reduced friction observed in ceramic oxides thanks to the formation of a lubricating film [162]. Considering the long-term excellent wear behavior, this type of bearing (OxZr) is being employed in young and active patients [163]. Clinical studies in TKA show no adverse effects associated with the use of OxZ femoral implants but no significant differences were found compared to CoCr components [164].

9.4.2   Carbon-based Coatings

Carbon-based materials are well known bioinert coatings. The interest of these coatings is their capacity to enhance the surface of medium- and heavy-load bearing components and wear parts, by imparting several significant material improvements, such as a higher wear resistance, elevated corrosion resistance, and much lower frictional effect. Moreover, carbon-based coatings have evidenced minimal protein adhesion and good blood compatibility due to the hydrophobicity of the surface. This section focuses on the use of pyrolytic carbon (PyC), nanocrystalline diamond, and diamond-like carbon (DLC) coatings in biomedical applications. Some of these applications are commercially available already while others are still under development.

PyC belongs to the family of turbostratic carbons, with a similar structure to graphite: sp^2-hybridized atoms form graphene sheets that are layered in a random manner. As was mentioned in Chapter 7, most important biomedical application of PyC is found in the prosthetic heart valve, due to its biocompatibility and thromboresistant qualities.

Usually, PyC coatings are deposited by thermal decomposition ("pyrolysis"), under inert N_2 or He atmosphere, of a hydrocarbon gas, such as propane, acetylene, or methane, at high temperature by using CVD. The process is carried out in a fluidized bed reactor [165, 166]. Since the coating process takes place at high temperatures (1000–2000 °C), the choice of substrates is limited, graphite being the substrate most commonly used. The preformed graphite substrate is placed within the fluidized bed in the vertical tube furnace where it is coated by the solid products from the pyrolysis reaction to the desired thickness.

Depending on the processing parameters (precursor, gas flow rate, temperature, and bed surface area) a variety of structures can be deposited: laminar, isotropic, columnar, or granular, with isotropic PyC the most interesting structure for biomedical applications. A careful control of the deposition conditions is crucial to obtain isotropic PyC coatings by this method [166].

Most artificial heart valves are lined with a thick PyC coating. At present, the most commonly implanted valves are bileaflet valves, such as St. Jude Medical prostheses (St. Jude Medical Inc.), On-X valves (On-X Life Technologies Inc.), or Medtronic Open Pivot (Medtronic, Inc). The biocompatibility of PyC has been well established through its successful use in heart valves and has been extrapolated to orthopedic applications. In this application it is currently marketed under the name PyroCarbon (Tornier, Inc., Integra Life

Sciences). These orthopedic implants are used in the replacement of small joints, such as metacarpophalangeal (knuckle) or trapezium bone (wrist joint), as well as in arthroplasty of the proximal interphalangeal joints.

In the early 1980s Sorin Biomedica Group developed and engineered a process for depositing a thin film of turbostratic carbon, Carbofilm^{TM}, on all types of substrates, including those employed in medical or heat-sensitive devices. The technique used is a PVD process carried out in high-vacuum conditions to avoid any chemical reaction. This company manufactures and markets artificial heart valves (Bicarbon products, Carbomedics) and cardiovascular stents coated with 0.5 μm thick Carbofilm^{TM} which improve hemocompatibility and reduce tissue overgrowth [167].

Diamond-related materials, including nanocrystalline diamond and DLC coatings, have remarkable properties and have also been considered for use as coatings on medical implants.

Nanodiamond (NCD) consists of sp^3-hybridized diamond bonds and exhibits grain sizes below 100 nm. The grain boundaries between crystallites contain sp^2 carbon atoms and hydrogenated carbon species.

NCD films are usually deposited by a CVD process, such as microwave plasma-enhanced CVD or hot filament CVD, from a carbon-containing gas and a carrier gas, such as hydrogen, argon, and nitrogen, at temperatures ranging from 450 to 900 °C [168–171]. It is possible to deposit relatively smooth high-quality NCD films of various thicknesses when very high nucleation densities and appropriate process parameters are used [169].

NCD films exhibit multiple features applicable to a broad range of functional devices (Chapter 7). NCD coatings combine a very low surface roughness with the outstanding properties of diamond, biological compatibility with blood, and high hydrophobicity, which are ideal features for applications in wear-resistant implants and cardiovascular medical devices. NCD may be used as a hard antibacterial coating to help lower the risk of infections. A recent study [172] shows that the bactericide and anti-adhesive properties of these coatings, tested against the *gram-negative Pseudomonas aeruginosa* bacteria, may be associated with their semiconducting properties: NCD surfaces react and establish chemical bonds with the cell wall or membrane, hindering bacterial adhesion and subsequent colonization on the NCD coating. On the other hand, ultra-NCD films are being explored as biocompatible functional electrodes for *in vivo* nerve stimulation, as well as encapsulation coatings for a Si microchip implantable inside the eye, as a component of an artificial retina, developed to restore sight to people blinded by the genetically-induced death of photoreceptors [173].

More experimental studies have been performed on *DLC* coatings, which are considered as a promising material for biomedical applications [167, 174, 175]. DLC is a group of materials based on metastable amorphous carbon coatings with a variety of properties due to the presence of different ratios of sp^2/sp^3-bonded carbon and different levels of hydrogen. As previously mentioned (Figure 7.5) different DLC coatings can be found as a function of their hydrogen content: non-hydrogenated amorphous carbon (a-C) and amorphous hydrogenated carbon (a-C:H). DLC containing high amounts of sp^3 fraction are tetrahedral amorphous carbon coatings (ta-C and ta-C:H).

The mechanical functionality of these DLC films is different because the sp^3/sp^{2}ratio and hydrogen amount determine the mechanical and tribological properties of a DLC coating.

Usually, films with a high content of sp^3-hybridized atoms are more similar to diamond and therefore very hard, but a high sp^3 content can lead to coating delamination. DLC coatings with a high proportion of sp^2 carbon atoms behave more like graphite and tend to be relatively soft. On the other hand, the presence of hydrogen in the film generates a larger number of C-H bonds which relieve the internal stress and produce a softer material. In this sense, a-C:H films with a high degree of hydrogenation have low friction and wear compared with a-C coatings.

A great variety of vacuum deposition technologies have been used to prepare DLC coatings on different substrates, with common methods based on PVD and variations on CVD: magnetron sputtering, PLD, IBAD, filtered cathodic vacuum arc (FCVA), plasma-assisted-CVD, and RF-plasma-enhanced CVD [176, 177]. Depending on the deposition techniques and source material, the chemistry composition and microstructure of the films may be different, and such differences may lead to variations in their properties, including biological response and antibacterial properties [178–180].

Solid carbon sources, such as graphite or glassy carbon, are used as source material for deposited coatings with low or no hydrogen, whereas for hydrogenated coatings, a hydrocarbon source (CH_4, C_2H_2, C_2H_4) is used.

The deposition technique used as well as the deposition parameters or the substrate characteristics determine both the sp^3/sp^2 ratio and the amount of hydrogen; therefore, different DLC coatings properties can be obtained. The sp^3/sp^2 ratio and hydrogen content in DLC coatings depend on the impact energy of the impinging atoms, which are controlled by plasma parameters, mainly the bias voltage and the gas pressure: at the lowest impact energies the precursor is not sufficiently decomposed, then films with higher hydrogen content are obtained. At intermediate energies, the amount of hydrogen decreases and sp^3-hybrydized carbons are favored. If the impact energies are too high, an increase in sp^2 bond carbons is obtained [174]. Usually, deposition techniques like PLD, magnetron sputtering, or FCVA are common techniques to prepare a-C coatings, whereas a-C:H films are prepared by plasma-enhanced CVD or plasma immersion ion implantation and deposition (PIII+D) [181]. Also, the DLC films prepared by sputter deposition may contain large amounts of sp^2-hybridized atoms and, therefore, they tend to be much softer than films prepared by using arc-PVD and PLD techniques [177].

On the other hand, DLC coatings can suffer from high levels of residual stress, causing poor substrate adhesion and delamination of the coating. One of the main concerns about DLC applications is the instability of DLC films in an aqueous environment, which leads to delamination of the coating as the adhesion of DLC films is degraded [182]. One of the main solutions for this poor surface adhesion is to deposit interlayers (buffer layers), such as Ti, CrC, and Si_3N_4, between the substrate and DLC film, or alternatively to dope the DLC coating with Ag, N, F, Ti, or Zr, in order to reduce the residual stresses in the coating [183, 184].

The DLC properties, such as the inert nature, hydrophobicity, and smoothness, are very important for their excellent blood compatibility, fundamentally reducing platelet activation in contact with the blood, which could subsequently trigger thrombosis. Also, DLC can serve as a protective coating in human blood environments, reducing the release of metal ions such as nickel, which is a common contact allergen, from the metallic implants. Several studies *in vivo* and *in vitro* have shown that DLC coatings prepared by different deposition techniques are quite biocompatible and do not induce any inflammatory reaction inside animal bodies [175, 181]. DLC coatings, deposited by PVD on coronary stent devices promote rapid and complete endothelization after seven days with very low platelets activation [185]. A correlation between composition and bonded carbons of DLC films and their blood compatibility is shown in a study performed by using the adsorption of two proteins, Fibrinogen and Human Serum Albumin, which play an important role in enhancing or inhibiting thrombus formation, respectively [186]. Also, DLC coatings have been investigated to strengthen and protect the surface of polymer-based medical products, providing additional rigidity to the plastic products and antibacterial activity [187]. Because of these features, DLC have found applications in coating different implant devices such as heart valves, cardiovascular stents, ophthalmic surgery needles, contact lenses, or medical wires. Plasma Chem GmbH manufactures coronary stents of stainless steel coated with 50–80 nm of DLC coating (BioDiamond Stents).

On the other hand, DLC can be used as a protective coating material for articulating joint replacements due to its high hardness, low wear and friction, high corrosion resistance, and smoothness. DLC films are used to reduce the coefficient of friction of a surface and improve wear resistance. For example, Diamonex DLC (Morgan Technical Ceramics) has been used to coat components of implantable joints. Renishaw Advanced Materials commercially offers DLC coatings, prepared by plasma-enhanced CVD, to coat knee prostheses for patients allergic to the materials in uncoated knee prostheses.

As previously mentioned, the presence of wear particles from joint surfaces is acknowledged as the main factor contributing to failure of joint arthroplasties. Numerous *in vitro* tests reporting on the friction and wear of DLC coatings on different substrate heads (Ti, Ti6Al4V, CoCrMo, or stainless steel) against UHMWPE have been performed (an overview of the different studies performed can be found in references [167, 184]). The *in vitro* experiments apparently showed contradicting results, mainly due to the different experimental set-ups: use of different testing methods (hip and knee simulator testing, pin on disc set-up), material substrate or test conditions (lubricant fluid, loading conditions). *In vitro* tests have shown low wear rates of Ti6Al4V femoral heads coated with DLC against UHMWPE, compared to conventional Al_2O_3 and CoCr, after hip joint simulator testing with a bovine calf serum lubricant [188]. However, DLC coatings show a very different performance *in vivo* [189]. Clinical studies were performed with more than 100 hip joint prostheses implanted consisting of DLC-coated femoral heads (Adamante®) articulating with PE. The coated implants showed only a 54% survival rate after 8.5 years. The surfaces of the explanted DLC-coated heads showed damage (numerous pits and delamination of the coating), which originated excessive wear of the PE counterpart and, in some cases, even of the metallic substrate of the heads. The disastrous results were associated with an inadequate adhesion of the coating.

Findings from clinical and laboratory-based tests on the appropriateness of DLC coating materials for joint replacement surfaces remain uncertain. Some results indicate DLC-coated materials to be promising whilst others suggest they are unsuitable [184]. Further experimental studies concerning the DLC coating composition, deposition techniques, or the presence of interlayers need to be carried out.

References

1. Geetha, M., Singh, A.K., Asokamani, R. and Gogia, A.K. (2009) Ti based biomaterials, the ultimate choice for orthopaedic implants-A review. *Prog. Mater Sci.*, **54**, 397–425.
2. Martin, P.M. (2010) *Handbook of Deposition Technologies for Films and Coatings*, Science, Applications and Technology, 3rd edn, Elsevier Inc.
3. Heimann, R.B. (1996) *Plasma-Spray Coating. Principles and Applications*, Wiley-VCH Verlag GmbH, Weinheim.
4. Liu, X., Chu, P.K. and Ding, C. (2004) Surface modification of titanium, titanium alloys, and related materials for biomedical applications. *Mater. Sci. Eng.*, **R47**, 49–121.
5. Surmenev, R.A. (2012) A review of plasma assisted methods for calcium phosphate-based coatings fabrication. *Surf. Coat. Technol.*, **206**, 2035–2056.
6. SreeHarsha, K.S. (2006) *Principles of Physical Vapor Deposition of Thin Films*, Elsevier, Oxford.
7. Hassa, D.D., Marciano, Y. and Wadley, H.N.G. (2004) Physical vapor deposition on cylindrical substrates. *Surf. Coat. Technol.*, **185**, 283–291.
8. Paital, S.R. and Dahotre, N.B. (2009) Calcium phosphate coatings for bio-implant applications: materials, performance factors, and methodologies. *Mater. Sci. Technol.*, **R66**, 1–70.
9. Krebs, H.U., Weisheit, M., Faupel, J. *et al.* (2003) in *Pulsed Laser Deposition (PLD)-A Versatile Thin Film Technique*, Advances in Solid State Physics, vol. **43** (ed B. Kramer), Springer-Verlag, Berlin, Heidelberg, pp. 505–518.
10. Rautray, T.R., Narayanan, R. and Kim, K.H. (2011) Ion implantation of titanium based biomaterials. *Prog. Mater. Sci.*, **56**, 1137–1177.
11. Choy, K.L. (2003) Chemical vapour deposition of coatings. *Prog. Mater. Sci.*, **48**, 57–170.
12. Brinker, C.J. and Scherer, G.W. (1990) *Sol-Gel Science. The Physics and Chemistry of Sol-Gel Processing*, Chapter 13, Academic Press, Inc, p. 787 Film formation.
13. Chiriac, A.P., Neamtu, I., Nita, L.E. and Nistor, M.T. (2010) Sol gel method performed for biomedical products implementation. *Mini Rev. Med. Chem.*, **10**, 990–1013.
14. Kandimalla, V.B. (2006) Immobilization of biomolecules in sol-gels: biological and analytical applications. *Crit. Rev. Anal. Chem.*, **36**, 73–106.
15. Boccaccini, A.R., Keim, S., Ma, R. *et al.* (2010) Electrophoretic deposition of biomaterials. *J. R. Soc. Interface*, **7**, S581–S613.
16. Hornberger, H., Virtanen, S. and Boccaccini, A.R. (2012) Biomedical coatings on magnesium alloys – a review. *Acta Biomater.*, **8**, 2442–2455.
17. Sonntag, A., Reinders, R. and Kretzer, J.P. (2012) What's next? Alternative materials for articulation in total joint replacement. *Acta Biomater.*, **8**, 2434–2441.
18. Heimann, RB. (2012) Structure, properties, and biomedical performance of osteoconductive bioceramic coatings. *Surf. Coat. Technol.* **233**.
19. DeGroot, K., Geesink, R.G., Klein, C.P. and Serekina, P. (1987) Plasma sprayed coatings of hydroxyapatite. *J. Biomed. Mater. Res.*, **21**, 1375–1381.

20. Sun, L., Berndt, C.C., Gross, K.A. and Kucuk, A. (2001) Material fundamentals and clinical performance of plasma-sprayed hydroxyapatite coatings: a review. *J. Biomed. Mater. Res., Part B*, **58**, 570–592.

21. Furlong, R.J. and Osborn, J.F. (1991) Fixation of hip prostheses by hydroxyapatite ceramic coatings. *J. Bone Joint Surg., Part B*, **73**, 741–745.

22. Geesink, R.G.T. (1989) Experimental and clinical experience with hydroxyapatite-coated hip implants. *Orthopedics*, **12**, 1239–1242.

23. Narayanan, R., Seshadri, S.K., Kwon, T.Y. and Kim, K.H. (2008) Calcium phosphate-based coatings on titanium and its alloys. *J. Biomed. Mater. Res., Part B*, **85**, 279–299.

24. ISO (2008) ISO 13779-2-2008. Implants for Surgery-Hydroxyapatite-Part 2: Coating of Hydroxyapatite, International Organization for Standardization.

25. Heimann, R.B. (2006) Thermal spraying of biomaterials. *Surf. Coat. Technol.*, **201**, 2012–2019.

26. Combes, C. and Rey, C. (2010) Amorphous calcium phosphates. Synthesis, properties and uses in biomaterials. *Acta Biomater.*, **6**, 3362–3378.

27. Tang, Q., Brooks, R., Rushton, N. and Best, S. (2010) Production and characterization of HA and SiHA coatings. *J. Mater. Sci. -Mater.*, **21**, 173–181.

28. Xu, J.L., Joguet, D., Cizek, J. *et al.* (2012) Synthesis and characterization on atmospheric plasma sprayed amorphous silica doped hydroxyapatite coatings. *Surf. Coat. Technol.*, **206**, 4659–4665.

29. Shimazaki, T., Miyamoto, H., Ando, Y., *et al.* (2010) *In vivo* antibacterial and silver-releasing properties of novel thermal sprayed silver-containing hydroxyapatite coating. *J. Biomed. Mater. Res., Part B* **92** : 386-389.

30. Yang, Y., Kim, K.H. and Ong, J.L. (2005) A review on calcium phosphate coatings produced using a sputtering process-an alternative to plasma spraying. *Biomaterials*, **26**, 327–33.7.

31. Van Dijk, K., Schaeken, H.G., Wolke, J.G.C. and Jansen, J.A. (1996) Influence of annealing temperature on RF magnetron sputtered calcium phosphate coatings. *Biomaterials*, **17**, 405–410.

32. Ong, J.L., Bessho, K., Cavin, R. and Carnes, D.L. (2002) Bone response to radio frequency sputtered calcium phosphate implants and titanium implants in vivo. *J. Biomed. Mater. Res.*, **59**, 184–190.

33. Hong, Z., Mello, A., Yoshida, T. *et al.* (2010) Osteoblast proliferation on hydroxyapatite coated substrates prepared by right angle magnetron sputtering. *J. Biomed. Mater. Res., Part A*, **93**, 878–885.

34. Surmenev, R.A., Ryabtseva, M.A., Shesterikov, E.V. *et al.* (2010) The release of nickel from nickel–titanium (NiTi) is strongly reduced by a sub-micrometer thin layer of calcium phosphate deposited by rf-magnetron sputtering. *J. Mater. Sci. Mater. Med.*, **21**, 1233–1239.

35. Wolke, J.G.C., van der Waerden, J.P.C.M., Schaeken, H.G. and Jansen, J.A. (2003) In vivo dissolution behavior of various RF magnetron-sputtered Ca-P coatings on roughened titanium implants. *Biomaterials*, **24**, 2623–2629.

36. Yan, Y.G., Wolke, J.G.C., Li, Y.B. and Jansen, J.A. (2006) Subcutaneous evaluation of RF magnetron-sputtered calcium pyrophosphate and hydroxyapatite-coated Ti implants. *J. Biomed. Mater. Res., Part A*, **97**, 815–822.

37. Thian, S., Huang, J., Barber, Z.H. *et al.* (2011) Surface modification of magnetron-sputtered hydroxyapatite thin films via silicon substitution for orthopaedic and dental applications. *Surf. Coat. Technol.*, **205**, 3472–3477.

38. Clèries, L., Martínez, E., Fernández-Pradas, J.M. *et al.* (2000) Mechanical properties of calcium phosphate coatings deposited by laser ablation. *Biomaterials*, **21**, 967–971.

39. Arias, J.L., Mayor, M.B., Pou, J. *et al.* (2003). Laser ablation rate of hydroxyapatite in different atmospheres. *Appl. Surf. Sci.*, **208–209**, 57–60.

40. Fernández-Pradas, J.M., Clèries, L., Martínez, E. *et al.* (2001) Influence of thickness on the properties of hydroxyapatite coatings deposited by KrF laser ablation. *Biomaterials*, **22**, 2171–2175.

41. Kim, H., Renato, P., Camata, B. *et al.* (2010) In vitro dissolution and mechanical behavior of c-axis preferentially oriented hydroxyapatite thin films fabricated by pulsed laser deposition. *Acta Biomater.*, **6**, 3234–3241.

42. Peraire, C., Arias, J.L., Bernal, D. *et al.* (2006) Biological stability and osteoconductivity in rabbit tibia of pulsed laser deposited hydroxyapatite coatings. *J. Biomed. Mater. Res., Part A*, **77**, 370–379.

43. Hontsu, S., Hashimoto, Y., Yoshikawa, Y. *et al.* (2012) Fabrication of hydroxyl apatite coating titanium web scaffold using pulsed laser deposition method. *J. Hard Tissue Biol.*, **21**, 181–187.

44. Choi, J.M., Kim, H.E. and Lee, I.S. (2000) Ion-beam-assisted deposition (IBAD) of hydroxyapatite coating layer on Ti-based metal substrate. *Biomaterials*, **21**, 469–473.

45. Bai, X., Sandukas, S., Appleford, M.R. *et al.* (2009) Deposition and investigation of functionally graded calcium phosphate coatings on titanium. *Acta Biomater.*, **5**, 3563–3572.

46. Lee, I., Zhao, B., Lee, G. *et al.* (2007) Industrial application of ion beam assisted deposition on medical implants. *Surf. Coat. Technol.*, **201**, 5132–5137.

47. Mendonça, G., Mendonça, D.B.S., Aragao, F.L.G. and Cooper, L.F. (2008) Advancing dental implant surface technology-From micron to nanotopography. *Biomaterials*, **29**, 3822–3835.

48. Coelho, P.G., Granjeiro, J.M., Romanos, G.E. *et al.* (2009) Basic research methods and current trends of dental implant surfaces. *J. Biomed. Mater. Res., Part B*, **88**, 579–596.

49. Allen, G.C., Ciliberto, E., Fragala, L. and Spoto, G. (1996) Surface and bulk study of calcium phosphate bioceramics obtained by metal organic chemical vapor deposition. *Nucl. Instrum. Methods B*, **116**, 457–460.

50. Cabañas, M.V. and Vallet-Regí, M. (2003) Calcium phosphate coatings deposited by aerosol chemical vapour deposition. *J. Mater. Chem.*, **13**, 1104–1107.

51. Hamdi, M. and Ide-Ektessabi, A. (2007) Dissolution behavior of simultaneous vapor deposited calcium phosphate coatings in vitro. *Mater. Sci. Eng., C Mater.*, **27**, 670–674.

52. Gan, L. and Pilliar, R. (2004) Calcium phosphate sol–gel-derived thin films on porous-surfaced implants for enhanced osteoconductivity. Part I: synthesis and characterization. *Biomaterials*, **25**, 5303–5312.

53. Choi, A.H. and Ben-Nissan, B. (2007) Sol-gel production of bioactive nanocoatings for medical applications. Part II: current research and development. *Nanomedicine*, **2**, 51–61.

54. Hijón, N., Cabañas, M.V., Izquierdo-Barba, I. and Vallet-Regí, M. (2004) Bioactive carbonate-hydroxyapatite coatings deposited onto Ti6Al4V substrate. *Chem. Mater.*, **16**, 1451–1455.

55. Hijón, N., Cabañas, M.V., Izquierdo-Barba, I. *et al.* (2006) Nanocrystalline bioactive apatite coatings. *Solid State Sci.*, **8**, 685–691.

56. Li, L.H., Kim, H.W., Lee, S.H. *et al.* (2005) Biocompatibility of titanium implants modified by microarc oxidation and hydroxyapatite coating. *J. Biomed. Mater. Res., Part A*, **73**, 48–54.

57. Kim, H.W., Kim, H.E. and Knowles, J.C. (2004) Fluor-hydroxyapatite sol–gel coating on titanium substrate for hard tissue implants. *Biomaterials*, **25**, 3351–3358.

58. Hijón, N., Cabañas, M.V., Peña, J. and Vallet-Regí, M. (2006) Dip coated silicon-substituted hydroxyapatite films. *Acta Biomater.*, **2**, 567–574.

59. Horkavcova, D., Oplistilova, R., Rohanova, D. *et al.* (2012) Deposition by sol-gel and characterisation of an antibacterial bioactive layer on a Ti substrate. *Glass Technol.: -Eur. J. Glass Sci. Technol., Part A*, **53**, 16–19.

60. Peña, J., Izquierdo-Barba, I., García, M.A. and Vallet-Regí, M. (2006) Room temperature synthesis of chitosan/apatite powders and coatings. *J. Eur. Ceram. Soc.*, **26**, 3631–3638.

61. Vila, M., Izquierdo-Barba, I., Bourgeois, A. and Vallet-Regí, M. (2011) Bimodal meso/macro porous hydroxyapatite coatings. *J. Sol-Gel Sci. Technol.*, **57**, 109–113.

62. Mendes, V.C., Moineddin, R. and Davies, J.E. (2009) Discrete calcium phosphate nanocrystalline deposition enhances osteoconduction on titanium-based implant surfaces. *J. Biomed. Mater. Res., Part A*, **90**, 577–585.

63. Junker, R., Dimakis, A., Thoneick, M. and Jansen, J.A. (2009) Effects of implant surface coatings and composition on bone integration: a systematic review. *Clin. Oral Implants Res.*, **20**, 185–206.

64. Xiao, X.F., Liu, R.F. and Tang, X.L. (2008) Electrophoretic deposition of silicon substituted hydroxyapatite coatings from n-butanol-chloroform mixture. *J. Mater. Sci. Mater. Med.*, **19**, 175–182.

65. Thair, L., Ismaeel, T., Ahmed, B. and Swadi, A.K. (2011) Development of apatite coatings on Ti-6Al-7Nb dental implants by biomimetic process and EPD: in vivo studies. *Surf. Eng.*, **27**, 11–18.

66. Dorozhkim, S.V. (2010) Nanosized and nanocrystalline calcium orthophosphates. *Acta Biomater.*, **6**, 715–734.

67. Bose, S. and Tarafder, S. (2012) Calcium phosphate ceramic systems in growth factor and drug delivery for bone tissue engineering: a review. *Acta Biomater.*, **8**, 1401–1421.

68. Liu, Y., Wu, G. and de Groot, K. (2010) Biomimetic coatings for bone tissue engineering of critical-sized defects. *J. R. Soc. Interface*, **7**, S631–S647.

69. Forsgren, J., Brohede, U., Strømme, M. and Engqvist, H. (2011) Co-loading of bisphosphonates and antibiotics to a biomimetic hydroxyapatite coating. *Biotechnol. Lett.*, **33**, 1265–1268.

70. Kokubo, T., Kushitani, H., Sakka, S. *et al.* (1990) Solutions able to reproduce *in vivo* surface structure changes in bioactive glass-ceramic A-W. *J. Biomed. Mater. Res.*, **24**, 721–734.

71. Barrere, F., van Blitterswijk, C.A., de Groot, K. and Layrolle, P. (2002) Influence of ionic strength and carbonate on the Ca-P coating formation from SBFx5 solution. *Biomaterials*, **23**, 1921–1930.

72. Bigi, A., Boanini, E., Bracci, B. *et al.* (2005) Nanocrystalline hydroxyapatite coatings on titanium: a new fast biomimetic method. *Biomaterials*, **26**, 4085–4089.

73. Wang, J., Boer, J. and de Groot, K. (2009) Proliferation and differentiation of osteoblast-like MC3T3-E1 cells on biomimetically and electrolytically deposited calcium phosphates coatings. *J. Biomed. Mater. Res., Part A*, **90**, 664–670.

74. Ballo, A.M., Xia, W., Palmquist, A. *et al.* (2012) Bone tissue reactions to biomimetic ion-substituted apatite surfaces on titanium implants. *J. R. Soc. Interface*, **9**, 1615–1624.

75. Barrere, F., Layrolle, P., Van Blitterswijk, C.A. and De Groot, K. (2001) Biomimetic coatings on titanium: a crystal growth study of octacalcium phosphate. *J. Mater. Sci. Mater. Med.*, **12**, 529–534.

76. Wang, J., Layrolle, P., Stigter, M. and de Groot, K. (2004) Biomimetic and electrolytic calcium phosphate coatings on titanium alloy: physicochemical characteristics and cell attachment. *Biomaterials*, **25**, 583–592.

77. Chen, Y.Y., Zhao, H.B., Wang, R. *et al.* (2012) Characterization and evaluation of titanium substrates coated with gelatin/hydroxyapatite composite for culturing rat bone marrow derived mesenchymal stromal cells. *Biomed. Eng. -Appl. Basis C*, **24**, 197–206.

78. Xia, Z., Yu, X. and Wei, M. (2012) Biomimetic collagen/apatite coating formation on Ti6Al4V substrates. *J. Biomed. Mater. Res., Part B*, **100**, 871–881.

79. Wang, X., Ito, A., Li, X. *et al.* (2011) Signal molecules–calcium phosphate coprecipitation and its biomedical application as a functional coating. *Biofabrication*, **3**, 022001.

80. Zitelli, J.P. and Higham, P. (1999) A novel method for solution deposition of hydroxyapatite onto three dimensionally porous metallic surfaces: peri-apatite Ha. *Mat. Res. Soc. Proc.*, **599**, 117. doi: 10.1557/PROC-599-117

81. Hansson, U., Ryd, L. and Toksvig-Larsen, S. (2008) A randomised RSA study of peri-apatite™ HA coating of a total knee prosthesis. *Knee*, **15**, 211–216.

82. Kuo, M.C. and Yen, S.K. (2002) The process of electrochemical deposited hydroxyapatite coatings on biomedical titanium at room temperature. *Mater. Sci. Eng., Part C*, **20**, 153–160.

83. Wang, S.H., Shih, W.J., Li, W.L. *et al.* (2005) Morphology of calcium phosphate coatings deposited on a Ti–6Al–4V substrate by an electrolytic method under 80 Torr. *J. Eur. Ceram. Soc.*, **25**, 3287–3292.

84. Kanamoto, K., Imamura, K., Kataoka, N. *et al.* (2009) Formation characteristics of calcium phosphate deposits on a metal surface by H_2O_2-electrolysis reaction under various conditions. *Colloids Surf., A*, **350**, 79–86.

85. Drevet, R., Benhayoune, H., Wortham, L. *et al.* (2010) Effects of pulsed current and H_2O_2 amount on the composition of electrodeposited calcium phosphate coatings. *Mater Charact*, **61**, 786–795.

86. López-Heredia, M.A., Weiss, P. and Layrolle, P. (2007) An electrodeposition method of calcium phosphate coatings on titanium alloy. *J. Mater. Sci. Mater. Med.*, **18**, 381–90.

87. Rakngarm, A. and Mutoh, Y. (2009) Electrochemical depositions of calcium phosphate film on commercial pure titanium and Ti–6Al–4V in two types of electrolyte at room temperature. *Mater. Sci. Eng. C-Mater.*, **29**, 275–283.

88. Narayanan, R., Seshadri, S.K., Kwon, T.Y. and Kim, K.H. (2007) Electrochemical nano-grained calcium phosphate coatings onTi–6Al–4V for biomaterial applications. *Scr. Mater.*, **56**, 229–232.

89. Drevet, R., Lemelle, A., Untereiner, V. *et al.* (2013) Morphological modifications of electrodeposited calcium phosphate coatings under amino acids effect. *Appl. Surf. Sci.*, **268**, 343–348.

90. Bøe, B.G., Röhrl, S.M., Heier, T. *et al.* (2011) A prospective randomized study comparing electro-chemically deposited hydroxyapatite and plasma-sprayed hydroxyapatite on titanium stems. *Acta Orthop.*, **82**, 13–19.

91. Hench, L.L. (2009) Genetic design of bioactive glass. *J. Eur. Ceram. Soc.*, **29**, 1257–1265.

92. Zhao, Y., Chen, C. and Wang, D. (2005) The current techniques for preparing bioglass coatings. *Surf. Rev. Lett.*, **12**, 505–13.

93. Sola, A. (2011) Bioactive glass coatings: a review. *Surf. Eng.*, **27**, 560–572.

94. Gómez-Vega, J.M., Saiz, E. and Tomsia, A.P. (1999) Glass-based coatings for titanium implant alloys. *J. Biomed. Mater. Res.*, **46**, 549–559.

95. López-Esteban, S., Saiz, E., Fujino, S. *et al.* (2003) Bioactive glass coatings for orthopedic metallic implants. *J. Eur. Ceram. Soc.*, **23**, 2921–2930.

96. Lee, T.M., Chang, E., Wang, B.C. and Yang, C.Y. (1996) Characteristics of plasma-sprayed bioactive glass coatings on Ti–6A1–4V alloy: an in vitro study. *Surf. Coat. Technol.*, **79**, 170–177.

97. Cannillo, V., Pierli, F., Sampath, S. and Siligardi, C. (2009) Thermal and physical characterisation of apatite/wollastonite bioactive glass–ceramics. *J. Eur. Ceram. Soc.*, **29**, 611–619.

98. Mardare, C.C., Mardare, A.I., Fernandes, J.R.F. *et al.* (2003) Deposition of bioactive glass ceramic thin-films by RF magnetron sputtering. *J. Eur. Ceram. Soc.*, **23**, 1027–30.

99. Stan, G.E., Pasuk, I., Husanu, M.A. *et al.* (2011) Highly adherent bioactive glass thin films synthetized by magnetron sputtering at low temperature. *J. Mater. Sci. Mater. Med.*, **22**, 2693–2710.

100. D'Alessio, L., Teghil, R., Zaccagnino, M. *et al.* (1999) Pulsed laser ablation and deposition of bioactive glass as coating material for biomedical applications. *Appl. Surf. Sci.*, **138–139**, 527–532.

101. Liste, S., Serra, J., González, P. *et al.* (2004) The role of the reactive atmosphere in pulsed laser deposition of bioactive glass films. *Thin Solid Films*, **453–454**, 224–228.

102. Comesaña, R., Quintero, F., Lusquiños, F. *et al.* (2010) Laser cladding of bioactive glass coatings. *Acta Biomater.*, **6**, 953–961.

103. Izquierdo-Barba, I., Asenjo, A., Esquivias, L. and Vallet-Regí, M. (2003) SiO_2-CaO Vitreous films deposited onto Ti6Al4V substrates. *Eur. J. Inorg. Chem.*, 1608–1613.

104. Fathi, M.H. and Doostmohammadi, A. (2009) Bioactive glass nanopowder and bio-glass coating for biocompatibility improvement of metallic implant. *J. Mater. Process Tech.*, **209**, 1385–1391.

105. Zhang, W., Chen, X., Liao, X. *et al.* (2011) Electrophoretic deposition of porous CaO–MgO–SiO$_2$ glass–ceramic coatings with B$_2$O$_3$ as additive on Ti–6Al–4V alloy. *J. Mater. Sci. Mater. Med.*, **22**, 2261–2271.

106. Cannillo, V. and Sola, A. (2010) Different approaches to produce coatings with bioactive glasses: enamelling vs plasma spraying. *J. Eur. Ceram. Soc.*, **30**, 2031–2039.

107. Bellucci, D., Cannillo, V. and Sola, A. (2010) An overview of the effects of thermal processing on bioactive glasses. *Sci. Sinter.*, **42**, 307–320.

108. Donald, I.W., Mallinson, P.M., Metcalfe, B.L. *et al.* (2011) Recent developments in the preparation, characterization and applications of glass- and glass–ceramic-to-metal seals and coatings. *J. Mater. Sci.*, **46**, 1975–2000.

109. Carvalho, F.L.S., Borges, C.S., Branco, J.R.T. and Pereira, M.M. (1999) Structural analysis of hydroxyapatite/bioactive glass composite coatings obtained by plasma spray processing. *J. Non-Cryst. Solid*, **247**, 64–68.

110. Gómez-Vega, J.M., Saiz, E., Tomsia, A.P. *et al.* (2000) Bioactive glass coatings with hydroxyapatite and Bioglass® particles on Ti-based implants. 1. Processing. *Biomaterials*, **21**, 2105–2111.

111. Balamurugan, A., Balossier, G., Michel, J. and Ferreira, J.M.F. (2009) Electrochemical and structural evaluation of functionally graded bioglass-apatite composites electrophoretically deposited onto Ti6Al4V alloys. *Electrochim. Acta*, **54**, 1192–1198.

112. Cannillo, V. and Sola, A. (2009) Potassium-based composition for a bioactive glass. *Ceram. Int.*, **35**, 3389–3393.

113. Peddi, L., Brow, R.K. and Brown, R.F. (2008) Bioactive borate glass coatings for titanium alloys. *J. Mater. Sci. Mater. Med.*, **19**, 3145–3152.

114. Goller, G. (2004) The effect of bond coat on mechanical properties of plasma sprayed bioglass-titanium coatings. *Ceram. Int.*, **30**, 351–355.

115. Kim, C.Y. and Lee, J.W. (2005) Surface bio-modification of titanium implants by an enamel process. *J. Ceram. Process. Res.*, **6**, 338–344.

116. Gómez-Vega, J.M., Saiz, E., Tomsia, A. *et al.* (2000) Novel bioactive functionally graded coatings on Ti6Al4V. *Adv. Mater.*, **12**, 894–898.

117. Mohamed, N., Rahaman, M.N., Li, Y. *et al.* (2008) Functionally graded bioactive glass coating on magnesia partially stabilized zirconia (Mg-PSZ) for enhanced bio-compatibility. *J. Mater. Sci. Mater. Med.*, **19**, 2325–2333.

118. Moritz, N., Rossi, S., Vedel, E. *et al.* (2004) Implants coated with bioactive glass by CO$_2$-laser, an in vivo study. *J. Mater. Sci. Mater. Med.*, **15**, 795–802.

119. Lee, J.H., Ryu, H.S., Lee, D. *et al.* (2005) Biomechanical and histomorphometric study on the bone–screw interface of bioactive ceramic-coated titanium screws. *Biomaterials*, **26**, 3249–3257.

120. Borrajo, J.P., Serra, J., González, P. *et al.* (2007) In vivo evaluation of titanium implants coated with bioactive glass by pulsed laser deposition. *J. Mater. Sci. Mater. Med.*, **18**, 2371–2376.

121. Foppiano, S., Marshall, S.J., Marshall, G.W. *et al.* (2007) Bioactive glass coatings affect the behavior of osteoblast-like cells. *Acta Biomater.*, **3**, 765–771.

122. Berbecaru, C., Stan, G.E., Pina, S. *et al.* (2012) The bioactivity mechanism of mag-netron sputtered bioglass thin films. *Appl. Surf. Sci.*, **258**, 9840–9848.
123. Mistry, S., Kundu, D., Datta, S. and Basu, D. (2011) Comparison of bioactive glass coated and hydroxyapatite coated titanium dental implants in the human jaw bone. *Aust. Dent. J.*, **56**, 68–75.
124. Kokubo, T., Miyaji, F., Kim, H. and Nakamura, T. (1996) Spontaneous formation of bone like apatite layer on chemically treated titanium metals. *J. Am. Ceram. Soc.*, **79**, 1127–1129.
125. Li, P., Ohtsuki, C., Kokubo, T. *et al.* (1994) The role of hydrated silica, titanium and alumina in inducing apatite on implants. *J. Biomed. Mater. Res.*, **28**, 7–15.
126. Kawai, T., Kizuki, T., Takadama, H. *et al.* (2010) Apatite formation on surface titanate layer with different Na content on Ti metal. *J. Ceram. Soc. Jpn.*, **118**, 19–24.
127. Kokubo, T., Matsushita, T., Takadama, H. and Kizuki, T. (2009) Development of bioactive materials based on surface chemistry. *J. Eur. Ceram. Soc.*, **29**, 1267–1274.
128. Takamada, H., Kim, H.M., Kokubo, T. and Nakamura, T. (2001) TEM-EDX study of mechanism of bone like apatite formation on bioactive titanium metal in simulated body fluid. *J. Biomed. Mater. Res.*, **57**, 441–448.
129. Kim, H.M., Himeno, T., Kawashita, M. *et al.* (2003) Surface potential change in bioactive titanium metal during the process of apatite formation in simulated body fluid. *Biomed. Mater. Res., Part A*, **67**, 1305–1309.
130. Kim, H.M., Kokubo, T., Fujibayashi, S. *et al.* (2000) Bioactive macroporous titanium surface layer on titanium substrate. *J. Biomed. Mater. Res.*, **52**, 553–557.
131. Nishio, K., Neo, M., Akiyama, H. *et al.* (2000) The effect of alkali- and heat-treated titanium and apatite-formed titanium on osteoblastic differentiation of bone marrow cells. *J. Biomed. Mater. Res.*, **2000** (52), 652–661.
132. Isaac, J., Loty, S., Hamdan, A. *et al.* (2009) Bone-like tissue formation on a biomimetic titanium surface in an explant model of osteoconduction. *J. Biomed. Mater. Res., Part A*, **89**, 585–593.
133. Nishiguchi, S., Fujibayashi, S., Kim, H.M. *et al.* (2003) Biology of alkali- and heat-treated titanium implants. *J. Biomed. Mater. Res., Part A*, **67**, 26–35.
134. Kawanabe, K., Ise, K., Goto, K. *et al.* (2009) Clinical device-related article: a new cementless total hip arthroplasty with bioactive titanium porous-coating by alkaline and heat treatment: average 4.8 year results. *J. Biomed. Mater. Res., Part B*, **90**, 476–481.
135. Zhao, C., Liang, K., Tan, J. *et al.* (2013) Bioactivity of porous titanium with hydrogen peroxide solution with or without tantalum chloride treatment at a low temperature. *Biomed. Mater.*, **8** (2), 025006. doi: 10.1088/1748-6041/8/2/025006
136. Pattanayak, D.K., Kawai, T., Matsushita, T. *et al.* (2009) Effect of HCl concentrations on apatite-forming ability of NaOH–HCl- and heat-treated titanium metal. *J. Mater. Sci.*, **20**, 2401–2411.
137. Kokubo, T., Pattanayak, D.K., Yamaguchi, S. *et al.* (2010) Positively charged bioac-tive Ti metal prepared by simple chemical and heat treatments. *J. R. Soc. Interface*, **7**, S503–S513.
138. Schliephak, H. and Scharnweber, D. (2008) Chemical and biological functionaliza-tion of titanium for dental implants. *J. Mater. Chem.*, **18**, 2404–2414.

139. Wieser, E., Tsyganow, I., Matz, W. *et al.* (1999) Modification of titanium by ion implantation of calcium and/or phosphorus. *Surf. Coat. Technol.*, **111**, 103–109.

140. Nayab, S., Jones, F.H. and Olsen, I. (2007) Effects of calcium ion-implantation of titanium on bone cell function *in vitro*. *J. Biomed. Mater. Res., Part A*, **83**, 296–302.

141. Rahaman, M.N., Yao, A., Bal, B.S. *et al.* (2007) Ceramics for prosthetic hip and knee joint replacement. *J. Am. Ceram. Soc.*, **90**, 1965–1988.

142. Rostlund, T., Albrektsson, B., Albrektsson, T. and McKellop, H. (1989) Wear of ion-implanted pure titanium against UHMWPE. *Biomaterials*, **10**, 176–181.

143. Lappalainen, R. and Santavirta, S.S. (2005) Potential of coatings in total hip replacement. *Clin. Orthop. Relat. Res.*, **430**, 72–79.

144. Abadias, G. (2008) Stress and preferred orientation in nitride-based PVD coatings. *Surf. Coat. Technol.*, **202**, 2223–2235.

145. Ramanuja, N., Levy, R.A., Dharmadhikari, S.N. *et al.* (2002) Synthesis and characterization of low pressure chemically vapor deposited titanium nitride films using $TiCl_4$ and NH_3. *Mater. Lett.*, **57**, 261–269.

146. Heide, N. and Schultze, J.W. (1993) Corrosion stability of TiN prepared by ion-implantation and PVD. *Nucl. Instrum. Methods B*, **80-1**, 467–1471.

147. Raimondi, M.T. and Pietrabissa, R. (2000) The in-vivo wear performance of prosthetic femoral heads with titanium nitride coating. *Biomaterials*, **21**, 907–913.

148. Turkan, U., Ozturk, O. and Eroglu, A.E. (2006) Metal ion release from TiN coated CoCrMo orthopedic implant material. *Surf. Coat. Technol.*, **200**, 5020–5027.

149. Galetz, M.C., Fleischmann, E.W., Konrad, C.H. *et al.* (2010) Abrasion resistance of oxidized zirconium in comparison with CoCrMo and titanium nitride coatings for artificial knee joints. *J. Biomed. Mater. Res., Part B*, **93**, 244–251.

150. Gotman, I., Gutmanas, E.Y. and Hunter, G. (2011) Wear-resistant ceramic films and coatings, in *Comprehensive Biomaterials* (ed P. Ducheyne) (ed.-in-Chief, Elsevier, Oxford, pp. 127–155.

151. Wilson, A.D., Leyland, A. and Matthews, A. (1999) A comparative study of the influence of plasma treatments, PVD coatings and ion implantation on the tribological performance of Ti-6Al-4V. *Surf. Coat. Technol.*, **114**, 70–80.

152. Venugopalan, R., Weimer, J.J., George, M.A. and Lucas, L.C. (2000) The effect of nitrogen diffusion hardening on the surface chemistry and scratch resistance of Ti-6Al-4V alloy. *Biomaterials*, **21**, 1669–1977.

153. Ma, S., Xu, K. and Jie, W. (2004) Wear behavior of the surface of Ti–6Al–4V alloy modified by treating with a pulsed d.c. plasma-duplex process. *Surf. Coat. Technol.*, **185**, 205–209.

154. Casadei, F., Pileggi, R., Valle, R. and Matthews, A. (2006) Studies on a combined reactive plasma sprayed/arc deposited duplex coating for titanium alloys. *Surf. Coat. Technol.*, **201**, 1200–1206.

155. Chen, L., Wang, S.Q., Du, Y. and Li, J. (2008) Microstructure and mechanical properties of gradient Ti(C, N) and TiN/Ti(C, N) multilayer PVD coatings. *Mater. Sci. Eng. A*, **478**, 336–339.

156. Maruyama, M., Capello, W.N., D'Antonio, J.A. *et al.* (2000) Effect of low friction ion treated femoral heads on polyethylene wear rates. *Clin. Orthop. Relat. Res.*, **370**, 183–191.

157. Fisher, J., Hu, X.Q., Stewart, T.D. *et al.* (2004) Wear of surface engineered metal-on-metal hip prostheses. *J. Mater. Sci. Mater. Med.*, **15**, 225–235.

158. Rieu, J. (1993) Ceramic formation on metallic surfaces (ceramization) for medical applications. *Clin. Mater.*, **12**, 227–235.

159. Evangelista, G.T., Fulkerson, E., Kummer, E. and Di Cesare, P.E. (2007) Surface damage to an oxinium femoral head prosthesis after dislocation. *J. Bone Joint Surg. Br., Part B*, **89**, 535–537.

160. Hobbs, L.W., Rosen, V.B., Mangin, S.P. *et al.* (2005) Oxidation microstructures and interfaces in the oxidized zirconium knee. *Int. J. Appl. Ceram. Technol.*, **2**, 221–246.

161. Ezzet, K.A., Hermida, J.C., Colwell, C.W. and D'Lima, D.D. (2004) Oxidized zirconium femoral components reduce polyethylene wear in a knee wear simulator. *Clin. Orthop. Rel. Res.*, **420**, 120–4.

162. Kop, A.M., Whitewood, C. and Johnston, D.J.L. (2007) Damage of oxinium femoral heads subsequent to hip arthroplasty dislocation – three retrieval case studies. *J. Arthroplasty*, **22**, 775–9.

163. Bourne, R.B., Barrack, R., Rorabeck, C.H. *et al.* (2005) Arthroplasty options for the young patient: oxinium on cross-linked polyethylene. *Clin. Orthop. Rel. Res.*, **441**, 159–67.

164. Hui, C., Salmon, L., Maeno, S. *et al.* (2011) Five-year comparison of oxidized zirconium and cobalt-chromium femoral components in total knee arthroplasty: a randomized controlled trial. *J. Bone Joint Surg. Am.*, **93**, 624–630.

165. Meadows, P.J., López-Honorato, E. and Xiao, P. (2009) Fluidized bed chemical vapor deposition of pyrolytic carbon–II. Effect of deposition conditions on anisotropy. *Carbon*, **47**, 251–262.

166. More, R.B., Haubold, A.D. and Bokros, J.C. (2013) in *Biomaterials Science: An Introduction to Materials in Medicine*, 3rd edn (eds B.D. Ratner, A.S. Hoffman and J.E. Lemons), Elsevier Academic Press, San Diego, CA, pp. 209–222.

167. Hauert, R. (2003) A review of modified DLC coatings for biological applications. *Diamond Relat. Mater.*, **12**, 583–589.

168. Narayan, R.J., Wei, W., Jin, C. *et al.* (2006) Microstructural and biological properties of nanocrystalline diamond coatings. *Diamond Relat. Mater.*, **15**, 1935–1940.

169. Butler, J.E. and Sumant, A.V. (2008) The CVD of nanodiamond materials. *Chem. Vap. Deposition*, **14**, 145–160.

170. Abreu, C.S., Amaral, M., Oliveira, F.J. *et al.* (2009) HFCVD nanocrystalline diamond coatings for tribo-applications in the presence of water. *Diamond Relat. Mater.*, **18**, 271–75.

171. Williams, O.A. (2011) Nanocrystalline diamond. *Diamond Relat. Mater.*, **20**, 621–640.

172. Medina, O., Nocua, J., Mendoza, F. *et al.* (2012) Bactericide and bacterial anti-adhesive properties of the nanocrystalline diamond surface. *Diamond Relat. Mater.*, **22**, 77–81.

173. Auciello, O. and Sumant, A.V. (2010) Status review of the science and technology of ultrananocrystalline diamond (UNCD™) films and application to multifunctional devices. *Diamond Relat. Mater.*, **19**, 699–718.

174. Dearneley, G. and Arps, J.H. (2005) Biomedical applications of diamond-like carbon (DLC) coatings: a review. *Surf. Coat. Technol.*, **200**, 2518–24.

175. Roy, R.K. and Lee, K.R. (2007) Biomedical applications of diamond-like carbon coatings: a review. *J. Biomed. Mater. Res., Part B*, **83**, 72–84.
176. Robertson, J. (2002) Diamond-like amorphous carbon. *Mater. Sci. Eng. R*, **37**, 129–181.
177. Erdemir, A. and Donnet, C. (2006) Tribology of diamond-like carbon films: recent progress and future prospects. *J. Phys. D: Appl. Phys.*, **39**, R311–R327.
178. Maa, W.J., Ruysa, A.J., Masonb, R.S. *et al.* (2007) DLC coatings: effects of physical and chemical properties on biological response. *Biomaterials*, **28**, 1620–1628.
179. Zhou, H., Xu, L., Ogino, A. and Nagatsu, M. (2008) Investigation into the antibacterial property of carbon films. *Diamond Relat. Mater.*, **17**, 1416–1419.
180. Ohgoe, Y., Hirakuri, K., Saitoh, H. *et al.* (2012) Classification of DLC films in terms of biological response. *Surf. Coat. Technol.*, **207**, 350–354.
181. Alaoski, A., Tiainen, V., Soininen, A. and Konttinen, Y.T. (2008) Load-bearing biomedical applications of diamond-like carbon coatings-current status. *Open Orthop. J.*, **2**, 43–50.
182. Park, S.J., Kwang-Ryeol, L., Ahn, S. and Kim, J. (2008) Instability of diamond-like carbon (DLC) films during sliding in aqueous environment. *Diamond Relat. Mater.*, **17**, 247–251.
183. Kumar, P., Babu, D., Mohan, L. *et al.* (2013) Wear and corrosion behavior of Zr-doped DLC on Ti-13Zr-13Nb biomedical alloy. *J. Mater. Eng. Perform.*, **22**, 283–293.
184. Love, C.A., Cook, R.B., Harvey, T.J. *et al.* (2013) Diamond like carbon coatings for potential application in biological implants-a review. *Tribol. Int.*, **63**, 141–150.
185. Castellino, M., Stolojan, V., Virga, A. *et al.* (2013) Chemico-physical characterisation and in vivo biocompatibility assessment of DLC-coated coronary stents. *Anal. Bioanal. Chem.*, **405**, 321–329.
186. Logothetidis, S. (2007) Haemocompatibility of carbon based thin films. *Diamond Relat. Mater.*, **16**, 1846–1857.
187. Grill, A. (2003) Diamond-like carbon coatings as biocompatible materials-an overview. *Diamond Relat. Mater.*, **12**, 166–170.
188. Affatato, S., Frigo, M. and Toni, A. (2000) An in vitro investigation of diamond-like carbon as a femoral head coating. *J. Biomed. Mater. Res., Part A*, **53B**, 221–226.
189. Taeger, G., Podleska, L.E., Schmidt, B. *et al.* (2003) Comparison of diamond-like-carbon and alumina-oxide articulating with polyethylene in total hip arthroplasty. *Mat.-Wiss. Werkstofftech.*, **34**, 1094–1100.

10

Scaffold Designing

Isabel Izquierdo-Barba

Departamento de Química Inorgánica y Bioinorgánica, Facultad de Farmacia, Universidad Complutense de Madrid, CIBER de Bioingeniería, Biomateriales y Nanomedicina (CIBER-BBN), Spain

10.1   Introduction

Scaffold design properties are a key factor in bone tissue engineering and represent more than just a passive component [1]. Scaffolds should provide a temporal structural support for cell growth during regeneration of the tissue, while eventually being reabsorbed, leaving only the newly-formed living tissue and the lesion fully healed [2]. Figure 10.1 displays the similarity between construction scaffolding and a 3D scaffold for bone tissue regeneration purposes.

According to Hutmacher [3], an ideal synthetic bone graft is a porous material that can act as a temporary template for bone growth in three dimensions. It should: (i) be biocompatible and bioresorbable with a controllable degradation and resorption rate to match bone cell/tissue growth *in vitro* and/or *in vivo*; (ii) have a suitable surface chemistry for cell attachment, proliferation, and differentiation; (iii) be three-dimensional and highly porous with an interconnected porous network for cell growth, vascularization, flow transport of nutrients, and metabolic waste; (iv) have mechanical properties to match those of the tissues at the site of implantation; (v) be made by a fabrication process that can be scaled up for mass production; and (vi) be sterilizable and meet regulatory requirements for clinical use. In general the ideal synthetic scaffold is expected to mimic porous cancellous bone, which fulfills most of the listed criteria [4, 5].

Figure 10.2 represents all steps implicated in the design of 3D scaffolds for bone tissue regeneration purposes. Ideally, the most modern processing techniques are based on the fabrication of 3D scaffolds based on medical imaging data which are "biodecorated" with stem cells and growth factors to accelerate bone regeneration. The 3D scaffold is implanted into the patient, where it undergoes reabsorption at the same time that the bone is newly

Bioceramics with Clinical Applications, First Edition. Edited by María Vallet-Regí.
© 2014 John Wiley & Sons, Ltd. Published 2014 by John Wiley & Sons, Ltd.

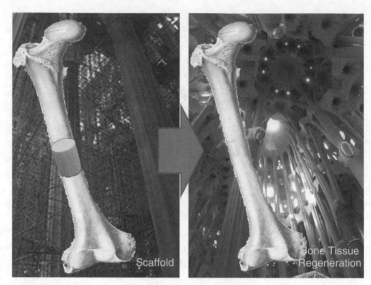

Figure 10.1 *Similarity between construction scaffolding and the 3D scaffold for bone tissue engineering. The image shows photographs of the Expiatory Temple of the Holy Family before and after its construction (Barcelona, Spain)*

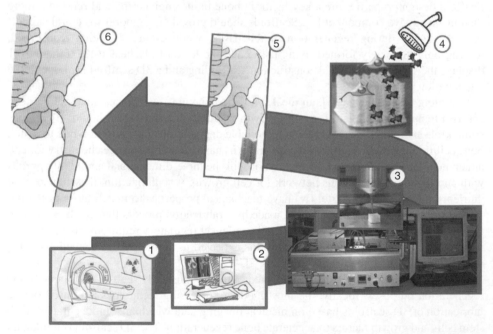

Figure 10.2 *Principle of scaffold-based tissue engineering. A tailored scaffold is prepared based on medical imaging data by rapid prototyping technologies (1–3), after incorporation of growth factors and cells (4), and implanted into the patient (5). In time, the scaffold is reabsorbed and the implanted construct is remodeled, to finally form functional tissue (6)*

formed, recovering finally its functionality [2]. Herein, an overview of the main requirements in the design of 3D scaffolds for bone tissue regeneration will be presented, as well
as the most relevant aspects of the different techniques of processing used to date.

10.2   Essential Requirements for Bone Tissue Engineering Scaffolds

Since Langer *et al.* pioneered the concept of reconstructing tissue using cells transplanted
on synthetic matrices in the early 1990s (Figure 10.2) [2], in the subsequent two decades,
research has "exploded" with the development of new materials and structures that serve as
scaffolds [6–8]. However, the penetration of these new scaffolding materials and structures
from the research laboratories to the clinic has been extremely limited. This limited penetration is likely due to a confluence of many factors, such as the need for a better understanding
of materials and design requirements, more complete understanding of material degradation *in vivo*, more comprehensive development of scalable methods for manufacturing,
integration of computational design techniques with scaffold manufacturing and more customized therapies to increase their acceptance. The integration of all these factors is needed
to reach from "concept to clinic" in the scaffolding issue [9]. To engineer successful scaffolds for a particular clinical application, one must understand the basic tissue structure and
metabolic requirements for the particular scaffold application. The design, material synthesis, and manufacture should be addressed taking into account the most relevant US Food
and Drug Administration (FDA) guidelines that govern this particular clinical application.
Therefore, four fundamental needs for tissue engineering scaffolds have been established:
Form, Function, Formation, and Fixation, which are denoted as the 4F requirements and
constitute the main pillars for the success of a scaffold [9]. *Form* is related to filling the
complex 3D defects, guiding the tissue shape to match the original 3D anatomy. *Function* is related to temporary support of everyday functional demands, typically mechanical,
within a defect until sufficient tissue has formed to take over these demands. *Formation*
is related to enhancing tissue regeneration through the delivery of appropriate biological
materials and by providing an appropriate mass-transport environment. *Fixation* is related
to the provision of all of this within a package that the surgeon can readily implant and
attach to tissues surrounding the defect.

It has been very well established that besides the material issue, which requires the
use of viable, non-mutagenic, non-antigen, non-carcinogenic, non-toxic, non-teratogenic
materials with high cell/tissue biocompatibility, the structural properties of 3D scaffolds
play an important role in the future success in clinical applications [10]. Hence, if the aim is
to mimic cancellous bone it is important to know in detail the bone structure. The reader is
encouraged to read Chapter 11 for further discussion related to bone structure. Figure 10.3
displays the hierarchical structure of natural bone which comprises different length scales
from macro- to nanoscale [11, 12]. Bone has a strong calcified outer compact layer
(Figure 10.3, A), which is comprised of many cylindrical Haversian systems or osteons
(Figure 10.3, B). The resident cells as osteocytes display cell membranes receptors which
respond to specific binding sites (Figure 10.3, C) and the well-defined nanoarchitecture
of the surrounding extracellular matrix (ECM) formed by nanocrystalline apatite crystals
organized into collagen fibers (Figure 10.3, D). Different studies have shown that the cells
are sensitive on these different levels to assure adequate tissue regeneration [7, 13–15].

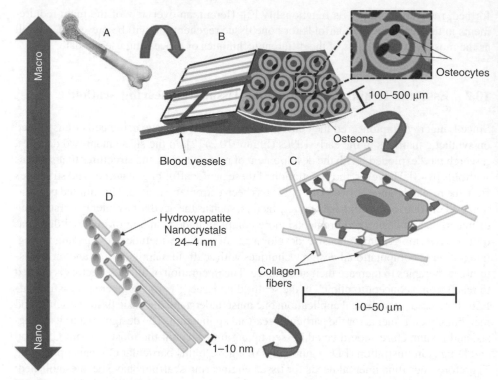

Figure 10.3 *Hierarchical organization of bone tissue at different length scales (See insert for color representation of the figure)*

Therefore, the main challenge is to maintain high levels of accurate control over these macro- (e.g., spatial form, mechanical strength, density, and porosity), micro- (e.g., pore size, pore distribution, pore interconnectivity), and nanostructural (e.g., surface area, roughness, topography) properties [1–7] to achieve adequate regeneration. Below are listed the main features to consider when designing scaffolds for bone tissue regeneration purposes:

1. *Macrostructure*: It is important to define the volume that will shape the regenerating tissue.
2. *Mechanical properties*: As a template to guide bone tissue regeneration, the scaffolds should have sufficient mechanical strength to provide temporary function in a defect while the tissue regenerates. If the native bone mechanical properties are used as a guideline in the scaffold designing, they must exhibit linear elastic properties with a moduli of hundreds of megapascals, with microstructures embedded that exhibit preferred orientations due to bone anisotropy. In addition, it is important to recognize that scaffold mechanical properties will decrease with the scaffold degradation. Thus, even if the scaffold has sufficient mechanical properties at the time of implantation, the way its mechanical properties change during the degradation could also affect function within the tissue defect [16, 17].
3. *Pore size, porosity, and interconnectivity*: While it is accepted that pore size is an important variable to stimulate cell ingrowth and new bone formation [3, 10], the interconnected porous network and porosity are critical in ensuring spatially uniform

cell distribution, cell survival, proliferation, and migration *in vitro*. Moreover, the scaffold's porosity (exceeding 60%) and degree of pore interconnectivity directly affect the diffusion of physiological nutrients and gases to, and the removal of metabolic waste and by-products from cells that have penetrated the scaffold. The size range of these pores is quite wide and at the moment there is a lack of consensus regarding the ideal porosity necessary for bone growth [18, 19]. It is well-known that pores with sizes smaller than 1 μm are appropriate to interact with proteins and are mainly responsible for inducing the formation of an apatite-like layer in contact with physiological fluids (bioactive behavior). Pores with sizes in the 1–20 μm range are important in cellular development, the type of cells attached, and the orientation and directionality of cellular in-growth. Moreover, pore sizes between 100 and 1000 μm are essential to assure nutrient supply, waste removal of cells, thus promoting the in-growth of bone cells. Finally, the presence of pores of sizes greater than 1000 μm will play an important role in the implant functionality [11, 20].

4. *Surface area and topography (nanoscale)*: Currently, different studies have established that the use of nanoscale structuring material to control cell behavior has important implications when designing new materials for tissue engineering [7, 21, 22]. Conventionally, scaffolds have been designed macroscopically to have mechanical properties similar to natural bone tissue, without the complexity and nanoscale details observed in real organs at the level of the cell matrix interaction (Figure 10.3). Different approaches through this nanoscale level may provide real benefits, since cells are inherently sensitive to local nanoscale changes. Some studies have shown that simply increasing the nanoscale roughness of the scaffold pore walls leads to an increase in cell attachment, proliferation, and expression of matrix components [23]. Figure 10.4 displays a schematic illustration concerning the influence of different length scales on the cellular spreading [7].

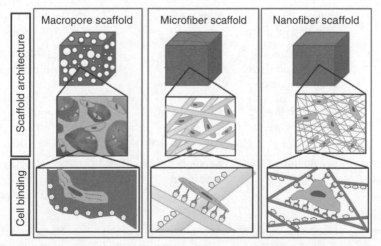

Figure 10.4   *Influence of pore scaffold architecture on cell adhesion and colonization. Cells binding to scaffolds with macroscale and microscale architectures flatten and spread as if cultured on flat surfaces. Scaffolds with nanoscale architectures exhibit larger surface areas to attach proteins, presenting many more binding sites to cell membrane receptors*

10.3　Scaffold Processing Techniques

There are a large variety of manufacturing methodologies to develop 3D porous scaffolds. These techniques range from conventional methods, such as deproteinization of bovine bone [24], solvent casting [25], particulate leaching [26], freeze-drying [27], gel casting replica [28–30], gas foaming [31], and their combinations [32], which are mainly focused on the introduction of open pores and inter-connected channels, to the most sophisticated technologies, such as rapid prototyping (RP) and electrospinning processing, which allow accurate control on the macro-nanoporosity scales [33, 34]. Table 10.1 summarizes the main features of different processing techniques for the manufacture of 3D scaffolds.

In general, conventional methodologies have shown limited reproducibility and low control over internal geometry. However, RP-assisted fabrications constitute a collection of techniques that can build an object in almost any shape by depositing layer by layer. The advantage of these techniques over conventional methods is that they allow production

Table 10.1　*Processing techniques for the preparation of 3D macroporous scaffolds*

Method	Via	Advantages	Disadvantages
Leaching	Particulates Fibers Meshes	Easy process Versatile Degradability and initial strength tailorable	Poor reproducibility Residual porogens Scarce or null interconnected
Freeze drying	Water freezing + sublimation	Highly porous structures	Energy and time consuming
Templates (replica)	Polymeric foams	High interconnectivity and porosity	Very low strength Crystal growth Limited porosity and connectivity
	Indirect rapid prototyping	Higher strength	Problems with residual solvents
Foaming	Gas generation from acid–base and decomposition reactions	Easy process	Scaffolds with low initial strength
	Surface active foaming agents	Easy process Injectable paste	Poor reproducibility Scaffolds with low initial strength
Rapid prototyping	SFF methodologies	Accurate control of the scaffold's architecture Reproducible and fast	Microarchitecture limited by particle's size
Electrospinning	Fiber deposition by electric forces	Accurate control of the nanostructure scaffold's architecture	Shape and macro architecture limited

of 3D scaffolds with defined and regular pore structures, which is dictated by a computer-aided design (CAD) file. Currently, electrospinning technology is being introduced in bone tissue regeneration applications due to its capability to obtain nanoscaled fibrous design of scaffolds that mimic functional collagen structures [22]. Because of the large amount of scaffold processing methods and to simplify things, this chapter will consider in detail three of the main scaffolding structures used for bone tissue regeneration purposes. These include foam type scaffolds and scaffolds derived by RP and electrospinning technologies, respectively. Figure 10.5 summarizes the main scaffold structures that will be discussed in this chapter.

10.3.1 Foam Scaffolds

Foam type scaffolds are macroporous networks that mimic the structure of cancellous bone as it has been demonstrated by X-ray microtomography [35]. These foam scaffolds exhibit

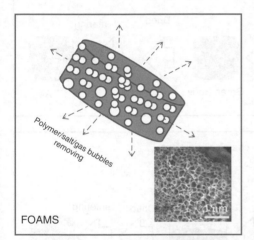

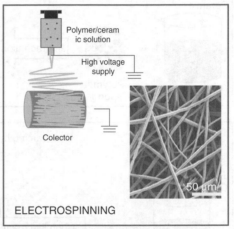

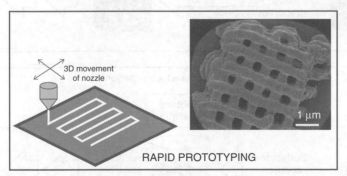

Figure 10.5 *Different technological approaches used to fabricate bioceramic scaffolds for bone tissue regeneration purposes. Highly porous scaffolds are fabricated by gas foaming or salt leaching (foams). Electrospinning is an ultrathin fiber-forming technique that uses an electrostatic force. Rapid prototyping allows production of 3D scaffolds with defined and regular pore structures*

an interconnectivity of open structures with porosity general higher than 90% and with pore sizes in the range of 200–700 μm. Different methodologies are employed to fabricate foam structures, which are summarized in Figure 10.6. These methodologies include indirect methods, such as the foam replica technique, and direct methods, such as gas forming chemical reactions and surface active foaming agents. These foam procedures are considered as effective routes to fabricate custom-made scaffolds with a given anatomical contour for hard tissue regeneration [36].

1. *Foam replica technique* is a widely used process in the fabrication of ceramic foams and was developed for the first time in 1963 [37]. It consists of the polymer-replication process in combination with gel casting methodology [28, 38]. The starting structure

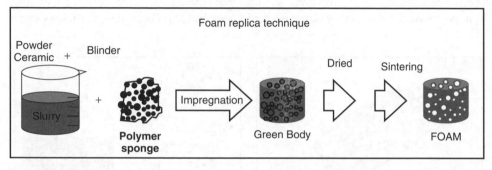

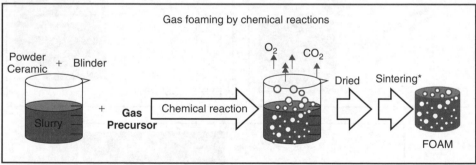

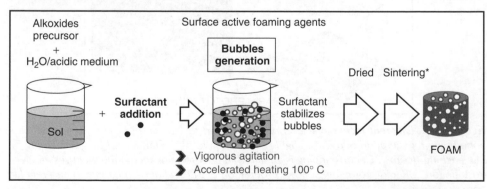

Figure 10.6 *Different methodologies to obtain foam-like scaffolds. (*) means that the sintering processing can be avoided by using biocompatible reactants*

(green body) is prepared by impregnation of polymer sponge (generally polyurethane PU sponge) with ceramic slurry. This polymer sponge exhibits the desired macrostructure and serves as a sacrificial template. After drying, the polymer sponge is slowly burned out at high temperature (> 450 °C) to minimize microstructure damage (microcracking) of the porous ceramics coating. Once the polymer sponge has been removed, the ceramic is sintered to the desired density. The foam replica technique offers a number of advantages over other scaffold fabrication techniques, such as the ability to produce foam with a highly porous structure with adjustable pore dimensions. Moreover, irregular shapes can be produced to match the size and shape of the bone defect. Nevertheless, as can easily be envisaged from this technique, positive replication renders very weak materials in terms of mechanical strength, often requiring the presence of a reinforcing phase [39]. The most important drawback of this technique is the need to eliminate the replica (polymer sponge). Often this step requires toxic solvents which if not thoroughly removed can then be harmful to the body.

2. *Gas foaming by chemical reactions.* Another strategy to obtain foam structures is through a foaming by gas generation via a chemical reaction of a ceramic slurry. The formation of gas bubbles is achieved by different chemical reactions, such as the addition of H_2O_2 followed by decomposition to O_2 [31], or through the acid–base reaction of NaH_2PO_4 with $NaHCO_3$ by formation of CO_2 bubbles [40, 41]. Such strategy allows control of the porosity content, pore dimensions, and pore shape by modulation of the processing conditions, such as the concentration of the reactants agents or temperature. In particular, gas foaming methods provide foams with higher interconnectivity rate and porosity than replica techniques. Therefore, these foaming methodologies have been widely used because they are straightforward, cost-effective, and easy to scale up [42]. Numerous scaffolds prepared from these techniques have resulted in successful clinical outcomes and are eventually expected to enter the commercial market [43, 44]. Moreover, foaming methodology offers the possibility of use in calcium phosphate cements [31, 45], which allow one to obtain *in situ* macroporosity [46, 47] (for more details see Chapter 8).

3. *Surface active foaming agents.* Another foam methodology is by using tensioactive molecules (surfactants) as gas bubble stabilizing agents. The incorporation of these surfactants can be used in both gel-cast foaming [48, 49] and sol–gel foaming processes [49]. In both techniques slurry/sol is foamed by different procedures in the presence of a surfactant which stabilizes the bubbles. Generally, to obtain gas bubbles vigorous stirring is used [50]. After that, a drying and an ageing period are necessary to increase the sol/slurry stability. Finally, a sintering process is carried out to remove the surfactant agent. Recently, this technology has been applied to obtain 3D foam systems of pure nanocrystalline hydroxyapatite (HA) with an open hierarchical interconnected macroporous network of between 1 and 400 μm and a crystal size similar to biological apatites, by a combination of the sol–gel route with the addition of a nonionic surfactant (Figure 10.7) [51, 52]. This combination allows the incorporation of a foaming step in the process by accelerated heating, resulting in the formation of an interconnected macroporous network, which can be easily tailored by controlling the amount of surfactant added [51]. Moreover this 3D interconnected architectural design of these HA foams allows excellent osteoblast internalization, proliferation, and differentiation, exhibiting adequate colonization over the entire scaffold surface with an appropriate degradation rate without any cytotoxic effects [53]. Furthermore, *in vivo*

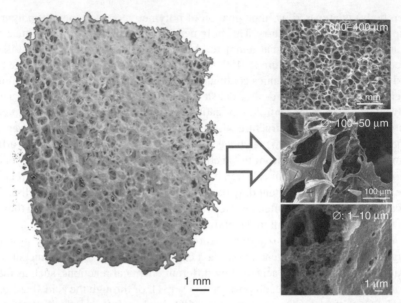

Figure 10.7   *Pure hydroxyapatite foam obtained by the gas foaming process in the presence of a surfactant. In this case the presence of surfactant allows one to obtain different porosity scales. Reproduced with permission from [51]. Copyright © 2010, The Royal Society of Chemistry*

studies have reveled high potential as bone regenerators and excellent osteointegration performance [54].

    On the other hand, gel-casting foaming methodology has also been prescribed to fabricate macroporous injectable calcium phosphate cements [45]. In this case the surfactant-foaming agent must be biocompatible and water soluble as the sintering process is omitted. Protein-based foaming agents, such as albumen or the protein mixture from egg white, have proved to be good foaming agents [55]. The most important advantage of using surface active foaming agents is that the injectability is preserved and, given the high stability of the foamed paste, macroporosity is maintained after injection.

4. *Strategies to enhance the mechanical properties.* Despite the macro/microstructural characteristics of foam structures being excellent for mimicing cancellous bone; their lack of initial strength is one of their important disadvantages, restricting their use to non-bearing situations [36]. Generally, there are different strategies to improve the mechanical stability, such as increasing the sintering temperature, coating with biodegradable polymer, producing a composite polymer/ceramic or synthesis of hybrids [6]. Polymer coating is a simple and cost-effective method which preserves the initial morphology and integrity, while improving handling, performance, and applicability [51, 54, 56–58]. The procedure is based on the immersion of foam in a biodegradable polymer solution, followed by drying at room temperature [51]. The polymer–ceramic composite foams are based on a mixture ceramic in the degradable polymer matrix in the starting slurry without sintering process [59]. The use of both polymer coatings and composite foam are obvious ways to induce toughness in the ceramic foam. However, the polymer phase masks the ceramic phase bioactivity/biocompatibility, providing islands with different cell response. Moreover, both components exhibit

different degradation rates. Ideally, the two phases should degrade congruently and at a controlled rate depending on the application. An alternative strategy is the synthesis of hybrid foams, where the degradable polymer is introduced into the sol stage of the sol–gel foaming process. Then a fine-scale interaction between the organic and inorganic materials results in this hybrid material as a single phase, leading to controlled congruent degradation and tailored mechanical properties [60]. Moreover, recent studies of template glasses (TGs) scaffolds have shown an increase in mechanical reinforcements through a biomimetic process. Thus, the outstanding bioactive behavior of TGs as well as their 3D mesopore structure allow the formation of apatite crystals not only in the external surface but also in the interior between the grains, which lead to a notable increase in the mechanical properties [61].

10.3.2 Rapid Prototyping Scaffolds

The RP techniques are referred to as solid free-form (SFF) manufacturing, which includes a group of techniques that are the most precise and reproducible avenues for controlling the internal pore size, porosity, pore interconnectivity, mechanical performance, and overall dimensions of tissue engineering scaffolds [1, 33]. RP techniques are defined as automated deposition of each tomographic layer sequence, based on programmed 3D images, into the desired architecture through an additive layer-by-layer method [62]. Thus, taking into account that one of the main requirements for translational applications is a high throughput and automated method, which can also produce patient-specific constructs, then such an RP-based method can potentially be used to fabricate such customized tissues [9].

Figure 10.8 shows the complete process in the design of a customized scaffold from RP techniques. The designed external and internal (pore) geometry of the scaffold can be

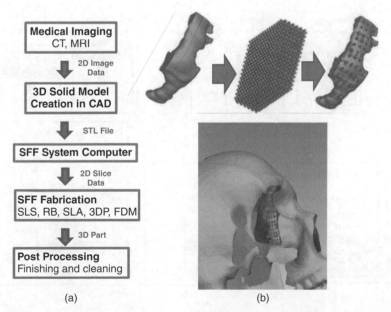

(a)   (b)

Figure 10.8 *(a) Typical RP process chain. (b) Computed tomography (CT)-scanning of the built structures allows assessment of the accuracy of the process, by comparing the scan data to the design. Reproduced with permission from [64]. Copyright © 2011, Elsevier*

built by using 3D drawing computer software or from clinical scanning data from magnetic resonance imaging (MRI) or tomography techniques [62]. Then a CAD-file can be obtained which describes the geometry and size of the parts to be built. For this, STL file format was developed; an STL file (acronym derived from stereolitography files) lists the coordinates of triangles that together make up the surface of the designed 3D structure. This designed structure is (virtually) sliced into layers. Finally, these data are uploaded to the RP-based instruments and the scaffold is fabricated (Figure 10.8) [63].

The most relevant RP methods in the design of 3D scaffolds for tissue engineering include: 3D printing (3DP), selective laser sintering (SLS), stereolithography (SLA), robocasting (RC), and fused deposition modeling (FDM). Figure 10.9 shows a scheme of the processing of each RP based-technique [64].

1. *3DP* is based primarily on the use of adhesive. The inkjet head prints droplets of a binder fluid onto a powder bed. This process is repeated for every layer until the 3D structure is printed and the remaining powder is removed.
2. *SLS* is based primarily on the use of heat. This technique employs a CO_2 laser beam to selectively sinter polymer or composite (polymer/ceramic, multiphase metal) powders to form material layers. The laser beam is directed onto the powder bed by a high precision laser scanning system.
3. *SLA* is based primarily on the use of light. The laser beam selectively initiates solidification in a thin layer of liquid photopolymer.
4. *RC* technique is based primarily on the use of a slurry. This technique consists of the robotic deposition of a highly concentrated colloidal suspension (inks) capable of fully

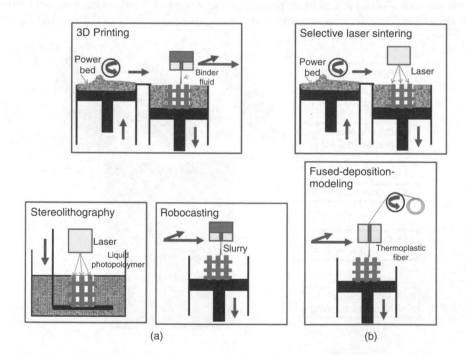

(a)   (b)

Figure 10.9   Main RP-based techniques relevant for tissue engineering applications

supporting their own weight during the assembly, thanks to their carefully tailored compositions and viscoelastic properties.

5. *FDM* uses filament material stock (generally thermoplastics) which is fed and melted inside a heated liquefier-head before being extruded layer by layer through a nozzle with a small orifice.

Table 10.2 summarizes the capabilities and limitations of each RP-based technique [64]. In general, SLA and FDM technologies have the disadvantage that they can be used for a very limited range of materials. SLA, despite being the most mature technology, has the disadvantage of a very low mechanical strength of the finished product. The high process temperatures given in the FDM methods can be inconvenient in some ceramics (glasses or nanostructured ceramics) or when biologically active molecules are added. Moreover, the occlusion of the pore edges limits the possibilities for using them in tissues technology. 3DP offers the advantage of high precision and rapid processing, and can be used for a large range of materials [63]. However, the use of toxic organic solvents needs a subsequent treatment to remove them before implantation, increasing costs and production times, as opposed to the desired final objectives. Meanwhile, SLS methods, despite having reduced accuracy compared to 3DP, remain within acceptable ranges, having the advantage of being solvent-free, thus requiring no further treatment except for sterilization prior to surgery. In

Table 10.2 *Main features of RP-assisted methods*

Method	Pore size/accuracy	Advantages	Disadvantages
3DP	45–500 μm/0.1	Broad material range No support structure needed Cost efficient	Small green strength Depowdering weak bonding between particles Powder can be trapped inside the body High temperature
SLS	45–100 μm/0.075	High mechanical properties No support structure needed	Powder can be trapped inside the body
SLA	400–1 μm/0.05	High accuracy	Material limitations Low mechanical properties Photopolymer needed Support structure needed
RC	250–1000 μm/0.1	High accuracy No support structure needed Combination of materials with two nozzles	Large build time Expensive process
FDM	250–1000 μm/0.05	No support structure needed No powder trapped	Material limitations High temperatures Thermoplast polymers required Mechanical anisotropy

addition, the final product has good mechanical strength properties. RC, 3DP, and SLS are the routes most used in the fabrication of ceramic scaffolds fabrication, as reported elsewhere [64].

RC technology also referred to as direct-write assembly, is unique among these processes because it allows one to build ceramic scaffolds using water-based inks with minimal organic content (< 1 wt%) and without the need for a sacrificial material or mold [65, 66]. The required temperature to remove this minimal organic content is low, around 300–400 °C. The colloidal inks (slurries) developed for RC must satisfy two important criteria [67]. First, their viscoelastic properties must allow them to flow through a deposition nozzle and then "set" immediately so that the shape is retained as additional layers are deposited or when they span gaps in the underlying structure. Second, the suspension must have a high solid volume concentration to minimize shrinkage during drying so that the particle network is able to resist the involved capillary stresses [68]. The stability of these high-solids-loading suspensions requires high dispersive forces between particles and, therefore, the role of dispersant is critical. The amount of dispersant has to be adjusted so as to efficiently coat the particles in the suspension, but an excess could cause flocculation of the particles due to the depletion effect [68, 69]. In order to avoid flocculation, polyethylenimine (PEI) is added, acting as a flocculating agent [68]. The resulting scaffolds exhibit a high percentage of easily controllable porosity and interconnectivity, easily with enhanced mechanical properties compared with conventional scaffold processing [70]. Moreover, by sintering processing it is possible to increase markedly these mechanical properties [71–73]. In addition, research has revealed improved compressive strength of bioceramic RC-assisted scaffolds by polymer infiltration [74].

On the other hand, the versatility of the RC slurries has allowed the incorporation of biodegradable polymers into the inks, abolishing the high temperature processing and thus enabling the incorporation of biologically active molecules during the scaffold processing [75–78]. Recently, polycaprolactone and nanocrystalline silicon-doped hydroxyapatite 3D scaffolds have been fabricated by the RP technique [79]. To enhance *in vivo* scaffolds efficiency demineralized bone matrix (DBM) was incorporated during the RP-assisted process. The *in vivo* studies using New Zealand rabbits have shown extraordinary regenerative properties, with new bone formation not only in the peripheral portions of the scaffolds but also within their macropores four months after implantation.

Furthermore, the real challenge in RC methodology has been the integration of nanoscale features into the micro-macrostructure matrices which, as has been commented above, is necessary to finely tune cellular responses [7, 9]. Therefore, owing to the nature of the sol–gel process, the sol can also be directly, printed prior to gelation if exhibiting adequate viscoelastic properties. Thus, 3D scaffolds can be produced with three scales of porosity: large pores (400–1000 μm) from the RP-assisted fabrication, ordered mesopores (5–10 nm) from the use of a copolymer template (F127) as a mesostructure directing agent, and additional macropores (1–80 μm) from the use of a sacrificial methyl cellulose template [80, 81]. The sol containing the templates was extruded onto a heated substrate using a RC-assisted device. Critical to success was the viscosity of the sol, which was controlled by the methyl cellulose content [82]. As an example, Figure 10.10 displays a 3D mesoporous glass in the SiO_2–P_2O_5 system with hierarchical pore structure composed of three different length scales of porosity: (i) highly ordered mesopores with diameters of about 4 nm; (ii) macropores with diameters in the 30–80 μm range with interconnections of about 2–4

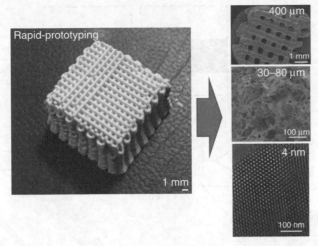

Figure 10.10 *Silica-based mesoporous 3D scaffolds obtaining by combination of two polymers such as Pluronic F127 and methyl cellulose using an RC-assisted technique*

and 8–9 μm; and (iii) ultra-large macropores of about 400 μm. This hierarchical porosity in the macro-micro and nanoscale of the resulting scaffolds makes them suitable for bone tissue engineering applications [81].

On the other hand, Park and coworkers have reported production of a dual-scale scaffold by combining the processes of RC and electrospinning technologies, which will be discussed in detail in the next section [83, 84]. In this process, the microfibrous layer of the scaffolds was first built via the RC process and then polymeric nanofibers were directly deposited onto the microfibrous layer by electrospinning. The subsequent microfibrous layers integrated with electrospun nanofibers were repeatedly laminated onto the previously integrated layers so that a 3D hybrid structure could be fabricated. The dually designed scaffold exhibited improved biological function in terms of human stem cell adhesion and proliferation [84].

10.3.3 Electrospinning Scaffolds

Electrospinning is a simple technique that uses electrical fields to produce nano to microfibers from polymer or ceramic/polymer solution. Solid fibers are obtained from electrified jets (using high voltage) that are continuously elongated due to the electrostatic repulsion between the surface charges and the evaporation solvent (Figure 10.11) [22].

One of the great advantages of this technique is the possibility of obtaining high surface area scaffolds which mimic the size scale of fibrous proteins found in the natural ECM [5, 13, 85]. The capacity to easily produce materials at the biological length scale has created a great interest in this methodology for tissue engineering and drug delivery applications [86].

The electrospinning technique is able to form non-woven fibrous mats which ensure fiber production from a broad range of precursor materials that include synthetic polymers, natural polymers, semiconductors, ceramics, or their combinations [69, 87]. However, despite numerous benefits, it has been extremely difficult to design macroscopically porous 3D

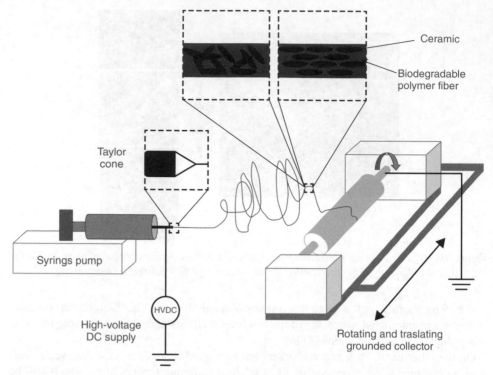

Figure 10.11 *Schematic of a typical electrospinning system. A melt ceramic polymer solution is forced through a needle using a syringe pump. The needle is connected to a high voltage DC supply, which injects charge of a certain polarity into a ceramic/polymer melt. If the electrostatic force created by the repulsion of similar charges is sufficient to overcome the surface tension of the polymeric melt solution the Taylor cone is formed and a fiber jet is emitted its apex. While the fiber jet is traveling toward the grounded collector it undergoes a chaotic whipping instability. The fiber jet is deposited on the collector which can be rotating and translating, as depicted here*

architectures by using nanofibers, which are characterized by entangled fibers and a densely packed membranous structure [87]. This limitation makes it suitable to use in combination with an RC-assisted process as mentioned above [83, 84].

As has been commented above, the great advantage of electrospinning technology is the accurate control of fiber size, porosity, and fiber shape by different processing variables, such as applied voltage, polymer melt flow rate, capillary–collector distance, polymer/ceramic concentration, solvent volatility, and solvent conductivity [22]. The capacity to adjust fiber size is one of the strengths of electrospinning, since fibers with diameters in the nanometer size range closely mimic the size scale of fibrous proteins found in the natural ECM, such as collagen. This ability of electrospun nanofibers to mimic the ECM is vital as previous studies have shown that both the size scale of the structure and the topography play important roles in cell proliferation and adhesion, respectively [13]. Also, non-woven fibrous mats comprised of nanofibers have a very high fraction of surface available to interact with cells, which makes them ideal for cell attachment. Additionally, the

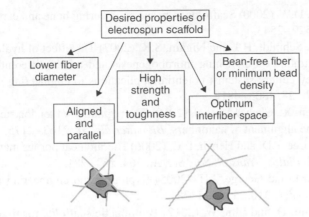

Figure 10.12   *Desired properties and required process parameters in electrospun scaffolds. Optimum interfiber space (supporting cell growth) and large interfiber space (hindering cell growth)*

porosity of electrospun mats aids in nutrient transport [22]. Figure 10.12 displays some of the desired properties of an electrospun scaffold for tissue engineering applications, as well as the importance of controling the interfiber space in order to achieve a better cell internalization and colonization.

The use of electrospun fibers and fiber meshes in tissue engineering applications often involves several considerations, including choice of material, fiber orientation, porosity, surface modification, and tissue application. By varying the processing and solution parameters, the fiber orientation (aligned vs. random) and porosity/pore size (cell infiltration) of the electrospun scaffold can be controlled and optimized for each individual application. After fabrication the surface of the scaffold can be modified with a high density of bioactive molecules, due to the relatively high scaffold surface area which will be discussed in the next chapter. Due to the flexibility in material selection as well as the ability to control the scaffold properties, electrospun scaffolds have been employed in a number of different tissue applications, including vascular, bone, neural, and tendon/ligament [88].

Despite the electrospinning technique having been widely developed with biodegradable polymers, recently, researchers have shown the possibility to obtain a bioceramic/biopolymer nanofibrous composite for biomedical applications including tissue regeneration [89–93]. In recent years, inorganic nanoparticles, such as HA, bioactive glass, and carbon nanotubes, have been widely co-electrospun into polymer nanofibers to enhance both the mechanical properties and the biocompatible response [93–98]. In this regard, the bioceramic needle oriented in polymer fiber is challenging but very important for enhancing the mechanical properties of these scaffolds.

References

1. Hollister, S.J. (2005) Porous scaffold design for tissue engineering. *Nat. Mater.*, **4**, 518–524.
2. Langer, R. and Vacanti, J.P. (1993) Tissue engineering. *Science*, **260**, 920–926.

3. Hutmacher, D.W. (2000) Scaffolds in tissue engineering bone and cartilage. *Biomaterials*, **21**, 2529–2543.

4. Rosines, E., Schmidt, H.J. and Nigam, S.K. (2007) The effect of hyaluronic acid size and concentration on branching morphogenesis and tubule differentiation in developing kidney culture systems: potential applications to engineering of renal tissues. *Biomaterials*, **28**, 4806–4817.

5. Yin, Z., Chen, X., Chen, J.L. *et al.* (2010) The regulation of tendon stem cell differentiation by the alignment of nanofibers. *Biomaterials*, **31**, 2163–2175.

6. Jones, J.R., Lee, P.D. and Hench, L.L. (2006) Hierarchical porous materials for tissue engineering. *Philos. Trans. R. Soc. Ser. A.*, **364**, 263–281.

7. Stevens, M.M. and George, J.H. (2005) Exploring and engineering the cell surface interface. *Science*, **310**, 1135–11138.

8. Kim, H., Shin, D. and Lim, W. (2012) Biomimetic scaffolds for tissue engineering. *Adv. Funct. Mater.*, **22**, 2446–2468.

9. Hollister, S.J. (2009) Scaffold Design and manufacturing from concept to clinic. *Adv. Mater.*, **21**, 3330–3342.

10. Karageorgiou, V. and Kaplan, D. (2005) Porosity of 3D biomaterial scaffolds and osteogenes. *Biomaterials*, **26**, 5474–5491.

11. Vallet-Regi, M. and Gonzalez-Calbet, J.M. (2004) Calcium phosphates as substitution of bone tissues. *Prog. Solid State Sci.*, **32**, 1–31.

12. Jones, J.R. (2009) New trends in bioactive scaffolds: the importance of nanostructure. *J. Eur. Ceram. Soc.*, **29**, 1275–1281.

13. Dalby, M., Gadegaard, N., Tare, R.A. *et al.* (2010) The control of human mesenchymal cell differentiation using nanoscale symmetry and disorder. *Nat. Mater.*, **6**, 997–1003.

14. Cima, L.G., Vacanti, J.P., Vacanti, C. *et al.* (1991) Tissue Engineering by cell transplantation using degradable polymer substrates. *J. Biomech. Eng.*, **113**, 143–151.

15. Jones, J.R. (2013) Review of bioactive glass: from Hench to hybrids. *Acta Biomater.*, **9**, 4457–4486.

16. Isaksson, H., Comas, O., van Donkelaar, C.C. *et al.* (2007) Bone regeneration during distraction osteogenesis: mechano-regulation by shear strain and fluid velocity. *J. Biomech.*, **40**, 2002–2011.

17. Vivanco, J., Aiyangar, A., Araneda, A. and Ploeg, H.-L. (2012) Mechanical characterization of injection-molded macro porous bioceramic bone scaffolds. *J. Mech. Behav. Biomed. Mater.*, **9**, 137–152.

18. Boyan, B.D., Hummert, T.W., Dean, D.D. and Schwratz, Z. (1996) Role of material surfaces in regulating bone and cartilage cell response. *Biomaterials*, **17**, 137–146.

19. Yoshikawa, T., Ohgushi, T. and Tamai, S. (1996) Immediate bone forming capability of prefabricated osteogenic hydroxyapatite. *J. Biomed. Mater. Res.*, **32**, 481–492.

20. Vallet-Regi, M., Colilla, M. and Izquierdo-Barba, I. (2008) Bioactive mesoporous silicas as controlled delivery systems: application in bone tissue regeneration. *J. Biomed. Nanotechnol.*, **4**, 1–15.

21. Place, E.S., Evans, N.D. and Stevens, M.M. (2009) Complexity in biomaterials for tissue engineering. *Nat. Mater.*, **8**, 457–470.

22. Sill, T.J. and Von Recum, H.A. (2008) Electrospinning: applications in drug delivery and tissue engineering. *Biomaterials*, **29**, 1989–2006.

23. Pattison, M.A., Wurster, S., Webster, T.J. and Haberstroh, K.M. (2005) Three-dimensional, nano-structured PLGA scaffolds for bladder tissue replacement applications. *Biomaterials*, **26**, 2491–2500.

24. Al Ruhaimi, K.A. (2001) Bone graft substitutes: a comparative qualitative histologic review of current osteoconductive grafting materials. *Int. J. Oral Maxillof. Implants*, **16**, 105–114.

25. Sanchez-Salcedo, S., Nieto, A. and Vallet-Regí, M. (2008) Hydroxyapatite/beta tricalcium phosphate/agarose macroporous scaffolds for bone tissue engineering. *Chem. Eng. J.*, **137**, 62–71.

26. Roy, T.D., Simon, J.L., Ricci, J.L. *et al.* (2003) Performance of degradable composite bone repair products made via three-dimensional fabrication techniques. *J. Biomed. Mater. Res. Part A*, **66**, 283–91.

27. Deville, S., Saiz, E., Nalla, R.K. and Tomsia, A.P. (2006) Freezing as a path to build complex composites. *Science*, **311**, 515–518.

28. Ramay, H.R. and Zhang, M.Q. (2003) Preparation of porous hydroxyapatite scaffolds by combination of the gel-casting and polymer sponge methods. *Biomaterials*, **24**, 2293–2303.

29. Padilla, S., Vallet-Regí, M., Ginebra, M.P. and Gil, F.J. (2005) Processing and mechanical properties of hydroxyapatite pieces obtained by gel casting method. *J. Eur. Ceram. Soc.*, **25**, 375–383.

30. Padilla, S., Sanchez-Salcedo, S. and Vallet-Regí, M. (2007) Bioactive glass as precursor of designed –architecture scaffolds for tissue engineering. *J. Biomed. Mater. Res.*, **81A**, 224–232.

31. Almirall, A., Larrecq, G., Delgado, J.A. *et al.* (2004) Fabrication of low temperature macroporous hydroxyapatite scaffolds by foaming and hydrolysis of an alpha-TCP paste. *Biomaterials*, **25**, 3671–3680.

32. Kim, S.S., Park, M.S., Jeon, O. *et al.* (2006) Poly(lactide-co-glycolide)/hydroxyapatite composite scaffolds for bone tissue engineering. *Biomaterials*, **27**, 1399–1409.

33. Yeong, W.Y., Chua, C.K., Leong, K.F. and Chandrasekaran, M. (2004) Rapid prototyping in tissue engineering: challenges and potential. *Trends Biotechnol.*, **22**, 643–652.

34. Pham, Q.P., Sharma, U. and Mikos, A.G. (2006) Electrospinning of polymeric nanofibers for tissue engineering applications: A review. *Tissue Eng.*, **12**, 1197–1211.

35. Martin, R.A., Yue, S., Hanna, J.V. *et al.* (2012) Characterizing the hierarchical structures of bioactive sol-gel silicate glass and scaffolds for bone regeneration. *Philos. Trans. R. Soc., Ser. A*, **370**, 1422–1443.

36. Yang, S.F., Leong, K.F., Du, Z.H. and Chua, C.K. (2001) The design of scaffolds for use in tissue engineering. *Tissue Eng. Part I*, **7**, 679–689.

37. Schwartzalder, K. and Somers, A.V. (1963) Method of making a porous shape of sintered refractory ceramic articles. US Patent 3090094.

38. Padilla, S., Roman, J. and Vallet-Regi, M. (2002) Synthesis of porous hydroxyapatites by combination of gelcasting and foams burn out methods. *J. Mater. Sci. -Mater. Med.*, **13**, 1193–1197.

39. Miao, X., Lim, W.K., Huang, X. and Chen, Y. (2005) Preparation and characterization of interpenetrating phase TCP/HA/PLGA composites. *Mater. Lett.*, **59**, 4000–4005.

40. del Real, R.P., Wolke, J.G.C., Vallet-Regi, M. and Jansen, J.A. (2002) A new method to produce macropores in calcium phosphate cements. *Biomaterials*, **23**, 3673–80.

41. del Real, R.P., Ooms, E., Wolke, J.G.C. *et al.* (2003) In vivo bone response to porous calcium phosphate cement. *J. Biomed. Mater. Res., Part A*, **65**, 30–6.

42. Yoon, J.J., Kim, J.H. and Park, T.G. (2003) Dexamethasone-releasing biodegradable polymer scaffolds fabricated by a gas-foaming/salt-leaching method. *Biomaterials*, **24**, 2323–2329.

43. Chen, J., Xu, J., Wang, A. and Zheng, M. (2009) Scaffolds for tendon and ligament repair: review of the efficacy of commercial products. *Expert Rev. Med. Devices*, **6**, 61–73.

44. Dorozhkin, S.V. (2010) Bioceramics of calcium orthophosphates. *Biomaterials*, **31**, 1465–1485.

45. Ginebra, M.P., Espanol, M., Montufar, E.B. *et al.* (2010) New processing approaches in calcium phosphate cements and their applications in regenerative medicine. *Acta Biomater.*, **6**, 2863–73.

46. Hesaraki, S. and Sharifi, D. (2007) Investigation of an effervescent additive as porogenic agent for bone cement macroporosity. *Biomed. Mater. Eng.*, **17**, 29–38.

47. Hesaraki, S., Zamanian, A. and Moztarzadeh, F. (2008) The influence of the acidic component of the gas-foaming porogen used in preparing an injectable porous calcium phosphate cement on its properties: acetic acid versus citric acid. *J. Biomed. Mater. Res., Part B*, **86**, 208–16.

48. Sepulveda, P. and Binner, J.G.P. (2001) Evaluation of the in situ polymerization kinetics for the gel casting of ceramic foams. *Chem. Mater.*, **13**, 3882–3887.

49. Wu, Z.Y., Hill, R.G., Yue, S. *et al.* (2011) Melt-derived glass scaffolds by a gel cast foaming technique. *Acta Biomater.*, **7**, 1807–17816.

50. Jones, J.R., Ehrenfried, L.M. and Hench, L.L. (2006) Optimizing bioactive glass scaffolds for bone tissue engineering. *Biomaterials*, **27**, 964–73.

51. Sanchez-Salcedo, S., Vila, M., Izquierdo-Barba, I. *et al.* (2010) Biopolymer-coated hydroxyapatite foams: a new antidote for heavy metal intoxication. *J. Mater. Chem.*, **20**, 6956–6961.

52. Vila, M., Sanchez-Salcedo, S., Cicuéndez, M. *et al.* (2011) Novel biopolymers-coated hydroxyapatite foams for removing heavy metals from polluted water. *J. Hazard. Mater.*, **192**, 71–77.

53. Cicuéndez, M., Izquierdo-Barba, I., Sánchez-Salcedo, S. *et al.* (2012) Biological performance of hydroxyapatite biopolymers foam: in vitro cell response. *Acta Biomater.*, **8**, 802–810.

54. Gil-Albarova, J., Vila, M., Badiola-Vargas, J. *et al.* (2012) In vivo osteointegration of three-dimensional crosslinked gelatin-coated hydroxyapatite foams. *Acta Biomater.*, **8**, 3777–3783.

55. Montufar, E.B., Gil, C., Traykova, T. *et al.* (2010) Foamed surfactant solution as a template for self-setting injectable hydroxyapatite scaffolds for bone regeneration. *Acta Biomater.*, **6**, 876–85.

56. Ruhé, P.Q., Hadberg, E.L., Padron, N.T. *et al.* (2006) Porous poly(DL-lactic-co-glycolic acid)/calcium phosphate cement composite for reconstruction of bone defects. *Tissue Eng.*, **12**, 789–800.

57. Chen, Q.Z. and Boccaccini, A.R. (2006) Poly(D,L-Lactic) coated 45S5 Bioglass ® -based scaffolds: processing and characterization. *J. Biomed. Mater. Res., Part A*, **77**, 445–457.

58. Bretcanu, O., Chen, Q.Z., Misra, S.K. *et al.* (2007) Biodegradable, polymer coated 45S5 bioglass-derived glass-ceramic scaffolds for bone tissue engineering. *Eur. J. Glass Sci. Technol., Part A*, **48**, 227–234.

59. Rezwan, K., Chen, Q.Z., Blaker, J.J. and Boccaccini, A.R. (2006) Biodegradable and bioactive porous polymer/inorganic composite scaffolds for bone tissue engineering. *Biomaterials*, **27**, 3413–3431.

60. Mohony, O., Tsigkou, O., Ionescu, C. *et al.* (2010) Silica-gelatin hybrids with tailorable degradation and mechanical properties for tissue regeneration. *Adv. Funct. Mater.*, **20**, 3835–3845.

61. Arcos, D., Vila, M., Lopez-Noriega, A. *et al.* (2011) Mesoporous bioactive glasses. Mechanical reinforcements by means of a biomimetic process. *Acta Biomater.*, **7**, 2952–2959.

62. Leong, K.F., Cheah, C.M. and Chua, C.K. (2003) Solid freeform fabrication of three-dimensional scaffolds for engineering replacement tissue and organs. *Biomaterials*, **24**, 2363–2378.

63. Butscher, A., Bohner, M., Hofmann, S. *et al.* (2011) Structural and material approaches to bone tissue engineering in powder-base three dimensional printing. *Acta Biomater.*, **7**, 907–920.

64. Butscher, A., Bohner, M., Roth, C. *et al.* (2012) Printability of calcium phosphate powder for three-dimensional printing of tissue engineering scaffolds. *Acta Biomater.*, **8**, 373–385.

65. Cesarano, J. and Carlvet, P. (2000) Free forming objects with low binder slurry. US Patent 6,027,326.

66. Cesarano, J., Segalman, R. and Calvert, P. (1998) Robocasting provides mold less fabrication from slurry deposition. *Ceram. Ind.*, **148**, 94–102.

67. Smay, J.E., Cesarano, J. and Lewis, J.A. (2002) Colloidal inks for directed assembly of 3-D periodic structures. *Langmuir*, **18**, 5429–5437.

68. Michna, S., Wu, W. and Lewis, J.A. (2005) Concentrated hydroxyapatite inks for direct-write assembly of 3-D periodic scaffolds. *Biomaterials*, **26**, 5632–5639.

69. Barnes, C.P., Sell, S.A., Boland, E.D. *et al.* (2007) Nanofiber technology: designing the next generation of tissue engineering scaffolds. *Adv. Drug Delivery Rev.*, **59**, 1413–1433.

70. Perera, F.H., Martinez-Vazquez, F.J., Miranda, P. *et al.* (2010) Clarifying the effect of sintering conditions on the microstructure and mechanical properties of beta-tricalcium phosphate. *Ceram. Int.*, **36**, 1929–1935.

71. Miranda, P., Pajares, A., Saiz, E. *et al.* (2008) Mechanical properties of calcium phosphate scaffolds fabricated by robocasting. *J. Biomed. Mater. Res., Part A*, **85**, 218–27.

72. Miranda, P., Saiz, E., Gryn, K. and Tomsia, A.P. (2006) Sintering and robocasting of betatricalcium phosphate scaffolds for orthopedic applications. *Acta Biomater.*, **2**, 457–466.

73. Miranda, P., Pajares, A., Saiz, E. *et al.* (2007) Fracture modes under uniaxial compression in hydroxyapatite scaffolds fabricated by robocasting. *J. Biomed. Mater. Res., Part A*, **83**, 646–655.

74. Martinez-Vazquez, F.J., Perera, F.H., Miranda, P. *et al.* (2010) Improving the compressive strength of bioceramic robocast scaffolds by polymer infiltration. *Acta Biomater.*, **6**, 4361–4368.

75. Franco, J., Hunger, P., Launey, M.E. *et al.* (2010) Direct write assembly of calcium phosphate scaffolds using a water-based hydrogel. *Acta Biomater.*, **6**, 218–28.
76. Jack, K.S., Velayudhan, S., Luckman, P. *et al.* (2009) The fabrication and characterization of biodegradable HA/PHBV nanoparticle–polymer composite scaffolds. *Acta Biomater.*, **5**, 2657–2667.
77. Zhao, L., Burguera, E.F., Xu, H.H.K. *et al.* (2010) Fatigue and human umbilical cord stem cell seeding characteristics of calcium phosphate–chitosan–biodegradable fiber scaffolds. *Biomaterials*, **31**, 840–847.
78. Billiet, T., Vandenhaute, M., Schelfhout, J. *et al.* (2012) A review of trends and limitations in hydrogel-rapid prototyping for tissue engineering. *Biomaterials*, **33**, 6020–6041.
79. Meseguer-Olmo, L., Vicente-Ortega, V., Alcaraz-Baños, M. *et al.* (2013) In vivo behavior of Si-Hydroxyapatite/ polycaprolactone/DMB scaffolds fabricated by 3D printing. *J. Biomed. Mater. Res., Part A*, **101**, 2038–2048. doi: 10.1002/jbn.a.34511.
80. Yun, H.S., Kim, S.E. and Hyeon, Y.T. (2007) Design and preparation of bioactive glasses with hierarchical pore networks. *Chem. Commun.*, 2139–2141.
81. García, A., Izquierdo-Barba, I., Colilla, M. *et al.* (2011) Preparation of 3-D scaffolds in the SiO_2–P_2O_5 system with tailored hierarchical meso-macroporosity. *Acta Biomater.*, **7**, 1265–1273.
82. Wu, C., Luo, Y., Cuniberti, G. *et al.* (2011) Three-dimensional printing of hierarchical and tough mesoporous bioactive glass scaffolds with a controllable pore architecture, excellent mechanical strength and mineralization ability. *Acta Biomater.*, **7**, 2644–2650.
83. Park, S.H., Kim, T.G., Kim, H.C. *et al.* (2008) Development of dual scale scaffolds via direct polymer melt deposition and electrospinning for applications in tissue regeneration. *Acta Biomater.*, **4**, 1198–1207.
84. Kim, T.G., Park, S.H., Chung, H.J. *et al.* (2010) Microstructure scaffolds coated with hydroxyapatite/collagen nanocomposite multilayer for enhance osteogenic induction of human mesenchymal stem cells. *J. Mater. Chem.*, **20**, 8927–8933.
85. Li, D. and Xia, Y.N. (2004) Electrospinning of nanofibers: reinventing the wheel? *Adv. Mater.*, **16**, 1151–1170.
86. Agarwal, S., Wendorff, J.H. and Greiner, A. (2009) Progress in the field of electrospinning for tissue engineering applications. *Adv. Mater.*, **21**, 3343–3351.
87. Teo, W.E. and Ramakrishna, S. (2006) A review on electrospinning design and nanofibre assemblies. *Nanotechnology*, **17**, R89–R106.
88. Pramnik, S., Pingguan-Murphy, B. and Osman, N. (2012) Progress of key strategies in development of electrospun scaffolds: bone tissue. *Sci. Technol. Adv. Mater.*, **13**, 043002(13 pp).
89. Kim, H.W., Song, J.H. and Kim, H.E. (2005) Nanofiber generation of gelatin-hydroxyapatite biomimetics for guided tissue regeneration. *Adv. Funct. Mater.*, **15**, 1988–1994.
90. Li, C.M., Vepari, C., Jin, H.J. *et al.* (2006) Electrospun silk-BMP-2 scaffolds for bone tissue engineering. *Biomaterials*, **27**, 3115–3124.
91. Tyagi, P., Catledge, S.A., Stanishevsky, A. *et al.* (2009) Nanomechanical properties of electrospun composite scaffolds based on polycaprolactone and hydroxyapatite. *J. Nanosci. Nanotechnol.*, **9**, 4839–4845.

92. Stanishevsky, A., Chowdhury, S., Chinoda, P. and Thomas, V. (2008) Hydroxyapatite nanoparticle loaded collagen fiber composites: Microarchitecture and nanoindentation study. *J. Biomed. Mater. Res., Part A*, **86**, 873–882.

93. Fu, S.Z., Wang, X.H., Guo, G. *et al.* (2010) Preparation and characterization of nano-hydroxyapatite/poly(ϵ-caprolactone) – poly(ethylene glycol) – poly(ϵ-caprolactone) composite fibers for tissue engineering. *J. Phys. Chem. C*, **2010** (114), 18372–18378.

94. Peng, F., Shaw, M.T., Olson, J.R. and Wei, M. (2011) Hydroxyapatite needle-shaped particles/poly(L-lactic Acid) electrospun scaffolds with perfect particle-along-nanofiber orientation and enhanced mechanical properties. *J. Phys. Chem. C*, **115**, 15743–15751.

95. Yu, C.-C., Chang, J.-J., Lee, Y.-H. *et al.* (2013) Electrospun scaffolds composing of alginate, chitosan, collagen and hydroxyapatite for applying in bone tissue engineering. *Mater. Lett.*, **93**, 133–136.

96. Lao, L., Wang, Y., Zhu, Y. *et al.* (2011) Poly(lactide-co-glycolide)/hydroxyapatite nanofibrous scaffolds fabricated by electrospinning for bone tissue engineering. *J. Mater. Sci. Mater. Med.*, **22**, 1873–1884.

97. Nie, H. and Wang, C.-H. (2007) Fabrication and characterization of PLGA/Hap composite scaffolds for delivery of BMP-2 plasmid DNA. *J. Controlled Release*, **120**, 111–121.

98. Franco, P.Q., Joao, C.F.C. and Borges, J.P. (2012) Electrospun hydroxyapatite fibers from a simple sol-gel system. *Mater. Lett.*, **67**, 233–236.

Part IV
Research on Future Ceramics

11

Bone Biology and Regeneration

Soledad Pérez-Amodio[1,2] *and Elisabeth Engel*[1,2,3]

[1]*Biomaterials for Regenerative Therapies, Institute for Bioengineering of Catalonia,
Spain*
[2]*Biomedical Research Networking center in Bioengineering, Biomaterials and
Nanomedicine (CIBER-BBN), Spain*
[3]*Department of Materials Science, Technical University
of Catalonia, Spain*

11.1   Introduction

Bone is considered a specialized connective tissue that functions as a mechanical support of the body. It is the only hard tissue (together with teeth) because its extracellular matrix is mineralized giving it its rigidity and strength. Bone provides internal support, protects vital internal organs, and allows skeletal motion due to attachment sites for muscles. Besides these mechanical functions, bone has relevant biological functions, being the body's principal mineral storage (regulating homeostasis) as well as blood (hematopoiesis), and mesenchymal stem cells (MSCs) reservoir.

Bone is made mostly of collagen, a protein that provides a soft framework, and calcium phosphate, a mineral that adds strength and hardens the framework.

This combination of collagen and calcium makes bone both flexible and strong, which in turn helps bone to withstand stress. More than 99% of the body's calcium is contained in the bones and teeth. The remaining 1% is found in the blood.

Throughout one's lifetime, old bone is removed (resorption) and new bone is added to the skeleton (formation). During childhood and teenage years, new bone is added faster than old bone is removed. As a result, bones become larger, heavier, and denser. Bone formation outpaces resorption until peak bone mass (maximum bone density and strength) is reached around age 30. After that time, bone resorption slowly begins to exceed bone formation.

For women, bone loss is fastest in the first few years after the menopause, and it continues into the postmenopausal years. Osteoporosis – which mainly affects women but may also

Bioceramics with Clinical Applications, First Edition. Edited by María Vallet-Regí.
© 2014 John Wiley & Sons, Ltd. Published 2014 by John Wiley & Sons, Ltd.

affect men – will develop when bone resorption occurs too quickly or when replacement occurs too slowly.

11.2   The Skeleton

Each bone of the skeleton constantly undergoes modeling during life to help it to adapt to changing biomechanical forces as well as remodeling to remove old bone and replace it with new, mechanically stronger bone to help preserve bone strength.

Bone grows in two ways in humans and other mammals. In endochondral bone growth, bone arises from columns of cartilage cells in growth plates which undergo a well-defined sequence of activities and changes in morphology. This results in a zone of provisional calcification which, by means of vascular invasion, then becomes bone (Figure 11.1). In membranous bone growth, bone does not arise from cartilage but rather from membrane, fibrous tissue, or mesenchyme.

Bones are classified into long bones, short bones, flat bones, and irregular bones. Long bones include the clavicles, humeri, radii, ulnae, metacarpals, femurs, tibiae, fibulae, metatarsals, and phalanges. Short bones include the carpal and tarsal bones, patellae, and sesamoid bones. Flat bones include the skull, mandible, scapulae, sternum, and ribs. Irregular bones include the vertebrae, sacrum, coccyx, and hyoid bone (Figure 11.2). Flat bones form by membranous bone formation, whereas long bones are formed by a combination of endochondral and membranous bone formation.

Bone is organized into cortical bone, also known as compact bone and trabecular bone, also known as cancellous or spongy bone. These two types are classified on the basis of porosity and the unit microstructure. Cortical bone is hard, with an average range of densities between 1.6 and 2.4 g cm^{-3} [1], and surrounds the marrow space. Trabecular bone is comprised of an arrangement of bony spicules called trabeculae. It is located at the interior of bone, including the ends of long bones, the interior of cuboidal bones and flat bones, and between the inner and outer layers of cortical bone in the skull (Figure 11.3).

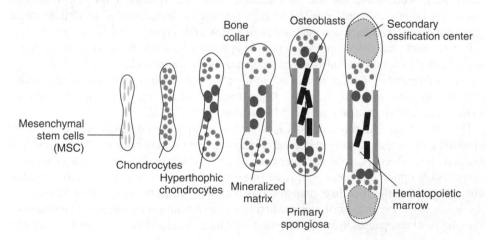

Figure 11.1   *Osteochondral bone growth*

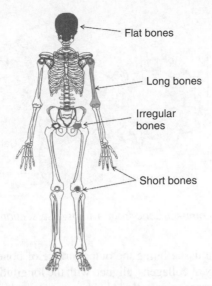

Flat bones

Long bones

Irregular bones

Short bones

Figure 11.2 *Different bone classification depending on their size and shape*

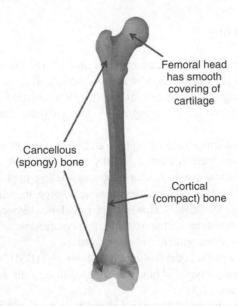

Femoral head has smooth covering of cartilage

Cancellous (spongy) bone

Cortical (compact) bone

Figure 11.3 *Bone classification depending on the structure*

Both cortical and trabecular bones are composed of osteons. Cortical osteons are called Haversian systems and represent the basic building block of cortical bone. An osteon is a cylindrical structure with a diameter of approximately 200 μm, aligned along the shaft of the long bones [2]. It consists of a central Haversian channel, circumferentially surrounded by an assembly of several layers of bone lamellae (Figure 11.4).

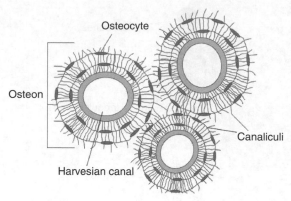

Osteocyte

Osteon

Canaliculi

Harvesian canal

Figure 11.4   *Diagram of compact bone from a transverse section of a typical long bone's cortex*

The periosteum is a thin tissue lining the outer surface of bone. The outer layer of the periosteum consists mostly of collagens, aligned with the longitudinal axis, and elastin [3], and is thought to have a mechanical role [4]. The periosteum's innermost layer consists mostly of progenitor cells that constantly build and repair bone [5].

11.3   Bone Remodeling

Bone remodeling is based on the concerted action of resorptive and formative cell populations in order to replace old bone with new bone and thus secure the integrity of the skeleton. Remodeling of bone is essential for (i) its adaptation to mechanical loading [6, 7], (ii) the maintenance of its strength [8], and (iii) the maintenance of calcium homeostasis.

In normal bone remodeling, bone resorption and bone formation are tightly regulated and maintained to ensure that, in mature healthy bone, there are no major net changes in bone mass or mechanical strength after each remodeling cycle. Adequate balance is controlled by the coupling of bone formation to bone resorption, which involves a number of coordinated signaling mechanisms. However, an imbalance between bone resorption and bone formation may occur under certain pathological conditions, which leads to abnormal bone remodeling and the development of bone disorders.

Bone remodeling takes place in the basic multicellular unit (BMU) and requires the coordinated activity of four major types of bone cells: bone-lining cells, osteocytes, osteoclasts, and osteoblasts [9] (Figure 11.5).

In a resting state, the surface of bone is covered by a monolayer of bone-lining cells, which belong to the osteoblast lineage [10]. Osteocytes also differentiate from osteoblasts and are embedded within the bone matrix [11]. Osteoclasts, the principle bone-resorbing cells, are multinucleated giant cells that differentiate from mononuclear cells upon stimulation by two important factors: the monocyte/macrophage-colony stimulating factor (M-CSF) and the receptor activator of nuclear factor κB (NF-κB) ligand (RANKL) [12]. Osteoblasts, the bone forming cells, are derived from MSCs through a multistep differentiation pathway. MSCs give rise to osteoprogenitors, which differentiate into preosteoblasts and then mature osteoblasts [13].

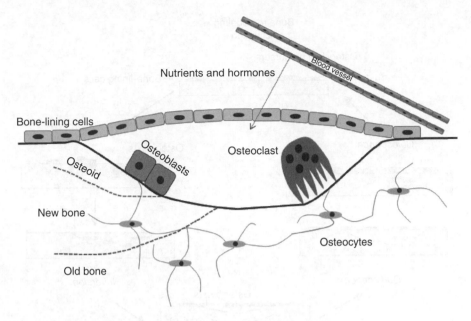

Figure 11.5   *Bone remodeling unit*

The remodeling process comprises four different but overlapping phases:

Phase 1: Initiation/activation of bone remodeling at a specific site
Phase 2: Bone resorption and current recruitment of MSCs and osteoprogenitors
Phase 3: Osteoblast differentiation and function
Phase 4: Mineralization of osteoid and completion of bone remodeling

The activation phase involves the recruitment and activation of mononuclear monocyte-macrophages from the circulation [14] that interact with bone lining cells that cover bone surfaces. Cells covering the bone surface produce M-CSF and RANKL [15, 16]. These factors are responsible for the induction of osteoclastogenesis. M-CSF binds to its receptor c-Fms in osteoclast precursors, providing the signals required for the survival and proliferation of the precursors [17]. RANK is expressed by osteoclast precursors and its activation by RANKL leads to expression of osteoclast specific genes such as cathepsin K, tartrate-resistant acid phosphatase (TRAP), calcitonin receptor and β3 integrin [18], and activation of resorption by mature osteoclasts.

After bone lining cells retract from the bone surface, osteoclast precursors attach to it and fuse to form osteoclasts. In response to activation of RANK by its ligand, multinucleated osteoclasts undergo internal structural changes that prepare them to resorb bone, such as the rearrangement of the actin cytoskeleton and formation of the sealing zone.

With the formation of osteoclasts, the remodeling process continues to Phase 2, in which bone resorption occurs, but the recruitment of MSCs/osteoprogenitors is also initiated. Phase 3 starts as osteoblast function (osteoid synthesis) begins to overtake bone resorption.

Phase 4 involves the mineralization of osteoid. Bone mineral consists of crystals of hydroxyapatite (HA) and non-collagenous proteins modulate the formation of these structures (Figure 11.6). Enzymes produced by osteoblasts, in particular alkaline

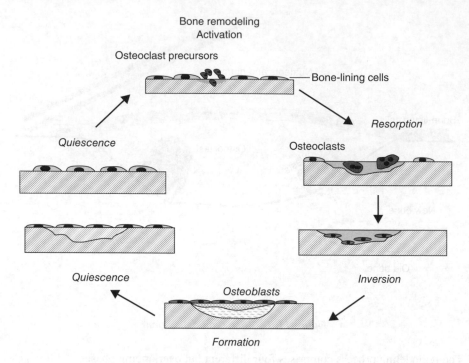

Figure 11.6   *Bone remodeling phases*

phosphatase (ALP), are essential for the process of mineralization. ALP regulates phosphoprotein dephosphorylation. By hydrolyzing phosphorylated substrates, this enzyme provides the inorganic phosphate necessary for HA crystallization [19].

11.4   Bone Cells

11.4.1   Bone Lining Cells

Bone lining cells belong to the osteoblast lineage. These cells are flattened and elongated and form a continuous layer, thereby helping to create and maintain a metabolic micro-environment. Ultrastructural studies have shown that these cells contain a relatively low number of mitochondria, free ribosomes, and rough endoplasmic reticulum. Bone lin- ing cells have cell processes that extend into the canaliculi and contact osteocyte cell process [20].

It has been proposed that bone lining cells play a role in bone remodeling by protecting bone surfaces from bone resorption. In addition, it is thought that the signals that initi- ate osteoclast formation may stimulate bone lining cells to predispose bone surfaces for bone resorption by digesting part of the non-mineralized bone matrix that covers the bone surface [21]. The removal of non-mineralized matrix by bone lining cells is mediated by matrix-metalloproteinase (MMP) activity, probably MMP-13 [22], revealing the mineral- ized matrix to osteoclasts [23].

Bone lining cells exert this resorbing activity by enwrapping collagen fibers protruding from the bone surface and subsequently digesting them in this secluded environment [24].

11.4.2   Osteoblasts

Osteoblasts originate from MSCs present in the bone marrow and in the periosteum. The commitment of MSCs to become osteoprogenitor cells is regulated by the tissue-specific transcription factor Runx2/Osterix [25, 26] and by molecules of the transforming growth factor-β (TGF-β) superfamily: the bone morphogenetic proteins (BMPs) [27, 28]. The BMPs as well as growth factors like fibroblast growth factors (FGFs), influence the proliferation, and differentiation of the osteoprogenitors into osteoblasts.

Osteoblast precursors change shape from spindle-shaped osteoprogenitors to large cuboidal differentiated osteoblasts on bone matrix. Pre-osteoblasts that are located near mature osteoblasts in the bone remodeling unit are usually recognizable because of their expression of ALP. Active mature osteoblasts that synthesize bone matrix have large nuclei, enlarged Golgi structures and, extensive endoplasmic reticulum. Figure 11.7 shows osteoblasts cultured on tissue culture plates.

Osteoblasts produce and secrete high amounts of bone matrix proteins, of which the bulk is formed by type I collagen. The network of type I collagen fibers provides the scaffold on which bone mineral is deposited. Osteoblasts also secrete non-collagenous proteins that together with type I collagen constitutes the osteoid. Non-collagenous proteins, such as osteonectin, osteopontin, bone sialoprotein, fibronectin, and osteocalcin, are abundantly produced by osteoblasts. After bone resorption is finished, osteoblast formation and function still continues to ensure a balance between bone removal and bone formation.

During bone formation, some osteoblasts become incorporated within the newly formed matrix and remain buried as osteocytes in spaces called lacunae [29, 30]. During the transformation from osteoblast to osteocyte the cuboidal osteoblast shape changes to a more dendritic-shaped cell, the osteocyte.

11.4.3   Osteocytes

Osteocytes are smaller than osteoblasts and have lost many of their cytoplasmic organelles [31]. These cells lie within lacunae in the newly formed bone matrix where they reside and ultimately undergo apoptosis. Osteocytes extend long filipodial extensions connecting them with each other as well as bone lining cells and osteoblasts of the bone surface [32].

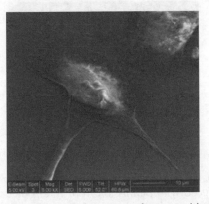

Figure 11.7   *Scanning electron microscope picture of an osteoblast cell isolated from trabecular bone*

Osteocytes have to remain in contact with other cells, and ultimately with the bone surface, to ensure the access of oxygen and nutrients.

The life span of osteocytes is probably determined by bone turnover, when osteoclasts resorb bone and "release" osteocytes. The fate of these osteocytes that are released by osteoclasts is currently unknown. There is little evidence that osteocytes may reverse their differentiation back into the osteoblastic state [33]. Some of them, only half released by osteoclast activity, may be re-embedded during new bone formation that follows bone resorption [34]. However, most of the osteocytes will probably die by apoptosis and become phagocytosed [35].

Osteocytes are mechano-sensors capable of transducing musculoskeletal-derived mechanical input into biological output [36]. They detect strain of mechanical loading, inducing the secretion of negative regulators of osteoblast bone formation, thereby enhancing bone formation. Osteocytes also detect microdamage (small defects in the bone matrix as a result of skeletal loading or pathologic conditions), initiating bone resorption via apoptotic bodies [37].

11.4.4 Osteoclasts

Osteoclasts form by the fusion of a mononuclear progenitor of the monocyte/macrophage family in a process called osteoclastogenesis [38].

Osteoclasts are the principle bone-resorbing cells constituting the smallest proportion of the bone cells, $1-3$ per μm^3 [39]. Osteoclasts exist in two functional states, the motile and the resorptive phases. During the motile state they migrate from the bone marrow to their resorptive site and in the resorptive phase they exert the bone resorption function. Motile osteoclasts are flattened, non-polarized cells. They are characterized by the presence of membrane protrusions, called lamellipodia and podosome complexes containing actin. Upon reaching the resorptive site, osteoclasts become polarized through cytoskeletal reorganization (Figure 11.8). As the osteoclast prepares to resorb bone it attaches to the bone surface through specific integrins and forms a specialized membrane area, the sealing zone. In the center of this sealing zone the osteoclast forms another unique membrane domain, the ruffled border. The actual degradation of bone takes place in this ruffled border

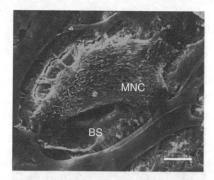

Figure 11.8 *Scanning electron microscope picture of a multinucleated cell isolated from peripheral blood. Reproduced with permission from (Pre-)Osteoclasts Induce Retraction of Osteoblasts Beofre their Fusion to Osteoclasts by S. perez-Amodio, W. Beertsen and V. Everts, Journal of Bone and Mineral Research, 19, 10, 1722–1731 Copyright (2004) American Society for Bone and Mineral Research*

area that is sealed off from the surrounding extracellular environment. The low pH in the resorption lacuna dissolves the mineral of the bone and proteinases secreted in the ruffled border digest the organic matrix.

Osteoclast differentiation and fusion is a multistep process that is regulated by several osteoclastogenic factors. Osteoclast precursors require factors produced by marrow stromal cells, osteoblasts, or T-lymphocytes to further differentiate [40]. M-CSF and RANKL, are the two factors responsible for the induction of osteoclastogenesis. M-CSF is produced by osteoblasts and stromal cells and binds to its receptor c-fms, on early osteoclast precursors, providing the signals required for the survival and proliferation of the precursors [41]. RANK is expressed by osteoclast precursors and its activation by RANKL leads to expression of specific osteoclast genes such as tartrate resistant acid phosphatase (TRAP), cathepsin K, calcitonin receptor, and β_3 integrin [18, 42]. Upon further stimulation with M-CSF and RANKL, osteoclast precursors fuse and form multinucleated cells. In response to activation of RANK by its ligand, mature osteoclasts undergo internal structural changes that prepare them to resorb bone, such as rearrangements of the actin cytoskeleton and formation of the healing zone (Figure 11.9).

The soluble decoy receptor osteoprotegerin (OPG) is the antagonist of RANKL [16]. OPG binds to RANKL and blocks the interaction with RANK, thereby inhibiting the differentiation of osteoclast precursors into mature osteoclasts [43]. OPG plays a pivotal role in the regulation of bone metabolism not only by inhibiting osteoclast differentiation and activation, but also by increasing osteoclast apoptosis [44].

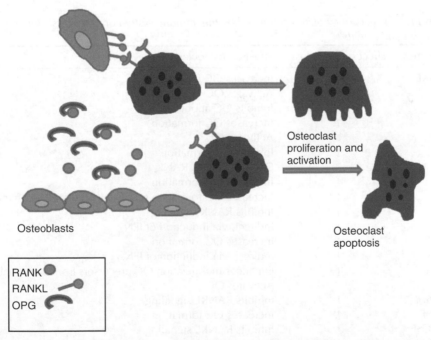

Figure 11.9 *OPG, produced by osteoblasts, binds to RANKL, and blocks the interaction with RANK (synthesized by osteoclasts), thereby inhibiting the differentiation of osteoclast precursors into mature osteoclasts and inducing their apoptosis*

Apart from the above-mentioned soluble factors, many cytokines are important regulators of osteoclast function. The majority of these cytokines are produced by cells of the immune system [45]. They regulate the expression of RANK and RANKL in osteoclasts and osteoblasts, respectively, or modulate the intracellular signaling mediated by RANK. The pro-inflammatory cytokines IL-1, TNF-α, and IL-6 might act synergistically on osteoclast formation and activity [46–48]. Osteoblasts and stromal cells are the main source of IL6 in bone; the effect of this cytokine in bone resorption is mediated primarily via stimulation of osteoclasts [49]. Also, IL-17, a cytokine that mediates autoimmune inflammatory arthritis is known to enhance osteoclastogenesis [50]. Other cytokines, such as IL-4, IL-10, and IL-13, inhibit osteoclastogenesis and osteoclast activation [51]. Table 11.1 shows the effects of several cytokines on bone metabolism and the possible underlying mechanisms of action.

Not all osteoclasts are the same and they can be categorized into subgroups depending on the matrix on which they are positioned. Osteoclasts present in long bones (e.g., the femur and the tibia) and the flat bones (e.g., the calvarium) are functionally different with respect to both acid secretion and proteases involved in degradation [52]. Data from mice deficient in the bicarbonate-anion exchanger Ae2 (Slc4a2) have shown that it is essential for bone resorption in long bones [53, 54], whereas it is not involved in bone resorption in calvaria [55], showing that distinct acid transport mechanisms are present in different subsets of osteoclasts. The proteolytic processes involved in the degradation of the two types of bone matrix are also different. Osteoclasts in flat bones preferentially engage MMP-mediated bone resorption, although cathepsin L also seems to be involved. Osteoclasts in long bones primarily depend on cysteine proteinases, in particular cathepsin K [56, 57]. Furthermore, TRAP appears to be involved in bone resorption, and more so in calvarial bone [58, 59].

Table 11.1   *Effects of cytokines produced by the immune system on osteoclast (OC) formation and activity*

Cytokines	Effect on bone loss	Effect on osteoclasts
RANKL	↑	Increases OC formation
		Activates OC
		Inhibits OC apoptosis
IL-1	↑	Increases OC formation
		Activates OC
IL-4	↓	Inhibits OC formation
		Down-regulates RANK
IL-6	↑	Increases OC formation
IL-7	↑	Increases OC formation (via T-cell mediated mechanism)
IL-10	↓	Inhibits RANK signaling
IL-12	↓	Indirect, via induction of IFN-γ
IL-17	↑	Increases OC formation
IL-18	↓	Indirect, via induction of IFN-γ
TNF-α	↑	Enhances mobilization OC precursors from bone marrow
		Activates OC
IFN-α/β	↓	Inhibits RANKL signaling
M-CSF	↑	Increases OC formation
GM-CSF	↓	Inhibits RANKL signaling
MCP-1	↑	Promotes OC fusion and activation

It has been suggested that bone matrix could play a role in the control of osteoclast activities, since compositional differences between long bone and flat matrices, including differences in the presence of putative cysteine proteinase inhibitors, have been reported [60].

11.5 Bone Extracellular Matrix

Bone tissue can be grossly divided into inorganic mineral material (mostly HA) and organic material from cells and the extracellular matrix. The inorganic phase of bone is a ceramic crystalline-type mineral that is an impure form of naturally occurring calcium phosphate, most often referred to as hydroxyapatite: $Ca_{10}(PO_4)_6(OH)_2$ [61]. Bone HA is not pure HA because the tiny apatite crystals (2–5 nm-thick, 15 nm-wide, 20–50 nm-long plates) contain impurities such as potassium, magnesium, strontium, and sodium (in place of the calcium ions), carbonate (in place of the phosphate ions), and chloride or fluoride (in place of the hydroxy ions). The bone mineral represents 80–90% of the volume in compact bones, decreasing to 15–25% in trabecular bones.

For the organic part, at the protein level collagens (mainly type I) represent 90% of the total bone protein content [62]. Trace amounts of type III and V and FACIT collagens are also present. FACIT collagens are members of the family of fibril-associated collagens with interrupted triple helices, a group of non-fibrillar collagens that serve as molecular bridges that are important for the organization and stability of extracellular matrices. Members of this family include collagens IX, XII, XIV, XIX, XX, and XXI.

Non-collagenous proteins comprise 10–15% of total bone protein. Approximately 25% of non-collagenous protein is exogenously derived, including serum albumin and α2-HS-glycoprotein, which bind to HA because of their acidic properties. The remaining exogenously derived non-collagenous proteins are composed of growth factors and a large variety of other molecules in small amounts that may affect bone cell activity.

The non-collagenous proteins are divided into several categories, including proteoglycans, glycosylated proteins with potential cell-attachment activities, and carboxylated proteins. These proteins regulate bone mineral deposition and turnover and regulate bone cell activity.

The main glycosylated protein present in bone is ALP. This protein is bound to the osteoblast cell surface via phosphoinositol linkage and it is also found free within the mineralized matrix.

Matrix maturation is associated with the expression of ALP and several non-collagenous proteins, such as osteocalcin, osteopontin, and bone sialoprotein. It is believed that these proteins help regulate order deposition of mineral by regulating the size of HA crystals formed.

11.6 Bone Diseases

Many bone diseases are disorders related to bone strength. These disorders are usually linked to deviations in minerals homeostasis (such as calcium or phosphorus), vitamin D, bone mass or bone structure, and some genetic diseases. Osteoporosis, rickets, osteomalacia, osteogenesis imperfecta (OI), marble bone disease (osteopetrosis), Paget's disease of bone, and fibrous dysplasia are some of them. Nonetheless in this chapter we will only mention some of them.

11.6.1   Osteoporosis

This is the most common metabolic bone disease and the main cause of bone fractures in post-menopausal women and the elderly in general. This represents an impact on society that affects the quality of life and the economy.

Osteoporosis is defined as a skeletal disorder characterized by compromised bone strength predisposing to an increased risk of fracture (NIH 1993 Consensus definition). The balance in bone remodeling is altered so that resorption predominates over formation and produces a net loss of bone which leads to osteoporosis. Around the world, one in three women and one in five men are at risk of an osteoporotic fracture. In fact, an osteoporotic fracture is estimated to occur every three seconds. The most common fractures occur at the hip, spine, and wrist. The likelihood of these fractures occurring, particularly at the hip and spine, increases with age in both women and men. The decrease in the sex hormones in men (testosterone and estrogen) and women (estrogen) are the main cause of bone loss in the elderly. For women, the rapid phase of bone loss is initiated by a dramatic decline in estrogen production by the ovaries at menopause. The loss of estrogen action on estrogen receptors in bone results in large increases in bone resorption. For aging men, sex steroid deficiency also appears to be a major factor in age-related osteoporosis. Although testosterone is the major sex steroid in men, some of it is converted by the aromatase enzyme into estrogen. As with women, the loss of sex steroid activity in men has an effect on calcium absorption and conservation, leading to progressive secondary increases in parathyroid hormone levels. As in older women, the resulting imbalance between bone resorption and formation results in slow bone loss that continues over life [63]. The three main mechanisms that cause osteoporosis are:

1. Lack of sufficient bone mass during the growth process.
2. Excessive bone resorption mediated by osteoclasts.
3. Inadequate new bone formation by osteoblasts during the bone turnover continuous process.

Menopause is the main cause of osteoporosis in women due to declining estrogen levels (Table 11.2). Estrogen loss by physiological menopause or the surgical removal of the

Table 11.2 *Besides age, other diseases, and environmental factors can contribute to bone loss*

Endocrine
Hyperthyroidism, hyperparathyroidism, Cushing's syndrome, and hypogonadism

Drugs
Corticosteroids, anticonvulsants, and lithium salts

Malignant tumor processes
Multiple myeloma and bone metastases from other tumors

Others
Malnutrition by anorexia nervosa, intestinal resections, ulcerative colitis, Crohn's disease, alcoholism, chronic renal failure, rheumatoid arthritis, prolonged immobilization, other causes of amenorrhea

ovaries causes rapid bone loss. Women, especially Caucasian and Asian, have lower bone mass than men. Bone loss results in a lower strength thereof, which easily leads to fractures of the wrist, spine, and hip.

11.6.2 Paget's Disease

Paget's disease is named after Sir James Paget who in 1876 first described it. This disease is also known as osteitis deformans, a designation that refers to inflammation of the bone and the secondary deformation that occurs with this disease. After osteoporosis, it is the most common bone disorder in our surrounding countries. The incidence of the disease increases with age and rarely occurs before age 40. It is more common in the Nordic countries, England, Western Europe, the United States, and New Zealand.

Paget's disease causes excessive growth and weakening of the bones. The most commonly affected sites are the pelvis (70%), femur (55%), tibia (32%), lumbar spine (53%), and skull (42%) [64]. The disease can affect one or several bones, but does not affect the entire skeleton.

Significant differences are apparent in the distribution and organization of collagen fibers due to the synthesis and deposition of fats by osteoblasts, and the lack of remodeling causes the loss of mechanical properties due to the decrease in strength.

The condition of this disease has been associated with measles and rubella. It is also known that there is an important genetic factor, with the disease often present in several members of the same family. All this makes one suspect transmission by a genetic mechanism.

11.6.3 Osteomalacia

Is the softening of the bones caused by a deficiency of vitamin D or by problems with the metabolism (breakdown and use) of this vitamin. These softer bones have a normal amount of collagen that gives structure to bones, but lack of calcium. In children, this condition is called rickets. Soft bones are more likely to bow and fracture than are harder and healthy bones.

11.6.4 Osteogenesis Imperfecta

OI is a genetic disorder in which bones are too brittle and break easily. It is often caused by a defect in a gene that produces type 1 collagen, a key pillar of the bone. There are many different defects that can affect this gene and the severity of the disease depends on the specific gene defect.

OI is an autosomal dominant, which means that you will have it if you have a copy of the gene. Most cases of OI are inherited from one parent, although some cases are the result of new genetic mutations.

11.7 Bone Mechanics

Biomechanically, bone can be considered as a biphasic composite material, with two phases: one the mineral and the other collagen. This combination confers better mechanical properties on the tissue than each component itself [65]. Bone has a tensile strength similar to cast iron, and yet is 3 times lighter and 20 times more flexible. The skeleton adapts to its specific function in the body by both its configuration and its microscopic structure. Nature tends to follow generally the law of the minimum so that the mechanical functions of loading and protection are achieved with minimum weight and maximum efficiency (Figure 11.10).

In 1982, Julius Wolff (a German anatomist and surgeon) described in his book "The Law of Bone transformation", how external applied forces could change the bone shape. This became what is nowadays known as Wolff's law: "The shape and structure of growing bones and adult bones depend on the stresses and strain to which they are subjected. By altering the lines of stress the shape of a bone could be changed". Although the rationale for the existence of Wolff's law has been challenged, many contemporary investigators still ascribe to the idea that there is a "Wolff's law" that states that bone models and remodels in response to the mechanical stresses it experiences so as to produce a minimal-weight structure that is "adapted" to its applied stresses [66, 67]. For instance, the racquet-holding arm bones of tennis players become much stronger than those of the other arm. Their bodies have strengthened the bones in their racquet-holding arm since it is routinely placed under higher than normal stresses players (Figure 11.10).

Due to different structural features mechanical properties differ in the two bone types. Macroscopically the cortical bone has much higher compressive stiffness (12–20 GPa vs 0.2–0.8 GPa) and strength (100–230 MPa vs 2–12 MPa) than the cancellous bone [68]. The various loads acting on the bones are related to the various activities of the individual, both compressive and tensile or shear. The cortical bone mainly works to compression, while trabecular bone must withstand forces of compression, tension, and shear. In

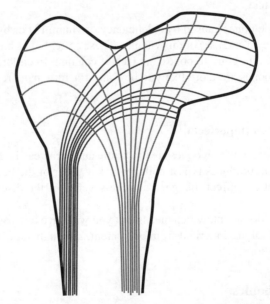

Figure 11.10 Wolff's law. The shape and structure of growing bones and adult bones depend on the stresses and strain to which they are subjected. By altering the lines of stress the shape of a bone could be changed

general, the bone mineral phase confers resistance to compression and shear, while collagen provides the tensile strength.

When performing mechanical tests bone properties have to be taken into account. Bone responds with a characteristic pattern to the forces applied on its surface, which depends on the strength, density, and composition of the tissue architecture. The first phase is elastic and generates a temporary deformation maintained while applying the acting force, and then regains its original shape. If the force increases, it enters a phase of plasticity and bone, although it, partially recovers, is deformed. Finally when the applied force exceeds the strength of the material fracture occurs (Figure 11.11).

Forces that act on bone tissue are tension, compression, and torsion. They can also be applied perpendicular to the bone surface, such as normal force, or obliquely, as a shear force. Bone is considered to have an anisotropic behavior, so the direction of the loading will exhibit different mechanical properties. Stiffness related to tension is maximal for axial loads and minimal for perpendicular loads (Figure 11.12).

Other factors, such as the strain rate, affect the mechanical properties of bone. The strain rate used in the traction assay affects strongly the stress–strain curve. Bone becomes more rigid and stronger the higher the strain rate applied. Physiological strain rates to which bones are normally subjected are 0.001 s^{-1} for a normal activity and 0.01 s^{-1} for strong one. The geometry of the bone will also influence its mechanical behavior. In tension and compression, the load to failure and stiffness are proportional to the cross-sectional area of the bone. The larger the area, the stronger and stiffer the bone.

Different parameters can affect bone mechanics: age, sex, genetics, lifestyle, and diet are some of them. Ageing is one of the main affecting factors, diminishing bone density, and producing architecture changes. As seen in Figure 11.13, it can be observed that bone loss

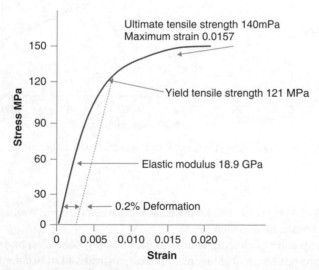

Figure 11.11   *Stress–strain curve of the bone in a traction assay. The yield tensile strength is produced at the yield point where the stress–strain curve deviates from the straight-line relationship. The slope of the line in this region where stress is proportional to strain is called the modulus of elasticity or Young's modulus*

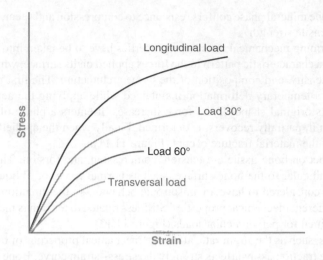

Figure 11.12   *Anisotropic behavior of the bone. (Adapted from [113].)*

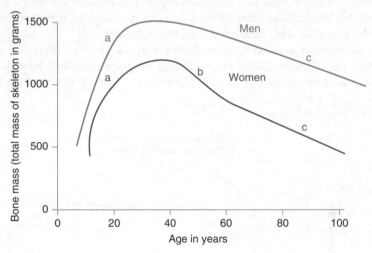

Figure 11.13   *Graph shows normal changes in bone mass with age, where a is peak bone mass, b is menopause, and c is ageing*

starts around 40–50 years old, with no differences between men and women, but it is more pronounced in women than in men.

Physical activity can also determine bone mass. Growing bones respond to low or moderate exercise through a significant increase in new cortical and trabecular bone. It is well known that astronauts lose bone density during their stay in space. It has been calculated that humans lose 1.5% of bone mass during each month in space. This loss occurs because the absence of the Earth's gravity disrupts the process of bone maintenance in its major function of supporting body weight.

11.8   Bone Tissue Regeneration

The increase in life expectancy and the ageing population in developed countries have caused a rise in the incidence of musculoskeletal diseases worldwide. According to a WHO report: "This increase will continue, especially in developing countries, due to the harmful effects of urban development and motorization" [69]. Musculoskeletal diseases have a huge economic impact of a magnitude equal to cardiovascular disease or cancer, and it is estimated that the total cost represents around 2.5% of the GDP of developed countries.

The treatment of large-bone defects caused by trauma or disease presents a significant clinical problem. Therapies include the use of autologous grafts, which is limited by availability and donor-site morbidity [70], and structural allografts, which is limited by lack of vascularization and high infection risk. These shortcomings have led to extensive research in bone tissue engineering.

Bone tissue engineering has emerged as the ideal strategy to solve this clinical problem. In the most common approach, a biomaterials scaffold with a well-defined architecture serves as a temporary structure for cells and guides their proliferation and differentiation into the desired tissue. Growth factors and other biomolecules can be incorporated into the scaffolds along with the cells, to guide the regulation of cellular functions during tissue regeneration [71].

As mentioned already in Chapter 10, biomaterials used to create scaffolds are designed to meet rigorous requirements that are essential for optimizing tissue formation [72]. Scaffolds for tissue engineering must (i) be biocompatible (not toxic) and be able to promote cell adhesion and proliferation; (ii) show, after *in vitro* tissue culture, mechanical properties that are comparable to those of the tissue to be replaced, (iii) have a porous three-dimensional (3D) architecture to permit proliferation of cells, vascularization, and diffusion of nutrients between the cells and the environment; (iv) degrade at a rate that matches the production of new tissue, into non-toxic compounds that can be resorbed or excreted by the body; and (v) be capable of being developed economically into anatomically relevant shapes and dimensions, and be sterilized for clinical use.

A variety of scaffolds have been developed for bone tissue engineering purposes and fabrication methods have already been described in Chapter 10. In this chapter we examine calcium phosphate bioceramics, bioactive glasses, and calcium phosphate cements for *in vivo* bone regeneration, and evaluate their efficacy as well as their drawbacks.

11.8.1   Calcium Phosphate and Silica-based Bioceramics

β-Tricalcium phosphates (β-TCPs) are highly osteoconductive and exert their effects via interaction with $\alpha2\beta1$ integrins on osteogenic cells and subsequent downstream activation of MAPK/ERK signaling pathways in these cells [73]. Scaffolds composed of β-TCP have been used in animal models to help to heal femoral defects [74], mandibular defects [75], and critical-size calvarial defects [76]. β-TCP seeded with adipose-derived stem cells (ASCs) has shown that this material has the capacity to induce stem cells to differentiate into osteoblastic cells [77].

Calcium phosphate scaffolds are a promising approach for bone regeneration because the normal process of mineral precipitation and ossification is dependent upon the extracellular concentration of both phosphate and calcium [78]. A major drawback to the use of β-TCP in the clinical setting is the observation that the resorption rate of the scaffold exceeds that of the formation of native bone [79].

Since the inorganic element of the extracellular matrix is composed of precipitated HA crystals, HA scaffolds designed with many different densities, porosities, sizes, and strength have been used in bone tissue engineering. This material is strongly osteoconductive but minimally osteogenic [80]. However, by seeding the scaffolds with osteogenic cells, the scaffold material can be made more osteogenic than on its own. HA has been used in animal models to help to correct radial [81], long bone [82], and cranial defects [83].

The combination of HA with other compounds, such as chitosan [84], collagen [85], and poly D, L-lactic-co-glycolic acid (PLGA) [86] for osteochondral repair has yielded scaffolds that are more useful than HA. These hybrid scaffolds are better promoters of osteoblast adhesion, proliferation, and differentiation than the pure HA scaffolds.

A common problem with HA scaffolds is their inability to maintain cells within the interior because of the lack of nutrient delivery to this area.

In an attempt to reduce the independent drawbacks of β-TCP and HA scaffolds, people have tried scaffolds composed of combinations of these two compounds, called biphasic calcium phosphate (BCP) scaffolds. BCP scaffolds have been used to regenerate bone in iliac wings defects in goats [87], in canine mandibular defects [88], and femoral head osteonecrosis cases [89].

BCP scaffolds have excellent cellular biocompatibility and have been demonstrated to be superior in inducing bone regeneration when compared with HA scaffolds [90].

Supramolecular chemistry has allowed the emergence of a new generation of advanced bioceramics, among which ordered mesoporous materials show potential features to be used as starting materials for the design and fabrication of scaffolds for bone tissue engineering purposes. Silica-based mesoporous materials are characterized by exhibiting high surface areas and a high surface density of silanol groups (Si−OH). It is well known that silanol groups, which are present in conventional bioactive glasses, are able to react with physiological fluids to produce nanometer-sized carbonated apatites [91, 92].

Their specific properties (high surface area, high pore volume, narrow pore size distribution in the mesopore region, and a pore channel system homogeneously organized in 2D or 3D arrangements) allow these nanomaterials to be used as controlled delivery systems of a wide range of biologically active molecules. These materials have shown high bioactive responses and biomimetic behavior not reported for any bioceramic up to date. Moreover, organic modification of mesoporous silica walls brings up many possibilities in molecule adsorption and release by promoting host−guest interactions [93].

11.8.2   Bioactive Glasses

Bioactive glasses are a versatile class of materials and structurally all silica-based glassed have the same basic building block SiO_4^{4-}. Porous glass ceramics made from bioglass 45S5 have been shown to induce bone formation when implanted in thigh muscles of dogs [94]. Formation of new bone using bioactive glasses has also been demonstrated using the rabbit radius osteotomy model [95], and when implanted in femoral condyles of rabbits in *in vivo* systems [96].

Seeding with MSCs has also been found to enhance tissue infiltration into bioactive glasses scaffold [97].

In a modified bone healing model of bone marrow ablation of rat tibia, it has been shown that filling of the intramedullary space with bioactive glass microspheres results in significant intramedullary new bone formation and high local bone turnover [98].

There are two mechanisms of bioactivity for bioactive glasses. Bone bonding is attributed to the formation of a hydroxycarbonate apatite layer, which interacts with collagen fibers of damaged bone to form a bond [99]. Bone bonding to the hydroxycarbonate apatite layer is thought to involve protein adsorption, incorporation of collagen fibrils, attachment of bone progenitor cells, cell differentiation, and the synthesis of extracellular matrix, followed by its mineralization [100].

11.8.3 Calcium Phosphate Cements

Calcium phosphate cement (CPC) has excellent biological properties, potential resorbability, molding capabilities, and easy manipulation. Recent *in vivo* studies have shown calcium phosphate cements to be promising materials for grafting applications. Studies in beagles demonstrated that calcium phosphate cements were replaced by new bone in alveolar bone defects [101, 102]. In another study, CPC was injected as a bone filler for gaps around oral implants placed on the femoral condyles of goats, excellent bone formation around the graft material was found [103].

Also, applying CPC to calvarial bone defects showed a progressive bone ingrowth, which was nearly replaced by host bone [104].

One of the main drawbacks in tissue engineering strategies that get even higher relevance in bone tissue engineering applications is vascularization. Bone, as you already know, is a highly vascularized tissue and rarely does a bone implant integrate optimally with the vascular network in the tissue. *In vivo* most cells do not survive more than a few hundred micrometers from the nearest capillary, due to diffusion limitations. Capillaries, and the vascular system, are required to supply essential nutrients, including oxygen, to remove waste products, and to provide biochemical communication [105]. Large bone defects regeneration remains as a great clinical problem due to the complexity in achieving a good vascularization network that could ensure the blood supply to allow rapid cell colonization.

Vascularization of new bioengineered tissue can be accomplished in three ways. Upon implantation of scaffolds that can release growth factors (such as vascular endothelial growth factor, VEGF) these will induce vessels from surrounding tissue to infiltrate the scaffold [106, 107]. This approach is problematic as blood vessels need time to infiltrate into tissues, and thereby this will predispose the tissue to ischemia as the cells wait for a nutrient supply to reach them. Vessels may also grow from within a scaffold seeded with endothelial cells to help enhance the rate of vascularization. Co-culture of endothelial cells with fibroblasts, osteoblasts, or myoblasts to study the interaction mechanisms also falls into this category [108–110]. This is based on the hypothesis that the transplanted cells will form capillary-like structures, join with inwards growing vessels from the host tissue, and combine with existing blood vessels to create a continuous vessel (anastomosis). This approach, like the previous one, is also associated with a time delay before vascular perfusion can occur, due to the time for the vessel network within the scaffold to join with the host vasculature. There is also a third method where matrices are prevascularized *in vivo*, extracted and then reimplanted at the site of interest, although this method involves serious limitations due to the need for a host (which is usually the same patient).

Recent results on the Ca^{2+} effect on angiogenesis and vascularization point to bioceramics being the most promising biomaterials capable of inducing vascularization for bone tissue regeneration [111, 112]. Such scaffolds, as themselves or as composites, are excellent systems that can gradually control degradation and, conversely, Ca^{2+} release.

11.9   Conclusions

Bone is a highly complex material with a very well organized and hierarchical structure responsible for its excellent mechanical properties due to the combination of collagen fibers and HA crystals in its composition. This metabolically active organ undergoes continuous remodeling throughout life. Bone remodeling involves the removal of mineralized bone by osteoclasts, followed by the formation of bone matrix through the osteoblasts that subsequently become mineralized. All these assets make bone a suitable model for regeneration, thus multiple studies have been performed in order to better understand the mechanism of bone formation and which biomaterial could be a good replacement, either temporal or permanent, that could undertake all the properties of bone.

It is clear that the ideal scaffold material will possess excellent biocompatible, osteocompatible, osteoinduction, osteoconduction, and neovasculogenic profiles.

Further advances should be made in order to answer questions such as: Can newly formed bone be self-supporting and renew itself? Is it necessary to use different types of scaffold and modulate their mechanical strength for different bone types? Can scaffold-derived bone support hematopoiesis?

It is almost certain that this knowledge and the capacity to correct bony defects will increase in the next few years.

References

1.   Bell, L.S., Cox, G. and Sealy, J. (2001) Determining isotopic life history trajectories using bone density fractionation and stable isotope measurements: a new approach. *Am. J. Phys. Anthropol.*, (146), 66–79.
2.   Rho, J.Y., Kuhn-Spearing, L. and Zioupos, P. (1998) Mechanical properties and the hierarchical structure of bone. *Med. Eng. Phys.*, **20** (2), 92–110.
3.   Allen, M.R., Hock, J.M. and Burr, D.B. (2004) Periosteum: biology, regulation, and response to osteoporosis therapies. *Bone*, **35**, 1003–1012.
4.   Yiannakopoulos, C.K., Kanellopoulos, A.D., Trovas, G.P. *et al.* (2008) The biomechanical capacity of the periosteum in intact long bones. *Arch. Orthop. Trauma Surg.*, **128**, 117–120.
5.   Colnot, C. (2009) Skeletal cell fate decisions within periosteum and bone marrow during bone regeneration. *J. Bone Miner. Res.*, **24**, 274–282.
6.   Frost, H.M. (2000) The Utah paradigm of skeletal physiology: an overview of its insights for bone, cartilage and collagenous tissue organs. *J. Bone Miner. Metab.*, **18**, 305–316.
7.   Vaananen, H.K. (1993) Mechanism of bone turnover. *Ann. Med.*, **25**, 353–359.
8.   Noble, B. (2003) Bone microdamage and cell apoptosis. *Eur. Cell. Mater.*, **6**, 46–55.

9. Eriksen, E.F. (2010) Cellular mechanisms of bone remodeling. *Rev. Endocr. Metab. Disord.*, **11**, 219–227.

10. Parfitt, A.M. (2001) The bone remodeling compartment: circulatory function for bone lining cells. *J. Bone Miner. Res.*, **16** (9), 1583–1585.

11. Santos, A., Bakker, A.D. and Klein-Nulend, J. (2009) The role of osteocytes in bone mechanotransduction. *Osteoporosis Int.*, **20**, 1027–1031.

12. Boyle William, J., Simonet, W.S. and Lacey, D.L. (2003) Osteoclasts differentiation and activation. *Nature*, **423** (6937), 337–342.

13. Ducy, P., Schinke, T. and Karsenty, G. (2000) The osteoblast: a sophisticated fibroblast under central surveillance. *Science*, **289** (5484), 1501–1504.

14. Roodman, G.D. (1999) Cell biology of the osteoclast. *Exp. Hematol.*, **27**, 1229–124.

15. Fuller, K., Wong, B., Fox, S. *et al.* (1998) TRANCE is necessary and sufficient for osteoblast-mediated activation of bone resorption in osteoclasts. *J. Exp. Med.*, **188**, 997–1001.

16. Lacey, D.L., Timms, E., Tan, H.L. *et al.* (1998) Osteoprotegerin ligand is a cytokine that regulates osteoclast differentiation and activation. *Cell*, **93**, 165–176.

17. Hamilton, J.A. (1997) CSF-1 signal transduction. *J. Leukoc. Biol.*, **62**, 145–155.

18. Teitelbaum, S.L. and Ross, F.P. (2003) Genetic regulation of osteoclast development and function. *Nat. Rev. Genet.*, **4**, 638–649.

19. Balcerzak, M., Hamade, E., Zhang, L. *et al.* (2003) The roles of annexins and alkaline phosphatase in mineralization process. *Acta Biochim. Pol.*, **50**, 1019–1038.

20. Miller, S.C. and Jee, W.S. (1987) The bone lining cell: a distinct phenotype? *Calcified Tissue Int.*, **41**, 1–5.

21. Chambers, T.J. and Fuller, K. (1985) Bone cells predispose bone surfaces to resorption by exposure of mineral to osteoclastic contact. *J. Cell Sci.*, **76**, 155–165.

22. Delaisse, J.M., Eeckhout, Y., Neff, L. *et al.* (1993) (Pro)collagenase (matrix metalloproteinase-1) is present in rodent osteoclasts and in the underlying bone-resorbing compartment. *J. Cell Sci.*, **106**, 1071–1082.

23. Chambers, T.J., Darby, J.A. and Fuller, K. (1985) Mammalian collagenase predisposes bone surfaces to osteoclastic resorption. *Cell Tissue Res.*, **241**, 671–675.

24. Everts, V., Delaisse, J.M., Korper, W. *et al.* (2002) The bone lining cell: its role in cleaning Howship's lacunae and initiating bone formation. *J. Bone Miner. Res.*, **17**, 77–90.

25. Ducy, P., Zhang, R., Geoffroy, V. *et al.* (1997) Osf2/Cbfa1: a transcriptional activator of osteoblast differentiation. *Cell*, **89**, 747–754.

26. Ducy, P. (2000) Cbfa1: a molecular switch in osteoblast biology. *Dev. Dyn.*, **219**, 461–471.

27. Asahina, I., Sampath, T.K. and Hauschka, P.V. (1996) Human osteogenic protein-1 induces chondroblastic, osteoblastic, and/or adipocytic differentiation of clonal murine target cells. *Exp. Cell Res.*, **222**, 38–47.

28. Katagiri, T. and Takahashi, N. (2002) Regulatory mechanisms of osteoblast and osteoclast differentiation. *Oral Dis.*, **8**, 147–159.

29. Baud, C.A. (1968) Submicroscopic structure and functional aspects of the osteocyte. *Clin. Orthop.*, **56**, 227–236.

30. Palumbo, C. (1986) A three-dimensional ultrastructural study of osteoid-osteocytes in the tibia of chick embryos. *Cell Tissue Res.*, **246**, 125–131.

31. Aubin, J.E. (1998) Advances in the osteoblast lineage. *Biochem. Cell Biol.*, **76**, 899–910.

32. Tasaka-Kamioka, K., Kamioka, H., Ris, H. and Lim, S.-S. (1998) Osteocyte shape is dependent on actin filament and osteocyte processes are unique actin-rich projections. *J. Bone Miner. Res.*, **13**, 1555–1568.

33. van der Plas, A., Aarden, E.M., Feijen, J.H. *et al.* (1994) Characteristics and properties of osteocytes in culture. *J. Bone Miner. Res.*, **9** (11), 1697–1704.

34. Suzuki, R., Domon, T. and Wakita, M. (2000) Some osteocytes released from their lacunae are embedded again in the bone and not engulfed by osteoclasts during bone remodeling. *Anat. Embryol (Berl).*, **202** (2), 119–128.

35. Elmardi, A.S., Katchburian, M.V. and Katchburian, E. (1990) Electron microscopy of developing calvaria reveals images that suggest that osteoclasts engulf and destroy osteocytes during bone resorption. *Calcif. Tissue Int.*, **46** (4), 239–245.

36. Cowin, S.C., Moss-Salentijn, L. and Moss, M.L. (1991) Osteocyte shape is dependent on actin filament and osteocyte processes are unique actin-rich projections. *J. Biomech. Eng.*, **113** (2), 191–197.

37. Kogianni, G., Mann, V. and Noble, B.S. (2008) Apoptotic bodies convey activity capable of initiating osteoclastogenesis and localized bone destruction. *J. Bone Miner. Res.*, **23** (6), 915–927.

38. Teitelbaum, S.L. (2000) Bone resorption by osteoclasts. *Science*, **289** (5484), 1504–1508.

39. Roodman, G.D. (1996) Advances in bone biology: the osteoclast. *Endocr. Rev.*, **17** (4), 308–332.

40. Udagawa, N., Takahashi, N., Akatsu, T. *et al.* (1990) Origin of osteoclasts: mature monocytes and macrophages are capable of differentiating into osteoclasts under a suitable microenvironment prepared by bone marrow-derived stromal cells. *Proc. Natl. Acad. Sci. U.S.A.*, **87** (18), 7260–7264.

41. Hodge, J.M., Kirkland, M.A. and Nicholson, G.C. (2007) Multiple roles of M-CSF in human osteoclastogenesis. *J. Cell. Biochem.*, **102** (3), 759–768.

42. Ejiri, S. (1983) The preosteoclast and its cytodifferentiation into the osteoclast: ultrastructural and histochemical studies of rat fetal parietal bone. *Arch. Histol. Jpn.*, **46**, 533–557.

43. Swarthout, J.T., D'Alonzo, R.C., Selvamurugan, N. and Partridge, N.C. (2002) Parathyroid hormone-dependent signaling pathways regulating genes in bone cells. *Gene*, **282**, 1–17.

44. Yasuda, H., Shima, N., Nakagawa, N. *et al.* (1998) Identity of osteoclastogenesis inhibitory factor (OCIF) and osteoprotegerin (OPG): a mechanism by which OPG/OCIF inhibits osteoclastogenesis in vitro. *Endocrinology*, **139**, 1329–1337.

45. Stashenko, P., Dewhirst, F.E., Peros, W.J. *et al.* (1987) Synergistic interactions between interleukin 1, tumor necrosis factor, and lymphotoxin in bone resorption. *J. Immunol.*, **138** (5), 1464–1468.

46. Lam, J., Takeshita, S., Barker, J.E. *et al.* (2000) TNF-alpha induces osteoclastogenesis by direct stimulation of macrophages exposed to permissive levels of RANK ligand. *J. Clin. Invest.*, **106** (12), 1481–1488.

47. Holt, I., Davie, M.W. and Marshall, M.J. (1996) Osteoclasts are not the major source of interleukin-6 in mouse parietal bones. *Bone*, **18** (3), 221–226.

48. Dinarello, C.A. (1997) Interleukin-1. *Cytokine Growth Factor Rev.*, **8** (4), 253–265.
49. Udagawa, N., Takahashi, N., Katagiri, T. *et al.* (1995) Interleukin (IL)-6 induction of osteoclast differentiation depends on IL-6 receptors expressed on osteoblastic cells but not on osteoclast progenitors. *J. Exp. Med.*, **182** (5), 1461–1468.
50. Kotake, S., Udagawa, N., Takahashi, N. *et al.* (1999) IL-17 in synovial fluids from patients with rheumatoid arthritis is a potent stimulator of osteoclastogenesis. *J. Clin. Invest.*, **103** (9), 1345–1352.
51. Moreno, J.L., Kaczmarek, M., Keegan, A.D. and Tondravi, M. (2003) IL-4 suppresses osteoclast development and mature osteoclast function by a STAT6-dependent mechanism: irreversible inhibition of the differentiation program activated by RANKL. *Blood*, **102** (3), 1078–1086.
52. Everts, V., de Vries, T.J. and Helfrich, M.H. (2009) Osteoclast heterogeneity: lessons from osteopetrosis and inflammatory conditions. *Biochem. Biophys. Acta*, **1792**, 757–765.
53. Josephsen, K., Praetorius, J., Frische, S. *et al.* (2009) Targeted disruption of the Cl-/HCO3- exchanger Ae2 results in osteopetrosis in mice. *Proc. Natl. Acad. Sci. U.S.A.*, **106** (5), 1638–1641.
54. Wu, J., Glimcher, L.H. and Aliprantis, A.O. (2008) HCO3-/Cl- anion exchanger SLC4A2 is required for proper osteoclast differentiation and function. *Proc. Natl. Acad. Sci. U.S.A.*, **105** (44), 16934–16939.
55. Jansen, I.D., Mardones, P., Lecanda, F. *et al.* (2009) Ae2(a,b)-deficient mice exhibit osteopetrosis of long bones but not of calvaria. *FASEB J.*, **23** (10), 3470–3481.
56. Everts, V., Korper, W., Jansen, D.C. *et al.* (1999) Functional heterogeneity of osteoclasts: matrix metalloproteinases participate in osteoclastic resorption of calvarial bone but not in resorption of long bone. *FASEB J.*, **13** (10), 1219–1230.
57. Shorey, S., Heersche, J.N. and Manolson, M.F. (2004) The relative contribution of cysteine proteinases and matrix metalloproteinases to the resorption process in osteoclasts derived from long bone and scapula. *Bone*, **35** (4), 909–917.
58. Perez-Amodio, S., Jansen, D.C., Schoenmaker, T. *et al.* (2006) Calvarial osteoclasts express a higher level of tartrate-resistant acid phosphatase than long bone osteoclasts and activation does not depend on cathepsin K or L activity. *Calcif. Tissue Int.*, **79** (4), 245–254.
59. Roberts, H.C., Knott, L., Avery, N.C. *et al.* (2007) Altered collagen in tartrate-resistant acid phosphatase (TRAP)-deficient mice: a role for TRAP in bone collagen metabolism. *Calcif. Tissue Int.*, **80** (6), 400–410.
60. van den Bos, T., Speijer, D., Bank, R.A. *et al.* (2008) Differences in matrix composition between calvaria and long bone in mice suggest differences in biomechanical properties and resorption: special emphasis on collagen. *Bone*, **43** (3), 459–468.
61. Lowenstam, H.A. and Weiner, S. (1989) *On Biomineralization*, Oxford University Press, New York.
62. Sommerfeldt, D.W. and Rubin, C.T. (2001) Biology of bone and how it orchestrates the form and function of the skeleton. *Eur. Spine J.*, **10** (Suppl 2), S86–S95.
63. Khosla, S., Melton, L.J. III, Atkinson, E.J. *et al.* (1998) Relationship of serum sex steroid levels and bone turnover markers with bone mineral density in men and women: a key role for bioavailable estrogen. *J. Clin. Endocrinol. Metab.*, **83**, 2266–2274.

64. Kanis, J.A. (1992) *Pathophysiology and Treatment of Paget's Disease of Bone*, 1st edn, Martin Dunita, London.
65. Bassett, C.A.L. (1965) Electrical effects in bone. *Sci. Am.*, **213**, 18.
66. Buckwalter, J.A. and Grodzinsky, A.J. (1999) Loading of healing bone, fibrous tissue, and muscle: implications for orthopaedic practice. *J. Am. Acad. Orthopedic Surgeons*, **7**, 291–299.
67. Goodship, A.E., Lanyon, L.E. and McFie, F. (1979) Functional adaptation of bone to increased stress. An experimental study. *J. Bone Joint Surg. Am.*, **61**, 539–546.
68. Hench, L.L. and Wilson, J. (eds) (1993) *Introduction to Bioceramics*, World Scientific, Singapore.
69. WHO Report (2009) http://www.who.int/ncd / cra (accessed 11 November 2013).
70. Mauffrey, C., Madsen, M., Bowles, R.J. and Seligson, D. (2012) Bone graft harvest site options in orthopaedic trauma: a prospective in vivo quantification study. *Injury*, **43** (3), 323–326.
71. Babensee, J.E., McIntire, L.V. and Mikos, A.G. (2000) Growth factor delivery for tissue engineering. *Pharm. Res.*, **17** (5), 497–504.
72. Hutmacher, D.W. (2001) Scaffold design and fabrication technologies for engineering tissues--state of the art and future perspectives. *J. Biomater. Sci. Polym. Ed.*, **12** (1), 107–124.
73. Lu, Z. and Zreiqat, H. (2010) Beta-tricalcium phosphate exerts osteoconductivity through alpha2beta1 integrin and down-stream MAPK/ERK signaling pathway. *Biochem. Biophys. Res. Commun.*, **394** (2), 323–329.
74. Jacobson, J.A., Yanoso-Scholl, L., Reynolds, D.G. *et al.* (2011) Teriparatide therapy and beta-tricalcium phosphate enhance scaffold reconstruction of mouse femoral defects. *Tissue Eng. Part A*, **17** (3-4), 389–398.
75. Nienhuijs, M.E., Walboomers, X.F., Briest, A. *et al.* (2010) Healing of bone defects in the goat mandible, using COLLOSS E and beta-tricalciumphosphate. *Biomed. Mater. Res. B Appl. Biomater.*, **92** (2), 517–524.
76. Hing, K.A., Wilson, L.F. and Buckland, T. (2007) Comparative performance of three ceramic bone graft substitutes. *Spine J.*, **7** (4), 475–490.
77. Liu, Q., Cen, L., Yin, S. *et al.* (2008) A comparative study of proliferation and osteogenic differentiation of adipose-derived stem cells on akermanite and beta-TCP ceramics. *Biomaterials*, **29** (36), 4792–4792.
78. Murshed, M., Harmey, D., Millán, J.L. *et al.* (2005) Unique coexpression in osteoblasts of broadly expressed genes accounts for the spatial restriction of ECM mineralization to bone. *Genes Dev.*, **19** (9), 1093–1104.
79. Hollinger, J.O. and Battistone, G.C. (1986) Biodegradable bone repair materials. Synthetic polymers and ceramics. *Clin. Orthop. Relat. Res.*, **207**, 290–305.
80. Patel, M., Dunn, T.A., Tostanoski, S. and Fisher, J.P. (2010) Cyclic acetal hydroxyapatite composites and endogenous osteogenic gene expression of rat marrow stromal cells. *J. Tissue Eng. Regen. Med.*, **4** (6), 422–436.
81. Niemeyer, P., Szalay, K., Luginbühl, R. *et al.* (2010) Transplantation of human mesenchymal stem cells in a non-autogenous setting for bone regeneration in a rabbit critical-size defect model. *Acta Biomater.*, **6** (3), 900–908.
82. Kasten, P., Vogel, J., Geiger, F. *et al.* (2008) The effect of platelet-rich plasma on healing in critical-size long-bone defects. *Biomaterials*, **29**, 3983–3992.

83. Lew, D., Farrell, B., Bardach, J. and Keller, J. (1997) Repair of craniofacial defects with hydroxyapatite cement. *J. Oral Maxillofac. Surg.*, **55** (12), 1441–1449.

84. Verma, D., Katti, K.S. and Katti, D.R. (2010) Osteoblast adhesion, proliferation and growth on polyelectrolyte complex-hydroxyapatite nanocomposites. *Philos. Trans. A Math. Phys. Eng. Sci.*, **368** (1917), 2083–2097.

85. Jones, G.L., Walton, R., Czernuszka, J. *et al.* (2010) Primary human osteoblast culture on 3D porous collagen-hydroxyapatite scaffolds. *J. Biomed. Mater. Res. A*, **94** (4), 1244–1250.

86. Xue, D., Zheng, Q., Zong, C. *et al.* (2010) Osteochondral repair using porous poly(lactide-co-glycolide)/nano-hydroxyapatite hybrid scaffolds with undifferentiated mesenchymal stem cells in a rat model. *J. Biomed. Mater. Res. A*, **94** (1), 259–270.

87. Kruyt, M.C., Stijns, M.M., Fedorovich, N.E. *et al.* (2004) Genetic marking with the DeltaLNGFR-gene for tracing goat cells in bone tissue engineering. *J. Orthop. Res.*, **22** (4), 697–702.

88. Peng, J., Wen, C., Wang, A. *et al.* (2011) Micro-CT-based bone ceramic scaffolding and its performance after seeding with mesenchymal stem cells for repair of load-bearing bone defect in canine femoral head. *J. Biomed. Mater. Res. B Appl. Biomater.*, **96** (2), 316–325.

89. Schopper, C., Ziya-Ghazvini, F., Goriwoda, W. *et al.* (2005) HA/TCP compounding of a porous CaP biomaterial improves bone formation and scaffold degradation-- a long-term histological study. *J. Biomed. Mater. Res. B: Appl. Biomater.*, **74** (1), 458–467.

90. Yuan, H., de Bruijn, J.D., Zhang, X. *et al.* (2001) Bone induction by porous glass ceramic made from bioglass (45S5). *J. Biomed. Mater. Res.*, **58** (3), 270–276.

91. Vallet-Regí, M., Ragel, C.V. and Salinas, A.J. (2003) Glasses with medical applications. *Eur. J. Inorg. Chem.*, **2003** (6), 1029–1042.

92. Li, P., Ohtsuki, C., Kokubo, T. *et al.* (1993) Effects of ions in aqueous media on hydroxyapatite induction by silica gel and its relevance to bioactivity of bioactive glasses and glass-ceramics. *J. Appl. Biomater.*, **4** (3), 221–229.

93. Salinas, A., Esbrit, P. and Vallet-Regí, M. (2013) A tissue engineering approach based on the use of bioceramics for bone repair. *Biomater. Sci.*, **1**, 40–51.

94. Wheeler, D.L., Stokes, K.E., Park, H.M. and Hollinger, J.O. (1997) Evaluation of particulate Bioglass in a rabbit radius ostectomy model. *J. Biomed. Mater. Res.*, **35** (2), 249–254.

95. Oonishi, H., Hench, L.L., Wilson, J. *et al.* (1999) Comparative bone growth behavior in granules of bioceramic materials of various sizes. *J. Biomed. Mater. Res.*, **44** (1), 31–43.

96. Fu, Q., Rahaman, M.N., Bal, B.S. *et al.* (2010) In vivo evaluation of 13-93 bioactive glass scaffolds with trabecular and oriented microstructures in a subcutaneous rat implantation model. *J. Biomed. Mater. Res. A*, **95** (1), 235–244.

97. Välimäki, V.V., Yrjans, J.J., Vuorio, E.I. and Aro, H.T. (2005) Molecular biological evaluation of bioactive glass microspheres and adjunct bone morphogenetic protein 2 gene transfer in the enhancement of new bone formation. *Tissue Eng.*, **11** (3-4), 387–394.

98. Välimäki, V.V., Yrjans, J.J., Vuorio, E. and Aro, H.T. (2005) Combined effect of BMP-2 gene transfer and bioactive glass microspheres on enhancement of new bone formation. *J. Biomed. Mater. Res. A*, **75** (3), 501–509.

99. Conejero, J.A., Lee, J.A. and Ascherman, J.A. (2007) Cranial defect reconstruction in an experimental model using different mixtures of bioglass and autologous bone. *J. Craniofac. Surg.*, **18** (6), 12.

100. Hench, L.L. and Polak, J.M. (2002) Third-generation biomedical materials. *Science*, **295** (5557), 1014–1017.

101. Fujikawa, K., Sugawara, A., Kusama, K. *et al.* (2002) Fluorescent labeling analysis and electron probe microanalysis for alveolar ridge augmentation using calcium phosphate cement. *Dent. Mater. J.*, **21** (4), 296–305.

102. Sugawara, A., Fujikawa, K., Kusama, K. *et al.* (2002) Histopathologic reaction of a calcium phosphate cement for alveolar ridge augmentation. *Biomed. Mater. Res.*, **61** (1), 47–52.

103. Comuzzi, L., Ooms, E. and Jansen, J.A. (2002) Injectable calcium phosphate cement as a filler for bone defects around oral implants: an experimental study in goats. *Clin. Oral Implants Res.*, **13** (3), 304–311.

104. Losee, J.E., Karmacharya, J., Gannon, F.H. *et al.* (2003) Reconstruction of the immature craniofacial skeleton with a carbonated calcium phosphate bone cement: interaction with bioresorbable mesh. *J. Craniofac. Surg.*, **14** (1), 117–124.

105. Ko, H.C., Milthorpe, B.K. and McFarland, C.D. (2007) Engineering thick tissues--the vascularisation problem. *Eur. Cell. Mater.*, **14**, 1–18; discussion 18–19.

106. Leach, J.K., Kaigler, D., Wang, Z. *et al.* (2006) Coating of VEGF-releasing scaffolds with bioactive glass for angiogenesis and bone regeneration. *Biomaterials*, **27**, 3249–3255.

107. Leach, J.K. and Mooney, D.J. (2008) Synthetic extracellular matrices for tissue engineering. *Pharm. Res.*, **25**, 1209–1211.

108. Unger, R.E., Sartoris, A., Peters, K. *et al.* (2007) Tissue-like self-assembly in cocultures of endothelial cells and osteoblasts and the formation of microcapillary-like structures on three-dimensional porous biomaterials. *Biomaterials*, **28**, 3965–3976.

109. Levenberg, S., Rouwkema, J., Macdonald, M. *et al.* (2005) Engineering vascularized skeletal muscle tissue. *Nat. Biotechnol.*, **23**, 879–884.

110. Choong, C.S., Hutmacher, D.W. and Triffitt, J.T. (2006) Co-culture of bone marrow fibroblasts and endothelial cells on modified polycaprolactone substrates for enhanced potentials in bone tissue engineering. *Tissue Eng.*, **12**, 2521–2531.

111. Aguirre, A., González, A., Planell, J.A. and Engel, E. (2010) Extracellular calcium modulates in vitro bone marrow-derived Flk-1+ CD34+ progenitor cell chemotaxis and differentiation through a calcium-sensing receptor. *Biochem. Biophys. Res. Commun.*, **2** (393), 156–161.

112. Aguirre, A., González Vázquez, A. *et al.* (2012) Control of microenvironmental cues with a smart biomaterial composite promotes endothelial progenitor cell angiogenesis. *Eur. Cells Mater.*, **01** (24), 90–106.

113. Nordin, M. and Frankel, V.H. (eds) (2001) *Basic Biomechanics of the Musculoskeletal System*, 3rd edn, Philadelphia, Lippincot Williams and Wilkins.

12

Ceramics for Drug Delivery

Miguel Manzano
Departamento de Química Inorgánica y Bioinorgánica, Facultad de Farmacia,
Universidad Complutense de Madrid, CIBER de Bioingeniería, Biomateriales y
Nanomedicina (CIBER-BBN), Spain

12.1 Introduction

The U.S. Food and Drug Administration (FDA) defines drugs as articles intended for use in the diagnosis, cure, mitigation, treatment, or prevention of disease and those articles also intended to affect the structure or any function of the body of man and other animals. In this sense, a medication or medicine is a drug employed to cure or attenuate any symptoms of a disease or medical condition. According to that, there are many different types of drugs for a great variety of purposes, such as analgesics, antibiotics, anti-inflammatories, chemotherapeutic agents, and so on.

However, although novel and more powerful pharmaceutical agents are being developed, more and more interest is being devoted to the methods in which these drugs are administered. In this sense, there is a clear need for prolonged and better control of drug administration. This has motivated a novel area of research called drug delivery technology that offers a great potential in the world of biomedicine. The classical drug administration routes are based on a systemic approach, where drugs are delivered to the bloodstream and then distributed all over the body. This approach presents some drawbacks, which include possible systemic toxicity with associated renal and liver problems and poor penetration into the targeted tissue. On the other hand, drug delivery systems (DDSs) allow delivery of the drug locally, reducing the side effects associated with the systemic administration and limiting the chance of overdose because the high drug concentration is located at the target tissue rather than throughout the body.

Bioceramics with Clinical Applications, First Edition. Edited by María Vallet-Regí.
© 2014 John Wiley & Sons, Ltd. Published 2014 by John Wiley & Sons, Ltd.

12.2   Drug Delivery

In conventional drug administration, such as oral tablets or intravenous injection, the drug concentration in the blood increases when the drug is taken, reaching a peak and then starting to decline (Figure 12.1). However, every drug presents a plasma concentration above which it is toxic and below which it is ineffective. In this sense, pharmaceutical agents should ideally maintain effective concentrations in plasma without reaching toxic levels. Although it is relatively easy to say, there are several reasons that impede reaching this goal with traditional therapies, such as, the great quantity of drugs that has to be injected to achieve the therapeutic effect over time; the low bioavailability of these drugs in areas with poor blood supply; and the difficulty of drug molecules to cross certain biological barriers [1].

Novel DDS, some of which are already commercially available, can maintain the drug concentration in plasma within the desired therapeutic range with just one dose [2]. Additionally, they can localize the delivery of the drug to certain tissues of the body, which reduces the systemic drug distribution throughout the whole body and potential side effects. In fact, DDS can be defined as those formulations that control the rate and period of drug delivery targeting specific areas of the body. DDS can also protect the drugs from degradation by the body, increasing the efficiency of the whole system and, therefore, increasing the patient's comfort.

Pharmacological features of traditional drugs have been improved when employed within DDS, which are engineered to enhance the pharmacokinetics and biodistribution of those drugs and/or work as drug reservoirs [3]. Both nanotechnology and biomaterials research areas have contributed widely to developing DDS that present (i) enhanced delivery of drugs that are poorly soluble in water; (ii) targeted delivery capabilities, so drugs can be sent to a cell or tissue in a specific manner; (iii) the possibility of achieving delivery of macromolecules to intracellular sites of action without previous degradation; (iv) the opportunity to deliver two or more drugs which allows the designing of combination therapies; (v) the possibility to combine therapies to allow the visualization of sites of drug delivery by the combination of drugs with imaging agents; and (vi) the reliability of real-time *in vivo* efficacy of the therapeutic agent [4]. Since the first drug delivery combination product approved by the FDA in the 1970s (an intrauterine device for contraception), over 70 controlled DDSs have been marketed (Figure 12.2) [5].

As it has been mentioned throughout this book, bioceramics are excellent candidates as materials for implantating into bone defects. However, those defects might have been

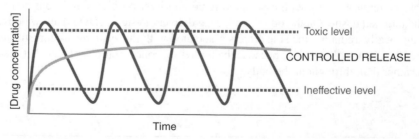

Figure 12.1 *Schematic representation of drug concentration versus time for traditional drug release systems (pulsatile) and a controlled drug release system. Adapted with permission from [5]. Copyright © 2012, Elsevier*

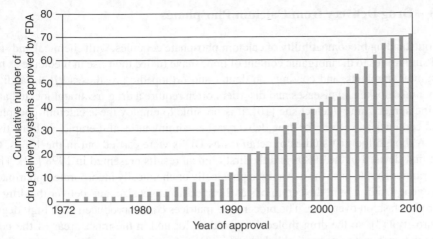

Figure 12.2 *Cumulative number of controlled drug delivery systems approved by the FDA per year [5]*

caused by many different situations or diseases, and in many situations the implantation is accompanied by administration of an appropriate drug. In this sense, implantable bioceramics can exploit the benefits of local drug delivery technology, overcoming the frequent problems of oral administration, such as the need for high dosage because of the poorly irrigated bone tissue. Thus, the synergetic effect of bioactive behavior and local drug delivery capabilities makes bioceramics exceptional materials for many different bone therapy purposes [6].

In general, any potential material to be employed as a drug or biomolecule delivery system must have the capacity to incorporate the drug or biomolecule, retain it without degrading or losing its activity, and deliver that drug progressively with time in the desired location. The method employed for loading the drug molecules into the carrier depends on the drug adsorbed but, especially, on the carrier material, since it is a process that should not modify massively the properties of the carrier. The release of the cargo from those matrices depends on many parameters, such as the micro/nanostructure of the carrier, the solubility of the adsorbed drug, the type of interaction between the host (matrix carrier) and the guest (adsorbed drug), and the mechanism of the degradation of the matrix. Those parameters and their influence on the release kinetics are normally evaluated *in vitro* following different experimental protocols before any *in vivo* application. The methods traditionally employed include the release on water, phosphate buffered saline (PBS) solution or simulated body fluid (SBF) at 37 °C, with or without magnetic stirring, and with or without refreshing the solution after a while [7]. Correlations of *in vitro* and *in vivo* drug release kinetics are uncommon because when dealing with *in vivo* situations there are many different external parameters that can influence the release kinetics [8]. However, the *in vitro* experiments are totally necessary because the obtained results only depend on the system, so they are a powerful tool to improve and implement the carrier system.

The development process of a DDS must cover different aspects. First, the drug loading must be verified. Secondly, the kinetics of the drug release have to be verified *in vitro*. In a final step, it is necessary to evaluate the clinical performance of the whole system.

12.3 Drug Delivery from Calcium Phosphates

The outstanding biocompatibility of calcium phosphate ceramics, with chemical and structural similarities to the inorganic content of bone, have fueled their use in medicine as bone substitutes, implants, and coatings on dental and orthopedic prostheses [9]. In addition, many musculoskeletal diseases and disorders often require a drug treatment located at the specific surgery/injury/defect site [10]. It is possible to employ these calcium-phosphate-based ceramics as vectors for drug delivery when aiming at local therapies in bone tissue [11]. Although the first studies of ceramics as DDSs were carried out in the 1930s, they were first devised in the 1980s and the first clinical results presented in 1998 [12]. Those bioceramics should have the capacity to physically or chemically incorporate a pharmaceutical agent, retain it until the ceramic is at the specific target size, and deliver the drug in a controlled fashion over time. The bioceramic matrices can be produced with high degrees of porosity [13], so the drug molecules could be located in the inner areas of the pores, or the therapeutic agents can be included in calcium phosphates by simple compaction of powered mixtures. However, in the former porous ceramics, the mechanical properties of the implant are deteriorated by the open structure, which limits these materials to fillers of small bone defects. In any case, the pores should be large enough to accommodate the drug molecules, which explains the efforts made by researchers in designing the pores of the bioceramics, controlling their number, size, shape, distribution, connectivity, and possible decoration, depending on the drug to be adsorbed [1].

Nowadays, one of the main areas of research in bone tissue engineering is centered on the design and processing of porous and biodegradable three-dimensional scaffolds made of bioceramics. There is also a growing trend for using these scaffolds as reservoirs of growth factors to be released once implanted, providing signals at local injury sites and stimulating cells to migrate and start the healing process. In an effort to enhance the osteoconductive and vascularization properties of bioceramic scaffolds, many biomolecules have been introduced to be then released, such as bone morphogenic proteins (BMPs), transforming growth factors (TGFs), basic fibroblast growth factors (bFGFs), and vascular endothelial growth factors (VEGFs) [14].

These scaffolds are becoming very popular because they provide not only the needed environment and structural integrity for bone regeneration, but also present the ability to control the dose and kinetics of the biomolecules release [15].

Traditional syntheses of calcium phosphate ceramics are normally based on thermal treatments, which means that the drug molecules must be loaded at the end of the process through impregnation technologies so they will not be degraded during the synthesis. However, more modern low-temperature synthetic procedures, such as precipitation methods, sol–gel chemistry, biomimetic processes, or supramolecular templating strategies, allow introduction of the drug molecules at different stages of the process, which represents an attractive alternative for employing calcium phosphates as DDSs.

12.3.1 Drug Delivery from Hydroxyapatite

Hydroxyapatite, $Ca_{10}(PO_4)_6(OH)_2$, is the most used calcium-phosphate-based bioceramic for implant fabrication because it is the most similar material to the inorganic component of bones [16]. Its potential use as a DDS has attracted significant interest because of the simultaneous use as bone substitute and drug delivery vector. For this type of application, hydroxyapatite ceramics need to present a porous structure and they are commonly produced using organic porogens that are burned out once they form the pores [17–20]. However, these porous bioceramics present mechanical properties below the necessary level, which limits their use to the filling of small bone defects. They have been successfully employed in the form of porous particulate products, which are commercially available, mainly in the oral surgery market, and have been investigated as biomolecules delivery systems [21–23].

Additionally, porous hydroxyapatite scaffolds have gained attention in the last few years because they can release drugs in a slow, local, and continuous fashion [14, 24, 25]. This technology is of particular interest when approaching diseases such as osteomyelitis, which is an inflammatory process accompanied by bone destruction caused by infective microorganisms. Conventional treatment of bone infection is based on surgical debridement and removal of possible foreign bodies, followed by antibiotic therapy. However, traditional routes of antibiotic administration consisting of intravenous injections and oral pills for long periods of time might cause systemic toxicity of the antibiotic. In this regard, local delivery of the anti-microbial agents can achieve elevated antibiotic concentrations at the site of infection without systemic toxicity. Using hydroxyapatite implants can fill the dead space after bone resection and guide its repair at the same time as locally releasing antibiotics to completely eradicate the infection. Additionally, the employment of hydroxyapatite avoids a second operation for the removal of the implant, as happens with other metallic materials. These hydroxyapatite implants should be porous, so the drug molecules can be adsorbed, to be then released once they are implanted. This approach, which was first evaluated using rats as animal models back in 1992 [26], has been explored with hydroxyapatite porous scaffolds implanted into osteomyelitis-induced rabbits [27], resulting in a great eradication of infection and new bone formation.

Besides the porous scaffolds morphology, many other variations have been produced through different powder processing techniques, such as sol–gel synthesis, solid state reactions, co-precipitation methods, hydrothermal reactions, microemulsion processing techniques, and mechanochemical synthesis [28]. Among those morphologies, hydroxyapatite nanoparticles can be employed as drug carriers [29], showing improved performances compared with conventional materials due to their large surface-to-volume ratios [30]. Hydroxyapatite nanoparticles can be produced through different methods that can be classified in two large groups: dry-synthesis (solid-state) and wet-synthesis [31]. Dry synthetic methods usually lead to stoichiometric and well-crystallized materials, although they require high temperatures and long treatment time. On the other hand, wet synthetic methods proceed at relatively low temperature but their crystallinity and Ca/P ratio are very low. In any case, hydroxyapatite nanoparticles have been investigated as a carrier of different biomolecules, such as growth factors [32], antibiotics [33–35], anticancer drugs [36–38], and even as gene carriers for the promising gene therapies [39]. These later therapies are based on the treatment of genetic diseases through manipulation of the genetic material, such as introducing genes into cells in a process called transfection. A variety of calcium phosphate nanoparticles have been observed to introduce DNA into the cells thanks to the great adherence of phosphate groups from nucleic acids to calcium phosphate [40–42].

Since hydroxyapatite is widely used as a bone substitute material, its combination with growth factors has been investigated to favor the bone regeneration process. It is possible to employ hydroxyapatite particles as carriers of growth factors, such as BMPs, bFGFs, and TGFs-β; in the same way as they can carry pharmaceutical agents. The release kinetics in this case would depend on the degradation of the hydroxyapatite particles, which could be regulated with the synthesis temperature [32].

12.3.2   Drug Delivery from Tricalcium Phosphates

Among the great diversity of calcium phosphates available, the Ca/P ratio is a very important parameter that strongly influences the ceramic properties. Tricalcium phosphate, TCP, $Ca_3(PO_4)_2$ show an atomic Ca/P ratio of 1.50, which makes it more reactive (more resorbable) than hydroxyapatite [16]. Traditionally, clinicians use mixtures of calcium phosphates formed by hydroxyapatite and β-TCPs, in the so-called β-TCP mixtures [43–47]. These calcium phosphates are being used to regenerate bone tissue throughout the body, covering all areas of the skeleton [48]. As has been mentioned throughout this book, these calcium phosphates are very versatile, because they can be employed as injectable materials, as coatings of other surfaces, as bulk materials in bone tissue engineering, or as bone defect fillers [49]. From the point of view of their potential application as DDSs, in this section we will refer to bulk TCPs, leaving a separate section for cements and coatings due to the clinical importance that they are gaining in the last few years.

12.3.3   Drug Delivery from Calcium Phosphate Cements

Calcium phosphate cements (CPCs) consist of a combination of calcium orthophosphates that when mixed with an aqueous solution forms a paste, which will set and harden after being implanted in an osseous defect. The use of CPCs as bone substitutes, that has been shown in the previous chapter of this book, together with the possibility of drug incorporation throughout the whole material volume, make these materials potential candidates as drug carriers for bone tissue engineering. Additionally, the self-setting ability at low temperature of these cements, detailed in a previous chapter, has powered their use as injectable delivery systems. It is for this reason that CPCs can be employed as osteoconductive bone grafts and local controlled DDSs at the same time (Figure 12.3). In fact, CPCs can be very useful in treatments of a variety of skeletal diseases, such as bone tumors, osteoporosis, or bone infections, because they can be employed as carriers of many pharmaceutical agents and growth factors [7, 50].

12.3.3.1   Drug Loading to Calcium Phosphate Cements

There are two important properties of CPCs that favor the incorporation of biologically active molecules: the hardening process takes place at room temperature, so there is no thermal degradation or loss of activity of the drugs, and the intrinsic porosity of the ceramic cements, which allows the incorporation of molecules within their network.

There are different methods to incorporate the drugs into the bioceramic cement that determine their distribution and interaction with the cement. They can be incorporated by: (i) blending the drug powder with the solid phase, which is the easiest way; (ii) dissolving the drug and adding the solution to the liquid phase of the cement, so a more homogeneous distribution of the drug within the cement will be achieved; (iii) dissolving the drug and

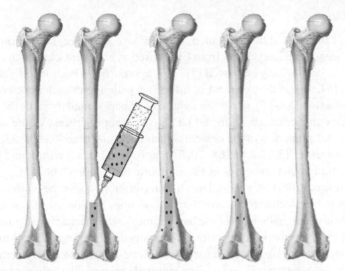

Figure 12.3 *Scheme of an injectable calcium phosphate cement with bone graft and drug delivery capabilities*

impregnating the solid blocks or granules of the cement; and (iv) loading the drug into polymer microspheres that can be added to the cement mixture, with the benefits of a better control over the drug release thanks to the polymer.

It is important to take into account that the drug addition can modify the setting kinetics of the cement, influencing its rheological properties and microstructural development, which, undoubtedly, would have an effect on the final mechanical properties of the cement [51]. However, it is impossible to predict that effect, since it depends on the chemistry and size of the adsorbed drug. Also the nature of the drug molecule will have a strong influence on the type of interaction with the calcium and phosphate ions from the cement matrix, such as inducing co-precipitation during setting [52] or complexing calcium ions, for example, when loading tetracyclines [53]. In a similar way, it is also important when adsorbing the drug to take into account how the variety of phenomena taking place during the cement setting, pH and ion concentration variation, could affect the functionality of the drug.

12.3.3.2   Drug Release from Calcium Phosphate Cements

The drug release from CPCs is mainly controlled by the diffusion of the drug molecules through the matrix, since the cement's degradation is too slow in comparison with the rate of drug release [50]. Besides, CPCs present connecting pores that would be permeated by the physiological media once implanted, and the drug release takes place by diffusion through the physiological fluids filling the pores. That diffusion depends on the porous structure of the material, so establishing mathematical models for release kinetics will depend on each type of CPC employed.

Regarding the type of biomolecules that can be loaded into CPCs, they are mainly related to the applicability of these cements as synthetic bone grafts, such as antibiotics, anti-inflammatories, anti-cancer drugs, anti-osteoporotics, proteins, and growth factors.

12.3.3.3   Antibiotics

The release of different antibiotics, such as aminoglycosides, cephalosporins, glycopeptides, quinolones, and tetracyclines, from CPCs used as bone replacement in traumatology or dentistry has been widely employed [7]. This approach has been investigated after surgical therapy [54, 55], in the treatment of infectious pathologies such as osteomyelitis and orthopedic infection [56–67], or for the treatment of periodontal diseases [68–70].

As has been commented above for all CPCs, the loading process of the antibiotic can modify important properties of the cements, such as setting time [53, 54, 58, 65, 69–75], mechanical properties [53, 62, 65, 68, 70, 76], porosity [71], microstructure [55, 58, 77], or even the antimicrobial properties of the antibiotic itself [60, 62, 66, 67].

Antibiotic release from CPCs should reach concentrations above the minimum inhibitory concentration to be effective for a certain period of time. The kinetic release of the drug is faster than the dissolution kinetics of the bioceramic, so the mechanism is governed by diffusion. In the case of orthopedic infection, commonly employed approaches to prevent and treat such infections are based on the use of antibiotic-impregnated polymethylmethacrylate. However, a second intervention is often required to remove the polymer implant. Using an absorbable material such as CPC will avoid that second intervention and also offer better elution of different antibiotics than from the polymer, maintaining higher antibiotic concentration for longer periods of time [63, 66]. However, when antibiotics are incorporated into CPCs there is a high probability of a very fast release (burst effect), especially when they are loaded in the solid state. A way to delay the antibiotic release so the effect can be prolonged over time and the burst release avoided has been explored through the incorporation of antibiotic into polymer spheres that are added to the cement formulation [56, 64].

In vivo animal models consisting of rabbits with induced osteomyelitis have demonstrated the efficiency of CPCs with gentamicin delivery capability. After three weeks of implanting those materials in the animals, the infection disappeared [62] or there was a significant reduction [61], while those animals treated systemically with gentamicin still presented infection.

12.3.3.4   Antiinflammatories

Inflammation is a process that commonly follows implantation, so introducing anti-inflammatory drugs into the implanted CPCs would help to reduce this problem. This approach has also been employed for the treatment of rheumatoid arthritis and chronic articular rheumatism [78, 79].

12.3.3.5   Anti-cancer Drugs

The treatment of bone tumors can be approached by surgery, radiation therapy or chemotherapy, or a combination of them. However, the systemic administration of anti-tumoral drugs shows a high toxicity for the rest of the body. The development of a carrier system that could be implanted and deliver those anti-cancer drugs locally rather than systemically will undoubtedly avoid the severe side effects of toxic oncologic chemotherapy. CPCs have been investigated as carriers for a variety of anti-cancer drugs, such as cisplatin [80, 81], doxorubicin [82], paclitaxel [83], methotrexate, or mercaptopurine [84, 85], at the same time as helping to reconstruct the bone defects after bone tumor resection. In this type of treatment, the delivery of the anti-tumoral drug right to the place where it is needed is even more important than in other diseases, and it is for this reason, together with some positive results in preliminary studies, that a great evolution of CPCs for the treatment of bone cancer is expected in the next few years.

12.3.3.6 Anti-osteoporotics

Osteoporosis is a bone disease that generates a decrease in bone mass and bone density, which results in an increment in the risk of fractures. The disease is caused by decreasing osteoblast activity together with increasing osteoclast activity. Traditional treatment methods are based on hormone replacement therapy, such as calcitonin [86] and anti-osteoporosis drugs, such as bisphosphonates, which are the only small molecular weight drug adsorbed into CPCs for the treatment of osteoporosis [87–92].

The main biotechnological application of CPCs as regeneration materials is based on their capacity to stimulate bone tissue regeneration. This usage has inspired their employment as antibiotics carriers, since they would prevent infections that might have been produced during surgical interventions, or in the treatment of bone infections.

It seems straightforward that the osteoconductivity features of these cements could be enhanced by the addition of certain molecules that upon local release could improve the bone growing abilities. Hence, growth factors, such as TGF-β1 and BMP, have been incorporated into CPCs, and the preclinical experiments, both *in vitro* and *in vivo*, have demonstrated an improved osteoconductivity, promoting the local bone ingrowth [93–95].

12.4 Drug Delivery from Silica-based Ceramics

The most relevant silica materials with medical applications are bioactive glasses, which are amorphous materials traditionally produced following the classic quenching of melts comprising SiO_2 and P_2O_5 as network formers and CaO and Na_2O as network modifiers [96]. However, the synthesis of bioactive materials has shifted toward the use of the sol–gel process because it allows a greater versatility in composition and structure [97]. This method allows the preparation of porous glasses and, consequently, has fueled their use as DDSs, as will be discussed in this section.

12.4.1 Drug Delivery from Glasses

In 1971, Hench proposed bioactive glasses as implantable materials for the first time [98]. These so-called bioactive glasses are highly reactive in physiological environments and can bond and integrate with living bone [99]. Additionally, they do not promote inflammation or toxicity while promoting the generation of bone matrix. For these reasons, bioactive glasses have been used clinically, mainly for filling osseous cavities, maxillofacial reconstruction, and dental applications [96].

The quick bioactive kinetics of glasses has inspired their use as bioactive accelerators of mineral apatites. However, their high reactivity when in contact with aqueous solution, together with their dense conformation, makes their application as DDSs very difficult. If a drug or biomolecule is anchored to the surface of the glass, the high reactivity of the glass in an aqueous medium would produce the quick release of those agents right after implantation. Biomolecules cannot be introduced into the network of the bulk material because the materials are normally produced at high temperature, at which the biomolecule or drug would degrade. Moreover, the fact that they are dense prevents loading the drug by impreganation after the synthesis. All these features inspired researchers to find an alternative synthetic method to produce bioglasses at a relatively low temperature, leading to the use of the sol–gel process.

12.4.1.1   Drug Delivery from Sol–Gel Glasses

As has been detailed before in this book, the sol–gel process is a chemical synthetic method where an oxide network is formed by polymerization reactions of chemical precursors [100]. From the perspective of drug delivery technology, the real advantage of this method is the low temperature of the process, since it allows introduction of organic molecules within the structure without thermal degradation. Additionally, thanks to the sol–gel technology, it is possible to control the textural properties, which allows the production of porous bioglasses, as will be detailed in the next section .

One of the major applications of bioglasses is in the treatment of bone defects because of their good osteoconductive properties. Moreover, their osteogenic potential can be enhanced through the incorporation of BMPs to induce osteoblast differentiation and, therefore, induce bone regeneration in critical defects. In this regard, the surface of bioglasses can be decorated with a variety of chemical groups to control the release of osteogenic agents. This approach has been used to release human BMP-2 to locally stimulate the production of bone, as an efficient alternative to systemic administration of TGFs [101]. A similar technology of bonding proteins to the surface of glasses has also been employed to covalently graft alkaline phosphatase, an enzyme involved in bone formation and mineralization, onto the surface of sol–gel glasses to enhance their bioactivity [102, 103]. Similarly, decoration of the glass surface with molecules to enhance their *in vivo* response can be done through the grafting of certain molecules that might help and promote cell attachment. This has been achieved by attaching collagen to the surface of a bioglass, improving cell attachment, and enhancing their applicability as tissue engineering materials [104].

Drug delivery technology has been applied in bioglasses to release antibiotics locally, that is, at the implantation site, in an effort to prevent implant-associated infections. When colonization by infectious microorganisms takes place, systemic administration of antibiotics does not always reach efficient concentrations at the infection site, mainly due to poor blood flow in the bone tissue. In that case, large antibiotic doses are needed, with the subsequent drawbacks to the rest of the body. Thus, the possibility of releasing antibiotics from the implant itself is a straightforward advantage because it is possible to obtain high drug concentrations where they are needed, that is, at the implant site. Further, this approach seems an important advance in orthopedics because it avoids the need for a second intervention to remove the implant because of infection. This approach was investigated with a bioglass as bone regenerator and gentamicin, a commonly employed antibiotic in traumatology, as the adsorbed drug [105]. The drug incorporation into the implant was carried out at room temperature, avoiding gentamicin degradation, through uniaxial and isostatic pressure. These implantable pieces were tested in an animal model using New Zealand rabbit femurs and in all cases the bone response was perfect osseous-integration, cortical, and sponge osseous tissue growth that allowed the osseous defect restoration and partial

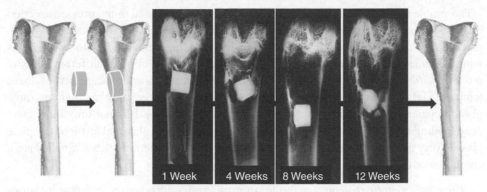

Figure 12.4 *Scheme of the implantation of a bioglass into rabbit femurs. Reproduced with permission from [105]. Copyright © 2002 Wiley Periodicals, Inc.*

resorption of the implant at a medium time length (Figure 12.4). Additionally, gentamicin levels in different organs (liver, kidney, and lungs) and in proximal and distal bone were evaluated, being above the minimal inhibitory concentration at osseous tissue. This level was observed to decrease over time, from 1 to 12 weeks, but was always above the minimal levels necessary to impede the microorganisms' development. Thus, the use of this implant technology avoids any possible infection process during and after material implantation while bone regeneration is achieved.

An innovative and promising approach that involves bioactive glasses with drug delivery potential has been presented recently, although to date it looks far from clinical use. The concept is based on a sol–gel glass with a double function: to favor the repair of bone tissue and provide *in situ* release of functionalized gold nanoparticles that can carry drugs attached to their surface [106]. The advantage of this system over surface-loaded sol–gel glasses is that while the latter release their cargo quickly due to the high reactivity of the glasses, the former offers a sustained release, since gold nanoparticles are located within the network of the glass, not just at its surface. Additionally, gold nanoparticles have been widely studied as DDSs [107], which adds additional control to this new and promising technology. These gold nanoparticles also allow the immobilization of different enzymes on the glass surface through the formation of self-assembled monolayers (SAMs) [108]. The enzyme evaluated so far has been Soybean peroxidase, and when comparing with merely adsorbed enzyme on the surface, a decrease in the oxidative stress of human osteoblasts cells was observed.

12.4.1.2 Drug Delivery from Templated Glasses

Although dense bioglasses have been investigated as DDSs, the generation of certain porosity into these glasses increases their potential, as happens in all bioceramics. In this sense, the combination of the sol–gel technology for the synthesis of glasses with the supramolecular chemistry of surfactants led to a novel type of highly ordered mesoporous materials for biomedical applications, both tissue engineering and drug delivery. As has been commented previously in this book, this new generation of materials, called ordered mesoporous bioactive glasses or templated glasses, show an enhanced bioactive behavior compared to conventional glasses [109]. Additionally, their excellent surface properties and high porosity together with their surface chemistry that allows easy organic modification, has popularized these templated glasses in the drug delivery world.

The problem of bone infection in orthopedic surgery mentioned in the previous section inspired the use of mesoporous bioglasses because of their regeneration capability and, particularly, their drug delivery potential. In fact, loading gentamicin into two glasses with the same chemical composition ($58SiO_2$-$37CaO$-$5P_2O_5$ mol%) but different porosity properties (dense and mesoporous), highlights the great adsorption capacity of these template glasses, which are able to load up to three times more gentamicin than the same glass prepared by conventional sol–gel procedures [110, 111]. It is not only their great drug loading capacity but also their variety of release kinetics that makes these template glasses very attractive for drug delivery technology. These different release kinetics are a consequence of:

1. *Their mesoporous structure*, with the drug molecules adsorbed at different locations of the porous mesostructure and, consequently, having different interactions with the material, leading to different release patterns:

   a. entrance of the pore with drug–glass surface interaction
   b. inner part of the pore with drug–pore wall interaction + diffusion through the pore channels
   c. inner part of the pore with drug–drug interaction + diffusion through the pore channels
   d. drugs located at the external surface of the materials with drug–glass surface interaction.

   Different textural properties would obviously result in different drug loading and release patterns. Thus, producing templated glasses using different surfactants, such as P123 or F127, would lead to a totally different mesostructure of the material and, therefore, to different loading capacities and release kinetics [112].

2. *The different chemical composition of both glass material and surface after a functionalization process*. The composition of the bioglass has influence on drug adsorption and release when the molecule adsorbed interacts with some of the glass components, as happens with tetracyclines that can be chelated by the CaO present in the glass [113]. When the percentage of CaO is increased, there would be an increase in tetracyclines adsorption on the material and also the release kinetics would be retarded. On the other hand, years of experience in drug delivery technologies based on both pure silica materials and mixed oxides has demonstrated that the organic modification of the matrix surface is the most influential parameter for both drug adsorption and release. Although this is absolutely true for pure silica materials, as will be described in the next section, the problem in the case of bioglasses is that decorating the surface of the material means a change in its composition, which can decrease its bioactive behavior [114]. Thus, it is necessary to carry out the functionalization process with special care in terms of percentage and composition. However, it has been observed that is possible to functionalize template glasses with different chemical groups, such as thiol, amino, hydroxy, and phenyl, without significantly decreasing the bioactive behavior, and increasing the control on drug, in this case ipriflavone, adsorption and, particularly, release kinetics [115].

12.4.2 Drug Delivery from Mesoporous Silica

Mesoporous silicas are ordered porous materials made of SiO_2, that are characterized by a high pore volume, narrow pore size distribution, and large surface area, as has been described in Chapter 5. They have become very popular in the biomedical world since they were proposed as DDSs for the first time back in 2001 by the Vallet-Regí research group [116]. The reasons for such great popularity are based on their properties, because mesoporous silicas fulfill most of the requirements for a matrix to be used as a DDS:

1. Materials made of SiO_2 with defined structures and surface properties are well known to be biocompatible [117]. Hence, silica has been employed to produce nanoparticles with biological applications [118, 119] and has been used as an additive in artificial implants to induce osteogenic properties [120]. Additionally, silica has been used in the formulation of several DDSs: magnetic nanoparticles [121], biopolymers [122], and micelles [123].
2. Mesoporous silica materials are characterized by a network of cavities that can be filled with different cargos, such as drugs, peptides, proteins, or even tiny nanoparticles. As mentioned above, the textural parameters of mesoporous silicas have led to these materials being widely used as DDSs:

   a. high pore volume, which allows the confinement of a large variety of pharmaceutical agents or biomolecules.
   b. large surface area, which gives a great potential for molecular adsorption.
   c. well-ordered mesopore distribution, which guarantees homogeneity and reproducibility of the adsorption and release mechanisms.

3. Ordered mesoporous materials present a surface that can be easily decorated with diverse organic moieties. By organic functionalization of the surface, both at the inner walls of the pores and the external surface, it is possible to control the host–guest interaction by selection of the organic group. Thus, it is possible to modulate the interaction between the drug and the matrix, which enables control of the adsorption and release rate of the drug, an important feature in DDSs.
4. The feasible functionalization of mesoporous silicas opens the possibility of closing the pore entrances with several organic groups. In this way, it is possible to load the pores with a certain drug and then close them up, avoiding any leaking and burst effect as a consequence of a quick release once implanted. This approach allows release of the cargo on demand, that is, when the pore gates are opened under certain stimuli, in the so-called smart delivery systems.
5. Ordered mesoporous silicas are relatively easy to produce and their architecture can be easily tuned by the adequate selection of the surfactant template. Additionally, ordered mesoporous silicas are thermally and chemically very stable, which offers a variety of different possibilities within drug delivery technologies.

12.4.2.1 *Loading Molecules into Mesoporous Silica*

The main parameters controlling the adsorption of molecules into the porous mesostructure are the textural properties of the matrix, as mentioned previously, that include pore size, type of mesostructure, surface area, and surface chemistry.

The process employed for loading the cargo into the network of cavities is normally based on impregnation techniques. After synthesis of mesoporous matrices, that has been described in detail in Chapter 5, including the surfactant removal, the pores have to be emptied to allow drug molecules to be adsorbed. The residual water that might be in the inner part of the pores is eliminated through dehydration in a vacuum oven. The materials are then soaked in a concentrated solution of the drug to be adsorbed. Although aqueous solvents are preferred to avoid traces of toxicity from certain organic solvents, the solvent should be chosen depending on the solubility of the drug to be adsorbed, reaching a concentration as high as possible to promote the maximum drug impregnation over the silica walls. Magnetic stirring may be applied to maximize the molecular diffusion into the channels.

The most important requirement of the process for loading into ordered mesoporous materials is the mesostructure survival, since any change in the structure will lead to variable release kinetics of the adsorbed drug, which takes place following a diffusion mechanism. Mesostructure is assessed by small angle X-ray diffraction (XRD) and N_2 adsorption, and both analyses should match those of the matrix before the loading process [124]. Additionally, it is possible to double check the pore structure prevalence by transmission electron microscopy (TEM) of the samples before and after adsorbing the drug.

Loading the drug into mesoporous materials can lead to adsorption of the molecules on the inner part of the pores and/or the external surface of the material. Thus, it is necessary to analyze where the drug is being adsorbed. This has been carried out traditionally through indirect methods, such as N_2 adsorption analysis, measuring the pore volume reduction after the loading process. Nowadays, the advances in electron microscopy and the development of scanning transmission electron microscopy (STEM) with a spherical aberration corrector (Cs), allow sufficient resolution to achieve atomic-level analysis. This technique was employed to analyze the location of zoledronate, a potent bisphosphonate, after loading it into SBA-15 material [125]. Using a high-angle annular dark-field detector, with the electron beam illuminating the wall, confirmed that the silica walls were composed of Si and O, as expected. On the other hand, when the electron beam was oriented in a direction parallel to the pores, the composition of the material inside the pores was found to be rich in nitrogen and carbon, which can only correspond to the drug molecules loaded into the inner part of the mesopores.

12.4.2.2   Pore Diameter

During the design of the carrier, the size of the molecule to be adsorbed into the pores should be taken into consideration. In fact, those dimensions determine the type of mesoporous host matrix chosen. The pore diameter is known to be a size-selective parameter; if the drug molecule is bigger than the pore entrance, the adsorption will take place only on the external surface of the matrix; however, if the molecule is smaller than the pore entrance, the adsorption would take place both inside the pores and on the external surface of the matrix. This is clearly exemplified when large molecules such as proteins are loaded into silica mesoporous materials. Small proteins (5.1 nm) were introduced into the channels of SBA-15 matrices, while large proteins, such as ovalbumin, serum albumin, or conalbumin, did not fit through the pore's entrance and were adsorbed only on the external surface of SBA-15 matrices [126]. There are many possible approaches to increase the size of the pores to be able to introduce large molecules into them, such as addition of swelling agents to the structure-directing agents. 1,3,5-Trimethylbenzene [127], triisopropylbenzene [128], and dodecane [129], enlarge the micelles during the synthetic process by up to 30%, although there is a thin red line on the employed amount because the addition of high concentrations might result in a loss of the long-range order and in the formation of mesocellular silica foams (MCFs). Other approaches include the use of a mixture of surfactant blends with long chains as structure directing agents [130], or adding a hydrothermal treatment step during the synthetic process of the mesoporous matrix and, thus, increase the size of the mesopores of SBA-15 up to 11.4 nm, which allowed to increase the loading of serum albumin protein from 15 to 27% [131].

12.4.2.3   Surface Area

Silica mesoporous materials are well-known for their high surface areas, about $1000\,m^2\,g^{-1}$, as has been mentioned several times in this chapter, justifying the interest of the catalysis industries in these materials. In terms of drug delivery technology it is one of the most important features to take into consideration, since molecular adsorption is a surface phenomenon, that is, the loaded molecules are retained because of their interaction with the matrix surface, both inside the pores and on the external surface of the matrix. The higher the surface area, the larger the amount of adsorbed molecules.

The relevance of the surface area in drug adsorption capability is evidenced when comparing the adsorption of the same drug molecule, alendronate, into matrices with different surface areas. MCM-41 and SBA-15 were employed for this investigation because they present the same mesostructure, 2D hexagonal with *p6mm* symmetry, but different surface areas, 1157 and $719\,m^2\,g^{-1}$, respectively [132]. In this study, the size-selective adsorption parameter was not a problem because the size of alendronate is about 0.6 nm and the pore entrances of the matrices are about 3 nm for MCM-41 and 9 nm for SBA-15. After proceeding with alendronate adsorption following the same experimental procedure for both materials, the loading percentage with respect to the total weight of the matrix was 14% for MCM-41 and 8% for SBA-15, that is, the matrix with the highest surface area, MCM-41, had almost double the loading capacity of SBA-15, highlighting the importance of surface area in the loading process. Similar observations were found in the loading of captopril, a hypertension drug, onto MCM-41 matrices produced with different surface areas by using templating surfactants with different alkyl chain length [133].

12.4.2.4   Pore Volume

The pore volume of mesoporous matrices is especially relevant when a large number of adsorbed molecules or particularly large molecules are loaded. Silica mesoporous materials are characterized by large mesopore volumes, about $1\,cm^3\,g^{-1}$, which allows the uptake of a large number of different molecules into their network of cavities. An approach to fill the pores completely with an adsorbed drug is based on consecutive impregnations of the material, MCM-41 in this case, in a concentrated solution of ibuprofen [134]. After every impregnation step, the ibuprofen content was analyzed, with an increase being observed on each cycle until the pores were completely filled. During the drug adsorption there is a first stage in which the molecules are retained by their interaction with the surface, and a second stage in which the loaded ibuprofen molecules are retained by ibuprofen–ibuprofen intermolecular interactions.

When large molecules, such as certain proteins, need to be adsorbed into the pores, the same strategies are used to increase the diameter as were used to increase the pore volume. Thus, the use of swelling agents during the synthesis process leads to the formation of MCFs [135], with a pore volume increase from $1.1 \, \text{cm}^3 \, \text{g}^{-1}$ in SBA-15 up to $1.9 \, \text{cm}^3 \, \text{g}^{-1}$ in MCFs. This increment in the pore volume allows the introduction of large proteins such as bovine serum albumin into the inside of the pores [136]. As can be expected, the higher the pore volume, the greater the protein adsorption. This can be observed when comparing the loading of bovine serum albumin into SBA-15 and MCF materials, with 15 and 24% protein loading, respectively [131].

12.4.2.5   Organic Functionalization

The surface of silica mesoporous materials is rich in Si-OH groups, so the interaction with drug molecules relies on the chemical interaction of the organic groups of the drugs with the silanol groups present on the surface of the matrices. Thus, it is possible to modulate the host–guest interaction through a variety of organic modifications of the silica walls before the drug loading. This is a very easy process based on the grafting of commercially available alkoxysilanes with organic groups at one end, and the chemical character of the silica walls can be modified as required. There are many characterization techniques to confirm the grafting of the organic moiety, such as ^{29}SiNMR and FTIR [137, 138], and to quantify the extent of organic functionalization, such as elemental analysis, thermogravimetry, or X-ray photoelectron spectroscopy (XPS) [138, 139]. The organic group is normally selected depending on the chemistry of the drug, to achieve a stronger attractive interaction, which will lead to greater drug adsorption and a more sustained release [140]. This effect was observed in the pioneering work on ibuprofen adsorbed into silica mesoporous materials, MCM-41 [116]. Before any kind of organic modification, the ruling host–guest interaction was SiOH-COOH. However, the functionalization of the silica walls with amine groups changed the interaction to a stronger NH_2-COOH attracting interaction, which led to an increase in the ibuprofen adsorption and delayed the ibuprofen release kinetics.

Although pore diameter is a size-selective parameter, the pore volume needs to be large enough to accommodate the maximum amount of drugs and the high surface area of mesoporous silica leads to great molecular adsorption, the organic modification of the silica walls has been found to be the most influential parameter in drug delivery technology from these materials. Additionally, the level of organic functionalization would also have an effect on the loading capacity of the matrix [141].

12.4.2.6   Releasing Molecules from Mesoporous Silica

Silica mesoporous materials loaded with drugs are expected, as any DDS, to release their cargo with acertain level of control. DDSs can be classified into two large groups: *passive systems*, where the molecules release rate is controlled by processes such as molecular diffusion (from stable porous mesostructures) or dissolution (from degradable matrices); and *active systems*, where the drug release takes place in response to a certain stimulus that acts as a trigger (so-called smart DDSs) [142]. In the case of silica mesoporous materials, the release is due to a diffusion mechanism, since the dissolution rate is too slow in comparison with the release kinetics. In the majority of the mesoporous matrices studied, there are two differentiated areas in the cumulative release plots: an initial burst due to the quick release of the particles adsorbed at the external surface of the matrix, and a more sustained and prolonged release due to the diffusion of the drug molecules throughout the mesopores.

12.4.2.7   Pore Diameter

Since the main factor controlling the release of the molecules, ruling out the initial burst effect, is the diffusion of the drug molecules through the mesopore channels, the textural properties of the host matrices play the most important role in the release kinetics. Among those properties, pore diameter is known to act as a drug release modulator, since a reduction in the pore diameter leads to a decrease in the drug release rates. This effect has been observed when producing a similar matrix, MCM-41, using surfactant templates with different alkyl chains, which led to different pore diameters. When loading the same drug, ibuprofen, in those matrices, a reduction in the release kinetics was observed with the smaller pore entrances [143].

12.4.2.8   Surface Area

In the same way that surface area is a very important factor to take into account in drug adsorption because the adsorption process itself is a surface phenomenon, it is reasonable to guess that the reverse process, release of molecules, would also strongly depend on the surface area. Thus, when the surface area of a matrix is very high, during the diffusion process the molecules find more available area to interact with, promoting extra host–guest interaction and, therefore, reducing the release kinetics. This effect has been observed when comparing the release rate of the same drug, alendronate, from matrices with similar mesostructures but different surface areas, MCM-41 and SBA-15, in which the kinetic release was slower in the matrix with the highest surface area [132].

12.4.2.9   Organic Functionalization

Organic modification of the mesoporous matrices is a key parameter in molecular adsorption because of the possibility of tuning the host–guest interactions, as commented above, so it is also a determining parameter of the release kinetics. Through the correct organic modification it is possible to control drug release, and this has been observed in all types of ordered mesoporous silicas, independent of the type of matrix and its mesostructure. Thus, in the previously described case of two matrices with the same mesostructure but different surface area, alendronate release was found to be dependent on the surface area of the matrix. However, an even bigger difference was observed when the surface of the materials was modified with amine groups, leading to a slower release rate than with the corresponding unmodified pure silica materials [132].

In drug delivery technologies, an important tool to save time and resources in drug carrier design is a mathematical model that can predict the release kinetics as a function of the textural parameters of the matrix. In the case of silica mesoporous materials, the prediction methods allow the design of delivery vectors with predicted release kinetics, as has been shown with the experimental and theoretical matching results of zoledronate release from SBA-15 [144].

Since the publication for the first time of the use of ordered mesoporous silicas eas DDSs, many different carrier systems have been designed to adsorb and release a great variety of biomolecules (Table 12.1 and Figure 12.5).

Table 12.1   *Silica mesoporous materials and biologically active molecules employed as drug carriers*

Mesoporous carrier	Drug/biomolecule	References
MCM-41$_{12}$	Ibuprofen	[116]
MCM-41$_{16}$	Ibuprofen	[116]
MCM-41$_{12}$-NH$_2$	Ibuprofen	[145]
MCM-41$_{16}$-NH$_2$	Ibuprofen	[145]
MCM-48	Ibuprofen	[146]
FDU-5	Ibuprofen	[146]
SBA-15	Ibuprofen	[147]
TDU-1	Ibuprofen	[148]
MCM-41-MS	Ibuprofen	[149]
MCM-41-DMS	Ibuprofen	[149]
MCM-41-TMS	Ibuprofen	[150]
MCM-41-Al	Naproxen	[150]
MCM-41-Al	Diflunisal	[151]
MCM-41	Pitoxicam	[152]
MCM-41	Aspirin	[153]
MCM-41-NH$_2$	Aspirin	[153]
SBA-15	Amoxicillin	[154]
SBA-15	Gentamicin	[124]
PLGA-SiO$_2$	Gentamicin	[155]
PLGA-SiO$_2$	Gentamicin	[155]
MCM-48	Erythromycin	[146]
FDU-5	Erythromycin	[146]
FDU-5-C$_8$	Erythromycin	[146]
SBA-15	Erythromycin	[156]
SBA-15-C$_8$	Erythromycin	[156]
SBA-15-C$_{18}$	Erythromycin	[156]
MCM-41	Erythromycin	[156]
SBA-15	Itraconazole	[157]
SBA-16	ZnNIA	[158]
SBA-16	ZnPCB	[158]
MCM-41$_{12}$	Captopril	[133]
MCM-41$_{16}$	Captopril	[133]
SBA-15	Captopril	[133]
MCM-41-TMS	Captopril	[159]
MCM-41	Sertraline	[160]
MSU-3-COOH	Famotidine	[161]
MSU-15-COOH	Famotidine	[162]
MSU-15-COOH-TMS	Famotidine	[162]
MCM-41	Alendronate	[132]

Table 12.1 *Silica mesoporous materials and biologically active molecules employed as drug carriers*

Mesoporous carrier	Drug/biomolecule	References
MCM-41-NH$_2$	Alendronate	[132]
SBA-15	Alendronate	[132]
SBA-15-NH$_2$	Alendronate	[132]
SBA-15	Bovine serum albumin	[131]
SBA-15-NH$_2$	Bovine serum albumin	[131]
SBA-15-C$_3$N$^+$Me	Tryptophan	[139]
SBA-15-C$_3$N$^+$Me$_2$C$_{18}$	Tryptophan	[139]
MSU-Tween-80	Pentapeptide	[163]
MCM-41	Cisplatin	[164]
MCM-41-NH$_2$	Cisplatin	[164]
MCM-41-NH$_2$-Fol	Cisplatin	[164]

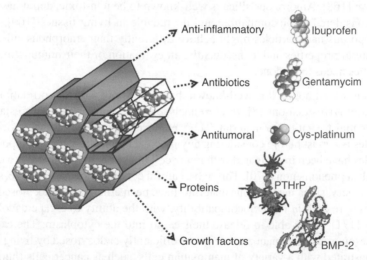

Figure 12.5 *Different drugs and biomolecules that have been loaded in ordered mesoporous silicas for drug delivery systems (See insert for color representation of the figure)*

12.4.2.10 Ordered Mesoporous Nanoparticles as Drug Delivery Systems

The production of silica mesoporous materials as nanoparticles, as described in Chapter 5, has been explored in the last few years as effective drug delivery platforms. In fact, they seem to be outstanding drug carriers related to the following features: straightforward synthesis, tunable pore architecture, thermal and chemical stability, relatively easy functionalization chemistry, low toxicity, and the potential to carry many different cargos (drugs, peptides, proteins, or nanoparticles) within their network of cavities [165–167]. These types of mesoporous nanoparticles are able to fulfill the necessary requirements to be employed as drug delivery nanoparticles, which in general are as follows: they should be biocompatible and able to evade the reticuloendothelial system, they should present a high loading capacity at the same time as protecting the cargo from degradation, and they should present a correct release pharmacokinetic together with zero premature release. Additionally, these silica mesoporous nanoparticles present an internal and external surface that can be organically modified independently [168], which means that it is possible to graft many different functionalities with multiple roles without affecting their capacity for drugs adsorption and release. Thus, silica mesoporous nanoparticles can be designed to present cell type or tissue specificity together with site-directing ability, which undoubtedly has fueled their research in the last few years.

12.4.2.11   Biocompatibility of Ordered Mesoporous Nanoparticles

The first things to take into consideration when designing any type of nanoparticle for drug delivery technology are the safety and toxicological issues. In this regard silica mesoporous materials present a similar chemical composition, SiO_2, and disordered structure at the atomic level as amorphous silica, which is approved by the EU regulations as a food additive [168]. Amorphous silica is well-known to be non-toxic, and it has traditionally been considered as biocompatible and degradable in living tissues [169]. However, silica mesoporous nanoparticles might behave differently than amorphous silica because of their textural properties and, consequently, an evaluation of their immunotoxicological characteristics needs to be done.

In vitro *Studies.*   Although it is well-known that there is not a direct correlation between *in vitro* and *in vivo* situations [8], *in vitro* analyses are an important tool to start analyzing the potential toxicity of a material. Thus, exploring the potential cytotoxicity of any nanoparticles is a must before considering any medical application, and mesoporous silica nanoparticles have been investigated with many cell lines concluding that it is a concentration dependent phenomenon [170]. Through a careful control of the particle size, shape, and surface chemistry, silica mesoporous materials have been efficiently endocytosed by living cells, resulting in high *in vitro* biocompatibility, with the ability to escape endolysosomal entrapment [171], being able to release their cargo into the cytoplasm. The experiments showing mesoporous silica nanoparticles to be efficiently endocytosed by living cells have been demonstrated with a variety of mammalian cells, such as cancer cells (human cervical cancer cells [172–176], pancreatic cancer cells [177], Chinese hamster ovarian cells [174], non-small-cell lung cancer epithelial cells [178]), non-cancer (neurological cells) [164], macrophages [178, 179], human mesenchymal stem cells [180, 181], human dendritic cells [182], and others [183]. From all these investigations, it has been concluded that particle shape and size are very important characteristics for a nanocarrier to be efficiently taken up by non-phagocytic cells [117]. One of the advantages of mesoporous silica nanoparticles over other types of nanocarriers is that they can be produced at the nanometric level with a precise control of the particle size. Moreover, they can be very easily modified on their external surface, which means that surface characteristics can be tuned on demand, such as grafting polyethylene glycol (PEG) chains, to give them stealth properties for the immune system and increase the circulation time in the bloodstream. In this sense, the possible toxicity of those silica mesoporous nanoparticles toward mammalian red blood cells has been evaluated [184, 185], presenting a good hemocompatibility and ensuring good behavior in the bloodstream.

In vivo *Studies.*   All the previously observed features that affected the *in vitro* toxicity and biocompatibility, that is, concentration, nanoparticle size and shape, surface area and surface chemistry, are factors that influence the *in vivo* response [186, 187]. Although there is some controversy about the *in vivo* toxicological results and more research is needed, there are some reports of good *in vivo* behavior. For instance, mesoporous silica nanoparticles of 100–130 nm in size have been well tolerated by mice as shown by serological, hematological, and histopathological examinations of blood samples and mouse tissues after intravenous administration of those nanoparticles [188]. Similarly, mesoporous silica nanoparticles have been employed as *in vivo* bioimaging systems without showing any toxic response in the short-term [189–192]. Finally, the administration route has been observed also as an important parameter when dealing with *in vivo* studies. Mesoporous silica nanoparticles showed good biocompatibility in rats when administered subcutaneously but resulted in death or euthanasia when intraperitoneal or intravenously injected [193]. Much more *in vivo* toxicological research is needed before any clinical phase, although the chemical versatility of mesoporous silica nanoparticles permits the development of materials with better blood circulation, biodegradability, and clearance from the body.

The different requirements for a DDS to be effective have been discussed in this chapter. However, one of the most important duties of a DDS, the Holy Grail, is to release the cargo in a suitable concentration at the desired target in a determined period of time. Mesoporous silica nanoparticles can fulfill these requirements through the so-called stimuli-responsive systems, in which the pores are closed once they have the cargo inside and then opened to release the drug molecules under certain stimuli. The gatekeeping concept was developed to close the pore entrances reversibly with diverse chemical entities to regulate the encapsulation and release of drug molecules [164]. Thanks to that concept, stimuli-responsive systems present the ability to allow cargo release on demand in response to different stimuli, such as temperature [194–198], pH [199–210], ultrasound [211], light [212–215], magnetic fields [175, 216–221], redox systems [164, 174, 175, 201, 222–227], peptides [228], enzymes [229, 230], or antibodies [231]. The major advantage of these capped drug nanocarriers is, apart from cargo protection, the zero premature release, which is very important when delivering a cytotoxic drug, as happens in cancer treatments. These systems are described in detail in Chapter 14.

12.5   Drug Delivery from Carbon Nanotubes

Carbon nanotubes (CNTs) present outstanding properties and a particular physicochemical architecture, as has been described in Chapter 7. Thanks to these outstanding properties, CNTs are being explored as an alternative platform for DDSs [232, 233]. CNTs present a considerable advantage over other drug carriers, because they can penetrate the cell membrane and, therefore, be taken up by cells (Figure 12.6). In fact, CNTs can travel through many cellular barriers and even enter the nucleus in some cases [234].

One of the major problems of CNTs for biomedical applications is their solubility in aqueous media. To counteract this, CNTs are normally oxidized under strong acidic conditions, but they can aggregate in the presence of salts present in most physiological media. Thus, further modifications are normally carried out, grafting hydrophilic polymers such as PEG [235]. A different method to stabilize CNTs is based on coating them with amphiphilic surfactant molecules adsorbed at the surface of the CNTs [236].

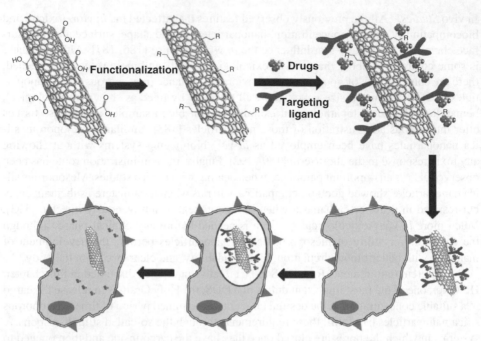

Figure 12.6 *Functionalization and drug loading on CNTs and internalization into cells to release their cargo into the cytoplasm*

Once the solubility task is solved, the next issue in their potential use as DDSs is their possible recognition as foreign particles in the bloodstream by the macrophages of the reticuloendothelial system. A proper functionalization with PEG can avoid that problem. In fact, *in vivo* experiments have demonstrated that PEG functionalization of CNTs decreases their toxicity [235].

CNTs can be functionalized to recognize a single cell, which improves the release efficiency since the carrier delivers the cargo only in the area where it is needed allowing reduction of the dosage [237, 238]. This approach is particularly interesting in the treatment of cancer, where the drugs can be encapsulated and delivered only in the tumor areas. Doxorubicin, a well-known anticancer drug usually administered intravenously in the form of a salt, has been adsorbed through non-covalent bonds onto the surface of PEG-CNTs [239–244]. Similarly, paclitaxel, an anticancer drug with very low water solubility, has been added to CNTs and *in vivo* models have shown a reduction in tumor volume in breast cancer and high tumor growth suppressing efficiency [245–248].

An imaginative approach is the encapsulation of drugs into the inner volume of the CNTs [249–252] then releasing these drugs after a chemical disintegration of the CNT cap [252]. The drug release can also be triggered by environmental changes, such as temperature and pH [249], or external stimuli, such as optical stimulation using near-infrared wavelengths (700–1100 nm), that do not affect biological structures [253].

However, one of the major drawbacks of CNTs as DDSs is their low drug loading capacity, which leads to the need to use high CNTs concentrations to achieve desired dosages, increasing the risk of possible toxicity. An imaginative solution to this problem can be by attaching to the surface of CNTs drug loaded liposomes, which have a high drug capacity [254–257]. These novel systems combine the efficient cell uptake of CNTs with the high drug-loading capacity of the liposomes. In addition, active targeting moieties can be grafted to these systems, promoting the release of drugs to diseased tissues while reducing cytotoxicity in healthy tissues [258].

Finally, these CNTs have also been employed in gene therapy, which involves the delivering of genetic materials into cells, due to their ability to cross cell and nucleus membranes and their capacity to load and release certain biologically active molecules, such as DNA [259–263].

12.6 Drug Delivery from Ceramic Coatings

Some traditional metallic implants present poor osteoconductivity and, as has been noted in Chapter 9, ceramic coatings of those implants can promote a better bioactive fixation. Additionally, these bioceramic coatings can present drug delivery capabilities, with drug molecules embedded into the coating and released locally at the implantation site (Figure 12.7) [10]. The release of antibiotics from the coated implant might prevent any type of infection after surgery, with the advantage of ensuring the release of antibiotics to the surrounding tissue at the target site, reducing associated toxicity to other areas of the body [264–267]. The effectiveness of releasing silver ions instead of antibiotics from bioceramics implants has also been proved in *in vitro* experiments [268].

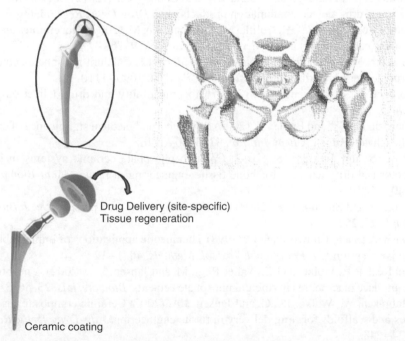

Drug Delivery (site-specific)
Tissue regeneration

Ceramic coating

Figure 12.7 *Schematic representation of metallic hip implants coated with a ceramic with drug delivery capabilities*

Bioceramic coatings of implants have also been also used to deliver biologically active molecules that can stimulate bone healing. Different growth factors, such as BMPs, have been loaded into calcium phosphate-coated orthopedic and dental implants to promote osseointegration [269]. In a similar approach, different bisphosphonates, such as zoledronate [270] or alendronate [271], have also been delivered locally from bioceramics coatings. Bisphosphonates are drugs employed in skeletal disorders inhibiting osteoclasts activity, but their systemic delivery by oral administration or intravenous injection normally generates serious side-effects and also leads to a very poor bioavailability [272]. Through the local delivery from bioceramic coatings of implants, straightforward benefits have been observed, inducing early stage osteogenesis and reducing the risk of bone fracture from osteoporosis through the inhibition of osteoclasts activity [270, 271].

References

1. Arcos, D. and Vallet-Regí, M. (2013) Bioceramics for drug delivery. *Acta Mater.*, **61**, 890–911.
2. Kost, J. and Langer, R. (2012) Responsive polymeric delivery systems. *Adv. Drug Deliv. Rev.*, **64** (Suppl.):327-341.
3. Allen, T.M. and Cullis, P.R. (2004) Drug delivery systems: entering the mainstream. *Science*, **303**, 1818–1822.
4. Farokhzad, O.C. and Langer, R. (2009) Impact of nanotechnology on drug delivery. *ACS Nano*, **3**, 16–20.
5. Couto, D.S., Perez-Breva, L., Saraiva, P. and Cooney, C.L. (2012) Lessons from innovation in drug-device combination products. *Adv. Drug Deliv. Rev.*, **64**, 69–77.
6. Vallet-Regi, M., Balas, F., Colilla, M. and Manzano, M. (2007) Bioceramics and pharmaceuticals: a remarkable synergy. *Solid State Sci.*, **9**, 768–776.
7. Ginebra, M.-P., Canal, C., Espanol, M. *et al.* (2012) Calcium phosphate cements as drug delivery materials. *Adv. Drug Deliv. Rev.*, **64**, 1090–1110.
8. Kohane, D.S. and Langer, R. (2010) Biocompatibility and drug delivery systems. *Chem. Sci.*, **1**, 441–446.
9. Dorozhkin, S.V. and Epple, M. (2002) Biological and medical significance of calcium phosphates. *Angew. Chem. Int. Ed.*, **41**, 3130–3146.
10. Bose, S. and Tarafder, S. (2012) Calcium phosphate ceramic systems in growth factor and drug delivery for bone tissue engineering: a review. *Acta Biomater.*, **8**, 1401–1421.
11. Paul, W. and Sharma, C.P. (2003) Ceramic drug delivery: a perspective. *J. Biomater. Appl.*, **17**, 253–264.
12. Dash, A.K. and Cudworth, G.C. (1998) Therapeutic applications of implantable drug delivery systems. *J. Pharmacol. Toxicol. Methods*, **40**, 1–12.
13. del Real, R.P., Wolke, J.G.C., Vallet-Regi, M. and Jansen, J.A. (2002) A new method to produce macropores in calcium phosphate cements. *Biomaterials*, **23**, 3673–3680.
14. Habraken, W., Wolke, J.G.C. and Jansen, J.A. (2007) Ceramic composites as matrices and scaffolds for drug delivery in tissue engineering. *Adv. Drug Deliv. Rev.*, **59**, 234–248.
15. Mourino, V. and Boccaccini, A.R. (2010) Bone tissue engineering therapeutics: controlled drug delivery in three-dimensional scaffolds. *J. R. Soc. Interface*, **7**, 209–227.

16. Vallet-Regi, M. and Gonzalez-Calbet, J.M. (2004) Calcium phosphates as substitution of bone tissues. *Prog. Solid State Chem.*, **32**, 1–31.

17. Orlovskii, V.P., Komlev, V.S. and Barinov, S.M. (2002) Hydroxyapatite and hydroxyapatite-based ceramics. *Inorg. Mater.*, **38**, 973–984.

18. Komlev, V.S. and Barinov, S.M. (2002) Porous hydroxyapatite ceramics of bi-modal pore size distribution. *J. Mater. Sci. Mater. Med.*, **13**, 295–299.

19. Li, Y.B., Tjandra, W. and Tam, K.C. (2008) Synthesis and characterization of nanoporous hydroxyapatite using cationic surfactants as templates. *Mater. Res. Bull*, **43**, 2318–2326.

20. Yao, J., Tjandra, W., Chen, Y.Z. *et al.* (2003) Hydroxyapatite nanostructure material derived using cationic surfactant as a template. *J. Mater. Chem.*, **13**, 3053–3057.

21. Paul, W. and Sharma, C.P. (1999) Development of porous spherical hydroxyapatite granules: application towards protein delivery. *J. Mater. Sci. -Mater. Med.*, **10**, 383–388.

22. Hong, M.H., Kim, K.M., Choi, S.H. *et al.* (2012) Fabrication of hollow hydroxyapatite spherical granules for hard tissue regeneration and alternative method for drug release test. *Micro Nano Lett.*, **7**, 634–636.

23. Paul, W., Nesamony, J. and Sharma, C.P. (2002) Delivery of insulin from hydroxyapatite ceramic microspheres: preliminary in vivo studies. *J. Biomed. Mater. Res.*, **61**, 660–662.

24. Sopyan, I., Mel, M., Ramesh, S. and Khalid, K.A. (2007) Porous hydroxyapatite for artificial bone applications. *Sci. Technol. Adv. Mater.*, **8**, 116–123.

25. Lebugle, A., Rodrigues, A., Bonnevialle, P. *et al.* (2002) Study of implantable calcium phosphate systems for the slow release of methotrexate. *Biomaterials*, **23**, 3517–3522.

26. Shinto, Y., Uchida, A., Korkusuz, F. *et al.* (1992) Calcium hydroxyapatite ceramic used as a delivery system for antibiotics. *J. Bone Joint Surg. -Br. Vol.*, **74**, 600–604.

27. Kundu, B., Soundrapandian, C., Nandi, S.K. *et al* (2010) Development of new localized drug delivery system based on ceftriaxone-sulbactam composite drug impregnated porous hydroxyapatite: a systematic approach for in vitro and in vivo animal trial. *Pharm. Res.*, **27**, 1659–1676.

28. Zhou, H. and Lee, J. (2011) Nanoscale hydroxyapatite particles for bone tissue engineering. *Acta Biomater.*, **7**, 2769–2781.

29. Uskokovic, V. and Uskokovic, D.P. (2011) Nanosized hydroxyapatite and other calcium phosphates: chemistry of formation and application as drug and gene delivery agents. *J. Biomed. Mater. Res. Part B -Appl. Biomater.*, **96**, 152–191.

30. Loo, S.C.J., Moore, T., Banik, B. and Alexis, F. (2010) Biomedical applications of hydroxyapatite nanoparticles. *Curr. Pharm. Biotechnol.*, **11**, 333–342.

31. Okada, M. and Furuzono, T. (2012) Hydroxylapatite nanoparticles: fabrication methods and medical applications. *Sci. Technol. Adv. Mater.*, **13**, 064103.

32. Matsumoto, T., Okazaki, M., Inoue, M. *et al* (2004) Hydroxyapatite particles as a controlled release carrier of protein. *Biomaterials*, **25**, 3807–3812.

33. Kano, S., Yamazaki, A., Otsuka, R. *et al.* (1994) Application of hydroxyapatite-sol as drug carrier. *Bio-Med. Mater. Eng.*, **4**, 283–290.

34. Venkatasubbu, G.D., Ramasamy, S., Ramakrishnan, V. and Kumar, J. (2011) Hydroxyapatite-alginate nanocomposite as drug delivery matrix for sustained release of ciprofloxacin. *J. Biomed. Nanotechnol.*, **7**, 759–767.

35. Uskokovic, V. and Desai, T.A. (2013) Phase composition control of calcium phosphate nanoparticles for tunable drug delivery kinetics and treatment of osteomyelitis. I. Preparation and drug release. *J. Biomed. Mater. Res. Part A*, **101**, 1416–1426.

36. Barroug, A. and Glimcher, M.J. (2002) Hydroxyapatite crystals as a local delivery system for cisplatin: adsorption and release of cisplatin in vitro. *J. Orthop. Res.*, **20**, 274–280.

37. Iafisco, M., Palazzo, B., Marchetti, M. *et al* (2009) Smart delivery of antitumoral platinum complexes from biomimetic hydroxyapatite nanocrystals. *J. Mater. Chem.*, **19**, 8385–8392.

38. Palazzo, B., Iafisco, M., Laforgia, M. *et al* (2007) Biomimetic hydroxyapatite–drug nanocrystals as potential bone substitutes with antitumor drug delivery properties. *Adv. Funct. Mater.*, **17**, 2180–2188.

39. Do, T.N.T., Lee, W.-H., Loo, C.-Y. *et al.* (2012) Hydroxyapatite nanoparticles as vectors for gene delivery. *Ther. Delivery*, **3**, 623–632.

40. Kumta, P.N., Sfeir, C., Lee, D.-H. *et al.* (2005) Nanostructured calcium phosphates for biomedical applications: novel synthesis and characterization. *Acta Biomater.*, **1**, 65–83.

41. Sokolova, V. and Epple, M. (2008) Inorganic nanoparticles as carriers of nucleic acids into cells. *Angew. Chem. Int. Ed.*, **47**, 1382–1395.

42. Sokolova, V., Rotan, O., Klesing, J. *et al* (2012) Calcium phosphate nanoparticles as versatile carrier for small and large molecules across cell membranes. *J. Nanopart. Res.*, **14**, 1–10.

43. Daculsi, G. (1998) Biphasic calcium phosphate concept applied to artificial bone, implant coating and injectable bone substitute. *Biomaterials*, **19**, 1473–1478.

44. Tancred, D.C., McCormack, B.A.O. and Carr, A.J. (1998) A synthetic bone implant macroscopically identical to cancellous bone. *Biomaterials*, **19**, 2303–2311.

45. Kivrak, N. and Tas, A.C. (1998) Synthesis of calcium hydroxyapatite-tricalcium phosphate (HA-TCP) composite bioceramic powders and their sintering behavior. *J. Am. Ceram. Soc.*, **81**, 2245–2252.

46. Sanchez-Salcedo, S., Werner, J. and Vallet-Regi, M. (2008) Hierarchical pore structure of calcium phosphate scaffolds by a combination of gel-casting and multiple tape-casting methods. *Acta Biomater.*, **4**, 913–922.

47. Sanchez-Salcedo, S., Arcos, D. and Vallet-Regi, M. (2008) Upgrading calcium phosphate scaffolds for tissue engineering applications. *Key Eng. Mater.*, **377**, 19–4242.

48. Dorozhkin, S.V. (2010) Bioceramics of calcium orthophosphates. *Biomaterials*, **31**, 1465–1485.

49. Grimandi, G., Weiss, P., Millot, F. and Daculsi, G. (1998) In vitro evaluation of a new injectable calcium phosphate material. *J. Biomed. Mater. Res.*, **39**, 660–666.

50. Ginebra, M.P., Traykova, T. and Planell, J.A. (2006) Calcium phosphate cements as bone drug delivery systems: a review. *J. Controlled Release*, **113**, 102–110.

51. Ginebra, M.P., Rilliard, A., Fernandez, E. *et al.* (2001) Mechanical and rheological improvement of a calcium phosphate cement by the addition of a polymeric drug. *J. Biomed. Mater. Res.*, **57**, 113–118.

52. Bigi, A., Bracci, B. and Panzavolta, S. (2004) Effect of added gelatin on the properties of calcium phosphate cement. *Biomaterials*, **25**, 2893–2899.

53. Ratier, A., Gibson, I.R., Best, S.M. *et al.* (2001) Setting characteristics and mechanical behaviour of a calcium phosphate bone cement containing tetracycline. *Biomaterials*, **22**, 897–901.

54. Takechi, M., Miyamoto, Y., Ishikawa, K. *et al* (1998) Effects of added antibiotics on the basic properties of anti-washout-type fast-setting calcium phosphate cement. *J. Biomed. Mater. Res.*, **39**, 308–316.

55. Hesaraki, S., Moztarzadeh, F. and Nezafati, N. (2009) Evaluation of a bioceramic-based nanocomposite material for controlled delivery of a non-steroidal anti-inflammatory drug. *Med. Eng. Phys.*, **31**, 1205–1213.

56. Bohner, M., Lemaitre, J., Merkle, H.P. and Gander, B. (2000) Control of gentamicin release from a calcium phosphate cement by admixed poly(acrylic acid). *J. Pharm. Sci.*, **89**, 1262–1270.

57. Hamanishi, C., Kitamoto, K., Tanaka, S. *et al.* (1996) A self-setting TTCP-DCPD apatite cement for release of vancomycin. *J. Biomed. Mater. Res.*, **33**, 139–143.

58. Bohner, M., Lemaitre, J., VanLanduyt, P. *et al.* (1997) Gentamicin-loaded hydraulic calcium phosphate bone cement as antibiotic delivery system. *J. Pharm. Sci.*, **86**, 565–572.

59. Sasaki, S. and Ishii, Y. (1999) Apatite cement containing antibiotics: efficacy in treating experimental osteomyelitis. *J. Orthop. Sci.: Off. J. Jpn. Orthop. Assoc.*, **4**, 361–369.

60. Ethell, M.T., Bennett, R.A., Brown, M.P. *et al.* (2000) In vitro elution of gentamicin, amikacin, and ceftiofur from polymethylmethacrylate and hydroxyapatite cement. *Veterinary Surg.*, **29**, 375–382.

61. Stallmann, H.P., Faber, C., Bronckers, A. *et al.* (2004) Osteomyelitis prevention in rabbits using antimicrobial peptide hLF1-11-or gentamicin-containing calcium phosphate cement. *J. Antimicrob. Chemother.*, **54**, 472–476.

62. Joosten, U., Joist, A., Frebel, T. *et al.* (2004) Evaluation of an in situ setting injectable calcium phosphate as a new carrier material for gentamicin osteomyelitis: studies in the treatment of chronic in vitro and in vivo. *Biomaterials*, **25**, 4287–4295.

63. Sasaki, T., Ishibashi, Y., Katano, H. *et al.* (2005) In vitro elution of vancomycin from calcium phosphate cement. *J. Arthroplasty*, **20**, 1055–1059.

64. Schnieders, J., Gbureck, U., Thull, R. and Kissel, T. (2006) Controlled release of gentamicin from calcium phosphate – poly(lactic acid-co-glycolic acid) composite bone cement. *Biomaterials*, **27**, 4239–4249.

65. Kisanuki, O., Yajima, H., Umeda, T. and Takakura, Y. (2007) Experimental study of calcium phosphate cement impregnated with dideoxy-kanamycin B. *J. Orthop. Sci.*, **12**, 281–288.

66. Urabe, K., Naruse, K., Hattori, H. *et al* (2009) In vitro comparison of elution characteristics of vancomycin from calcium phosphate cement and polymethylmethacrylate. *J. Orthop. Sci.*, **14**, 784–793.

67. Jiang, P.-J., Patel, S., Gbureck, U. *et al.* (2010) Comparing the efficacy of three bioceramic matrices for the release of vancomycin hydrochloride. *J. Biomed. Mater. Res. Part B -Appl. Biomater.*, **93B**, 51–58.

68. Ratier, A., Freche, M., Lacout, J.L. and Rodriguez, F. (2004) Behaviour of an injectable calcium phosphate cement with added tetracycline. *Int. J. Pharm.*, **274**, 261–268.

69. Tamimi, F., Torres, J., Bettini, R. *et al* (2008) Doxycycline sustained release from brushite cements for the treatment of periodontal diseases. *J. Biomed. Mater. Res. Part A*, **85**, 707–714.

70. Hamdan Alkhraisat, M., Rueda, C., Cabrejos-Azama, J. *et al* (2010) Loading and release of doxycycline hyclate from strontium-substituted calcium phosphate cement. *Acta Biomater.*, **6**, 1522–1528.

71. Ishikawa, K., Miyamoto, Y., Takechi, M. *et al* (1997) Non-decay type fast-setting calcium phosphate cement: hydroxyapatite putty containing an increased amount of sodium alginate. *J. Biomed. Mater. Res.*, **36**, 393–399.

72. Takechi, M., Miyamoto, Y., Momota, Y. *et al* (2002) The in vitro antibiotic release from anti-washout apatite cement using chitosan. *J. Mater. Sci. Mater. Med.*, **13**, 973–978.

73. Tamimi Marino, F., Torres, J., Tresguerres, I. *et al.* (2007) Vertical bone augmentation with granulated brushite cement set in glycolic acid. *J. Biomed. Mater. Res. Part A*, **81A**, 93–102.

74. Hesaraki, S., Nemati, R. and Nosoudi, N. (2009) Preparation and characterisation of porous calcium phosphate bone cement as antibiotic carrier. *Adv. Appl. Ceram.*, **108**, 231–240.

75. Chen, C.-H.D., Chen, C.-C., Shie, M.-Y. *et al.* (2011) Controlled release of gentamicin from calcium phosphate/alginate bone cement. *Mater. Sci. Eng., C -Mater. Biol. Appl.*, **31**, 334–341.

76. Hofmann, M.P., Mohammed, A.R., Perrie, Y. *et al.* (2009) High-strength resorbable brushite bone cement with controlled drug-releasing capabilities. *Acta Biomater.*, **5**, 43–49.

77. Hesaraki, S., Zamanian, A. and Moztarzadeh, F. (2008) The influence of the acidic component of the gas-foaming porogen used in preparing an injectable porous calcium phosphate cement on its properties: acetic acid versus citric acid. *J. Biomed. Mater. Res. Part B Appl. Biomater.*, **86**, 208–216.

78. Otsuka, M., Nakahigashi, Y., Matsuda, Y. *et al.* (1994) A novel skeletal drug-delivery system using self-setting calcium-phosphate cement. 7. Effect of biological factors on indomethacin release from the cement loaded on bovine bone. *J. Pharm. Sci.*, **83**, 1569–1573.

79. Otsuka, M., Nakahigashi, Y., Matsuda, Y. *et al.* (1997) A novel skeletal drug delivery system using self-setting calcium phosphate cement VIII: the relationship between in vitro and in vivo drug release from indomethacin-containing cement. *J. Controlled Release*, **43**, 115–122.

80. Tahara, Y. and Ishii, Y. (2001) Apatite cement containing cis-diamminedichloroplatinum implanted in rabbit femur for sustained release of the anticancer drug and bone formation. *J. Orthop. Sci.: Off. J. Jpn. Orthop.Assoc.*, **6**, 556–565.

81. Tanzawa, Y., Tsuchiya, H., Shirai, T. *et al* (2011) Potentiation of the antitumor effect of calcium phosphate cement containing anticancer drug and caffeine on rat osteosarcoma. *J. Orthop. Sci.*, **16**, 77–84.

82. Tani, T., Okada, K., Takahashi, S. *et al.* (2006) Doxorubicin-loaded calcium phosphate cement in the management of bone and soft tissue tumors. *In Vivo*, **20**, 55–60.
83. Lopez-Heredia, M.A., Kamphuis, G.J.B., Thune, P.C. *et al.* (2011) An injectable calcium phosphate cement for the local delivery of paclitaxel to bone. *Biomaterials*, **32**, 5411–5416.
84. Otsuka, M., Matsuda, Y., Suwa, Y. *et al.* (1994) A novel skeletal drug-delivery system using a self-setting calcium-phosphate cement. 5. Drug-release behavior from a heterogeneous drug-loaded cement containing an anticancer drug. *J. Pharm. Sci.*, **83**, 1565–1568.
85. Otsuka, M., Matsuda, Y., Fox, J.L. and Higuchi, W.I. (1995) Novel skeletal drug-delivery system using self-setting calcium-phosphate cement. 9. Effects of the mixing solution volume on anticancer drug-release from homogeneous drug-loaded cement. *J. Pharm. Sci.*, **84**, 733–736.
86. Li, D.X., Fan, H.S., Zhu, X.D. *et al* (2007) Controllable release of salmon-calcitonin in injectable calcium phosphate cement modified by chitosan oligosaccharide and collagen polypeptide. *J. Mater. Sci. Mater. Med.*, **18**, 2225–2231.
87. Panzavolta, S., Torricelli, P., Bracci, B. *et al.* (2009) Alendronate and pamidronate calcium phosphate bone cements: setting properties and in vitro response of osteoblast and osteoclast cells. *J. Inorg. Biochem.*, **103**, 101–106.
88. Jindong, Z., Hai, T., Junchao, G. *et al.* (2010) Evaluation of a novel osteoporotic drug delivery system in vitro: alendronate-loaded calcium phosphate cement. *Orthopedics*, **33**, 561.
89. Panzavolta, S., Torricelli, P., Bracci, B. *et al.* (2010) Functionalization of biomimetic calcium phosphate bone cements with alendronate. *J. Inorg. Biochem.*, **104**, 1099–1106.
90. Giocondi, J.L., El-Dasher, B.S., Nancollas, G.H. *et al.* (2010) Molecular mechanisms of crystallization impacting calcium phosphate cements. *Philos. Trans. R. Soc. A - Math. Phys. Eng. Sci.*, **368**, 1937–1961.
91. Schnitzler, V., Fayon, F., Despas, C. *et al.* (2011) Investigation of alendronate-doped apatitic cements as a potential technology for the prevention of osteoporotic hip fractures: critical influence of the drug introduction mode on the in vitro cement properties. *Acta Biomater.*, **7**, 759–770.
92. Su, K., Shi, X., Varshney, R.R. and Wang, D.-A. (2011) Transplantable delivery systems for in situ controlled release of bisphosphonate in orthopedic therapy. *Expert Opin. Drug Deliv.*, **8**, 113–126.
93. Blom, E.J., Klein–Nulend, J., Klein, C.P. *et al.* (2000) Transforming growth factor-β1 incorporated during setting in calcium phosphate cement stimulates bone cell differentiation in vitro. *J. Biomed. Mater. Res.*, **50**, 67–74.
94. Szivek, J.A., Nelson, E.R., Hajdu, S.D. *et al.* (2004) Transforming growth factor-β1 accelerates bone bonding to a blended calcium phosphate ceramic coating: a dose-response study. *J. Biomed. Mater. Res. Part A*, **68**, 537–543.
95. Haddad, A.J., Peel, S.A.F., Clokie, C.M.L. and Sandor, G.K.B. (2006) Closure of rabbit calvarial critical-sized defects using protective composite allogeneic and alloplastic bone substitutes. *J. Craniofacial Surg.*, **17**, 926–934.
96. Vallet-Regi, M., Ragel, C.V. and Salinas, A.J. (2003) Glasses with medical applications. *Eur. J. Inorg. Chem.*, **2003**, 1029–1042.

97. Li, R., Clark, A.E. and Hench, L.L. (1991) An investigation of bioactive glass powders by sol-gel processing. *J. Appl. Biomater.*, **2**, 231–239.

98. Hench, L.L., Splinter, R.J., Allen, W.C. and Greenlee, T.K. (1971) Bonding mechanisms at the interface of ceramic prosthetic materials. *J. Biomed. Mater. Res.*, **5**, 117–141.

99. Hench, L.L. (1991) Bioceramics – from concept to clinic. *J. Am. Ceram. Soc.*, **74**, 1487–1510.

100. Hench, L.L. and West, J.K. (1990) The sol-gel process. *Chem. Rev.*, **90**, 33–72.

101. Verne, E., Vitale-Brovarone, C., Bui, E. *et al.* (2009) Surface functionalization of bioactive glasses. *J. Biomed. Mater. Res. Part A*, **90**, 981–992.

102. Verne, E., Ferraris, S., Vitale-Brovarone, C. *et al* (2010) Alkaline phosphatase grafting on bioactive glasses and glass ceramics. *Acta Biomater.*, **6**, 229–240.

103. Verne, E., Ferraris, S., Brovarone, C.V. *et al* (2010) Enzyme grafting to bioactive glasses, in *Advances in Bioceramics and Biotechnologies* (eds R. Narayan and M.J. Westerville), American Ceramic Society, pp. 133–137.

104. Chen, Q.Z., Ahmed, I., Knowles, J.C. *et al.* (2008) Collagen release kinetics of surface functionalized 4555 Bioglass (R)-based porous scaffolds. *J. Biomed. Mater. Res. Part A*, **86**, 987–995.

105. Meseguer-Olmo, L., Ros-Nicolas, M.J., Clavel-Sainz, M. *et al.* (2002) Biocompatibility and in vivo gentamicin release from bioactive sol-gel glass implants. *J. Biomed. Mater. Res.*, **61**, 458–465.

106. Aina, V., Marchis, T., Laurenti, E. *et al.* (2010) Functionalization of sol gel bioactive glasses carrying Au nanoparticles: selective Au affinity for amino and thiol ligand groups. *Langmuir*, **26**, 18600–18605.

107. Sperling, R.A., Rivera Gil, P., Zhang, F. *et al.* (2008) Biological applications of gold nanoparticles. *Chem. Soc. Rev.*, **37**, 1896–1908.

108. Aina, V., Ghigo, D., Marchis, T. *et al.* (2011) Novel bio-conjugate materials: soybean peroxidase immobilized on bioactive glasses containing Au nanoparticles. *J. Mater. Chem.*, **21**, 10970–10981.

109. Yan, X.X., Yu, C.Z., Zhou, X.F. *et al.* (2004) Highly ordered mesoporous bioactive glasses with superior in vitro bone-forming bioactivities. *Angew. Chem. Int. Ed.*, **43**, 5980–5984.

110. Xia, W. and Chang, J. (2006) Well-ordered mesoporous bioactive glasses (MBG): a promising bioactive drug delivery system. *J. Controlled Release*, **110**, 522–530.

111. Xia, W. and Chang, J. (2008) Preparation, in vitro bioactivity and drug release property of well-ordered mesoporous 58S bioactive glass. *J. Non-Cryst. Solids*, **354**, 1338–1341.

112. Zhao, Y.F., Loo, S.C.J., Chen, Y.Z. *et al.* (2008) In situ SAXRD study of sol-gel induced well-ordered mesoporous bioglasses for drug delivery. *J. Biomed. Mater. Res. Part A*, **85A**, 1032–1042.

113. Zhao, L., Yan, X., Zhou, X. *et al.* (2008) Mesoporous bioactive glasses for controlled drug release. *Microporous Mesoporous Mater.*, **109**, 210–215.

114. Sun, J., Li, Y.S., Li, L. *et al.* (2008) Functionalization and bioactivity in vitro of mesoporous bioactive glasses. *J. Non-Cryst. Solids*, **354**, 3799–3805.

115. Lopez-Noriega, A., Arcos, D. and Vallet-Regi, M. (2010) Functionalizing meso-porous bioglasses for long-term anti-osteoporotic drug delivery. *Chem. Eur. J.*, **16**, 10879–10886.
116. Vallet-Regi, M., Ramila, A., del Real, R.P. and Perez-Pariente, J. (2001) A new property of MCM-41: drug delivery system. *Chem. Mater.*, **13**, 308–311.
117. Slowing, I.I., Vivero-Escoto, J.L., Wu, C.W. and Lin, V.S.Y. (2008) Mesoporous silica nanoparticles as controlled release drug delivery and gene transfection carriers. *Adv. Drug Deliv. Rev.*, **60**, 1278–1288.
118. Gerion, D., Herberg, J., Bok, R. *et al.* (2007) Paramagnetic silica-coated nanocrystals as an advanced MRI contrast agent. *J. Phys. Chem. C*, **111**, 12542–12551.
119. Bottini, M., D'Annibale, F., Magrini, A. *et al.* (2007) Quantum dot-doped silica nanoparticles as probes for targeting of T-lymphocytes. *Int. J. Nanomed.*, **2**, 227–233.
120. Areva, S., Aaritalo, V., Tuusa, S. *et al.* (2007) Sol-gel-derived TiO_2-SiO_2 implant coatings for direct tissue attachment. Part II: evaluation of cell response. *J. Mater. Sci. Mater. Med.*, **18**, 1633–1642.
121. Dormer, K., Seeney, C., Lewelling, K. *et al.* (2005) Epithelial internalization of super-paramagnetic nanoparticles and response to external magnetic field. *Biomaterials*, **26**, 2061–2072.
122. Allouche, J., Boissiere, M., Helary, C. *et al.* (2006) Biomimetic core-shell gela-tine/silica nanoparticles: a new example of biopolymer-based nanocomposites. *J. Mater. Chem.*, **16**, 3120–3125.
123. Huo, Q.S., Liu, J., Wang, L.Q. *et al.* (2006) A new class of silica cross-linked micellar core-shell nanoparticles. *J. Am. Chem. Soc.*, **128**, 6447–6453.
124. Doadrio, A.L., Sousa, E.M.B., Doadrio, J.C. *et al.* (2004) Mesoporous SBA-15 HPLC evaluation for controlled gentamicin drug delivery. *J. Controlled Release*, **97**, 125–132.
125. Vallet-Regi, M., Manzano, M., Gonzalez-Calbet, J.M. and Okunishi, E. (2010) Evidence of drug confinement into silica mesoporous matrices by STEM spherical aberration corrected microscopy. *Chem. Commun.*, **46**, 2956–2958.
126. Yiu, H.H.P., Botting, C.H., Botting, N.P. and Wright, P.A. (2001) Size selective protein adsorption on thiol-functionalised SBA-15 mesoporous molecular sieve. *Phys. Chem. Chem. Phys.*, **3**, 2983–2985.
127. Kruk, M., Jaroniec, M. and Sayari, A. (2000) New insights into pore-size expansion of mesoporous silicates using long-chain amines. *Microporous Mesoporous Mater.*, **35-6**, 545–553.
128. Kimura, T., Sugahara, Y. and Kuroda, K. (1998) Synthesis of mesoporous aluminophosphates using surfactants with long alkyl chain lengths and triisopropyl-benzene as a solubilizing agent. *Chem. Commun.*, 559–560.
129. Blin, J.L., Otjacques, C., Herrier, G. and Su, B.L. (2000) Pore size engineering of mesoporous silicas using decane as expander. *Langmuir*, **16**, 4229–4236.
130. Smarsly, B., Polarz, S. and Antonietti, M. (2001) Preparation of porous silica materials via sol-gel nanocasting of nonionic surfactants: a mechanistic study on the self-aggregation of amphiphiles for the precise prediction of the mesopore size. *J. Phys. Chem. B*, **105**, 10473–10483.

131. Vallet-Regi, M., Balas, F., Colilla, M. and Manzano, M. (2008) Bone-regenerative bioceramic implants with drug and protein controlled delivery capability. *Prog. Solid State Chem.*, **36**, 163–191.

132. Balas, F., Manzano, M., Horcajada, P. and Vallet-Regi, M. (2006) Confinement and controlled release of bisphosphonates on ordered mesoporous silica-based materials. *J. Am. Chem. Soc.*, **128**, 8116–8117.

133. Qu, F.Y., Zhu, G.S., Huang, S.Y. *et al* (2006) Controlled release of Captopril by regulating the pore size and morphology of ordered mesoporous silica. *Microporous Mesoporous Mater.*, **92**, 1–9.

134. Charnay, C., Begu, S., Tourne-Peteilh, C. *et al.* (2004) Inclusion of ibuprofen in mesoporous templated silica: drug loading and release property. *Eur. J. Pharm. Biopharm.*, **57**, 533–540.

135. Lettow, J.S., Han, Y.J., Schmidt-Winkel, P. *et al.* (2000) Hexagonal to mesocellular foam phase transition in polymer-templated mesoporous silicas. *Langmuir*, **16**, 8291–8295.

136. Schmidt-Winkel, P., Lukens, W.W., Zhao, D.Y. *et al.* (1999) Mesocellular siliceous foams with uniformly sized cells and windows. *J. Am. Chem. Soc.*, **121**, 254–255.

137. Horcajada, P., Ramila, A., Gerard, F. and Vallet-Regi, M. (2006) Influence of superficial organic modification of MCM-41 matrices on drug delivery rate. *Solid State Sci.*, **8**, 1243–1249.

138. Menaa, B., Miyagawa, Y., Takahashi, M. *et al.* (2009) Bioencapsulation of apomyoglobin in nanoporous organosilica sol-gel glasses: influence of the siloxane network on the conformation and stability of a model protein. *Biopolymers*, **91**, 895–906.

139. Balas, F., Manzano, M., Colilla, M. and Vallet-Regi, M. (2008) L-Trp adsorption into silica mesoporous materials to promote bone formation. *Acta Biomater.*, **4**, 514–522.

140. Manzano, M. and Vallet-Regi, M. (2010) New developments in ordered mesoporous materials for drug delivery. *J. Mater. Chem.*, **20**, 5593–5604.

141. Nieto, A., Balas, F., Colilla, M. *et al.* (2008) Functionalization degree of SBA-15 as key factor to modulate sodium alendronate dosage. *Microporous Mesoporous Mater.*, **116**, 4–13.

142. Arruebo, M., Vilaboa, N. and Santamaria, J. (2010) Drug delivery from internally implanted biomedical devices used in traumatology and in orthopedic surgery. *Expert Opin. Drug Deliv.*, **7**, 589–603.

143. Horcajada, P., Ramila, A., Perez-Pariente, J. and Vallet-Regi, M. (2004) Influence of pore size of MCM-41 matrices on drug delivery rate. *Microporous Mesoporous Mater.*, **68**, 105–109.

144. Manzano, M., Lamberti, G., Galdi, I. and Vallet-Regi, M. (2011) Anti-osteoporotic drug release from ordered mesoporous bioceramics: experiments and modeling. *AAPS Pharm.*, **12**, 1193–1199.

145. Munoz, B., Ramila, A., Perez-Pariente, J. *et al.* (2003) MCM-41 organic modification as drug delivery rate regulator. *Chem. Mater.*, **15**, 500–503.

146. Izquierdo-Barba, I., Martinez, A., Doadrio, A.L. *et al.* (2005) Release evaluation of drugs from ordered three-dimensional silica structures. *Eur. J. Pharm. Sci.*, **26**, 365–373.

147. Song, S.W., Hidajat, K. and Kawi, S. (2005) Functionalized SBA-15 materials as carriers for controlled drug delivery: influence of surface properties on matrix-drug interactions. *Langmuir*, **21**, 9568–9575.
148. Heikkila, T., Salonen, J., Tuura, J. *et al* (2007) Mesoporous silica material TUD-1 as a drug delivery system. *Int. J. Pharm.*, **331**, 133–138.
149. Tang, Q.L., Xu, Y., Wu, D. and Sun, Y.H. (2006) Hydrophobicity-controlled drug delivery system from organic modified mesoporous silica. *Chem. Lett.*, **35**, 474–475.
150. Tang, Q.L., Xu, Y., Wu, D. *et al* (2006) Studies on a new carrier of trimethylsilyl-modified mesoporous material for controlled drug delivery. *J. Controlled Release*, **114**, 41–46.
151. Cavallaro, G., Pierro, P., Palumbo, F.S. *et al* (2004) Drug delivery devices based on mesoporous silicate. *Drug Deliv.*, **11**, 41–46.
152. Ambrogi, V., Perioli, L., Marmottini, F. *et al.* (2007) Improvement of dissolution rate of piroxicam by inclusion into MCM-41 mesoporous silicate. *Eur. J. Pharm. Sci.*, **32**, 216–222.
153. Zeng, W., Qian, X.F., Zhang, Y.B. *et al.* (2005) Organic modified mesoporous MCM-41 through solvothermal process as drug delivery system. *Mater. Res. Bull*, **40**, 766–772.
154. Vallet-Regi, M., Doadrio, J.C., Doadrio, A.L. *et al.* (2004) Hexagonal ordered mesoporous material as a matrix for the controlled release of amoxicillin. *Solid State Ion.*, **172**, 435–439.
155. Xue, J.M. and Shi, M. (2004) PLGA/mesoporous silica hybrid structure for controlled drug release. *J. Controlled Release*, **98**, 209–217.
156. Doadrio, J.C., Sousa, E.M.B., Izquierdo-Barba, I. *et al.* (2006) Functionalization of mesoporous materials with long alkyl chains as a strategy for controlling drug delivery pattern. *J. Mater. Chem.*, **16**, 462–466.
157. Mellaerts, R., Aerts, C.A., Van Humbeeck, J. *et al.* (2007) Enhanced release of itraconazole from ordered mesoporous SBA-15 silica materials. *Chem. Commun.*, 1375–1377.
158. Zelenak, V., Hornebecq, V. and Llewellyn, P. (2005) Zinc(II)-benzoato complexes immobilised in mesoporous silica host. *Microporous Mesoporous Mater.*, **83**, 125–135.
159. Qu, F.Y., Zhu, G.S., Huang, S.Y. *et al.* (2006) Effective controlled release of captopril by silylation of mesoporous MCM-41. *ChemPhysChem*, **7**, 400–406.
160. Nunes, C.D., Vaz, P.D., Fernandes, A.C. *et al.* (2007) Loading and delivery of sertraline using inorganic micro and mesoporous materials. *Eur. J. Pharm. Biopharm.*, **66**, 357–365.
161. Tang, Q.L., Xu, Y., Wu, D. and Sun, Y.H. (2006) A study of carboxylic-modified i-nesoporous silica in controlled delivery for drug famotidine. *J. Solid State Chem.*, **179**, 1513–1520.
162. Xu, W., Gao, Q., Xu, Y. *et al* (2008) Controlled drug release from bifunctionalized mesoporous silica. *J. Solid State Chem.*, **181**, 2837–2844.
163. Tourne-Peteilh, C., Lerner, D.A., Charnay, C. *et al.* (2003) The potential of ordered mesoporous silica for the storage of drugs: the example of a pentapeptide encapsulated in a MSU-Tween 80. *ChemPhysChem*, **4**, 281–286.

164. Lai, C.Y., Trewyn, B.G., Jeftinija, D.M. *et al.* (2003) A mesoporous silica nanosphere-based carrier system with chemically removable CdS nanoparticle caps for stimuli-responsive controlled release of neurotransmitters and drug molecules. *J. Am. Chem. Soc.*, **125**, 4451–4459.

165. Singh, N., Karambelkar, A., Gu, L. *et al.* (2011) Bioresponsive mesoporous silica nanoparticles for triggered drug release. *J. Am. Chem. Soc.*, **133**, 19582–19585.

166. Li, Z., Barnes, J.C., Bosoy, A. *et al.* (2012) Mesoporous silica nanoparticles in biomedical applications. *Chem. Soc. Rev.*, **41**, 2590–2605.

167. Tang, F., Li, L. and Chen, D. (2012) Mesoporous silica nanoparticles: synthesis, bio-compatibility and drug delivery. *Adv. Mater.*, **24**, 1504–1534.

168. Garcia-Bennett, A.E. (2011) Synthesis, toxicology and potential of ordered meso-porous materials in nanomedicine. *Nanomedicine*, **6**, 867–877.

169. Martin, K.R. (2007) The chemistry of silica and its potential health benefits. *J. Nutr. Health Aging*, **11**, 94–98.

170. Vivero-Escoto, J.L., Slowing, I.I., Trewyn, B.G. and Lin, V.S.Y. (2010) Mesoporous silica nanoparticles for intracellular controlled drug delivery. *Small*, **6**, 1952–1967.

171. Lin, Y.S., Tsai, C.P., Huang, H.Y. *et al.* (2005) Well-ordered mesoporous silica nanoparticles as cell markers. *Chem. Mater.*, **17**, 4570–4573.

172. Slowing, I.I., Trewyn, B.G. and Lin, V.S.Y. (2007) Mesoporous silica nanoparticles for intracellular delivery of membrane-impermeable proteins. *J. Am. Chem. Soc.*, **129**, 8845–8849.

173. Slowing, I., Trewyn, B.G. and Lin, V.S.Y. (2006) Effect of surface functionalization of MCM-41-type mesoporous silica nanoparticles on the endocytosis by human cancer cells. *J. Am. Chem. Soc.*, **128**, 14792–14793.

174. Radu, D.R., Lai, C.Y., Jeftinija, K. *et al.* (2004) A polyamidoamine dendrimer-capped mesoporous silica nanosphere-based gene transfection reagent. *J. Am. Chem. Soc.*, **126**, 13216–13217.

175. Giri, S., Trewyn, B.G., Stellmaker, M.P. and Lin, V.S.Y. (2005) Stimuli-responsive controlled-release delivery system based on mesoporous silica nanorods capped with magnetic nanoparticles. *Angew. Chem. Int. Ed.*, **44**, 5038–5044.

176. Lu, F., Wu, S.H., Hung, Y. and Mou, C.Y. (2009) Size effect on cell uptake in well-suspended, uniform mesoporous silica nanoparticles. *Small*, **5**, 1408–1413.

177. Lu, J., Liong, M., Zink, J.I. and Tamanoi, F. (2007) Mesoporous silica nanoparticles as a delivery system for hydrophobic anticancer drugs. *Small*, **3**, 1341–1346.

178. Yu, T., Malugin, A. and Ghandehari, H. (2011) Impact of silica nanoparticle design on cellular toxicity and hemolytic activity. *ACS Nano*, **5**, 5717–5728.

179. Witasp, E., Kupferschmidt, N., Bengtsson, L. *et al.* (2009) Efficient internalization of mesoporous silica particles of different sizes by primary human macrophages without impairment of macrophage clearance of apoptotic or antibody-opsonized target cells. *Toxicol. Appl. Pharmacol.*, **239**, 306–319.

180. Chung, T.H., Wu, S.H., Yao, M. *et al.* (2007) The effect of surface charge on the uptake and biological function of mesoporous silica nanoparticles 3T3-L1 cells and human mesenchymal stem cells. *Biomaterials*, **28**, 2959–2966.

181. Huang, D.-M., Hung, Y., Ko, B.-S. *et al.* (2005) Highly efficient cellular labeling of mesoporous nanoparticles in human mesenchymal stem cells: implication for stem cell tracking. *FASEB J.*, **19**, 2014–2016.

182. Vallhov, H., Gabrielsson, S., Stromme, M. *et al.* (2007) Mesoporous silica particles induce size dependent effects on human dendritic cells. *Nano Lett.*, **7**, 3576–3582.

183. Tao, Z.M., Morrow, M.P., Asefa, T. *et al.* (2008) Mesoporous silica nanoparticles inhibit cellular respiration. *Nano Lett.*, **8**, 1517–1526.

184. Slowing, I.I., Wu, C.W., Vivero-Escoto, J.L. and Lin, V.S.Y. (2009) Mesoporous silica nanoparticles for reducing hemolytic activity towards mammalian red blood cells. *Small*, **5**, 57–62.

185. Lin, Y.S. and Haynes, C.L. (2010) Impacts of mesoporous silica nanoparticle size, pore ordering, and pore integrity on hemolytic activity. *J. Am. Chem. Soc.*, **132**, 4834–4842.

186. Lopez, T., Basaldella, E.I., Ojeda, M.L. *et al.* (2006) Encapsulation of valproic acid and sodic phenytoin in ordered mesoporous SiO_2 solids for the treatment of temporal lobe epilepsy. *Opt. Mater.*, **29**, 75–81.

187. Wu, S.-H., Lin, Y.-S., Hung, Y. *et al.* (2008) Multifunctional mesoporous silica nanoparticles for intracellular labeling and animal magnetic resonance imaging studies. *ChemBioChem*, **9**, 53–57.

188. Lu, J., Liong, M., Li, Z.X. *et al.* (2010) Biocompatibility, biodistribution, and drug-delivery efficiency of mesoporous silica nanoparticles for cancer therapy in animals. *Small*, **6**, 1794–1805.

189. Liu, H.M., Wu, S.H., Lu, C.W. *et al.* (2008) Mesoporous silica nanoparticles improve magnetic labeling efficiency in human stem cells. *Small*, **4**, 619–626.

190. Hsiao, J.K., Tsai, C.P., Chung, T.H. *et al.* (2008) Mesoporous silica nanoparticles as a delivery system of gadolinium for effective human stem cell tracking. *Small*, **4**, 1445–1452.

191. Kim, J., Kim, H.S., Lee, N. *et al.* (2008) Multifunctional uniform nanoparticles composed of a magnetite nanocrystal core and a mesoporous silica shell for magnetic resonance and fluorescence imaging and for drug delivery. *Angew. Chem. Int. Ed.*, **47**, 8438–8441.

192. Taylor, K.M.L., Kim, J.S., Rieter, W.J. *et al.* (2008) Mesoporous silica nanospheres as highly efficient MRI contrast agents. *J. Am. Chem. Soc.*, **130**, 2154–2155.

193. Hudson, S.P., Padera, R.F., Langer, R. and Kohane, D.S. (2008) The biocompatibility of mesoporous silicates. *Biomaterials*, **29**, 4045–4055.

194. Schlossbauer, A., Warncke, S., Gramlich, P.M.E. *et al.* (2010) A programmable DNA-based molecular valve for colloidal mesoporous silica. *Angew. Chem. Int. Ed.*, **49**, 4734–4737.

195. You, Y.Z., Kalebaila, K.K., Brock, S.L. and Oupicky, D. (2008) Temperature-controlled uptake and release in PNIPAM-modified porous silica nanoparticles. *Chem. Mater.*, **20**, 3354–3359.

196. Chang, J.H., Shim, C.H., Kim, B.J. *et al* (2005) Bicontinuous, thermoresponsive, L-3-phase silica nanocomposites and their smart drug-delivery applications. *Adv. Mater.*, **17**, 634–637.

197. Fu, Q., Rao, G.V.R., Ista, L.K. *et al.* (2003) Control of molecular transport through stimuli-responsive ordered mesoporous materials. *Adv. Mater.*, **15**, 1262–1266.

198. Zhou, Z.Y., Zhu, S.M. and Zhang, D. (2007) Grafting of thermo-responsive polymer inside mesoporous silica with large pore size using ATRP and investigation of its use in drug release. *J. Mater. Chem.*, **17**, 2428–2433.

199. Meng, H.A., Xue, M., Xia, T.A. *et al.* (2010) Autonomous in vitro anticancer drug release from mesoporous silica nanoparticles by ph-sensitive nanovalves. *J. Am. Chem. Soc.*, **132**, 12690–12697.

200. Liu, R., Zhang, Y., Zhao, X. *et al.* (2010) pH-responsive nanogated ensemble based on gold-capped mesoporous silica through an acid-labile acetal linker. *J. Am. Chem. Soc.*, **132**, 1500–1501.

201. Park, C., Oh, K., Lee, S.C. and Kim, C. (2007) Controlled release of guest molecules from mesoporous silica particles based on a pH-responsive polypseudorotaxane motif. *Angew. Chem. Int. Ed.*, **46**, 1455–1457.

202. Fisher, K.A., Huddersman, K.D. and Taylor, M.J. (2003) Comparison of micro- and mesoporous inorganic materials in the uptake and release of the drug model fluorescein and its analogues. *Chem. Eur. J.*, **9**, 5873–5878.

203. Yang, Q., Wang, S.H., Fan, P.W. *et al.* (2005) pH-responsive carrier system based on carboxylic acid modified mesoporous silica and polyelectrolyte for drug delivery. *Chem. Mater.*, **17**, 5999–6003.

204. Leung, K.C.F., Nguyen, T.D., Stoddart, J.F. and Zink, J.I. (2006) Supramolecular nanovalves controlled by proton abstraction and competitive binding. *Chem. Mater.*, **18**, 5919–5928.

205. Nguyen, T.D., Leung, K.C.F., Liong, M. *et al.* (2006) Construction of a pH-driven supramolecular nanovalve. *Org. Lett.*, **8**, 3363–3366.

206. Xu, W.J., Gao, Q., Xu, Y. *et al.* (2009) pH-Controlled drug release from mesoporous silica tablets coated with hydroxypropyl methylcellulose phthalate. *Mater. Res. Bull.*, **44**, 606–612.

207. Song, S.W., Hidajat, K. and Kawi, S. (2007) pH-Controllable drug release using hydrogel encapsulated mesoporous silica. *Chem. Commun.*, 4396–4398.

208. Zhu, Y.F., Shi, J.L., Shen, W.H. *et al.* (2005) Stimuli-responsive controlled drug release from a hollow mesoporous silica sphere/polyelectrolyte multilayer core-shell structure. *Angew. Chem. Int. Ed.*, **44**, 5083–5087.

209. Casasus, R., Marcos, M.D., Martinez-Manez, R. *et al.* (2004) Toward the development of ionically controlled nanoscopic molecular gates. *J. Am. Chem. Soc.*, **126**, 8612–8613.

210. Zheng, H.Q., Wang, Y. and Che, S.N. (2011) Coordination bonding-based mesoporous silica for pH-responsive anticancer drug doxorubicin delivery. *J. Phys. Chem. C*, **115**, 16803–16813.

211. Kim, H.J., Matsuda, H., Zhou, H.S. and Honma, I. (2006) Ultrasound-triggered smart drug release from a poly(dimethylsiloxane)-mesoporous silica composite. *Adv. Mater.*, **18**, 3083–3088.

212. Ferris, D.P., Zhao, Y.L., Khashab, N.M. *et al.* (2009) Light-operated mechanized nanoparticles. *J. Am. Chem. Soc.*, **131**, 1686–1688.

213. Knezevic, N.Z., Trewyn, B.G. and Lin, V.S.Y. (2011) Functionalized mesoporous silica nanoparticle-based visible light responsive controlled release delivery system. *Chem. Commun.*, **47**, 2817–2819.

214. Mal, N.K., Fujiwara, M. and Tanaka, Y. (2003) Photocontrolled reversible release of guest molecules from coumarin-modified mesoporous silica. *Nature*, **421**, 350–353.

215. Lu, J., Choi, E., Tamanoi, F. and Zink, J.I. (2008) Light-activated nanoimpeller-controlled drug release in cancer cells. *Small*, **4**, 421–426.

216. Ruiz-Hernandez, E., Baeza, A. and Vallet-Regi, M. (2011) Smart drug delivery through DNA/magnetic nanoparticle gates. *ACS Nano*, **5**, 1259–1266.

217. Kong, S.D., Zhang, W.Z., Lee, J.H. *et al.* (2010) Magnetically vectored nanocapsules for tumor penetration and remotely switchable on-demand drug release. *Nano Lett.*, **10**, 5088–5092.

218. Ruiz-Hernandez, E., Lopez-Noriega, A., Arcos, D. *et al.* (2007) Aerosol-assisted synthesis of magnetic mesoporous silica spheres for drug targeting. *Chem. Mater.*, **19**, 3455–3463.

219. Lopez-Noriega, A., Ruiz-Hernandez, E., Stevens, S.M. *et al.* (2009) Mesoporous microspheres with doubly ordered core-shell structure. *Chem. Mater.*, **21**, 18–20.

220. Arruebo, M., Galan, M., Navascues, N. *et al.* (2006) Development of magnetic nanostructured silica-based materials as potential vectors for drug-delivery applications. *Chem. Mater.*, **18**, 1911–1919.

221. Baeza, A., Guisasola, E., Ruiz-Hernández, E. and Vallet-Regí, M. (2012) Magnetically triggered multidrug release by hybrid mesoporous silica nanoparticles. *Chem. Mater.*, **24**, 517–524.

222. Liu, R., Zhao, X., Wu, T. and Feng, P.Y. (2008) Tunable redox-responsive hybrid nanogated ensembles. *J. Am. Chem. Soc.*, **130**, 14418–14419.

223. Hernandez, R., Tseng, H.-R., Wong, J.W. *et al.* (2004) An operational supramolecular nanovalve. *J. Am. Chem. Soc.*, **126**, 3370–3371.

224. Nguyen, T.D., Tseng, H.R., Celestre, P.C. *et al* (2005) A reversible molecular valve. *Proc. Natl. Acad. Sci. U.S.A.*, **102**, 10029–10034.

225. Nguyen, T.D., Liu, Y., Saha, S. *et al.* (2007) Design and optimization of molecular nanovalves based on redox-switchable bistable rotaxanes. *J. Am. Chem. Soc.*, **129**, 626–634.

226. Torney, F., Trewyn, B.G., Lin, V.S.Y. and Wang, K. (2007) Mesoporous silica nanoparticles deliver DNA and chemicals into plants. *Nat. Nanotechnol.*, **2**, 295–300.

227. Gruenhagen, J.A., Lai, C.Y., Radu, D.R. *et al.* (2005) Real-time imaging of tunable adenosine 5-triphosphate release from an MCM-41-type mesoporous silica nanosphere-based delivery system. *Appl. Spectrosc.*, **59**, 424–431.

228. Porta, F., Lamers, G.E.M., Zink, J.I. and Kros, A. (2011) Peptide modified mesoporous silica nanocontainers. *Phys. Chem. Chem. Phys.*, **13**, 9982–9985.

229. Bernardos, A., Mondragon, L., Aznar, E. *et al.* (2010) Enzyme-responsive intracellular controlled release using nanometric silica mesoporous supports capped with "Saccharides". *ACS Nano*, **4**, 6353–6368.

230. Schlossbauer, A., Kecht, J. and Bein, T. (2009) Biotin-avidin as a protease-responsive cap system for controlled guest release from colloidal mesoporous silica. *Angew. Chem. Int. Ed.*, **48**, 3092–3095.

231. Climent, E., Bernardos, A., Martinez-Manez, R. *et al.* (2009) Controlled delivery systems using antibody-capped mesoporous nanocontainers. *J. Am. Chem. Soc.*, **131**, 14075–14080.

232. Peretz, S. and Regev, O. (2012) Carbon nanotubes as nanocarriers in medicine. *Curr. Opin. Colloid Interface Sci.*, **17**, 360–368.

233. Fisher, C., Rider, A.E., Han, Z.J. *et al.* (2012) Applications and nanotoxicity of carbon nanotubes and graphene in biomedicine. *J. Nanomater.*, **2012**, 19pp.

234. Mu, Q., Liu, W., Xing, Y. *et al* (2008) Protein binding by functionalized multiwalled carbon nanotubes is governed by the surface chemistry of both parties and the nanotube diameter. *J. Phys. Chem. C*, **112**, 3300–3307.

235. Bottini, M., Rosato, N. and Bottini, N. (2011) PEG-modified carbon nanotubes in biomedicine: current status and challenges ahead. *Biomacromolecules*, **12**, 3381–3393.

236. Chen, R.J., Zhang, Y.G., Wang, D.W. and Dai, H.J. (2001) Noncovalent sidewall functionalization of single-walled carbon nanotubes for protein immobilization. *J. Am. Chem. Soc.*, **123**, 3838–3839.

237. Bianco, A., Kostarelos, K. and Prato, M. (2005) Applications of carbon nanotubes in drug delivery. *Curr. Opin. Chem. Biol.*, **9**, 674–679.

238. Hilder, T.A. and Hill, J.M. (2009) Modeling the loading and unloading of drugs into nanotubes. *Small*, **5**, 300–308.

239. Nakashima, N. (2006) Solubilization of single-walled carbon nanotubes with condensed aromatic compounds. *Sci. Technol. Adv. Mater.*, **7**, 609–616.

240. Liu, Z., Sun, X., Nakayama-Ratchford, N. and Dai, H. (2007) Supramolecular chemistry on water-soluble carbon nanotubes for drug loading and delivery. *ACS Nano*, **1**, 50–56.

241. Liu, Z., Fan, A.C., Rakhra, K. *et al.* (2009) Supramolecular stacking of doxorubicin on carbon nanotubes for in vivo cancer therapy. *Angew. Chem. Int. Ed.*, **48**, 7668–7672.

242. Heister, E., Neves, V., Tilmaciu, C. *et al.* (2009) Triple functionalisation of single-walled carbon nanotubes with doxorubicin, a monoclonal antibody, and a fluorescent marker for targeted cancer therapy. *Carbon*, **47**, 2152–2160.

243. Zhang, X., Meng, L., Lu, Q. *et al.* (2009) Targeted delivery and controlled release of doxorubicin to cancer cells using modified single wall carbon nanotubes. *Biomaterials*, **30**, 6041–6047.

244. Fabbro, C., Ali-Boucetta, H., Da Ros, T. *et al.* (2012) Targeting carbon nanotubes against cancer. *Chem. Commun.*, **48**, 3911–3926.

245. Liu, Z., Robinson, J.T., Tabakman, S.M. *et al.* (2011) Carbon materials for drug delivery and cancer therapy. *Mater. Today*, **14**, 316–323.

246. Liu, Z., Chen, K., Davis, C. *et al.* (2008) Drug delivery with carbon nanotubes for in vivo cancer treatment. *Cancer Res.*, **68**, 6652–6660.

247. Berlin, J.M., Leonard, A.D., Pham, T.T. *et al.* (2010) Effective drug delivery, in vitro and in vivo, by carbon-based nanovectors noncovalently loaded with unmodified paclitaxel. *ACS Nano*, **4**, 4621–4636.

248. Berlin, J.M., Pham, T.T., Sano, D. *et al.* (2011) Noncovalent functionalization of carbon nanovectors with an antibody enables targeted drug delivery. *ACS Nano*, **5**, 6643–6650.

249. Hilder, T.A. and Hill, J.M. (2008) Encapsulation of the anticancer drug cisplatin into nanotubes. Proceeding of the international Conference on Nanoscience and Nanotechnology ICONN, pp. 109–112.

250. Son, S.J., Bai, X. and Lee, S.B. (2007) Inorganic hollow nanoparticles and nanotubes in nanomedicine. Part 1. Drug/gene delivery applications. *Drug Discov. Today*, **12**, 650–656.

251. Prakash, R., Washburn, S., Superfine, R. *et al.* (2003) Visualization of individual carbon nanotubes with fluorescence microscopy using conventional fluorophores. *Appl. Phys. Lett.*, **83**, 1219–1221.

252. Ritschel, M., Leonhardt, A., Elefant, D. *et al.* (2007) Rhenium-catalyzed growth carbon nanotubes. *J. Phys. Chem. C*, **111**, 8414–8417.

253. Moon, H.K., Lee, S.H. and Choi, H.C. (2009) In vivo near-infrared mediated tumor destruction by photothermal effect of carbon nanotubes. *ACS Nano*, **3**, 3707–3713.

254. Gao, Y. and Kyratzis, I. (2008) Covalent immobilization of proteins on carbon nanotubes using the cross-linker 1-ethyl-3-(3-dimethylaminopropyl)carbodiimide-a critical assessment. *Bioconjugate Chem.*, **19**, 1945–1950.

255. Gabizon, A., Shmeeda, H. and Barenholz, Y. (2003) Pharmacokinetics of pegylated liposomal doxorubicin – review of animal and human studies. *Clin. Pharmacokinetics*, **42**, 419–436.

256. Dhar, S., Liu, Z., Thomale, J. *et al.* (2008) Targeted single-wall carbon nanotube-mediated Pt(IV) prodrug delivery using folate as a homing device. *J. Am. Chem. Soc.*, **130**, 11467–11476.

257. Zucker, D., Andriyanov, A.V., Steiner, A. *et al.* (2012) Characterization of PEGylated nanoliposomes co-remotely loaded with topotecan and vincristine: relating structure and pharmacokinetics to therapeutic efficacy. *J. Controlled Release*, **160**, 281–289.

258. Maruyama, K., Ishida, O., Takizawa, T. and Moribe, K. (1999) Possibility of active targeting to tumor tissues with liposomes. *Adv. Drug Deliv. Rev.*, **40**, 89–102.

259. Abu-Salah, K.M., Ansari, A.A. and Alrokayan, S.A. (2010) DNA-based applications in nanobiotechnology. *J. Biomed. Biotechnol.*, **2010**, 715295–715295.

260. Liu, Z., Winters, M., Holodniy, M. and Dai, H. (2007) siRNA delivery into human T cells and primary cells with carbon-nanotube transporters. *Angew. Chem. Int. Ed.*, **46**, 2023–2027.

261. Wu, Y., Phillips, J.A., Liu, H. *et al.* (2008) Carbon nanotubes protect DNA strands during cellular delivery. *ACS Nano*, **2**, 2023–2028.

262. Pantarotto, D., Singh, R., McCarthy, D. *et al.* (2004) Functionalized carbon nanotubes for plasmid DNA gene delivery. *Angew. Chem. Int. Ed.*, **43**, 5242–5246.

263. Wang, X., Ren, J. and Qu, X. (2008) Targeted RNA interference of cyclin A(2) mediated by functionalized single-walled carbon nanotubes induces proliferation arrest and apoptosis in chronic myelogenous leukemia K562 cells. *ChemMedChem*, **3**, 940–945.

264. Acharya, G. and Park, K. (2006) Mechanisms of controlled drug release from drug-eluting stents. *Adv. Drug Deliv. Rev.*, **58**, 387–401.

265. Radin, S., Campbell, J.T., Ducheyne, P. and Cuckler, J.M. (1997) Calcium phosphate ceramic coatings as carriers of vancomycin. *Biomaterials*, **18**, 777–782.

266. Yang, Z., Han, J., Li, J. *et al.* (2009) Incorporation of methotrexate in calcium phosphate cement: behavior and release in vitro and in vivo. *Orthopedics*, **32**, 27–27.

267. Stigter, M., Bezemer, J., de Groot, K. and Layrolle, P. (2004) Incorporation of different antibiotics into carbonated hydroxyapatite coatings on titanium implants, release and antibiotic efficacy. *J. Controlled Release*, **99**, 127–137.

268. Roy, M., Bandyopadhyay, A. and Bose, S. (2009) In vitro antimicrobial and biological properties of laser assisted tricalcium phosphate coating on titanium for load bearing implant. *Mater. Sci. Eng. C-Mater. Biol. Appl.*, **29**, 1965–1968.

269. Liu, Y., de Groot, K. and Hunziker, E.B. (2005) BMP-2 liberated from biomimetic implant coatings induces and sustains direct ossification in an ectopic rat model. *Bone*, **36**, 745–757.
270. Peter, B., Pioletti, D.P., Laib, S. *et al.* (2005) Calcium phosphate drug delivery system: influence of local zoledronate release on bone implant osteointegration. *Bone*, **36**, 52–60.
271. Garbuz, D.S., Hu, Y., Kim, W.Y. *et al.* (2008) Enhanced gap filling and osteoconduction associated with alendronate-calcium phosphate-coated porous tantalum. *J. Bone Joint Surg. -Am.*, **90A**, 1090–1100.
272. Ezra, A. and Golomb, G. (2000) Administration routes and delivery systems of bisphosphonates for the treatment of bone resorption. *Adv. Drug Deliv. Rev.*, **42**, 175–195.

13
Ceramics for Gene Transfection

Blanca González[1,2]

[1]*Departamento de Química Inorgánica y Bioinorgánica, Facultad de Farmacia,
Universidad Complutense de Madrid, Spain*
[2]*Centro de Investigación Biomédica en Red en Bioingeniería,
Biomateriales y Nanomedicina (CIBER-BBN), Spain*

13.1 Gene Transfection

The process by which foreign nucleic acids sequences are introduced into the interior of a cell is known as *gene transfection* [1, 2].

The incorporation of genetic material into a host genome requires delivery of the genes, DNA fragments, or plasmid DNA (pDNA) into the cell nucleus. If the gene is safely introduced into the cell, reaches the nucleus and is inserted into the nuclear DNA it would be finally expressed as the protein which it codifies for. Distinction can be made between a transient transfection, where DNA does not integrate into the host chromosome, and a stable transfection, where the foreign DNA is integrated into the chromosome and passed over to the next generation.

Another alternative involves *RNA interference (RNAi)*, a mechanism of action in which small or short interfering RNA (siRNA) sequences bind to targeted messenger RNA (mRNA) and initiate its degradation, leading to gene silencing [3]. In this case of RNA targeting the gene delivery into the nucleus is not generally required. The RNAi technique is an effective method to inhibit protein expression by targeted cleavage of mRNA and has made substantial progress since the first demonstration of gene knockdown in mammalian cells [4]. Hence, siRNA-based therapeutics have great potential to suppress pathogenic gene expressions [5–7].

The two main nucleic acid delivery approaches, DNA delivery and siRNA delivery, are conceptually different and will be referred to throughout this chapter.

Bioceramics with Clinical Applications, First Edition. Edited by María Vallet-Regí.
© 2014 John Wiley & Sons, Ltd. Published 2014 by John Wiley & Sons, Ltd.

However, the delivery of DNA/siRNA into the interior of a cell is not an easy task and needs to overcome several biological obstacles. For instance, the lipid bilayer of the cell membrane acts as a biological barrier against foreign and/or pathogenic nucleic acids, hence preventing therapeutic delivery of genes, siRNA, and pDNA. To get around this problem there are physical, chemical, and biological methods that notably enhance the transfection efficiency (Figure 13.1) [8].

On the one hand the *physical methods* are microinjection, electroporation, which consists in the generation of micropores in the cellular membrane with controlled pulsed electric fields of submicrosecond duration that facilitates the capture of the exogenous DNA molecules, or bombardment of the cells with microparticles that carry adsorbed DNA fragments. These methods are limited in their ability to treat large quantities of cells and DNA and, moreover, cell damaging may occur.

On the other hand, the *chemical and biological methods* make use of effective gene carrier vectors for the gene delivery. These must be vehicles able to compact and protect oligonucleotides and to actively cross the lipid membrane, delivering nucleic acid cargos with efficiency and limited toxicity. Research efforts focused on designing effective carrier vectors have led to the development of two main strategies: *viral and nonviral gene delivery*.

Viral systems exhibit high efficiency at delivering both DNA and RNA to numerous cell lines, since they harness the infection mechanism of natural viruses to introduce themselves into the cells. Then, viral carriers are evidently most effective but possess safety issues and the possibility of gene recombination [9]. The main virus families used are retrovirus, adenovirus, adeno-associated virus, herpes virus, and Sendai virus, among others.

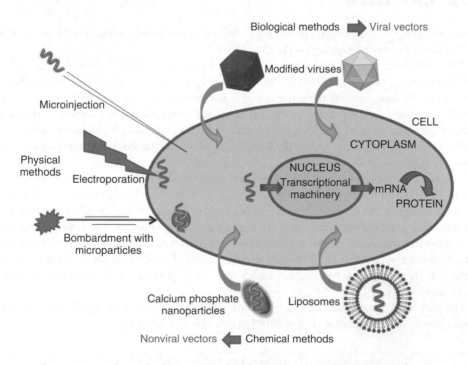

Figure 13.1   *Physical, chemical, and biological methods for gene transfection*

Although viral particles are engineered to eliminate the viral pathogenicity, biosafety concerns, including toxicity, immunogenicity, and limitations regarding scale-up procedures, still exist.

Therefore, the current research is focused on the design of synthetic nonviral nanovectors, as a safer and more biocompatible alternative to viral vectors [10–12]. However, an emerging challenge of nonviral carriers is their low transfection performance since they are hindered by numerous extra- and intracellular obstacles. Based on different chemical building blocks, a broad variety of nonviral vectors have been developed from biocompatible nanostructured materials. These systems are, for example, cationic polymers [13], polymeric nanocapsules [14], liposomes [15], dendrimers [16, 17], and inorganic nanoparticles [18–21].

Nowadays the applications of gene transfection are mainly in the field of biotechnology, since gene transfection performed *in vitro* opens huge possibilities for the DNA reprogramming of bacteria and eukaryotic cells [22]. The manufacture of drugs using transgenic plants or animals is an example of biotechnological applications regarding human health care. Perhaps the most promising application is gene therapy, which uses genes as a medicine and is currently being tested as a therapeutic option for some genetic diseases. At the moment there are several clinical trials using viral vectors for the treatment of genetic diseases [23]. Although gene therapy is a promising treatment option for a number of diseases (including inherited disorders, some types of cancer, and certain viral infections), the technique remains risky, and is still under study to make sure that it will be safe and effective.

Gene therapy can be defined as the treatment or prevention, in the case of DNA vaccines, of genetic disorders by the transfer of therapeutic genetic material (DNA or RNA) into a group of cells (a tissue or organ) [24–27]. The introduced genetic material has the ability to express the right version of a protein in cells where the endogenous gene is deficient. Currently, gene therapy is also being developed as an alternative or complementary therapy in cancer treatment [28] and other acquired diseases, such as cardiovascular and neurodegenerative ones and some immuno-deficiencies.

The therapeutic delivery of nucleic acids is intended to correct or modify the expression of the gene influencing the disease process. This can be achieved by means of replacing a mutated gene that causes disease with a healthy copy of the gene, inactivating (or "knocking out") a mutated gene that is functioning improperly, up or down regulating the expression of target genes, introducing a new gene into the body to help fight a disease, or even inducing the malignant cell apoptosis in the case of cancer.

However, gene delivery is one of the biggest challenges in the field of gene therapy. In addition, there are other stages for a successful gene delivery and expression. In practice, good candidate diseases for gene therapy are those originated by defects in only one gene. Moreover, it is mandatory to know which gene is mutated in the disorder and to have its normal version available. It is important to target the right cells and activate their transcription and translation machinery properly. The insertion of the gene in the host genome is not at random, since insertions in oncogenes can generate their deregulation and produce tumor cells, therefore the vector should possess an insertion system to avoid this problem. Another requirement is avoiding harmful side effects, since there is a risk that an unfamiliar biological substance introduced into the body may be toxic or the body may produce an immune response against it.

In gene therapy the way of bringing the target cells and the gene delivery vectors into contact may be *ex vivo*, by removing a sample of the patient's cells and exposing them to the gene vector in a laboratory setting (*in vitro*). Then, the cells containing the vector are subsequently returned to the patient. Alternatively, the gene vector can be injected or given intravenously directly into a specific tissue in the organism *in vivo*, where it is taken up by individual cells. The treatment *ex vivo* optimizes the transfection to the targeted cells, although it can alter their characteristics due to the *in vitro* manipulation. The *in vivo* treatment does not alter the target cells or tissue but hinders the vector from reaching the site of interest and increases its possible toxic effects.

13.2 Gene Transfection Based on Nonviral Vectors

Due to *in vitro* and *in vivo* barriers for gene delivery, a nonviral vector should fulfill several requirements in order to deal with those biological barriers and improve the efficiency of the gene transfer (see Figure 13.2a) [11, 29].

1. During systemic circulation and at the extracellular space, free oligonucleotides and DNA are rapidly degraded by serum exonucleases in the blood. The vector must condense DNA to protect it from nuclease degradation by means of compaction of DNA into spherical complexes. Complexation of DNA via electrostatic interactions between the negatively charged phosphate backbone of DNA and cationic molecules leads to charge neutralization and a compaction of the nucleotide fragments.
2. Internalization or cellular uptake of naked DNA via plasma membrane permeation is hindered by the size and negative charge of the DNA, that is, the negative potential of the cell membrane impedes the incorporation of negatively charged phosphate-containing DNA. Therefore, the vector must exhibit a positive net electric surface charge at physiological pH to favor vector wrapping by the membrane and cellular uptake. Furthermore, vectors derivatized with targeting ligands are internalized more rapidly and efficiently by receptor-mediated endocytosis, whereas plain or non-targeted vectors are taken up by non-specific endocytosis (for a description of the endocytosis mechanism see Section 14.3.1).
3. Following uptake into endocytic vesicles, gene delivery vehicles must possess a mechanism to protect DNA from the acid environment inside endosomes and perform endosomal escape. In this respect, macromolecules that have amine groups with low pK_a values exhibit a "proton sponge" behavior. They are able to buffer the endosomal vesicle, which leads to endosomal swelling and lysis, thus releasing the DNA or DNA/vector complex into the cytoplasm. Once released into the cytoplasm, DNA, or DNA/vector complexes en route to the nucleus must overcome additional barriers in the cytosol that hamper delivery of the DNA into the nucleus of the host cell. The cytoplasm is abundant in endonucleases and various organelles that cause degradation and hinder vector movements.
4. The nuclear envelope is the ultimate obstacle to the nuclear entry of DNA. There is a higher transfectability in dividing cells since disassembly of the nuclear membrane assists vector entry into the nuclei during mitotic cell division [30]. In non-dividing

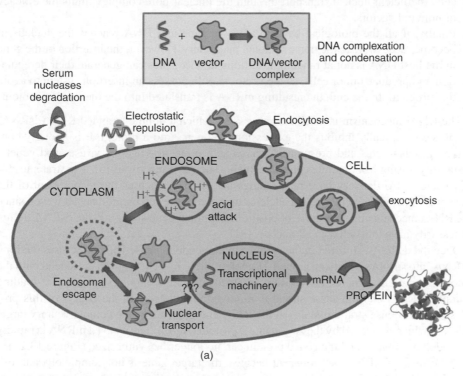

(a)

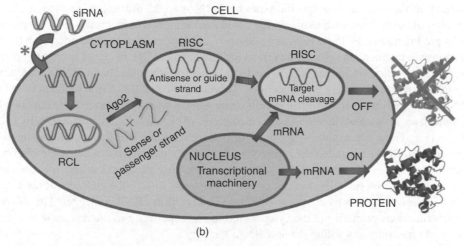

(b)

Figure 13.2 (a) Cellular trafficking of nonviral vectors. (b) Simplified model of the RNAi mechanism in mammalian cells (See insert for color representation of the figure)

cells inefficient nuclear transport through the nuclear pore complex limits the efficacy of nonviral vectors.

5. Finally, if all the biological barriers are overcome and DNA reaches the nucleus, it becomes accessible to the transcriptional machinery. However, the insertion of the gene in the host DNA is not at random, insertions in oncogenes can generate their deregulation and produce tumor cells, so the vector should possess an insertion system to avoid this problem. In the end, the resulting mRNA is translated into the therapeutic protein.

The RNAi mechanism is a cellular process by which a double-stranded RNA (dsRNA) sequence specifically inhibits the expression of a gene into a protein. It is strongly conserved in eukaryotes and presumably serves as a protection against viruses and genetic instability arising from mobile genetic elements such as transposons [5]. The strategies for DNA and RNAi delivery have many things in common, such as the requirement of the transport of the negatively charged nucleic acids across the cell membrane. However, since the RNAi mechanism is a process of posttranscriptional gene silencing, the dsRNA do not necessarily have to reach the cell nucleus [31].

Once the dsRNA is located in the cytoplasm, the endonuclease Dicer processes the long dsRNA into siRNAs which are around 21 nucleotides long. Then, siRNA is loaded onto the ribonucleoprotein complex RISC (RNA-induced silencing complex) by the RISC-loading complex (RLC) and the sense strand is removed and cleaved in the course of this process. The antisense strand remains in the RISC and guides it to the complementary target mRNA, which is cleaved by the Argonaute 2 protein (Ago2). The cleaved mRNA is rapidly degraded by RNases and the coded protein can no longer be synthesized (Figure 13.2b).

The effects of siRNAs are transient because the target gene is not completely shut off, which is why RNAi is referred to as a "knockdown" technology as opposed to "knockout" in the case of transgene animals created by homologous recombination. The technique of turning off the expression of specific genes by dsRNAs could, initially, be applied to a large number of eukaryotes, such as plants, but could not be applied to mammals since long dsRNAs are interpreted by these cells as a pathogen and trigger an unspecific interferon (INF) response in mammalian cells. However, on the one hand the INF response is only triggered by dsRNAs which are longer than 30 nucleotides and, on the other hand, RNAi is induced by RNAs of approximately 21 nucleotides. Therefore, the solution to the problem of using RNAi for experiments in mammalian cells came when it was demonstrated that chemically synthesized 21-nucleotide siRNAs also mediate RNAi in cultured mammalian cells [32]. This created new opportunities, not only for research, but also for therapeutic treatments. In addition, RNAi expanded the repertoire of the already well known antisense oligonucleotide strategies. These strategies differ since antisense oligonucleotides are modified single-strand DNA (ssDNA) molecules which primarily induce the cleavage of the target RNA in the cell nucleus by activation of RNase H. In contrast, RNAi is triggered by dsRNA which functions primarily in the cytoplasm, but the experience in the antisense field has allowed rapid progress within the new RNAi strategy.

13.3   Ceramic Nanoparticles for Gene Transfection

The use of nanoparticles in medicine is an emerging field of nanobiotechnology [33]. As a result of their small size, nanoparticles can penetrate the cell wall and deliver drugs or biomolecules into living systems, usually for a therapeutic purpose. Regarding their size, the upper limit for an efficient uptake through the cell membrane appears to be around 100 nm. Besides viral, polymeric, and liposomal vectors, inorganic nanoparticles are especially suitable for gene transfection because they can be engineered and surface functionalized in many different ways [20]. Their description in this chapter will be restricted, following the subject of this book, to the next ceramic, and inorganic nanoparticles: calcium phosphate, mesoporous silica, carbon allotropes, and iron oxide nanoparticles (IONPs).

Most of the systems described herein relate to basic research and indeed significant advances have been achieved from *in vitro* experiments and also some *in vivo* tests. Nevertheless, future research efforts are needed to transfer the application of these synthetic nanovectors to clinical nanomedicine.

Inorganic nanoparticles offer many ways to prepare systems with a defined particle size, surface functionalization, nucleic acid protection, and biocompatibility. It is possible to fine tune their nanostructure, for example, by coating them with different layers or by generating internal nanopores and, therefore, their use as carriers can be extended. The surface functionalization of nanoparticles (shell) is important for uptake and short-term cellular interaction, whereas the chemical composition of the interior (core) is important for long-term biodegradability and biocompatibility.

As a requisite for DNA/siRNA complexation and protection, most of the inorganic NPs need the introduction of cationic charge in their surface. The general method to achieve this is functionalization with amine polymeric structures, thus resulting in organic–inorganic hybrid materials [34]. Amine polymeric compounds possess primary and tertiary amines among other functional groups in their structures. At physiological pH (pH 7.4) primary amines (pK_a 9–11) are positively charged ammonium groups, complexing the DNA fragment through ionic interaction with negatively charged phosphate groups of nucleic acids. The entry into the cell via endocytosis is optimized since protonated residues on the resulting compact complexes favor the binding to the negatively charged cell surface. Thereafter, tertiary amines of the polymeric framework (pK_a 5–8) are responsible for a "proton sponge" behavior at the intracellular level. Their protonation protects the DNA from degradation in an endosomal compartment (pH 5–6). Both, DNA complexation for the endocytosis process and proton sponge behavior, would explain the efficiency found with these polyamine systems.

Cationic polymers such as polyethylenimine (PEI) and poly(L-lysine) (PLL), as well as cationic dendrimers such as polyamidoamine (PAMAM) and poly(propyleneimine) (PPI) dendrimers are effective gene transfection agents by themselves (Figure 13.3) [35–37]. In general, the high molecular weight (HMW) versions of these polymers and the higher generations of the dendrimers show high transfection efficiency, because of an enhanced cellular uptake due to the high load of positively charged amines on their surfaces, and a high level of endosomal escape due to a proton buffering effect over a broad pH range [38].

However, cytotoxicity of these polyamine systems occurs via a proton sponge effect that leads to heightened proton pump activity inside the cell, osmotic swelling of the endocytic compartment, endosomal rupture, and ultimately cell death by a mitochondrial mediated mechanism [39]. Therefore, for these polyamine systems the transfection efficiency has been found to be usually contradicted by the biocompatibility [40].

Moving to lower molecular weight (LMW) polymers or generation dendrimers it is possible to obtain nontoxic products, but the gene transfection efficiency sinks quite drastically [41]. To overcome this problem inorganic nanoparticles are generally used as an inorganic support to covalently attach LMW polymers or generation dendrimers. As a result, the

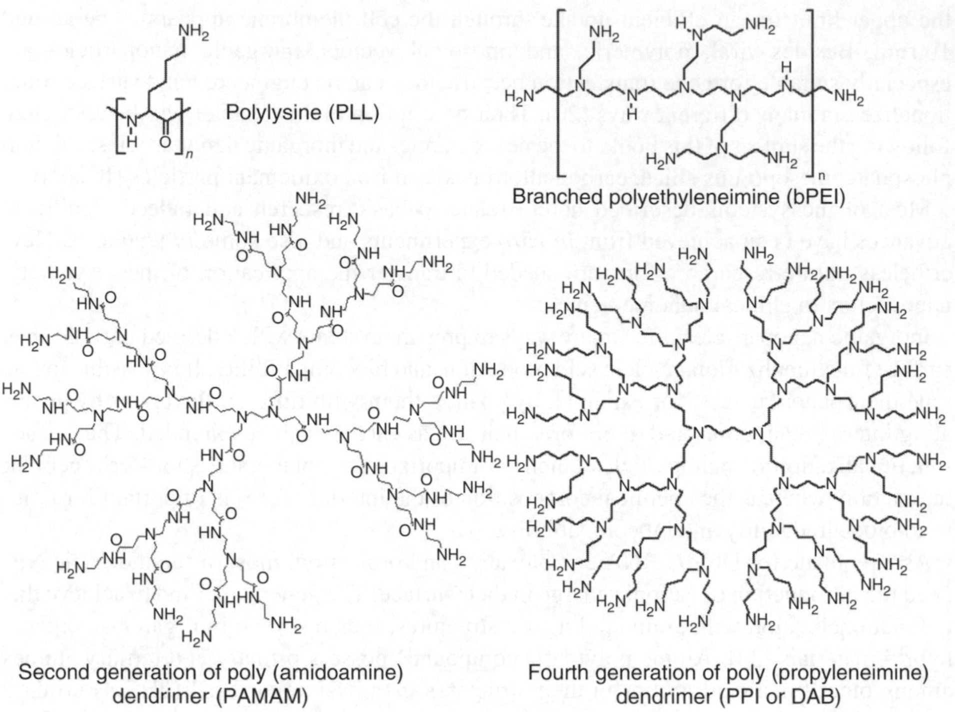

Polylysine (PLL)

Branched polyethyleneimine (bPEI)

Second generation of poly (amidoamine) dendrimer (PAMAM)

Fourth generation of poly (propyleneimine) dendrimer (PPI or DAB)

Figure 13.3   *Structures of cationic polymers and dendrimers used as gene transfection agents*

global surface charges of the hybrid material increase and, therefore, the system enables an optimal pDNA complexation. For instance, LMW branched polyethylenimine (bPEI) exhibits enhanced cellular uptake and transfection efficiency with low cytotoxicity when it is conjugated to inorganic nanomaterials such as gold, silica, carbon materials, and iron oxide [42–46].

Engineered nanomaterials offer the opportunity to design multimodal approaches for the treatment of diseases, including cancer. Considering the use of a nanomaterial in which several properties are integrated, targeting, imaging, and therapy can be combined. In addition, a synergistic effect in a combined therapy is highly desirable. Regarding gene delivery, this synergistic effect can be achieved in two main ways:

- Combined hyperthermia and gene therapies, which can be fulfilled when the vector is composed of graphene oxide (GO) which possesses a strong near-infrared (NIR) optical absorption ability used for photothermal therapy of cancer cells, and also in the case of IONPs, that respond to a magnetic field exerting magnetic hyperthermia.
- Sequential delivery or co-delivery of siRNA and anticancer drugs. This combined therapy is mainly aimed at overcoming drug resistance in the treatment of cancer.

The term *multidrug resistance (MDR)* is used to define a resistance phenotype where cancer cells become resistant simultaneously to multiple drugs [47, 48]. This effect can be divided into two distinct classes, pump and nonpump resistance [49]. The pump resistance is caused by certain proteins that form membrane-bound adenosine triphosphate (ATP)-dependent active drug efflux pumps, which significantly decrease the intracellular concentration of the drug and thereby the efficacy of the treatment. Membrane proteins, P-glycoprotein (Pgp) and MDR-associated proteins have been shown to be the main players for pump resistance to a broad range of structurally and functionally distinct cyto-toxic agents [50]. The main mechanism of nonpump resistance is an activation of cellular antiapoptotic defense, mainly by Bcl-2 protein [51, 52]. Most of the anticancer drugs trigger apoptosis and simultaneously activate both pump and nonpump cellular defense of MDR, which prevents cell death. Therefore, to effectively suppress the overall resistance to chemotherapy, it is essential to simultaneously inhibit both pump and nonpump mecha-nisms of cellular resistance by targeting all the intracellular molecular targets [49, 53–55].

Since siRNA induces specific silencing of targeted protein, this technique offers signifi-cant potential in overcoming MDR of cancer cells [32, 56]. Special sequences of siRNAs targeted against mRNA encoding major proteins responsible for pump and nonpump cellu-lar defense have been developed and shown a substantial efficacy *in vitro* [54, 55, 57, 58]. Therefore, efficient co-delivery methods to cancer cells of siRNAs targeted against mRNA encoding major proteins responsible for pump and nonpump cellular defense, simultane-ously with a traditional anticancer drug for enhanced chemotherapy efficacy, have become an emerging objective of research.

A common strategy to validate or to test a nanomaterial as an effective gene carrier is the use of a reporter gene to measure the transfection efficiency. The most frequent genes used as a model are a pDNA which codes for the enhanced green fluorescent protein (pEGFP) and another one for the luciferase (pLuc). In both cases transfection efficiency is easy to quantify by measuring the expression of the resulting protein by flow cytometry.

13.3.1 Calcium Phosphate Nanoparticles

Calcium phosphate (CaP) bioceramics are the inorganic component of biological hard tis-sues, for example, bone and teeth, where they occur as carbonated hydroxyapatite (HAp). With the exception of enamel, CaP in the biological domain almost exclusively exists in the form of nanoparticles (see Section 3.7). Hence, calcium phosphate nanoparticles are non-toxic for cells and, therefore, they are biocompatible and biodegradable due to the absence of both foreign body response and toxic by-products. Both Ca^{2+} and PO_4^{3-} are naturally found in the body in the form of amorphous and crystalline solids, in hard tissues as well as in physiological fluids (in concentrations of 1–5 mM). The only objection is that the degradation of CaP might have a harmful effect on cells if the release of Ca^{2+} increases the intracellular level of calcium and initiates protein aggregation. Furthermore, CaP NPs are easy to prepare with low manufacturing costs and excellent storage abilities. Therefore, nanosized CaP particles are one of the most viable options for gene delivery, and they have been extensively used as *in vitro* gene delivery vectors for over 35 years [20, 59, 60]. They have even been denoted as "second-generation nonviral vectors in gene therapy" [61, 62].

DNA molecules present a good adherence to calcium phosphate nanoparticles, probably due to the binding affinity of Ca^{2+} ions on the surface of CaP to the helical phosphate groups in nucleic acids [63]. The encapsulation of DNA into CaP nanoparticles protects the nucleic acid from both the extracellular and cytoplasmatic environment, enabling its efficient deliv-ery into the cell nucleus. The cell entry of the DNA/CaP complex follows an endocytic pathway. Once the nanoparticles are endocytosed, and the endosome starts maturing into lysosome, proton pumps in the lysosomal membrane induce a pH drop. The CaP phase dissolves due to the low pH, resulting in large quantities of Ca^{2+} and PO_4^{3-} ions being released into the endosomal mixture inside the vesicle. As a result, the osmotic imbalance leads to a massive influx of H_2O into the vesicle which destabilizes the endosomal membrane and ruptures the vesicle, so the DNA/CaP complexes and/or DNA dissociated from CaP vectors, are freely released into the cytosol. Calcium ions also enhance the nuclear uptake of DNA through the nuclear pore complex due to the involvement of calcium in intracellular signaling pathways. Olton *et al.* [64] revealed that the cellular uptake of the DNA/CaP complexes, as well as the subsequent gene expression, in HeLa and Cos-7 cells is mediated by both clathrin- and caveolae-mediated endocytosis, unlike the traditional belief that clathrin-mediated endocytosis process is solely responsible for the cell internalization of the DNA/CaP complexes. Furthermore, it was confirmed that the involvement of the caveolae-mediated endocytic pathway was much stronger (for a description of the endocytosis mechanism see Section 14.3.1).

The original method for calcium phosphate transfection was discovered by Graham and van der Eb in 1973 [65]. According to this method, the formation of nanoparticles occurs on the DNA backbone. For the preparation of the DNA/CaP complex for transfection, a calcium chloride solution is mixed with the DNA and a subsequent addition of phosphate-buffered saline solution results in the formation of calcium phosphate with DNA fine precipitates (nano- and microparticles). This dispersion is added to a cell suspension, and the nanoparticles are taken up by the cells.

In the last years, another method for nanoparticle preparation has been described by the group of Epple [66]. A dispersion of calcium phosphate nanoparticles is prepared by rapid mixing of aqueous solutions of calcium nitrate and diammonium hydrogen phosphate at pH 9 under constant stirring. Immediately after mixing, the nanoparticles are functionalized with the aqueous solution of nucleic acid. Such nanoparticles are very small, in the range of 10–20 nm, and well defined, consisting of a calcium phosphate core and an outer shell of nucleic acid, whereas the precipitates of the standard method consist of a poorly defined DNA/CaP precipitate. The interactions between calcium phosphate and nucleic acid are electrostatic. To enhance the DNA protection against attack of nucleases and, therefore, improve the transfection efficiency, the single-shell nanoparticles were modified by the addition of additional layers of calcium phosphate (for protection) and nucleic acid (for colloidal stabilization), leading to multi-shell NPs [67]. A further improvement of this system was accomplished by the introduction of an outer shell of PEI. The outermost layer of PEI on the CaP gives a positive charge on the nanoparticle surface, which leads to electrostatic colloidal stabilization and is helpful for the penetration of NPs through the negatively charged cell membrane [68, 69].

Other authors have systematically studied several factors and properties of the CaP nanoparticles affecting the ability of the NPs to effectively bind, package, and condense DNA, which ultimately affect the DNA protection, release and, therefore, the transfection efficiency. In these studies several parameters have been optimized: the stoichiometry (Ca/P ratio) and the mode in which the precursor solutions are mixed, the CaP phase, which can affect or modulate the dissolution rate, and the optimal nanosized range [70, 71].

Recent strategies of preparation of CaP NPs for gene transfection have in common the use of a polymer such as poly(ethylene glycol) (PEG) to impart stability to the NPs, as well as stealth properties by minimizing proteins adsorption (opsonization) in physiological fluids. The use of a bisphosphonate (bp)-modified PEG enables a strong binding of this polymer to the CaP nanoparticle surface through the bp group. Then, stable DNA/CaP nanoparticles have been produced by using this PEG-bp derivative [72]. In another example, the design of a block copolymer possessing a PEG block and a phosphate-based pH-dependent block, allows the preparation of pH-sensitive CaP NPs as gene carriers. The PEG block forms a hydrophilic shell that controls the size of the CaP particles to approximately 200 nm. The pH-sensitive block has many phosphomonoester moieties which strongly interact with the calcium ions within the CaP core, therefore stabilizing the CaP NPs at neutral pH. Moreover, since the phosphomonoester moieties have two distinct pK_a values (~ 3 and ~ 7) the CaP NPs can be destabilized by gradual protonation of the phosphomonoester moieties during acidification in the early endosome, allowing the release of encapsulated nucleic acids [73]. Both systems exhibit great potential as efficient and safe gene delivery vectors for future gene therapy applications.

A very recent work describes a method for the spatial application of CaP nanoparticle-based gene delivery. The coating of implants with CaP NPs could be applied for localized gene delivery in bone diseases, avoiding the systemic application of gene carriers in nanoparticulate form. To achieve this aim, conductive surfaces, such as titanium and ITO (indium tin oxide-coated glass, electrically conductive, and transparent), were electrophoretically coated with DNA-functionalized CaP NPs. The resulting bioactive substrates were assayed in gene transfection experiments on NIH3T3 and HeLa cell lines, using the reporter gene for the enhanced green fluorescence protein (EGFP). The transfection was also followed by live cell imaging in the transparent ITO substrate [74].

13.3.2   Mesoporous Silica Nanoparticles

In addition to their application for the controlled release of drugs, mesoporous silica nanoparticles (MSNPs) have the potential to act as delivery platforms of genetic material (see Sections 5.7, 5.8, and 12.3.2). For this purpose, the external silica surface has to be provided with a positive charge in order to maximize the binding of negatively charged nucleic acids and their delivery into cells. The introduction of cationic charge on the surface of silica materials for DNA complexation is usually performed via functionalization with amine polymeric structures [75–77], thus resulting in organic–inorganic hybrid materials with a porous structure. Hence, it is possible to achieve a double delivery of both chemical therapeutic agents, allocated within the mesopores, and genes, complexed in the outer surface of the nanodevice. This ability would be especially useful in cancer treatments for achieving synergistic or additive effects in combined therapies. Current combined therapies lack simultaneous targeted drug and nucleic acid delivery to tumors. As just mentioned, MSNPs are envisioned systems capable of accomplishing targeted delivery of drug/gene and drug/siRNA combinations to the same population of tumor cells in a coordinated manner.

In a pioneering work, the group of Lin developed a gene transfection reagent based on MCM-41-type mesoporous silica nanospheres functionalized with the second generation of PAMAM dendrimers [78]. Low generation of PAMAM dendrimers, such as G2 in this case, covalently attached to MSNPs as an inorganic support results in an increase in the global surface charge of the hybrid material and, therefore, the system enables an optimal pDNA complexation while using nontoxic dendrimer generations. The incorporation of dendritic macromolecules in the outer surface of the MSNPs was achieved following a synthetic approach in which the silica surface was previously functionalized with a reactive linker to covalently bond the dendrimers in a second step. Isocyanatopropyltriethoxysilane was grafted onto the silica surface to yield isocyanatopropyl-functionalized MSNPs. At this stage the material was loaded with the fluorescent dye Texas Red. Then the amine-terminated second generation of PAMAM dendrimers were covalently bonded to the silica nanoparticles through urea linkages. The fast kinetics of the reaction in the second step, as well as the ratio between the size of the dendrimer generation (about 2.9 nm, hydrodynamic diameter) and the pore diameter (2.7 nm), favor the distribution of the dendrimers in the outer surface of the MSNP material, blocking the entrance of the pores of the MCM-41 type. With this strategy the bioceramic was provided with a capping dendrimer to store Texas Red loaded in the pores with the aim of tracking the distribution of the MSNP material once internalized into the cells. Furthermore, the main role of the attached dendrimers is to efficiently complex DNA at physiological pH, where the dendrimers are positively charged, and to protect the DNA from enzymatic cleavage. The transfection efficiency of the system was evaluated by using a pDNA which codes for the enhanced green fluorescent protein (pEGFP-C1) and the successful delivery of the plasmid to the cell nucleus was confirmed with the observation of significant expression of the green fluorescent protein in the transfected cells.

Inspired by the group of Lin, which reported as well the finding that MSNPs can simultaneously deliver DNA and chemicals into plant cells [19, 79], other authors have reported co-delivery of nucleic acids and chemotherapeutic agents or drugs.

MSNPs functionalized with amine-terminated PAMAM dendrimers to bind Bcl-2-targeted siRNA, using the method reported by Lin's group [78], were loaded with doxorubicin (Dox). The system efficiently delivered simultaneously the siRNA targeted against mRNA encoding the antiapoptotic protein Bcl-2 and the hydrophobic drug Dox [80]. The combined drug/siRNA delivery resulted in a substantially enhanced activity against multidrug resistant human ovarian cancer cells *in vitro*. The anticancer efficacy of Dox increased 132 times compared to free Dox, mainly because the simultaneously delivered siRNA significantly silenced the Bcl-2 mRNA, suppressing the nonpump resistance, and substantially enhanced the anticancer action of Dox, efficiently overcoming the MDR of the cancer cell line used. The data also suggested that the Dox delivered by MSNs had minimal premature release in the extracellular environment, which can greatly eliminate side effects of Dox.

The group of Zink has studied the effects of several different PEI polymer sizes as coating on MSNPs in terms of cellular uptake, cellular toxicity, and efficiency of drug and nucleic acid delivery [44]. PEIs ranging from a molecular weight of 0.6 to 25 kDa were explored for achieving cellular delivery versus reduction in toxicity. The MSNPs surface was modified with PEI via electrostatic interaction. To prepare the hybrid PEI-MSNPs, phosphonate-modified MSNPs were first synthesized and the coating with PEI was subsequently performed by mixing them with different molecular weight PEIs (0.6, 1.2, 1.8, 10, and 25 kDa). The stability of PEI attached to the MSNP surface was confirmed by using two different fluorescent labels on both the MSNPs and the PEIs. Then, confocal microscopy confirmed that both labels co-localize in the cell. The authors show that the cellular uptake of PEI-coated MSNPs, irrespective of polymer size, is considerably enhanced compared to unmodified MSNPs (silanol surface). Moreover, the authors demonstrate that the reduction of the polymer size is capable of scaling back the cytotoxic effect, still enabling efficient cellular delivery of nucleic acids with non-toxic MSNPs coated with LMW PEI. The dual function of these devices was further demonstrated in this study since the facilitated cellular uptake of cationic particles enhances the ability of the MSNPs to deliver the hydrophobic chemotherapeutic agent, paclitaxel, to pancreatic cancer cells.

Once having studied the delivery of siRNA delivery in mammalian cells [81], the same group further demonstrated dual drug and siRNA delivery by an MSNPs platform to overcome multidrug resistance in cancer cells [82]. The co-delivery strategy utilizes a siRNA to silence the expression of a protein that contributes to the formation of a drug efflux pump that prevents the intracellular build-up of chemotherapeutic agents, together with an appropriate anticancer drug for drug-resistant cells. To test the utility of MSNP as a dual delivery platform, the authors used the drug-resistant squamous carcinoma cell line KB-V1, which exhibits MDR as a result of Pgp overexpression, to see if Pgp knockdown restores Dox sensitivity. The authors focused the study in engineering the MSNPs by means of surface functionalization. The attachment of negative (phosphonate) as well as positive (PEI) surface groups makes the MSNPs functionally effective. The functionalization of the particle surface with a phosphonate group allows electrostatic binding of Dox to the porous interior, from where the drug could be released by acidification of the medium, as occurs during the lysosomal processing when the system enters the cell. In addition, phosphonate modification also allows exterior coating with the cationic polymer, PEI, which enables the MSNPs to contemporaneously deliver Pgp siRNA. This study demonstrates the feasibility of the MSNP platform to improve the cytotoxicity of Dox by co-delivery of Pgp siRNA as proof-of-principle. In addition the authors interestingly point out that MDR phenotype is quite complex, involving several genes up- or down-regulated, as well as a combination of drug resistance mechanisms. The understanding of the MDR mechanisms may provide researchers with new opportunities to develop more therapeutic components or combinations to achieve better therapeutic effects.

An enhanced gene and siRNA delivery has been reported by Oupicky and coworkers [83] using a system of polycation-modified MSNPs loaded with chloroquine, a lysosomotropic agent. The MSNPs modified with the polycations poly(2-(dimethylamino)ethylmethacrylate) or poly (2-(diethylamino)ethylmethacrylate), were further stabilized by surface modification with PEG to prevent aggregation. The biological studies showed that polycation-modified MSNPs were able to simultaneously deliver chloroquine with DNA and siRNA in cell culture. The codelivery of chloroquine and the nucleic acids leads to a significantly increased transfection and silencing activity of the complexes compared with MSNPs not loaded with chloroquine, since the endosomal escape of the nucleic acids is aided by the endosome disrupting agent.

Recent research trends are focused on the possibility to host the genetic material inside the mesopores of the MSNPs, instead of carrying it on the outer surface. With this aim, the synthesis of MSNPs with very large pores to provide more inner space has to be carried out, while maintaining a narrow particle size distribution, of about 70–300 nm, of the NPs diameter. One of the advantages brought up in this strategy is an effective protection against nucleases without needing cationic polymers in the surface which can increase the cytotoxicity of the whole carrier. The research work developed by Min and coworkers [84] is the first to report a facile synthesis of monodispersed MSNPs with an overall diameter of about 250 nm possessing ultralarge pores of about 23 nm, and to demonstrate their application as a gene delivery carrier with low cytotoxicity. To prepare the positively charged surface for loading pDNA through electrostatic interaction, primary amine groups were introduced using 3-aminopropyltriethoxysilane (APTES) through a postgrafting method. The aminated MSNPs with large pores provide cationic pores large enough to encapsulate plasmids, at least partially, without supplementary cationic polymers. Moreover, the system readily enters the cell, showing an efficient cellular uptake, and can efficiently protect pDNAs from nuclease-mediated degradation and shows much higher transfection efficiency, maintaining a low cytotoxicity, compared to the same aminated system with small pores (about 2 nm).

The potency of using large pore mesoporous silica nanoparticles (LP-MSNs) as carriers for siRNA delivery has been exploited in a following study by Qiao and coworkers [85]. First, large pore MSNPs, 100–200 nm in diameter and with a 3D cubic mesostructure, were synthesized following a method previously reported [86]. To act as a host for nucleic acids, the surface of the silica requires modification in order to generate sufficient binding affinity for the negatively charged nucleic acids. In this case, the large pore MSNPs were functionalized by grafting epoxysilane (3-glycidoxypropyl trimethoxysilane), and subsequently reacted with PLL. The same materials amine functionalized with APTES were also synthesized for comparison. After a deep characterization study of the hybrid materials, the influences of the different surface functionalization (PLL, amine, and unfunctionalized) on DNA adsorption, cell uptake, cytotoxicity, and biological function were then substantiated. PLL-functionalized MSNPs were proven to be superior as gene carriers compared to amino-functionalized and the native MSNPs. The system was tested to deliver functional siRNA against minibrain-related kinase and polo-like kinase 1 in osteosarcoma cancer cell line KHOS cells. These two oncogenic genes are highly expressed in osteosarcoma cells and silencing them with siRNAs can inhibit KHOS cell growth. Then, the functionalized particles demonstrated great potential for efficient gene transfer into cancer cells as a decrease in the cellular viability of the osteosarcoma cancer cells was induced. Moreover, the PLL-modified silica nanoparticles also exhibited a high biocompatibility, with low cytotoxicity observed up to 100 $\mu g\,ml^{-1}$.

This technique of using MSNPs with large pores to accommodate siRNAs has been reported by the group of Brinker, employing "protocells" as a delivery platform with unique attributes for therapeutic oligonucleotides [87]. Protocells consist of MSNP-supported lipid bilayers and are formed via fusion of liposomes to porous silica nanoparticles and synergistically combines the features of both components. The high pore volume and surface area of the spherical mesoporous silica core allows high-capacity encapsulation of a spectrum of cargos. The supported lipid bilayer, whose composition can be modified for specific biological applications, serves as a modular and reconfigurable scaffold, allowing the attachment of a variety of molecules that provide cell-specific targeting and controlled intracellular trafficking. To form protocells loaded with therapeutic RNA and targeted to hepatocellular carcinomas (HCCs), silica NPs with an average diameter of 165 nm and multimodal pore morphology composed of large, 23–30 nm, surface accessible pores interconnected by 3–13 nm pores, are modified with an amine-containing silane and then loaded with the siRNA. Subsequently, PEGylated liposomes were fused to the siRNA-loaded cores and the resulting supported lipid bilayer was then chemically conjugated with a targeting peptide (SP94) that binds to HCC and an endosomolytic peptide (H5WYG) that promotes endosomal/lysosomal escape of internalized protocells. In terms of loading capacity, stability, targeting, and internalization efficiency, compared with the existing analogous liposomes and MSNPs, protocells might address many of the deficiencies that currently limit the clinical use of these nanosystems as delivery platform for therapeutic oligonucleotides.

13.3.3   Carbon Allotropes (Fullerenes, CNTs, Graphene Oxide)

13.3.3.1   *Fullerenes*

Among the allotropic form of carbon fullerenes, C_{60} is the most representative of the fullerene family. *Fullerene C_{60}* has 60 carbon atoms arranged in a spherical closed structure in which every carbon atom is covalently bonded to three other adjacent atoms with sp^2 hybridization (see Section 7.1.3). Although their chemical and physical unique features make them very attractive for applications and use in many different fields, either in biological and materials chemistry, the well-known lack of solubility in polar solvents and the consequent formation of aggregates in aqueous solutions, is a negative aspect of these molecules for use in medicinal chemistry. This drawback due to the hydrophobicity of the sphere can be solved by means of supramolecular or chemical approaches. The most versatile methodology to increase the hydrophilicity is based on the chemical modification of the sphere with functional groups, such as amino, amino acid, or carboxylic acid, which leads to a wide variety of C_{60} derivatives. However, synthesis of functionalized fullerenes needs indeed a laborious synthetic effort.

In the field of nucleic acid delivery, Nakamura and coworkers pioneered the potential use of fullerene derivatives as nanocarriers, showing that cationic tetraamino fullerene derivatives exhibit a practically useful level of transfection efficiency [88, 89]. In a systematic structure and activity relationship investigation performed on a library of 22 fullerene cationic derivatives, these researches have also shown that an appropriate hydrophobic–hydrophilic balance appears to be essential to form fullerene–DNA nanostructures capable of crossing the cell membrane and thus to release DNA for gene expression [90].

Poly-*N,N*-dimethylfulleropyrrolidonium adducts have been prepared by multifunctionalizing C_{60} via multiple 1,3-dipolar cycloaddition reactions of azomethine ylides. These cationic fullerenes developed by Prato and coworkers have the advantage of being totally water soluble and are able to tightly complex pDNA [91].

Another class of water-soluble fullerene derivatives prepared using Hirsch–Bingel chemistry has been evaluated as transfecting agents [92]. The different C_{60} derivatives yield either positively charged, negatively charged, or neutral chemical functionalities under physiological conditions. The authors made an effort to elucidate the relationship between the hydrophobicity of the fullerene core, the hydrophilicity of the water-solubilizing groups, and the overall charge state of the C_{60} vectors in gene delivery and expression. Although all C_{60} derivatives showed statistically significant transfection efficiency compared to naked DNA, only two positively charged C_{60} derivatives showed efficient *in vitro* transfection, an octa-amino-, and a dodeca-amino-derivatized C_{60}.

Indeed, a high number of protonable amino groups are needed in a fullerene derivative for a good DNA condensation capacity. This fact has been recently investigated by Nieregarten and coworkers [93]. Taking advantage of a synthetic strategy based on the post-functionalization of a readily available fullerene hexakis-adduct core, under the

Cu-mediated Huisgen reaction conditions, the authors have prepared the first, second, and third generations of fullerenodendrimers with amine terminal groups. Only the second and third generation, 48 and 96 amino groups, respectively, have enough amino groups to ensure DNA compaction into stable and positively charged polyplexes that fruitfully deliver DNA into cells. These spherical fullerenodendrimers also exhibit high transfection efficiency, while maintaining low toxicity.

13.3.3.2   Carbon Nanotubes (CNTs)

Carbon nanotubes (CNTs) are molecular-scale tubes of graphitic carbon with unique size- and structure-dependent optical, electronic, magnetic, thermal, chemical, and mechanical characteristics. CNTs consist of carbon atoms arranged in sheets of graphene rolled up into cylindrical hollow shapes. Their lengths may vary from several hundred nanometers to several micrometers with diameters of 0.4–2 nm for single-walled carbon nanotubes (SWCNTs) and 2–100 nm for multi-walled carbon nanotubes (MWCNTs) (see Section 7.1.4).

Pristine CNTs have very poor solubility in water, which is a basic requirement to be used in biological applications. However, development in the functionalization chemistry of CNTs [94, 95] has led to their enhanced dispersibility in aqueous physiological media, by means of adding polar or charged groups. In addition to increased solubility, introduction of positively charged groups at the surface of CNTs indeed broadens the spectrum for their potential biological applications, including the improvement of their affinity toward DNA/siRNA for gene delivery [96].

Furthermore, CNTs also have potential applications as carriers for delivery of nucleic acids into cells because of another unique property: the ability to cross cell membranes. Recently, it has been demonstrated that covalently functionalized CNTs can enter inside different cell types through energy dependent or independent methods, that is, endocytosis or a "*nanoneedle*" mechanism, respectively [97].

Various types of functionalization of CNTs have been described for DNA/siRNA delivery: a method based on the 1,3-dipolar cycloaddition of azomethine ylides, a covalent linkage approach through amidation of carboxylic acids and a noncovalent approach by using cationic phospholipids adsorbed onto the CNT surface.

The covalent functionalization of CNTs by using a method based on the 1,3-dipolar cycloaddition of azomethine ylides (the Prato reaction) was used in the first employment of CNTs as a novel gene delivery vector system, reported by Bianco *et al.* [98]. Both SWCNTs and MWCNTs were functionalized with a pyrrolidine ring bearing a free amine-terminal oligoethylene glycol moiety attached to the nitrogen atom ($-CH_2CH_2OCH_2CH_2OCH_2CH_2NH_3^+Cl^-$). The authors reported that pDNA was able to associate in a condensed globular conformation through electrostatic interactions onto the surface of functionalized-CNTs at specific charge ratios. The results showed 5–10 times higher levels of gene expression compared to naked pDNA and the lack of cytotoxicity of the complex was also demonstrated.

In another study, the same authors reported the gene transfer efficiency between the next different types of CNTs chemically functionalized with cationic head groups: $SWNT-NH_3^+$, $MWNT-NH_3^+$, and $SWNT-Lys-NH_3^+$ (Lys = lysine). They concluded that the surface area, length, width, and charge density of the functionalized-CNTs were the critical parameters for the interaction with pDNA and the consequent formation of a biologically active complex [99].

In an advanced *in vivo* study, these authors used the MWNTs functionalized with ammonium groups at the end of a triethylene glycol chain, for the delivery of a siRNA toxic sequence and achieved therapeutic silencing leading to tumor-growth arrest and prolonged survival of animals bearing a human lung tumor xenograft. Interestingly, that study included a comparison between the gene-silencing capacities of cationic CNTs and cationic liposomes. In this evaluation, a statistically significant benefit from *in vivo* local (intratumoral) administration of siRNA:CNT complexes instead of siRNA:liposomes was found [100].

This kind of MWCNTs, functionalized with ammonium groups at the end of a tri-ethylene glycol chain through a pyrrolidine ring, disperse well in water and allow the formation of stable complexes with the phosphate ions of siRNA. In an attempt to increase in a controllable manner the number of cationic groups at the surface of the CNTs, the authors reported the synthesis of successive generations of PAMAM dendrons linked to the surface of MWNTs. Starting from pristine MWNTs, initially functionalized using 1,3-dipolar cycloaddition of azomethine ylides, a stepwise divergent synthetic process was followed to achieve growth of the PAMAM dendrons on the MWNTs up to the second-generation. The final dendron–MWNT adducts were complexed with a noncoding siRNA sequence, and the complexes were successfully used *in vitro* to transfect mammalian cells. The cytoplasmic delivery of siRNA remarkably increased on going from dendron generation G0 to generation G2 in comparison with the delivery by the precursor ammonium-functionalized $MWNT-NH_3^+$ [101]. The electrophoretic motility on agarose gel of these complexes showed that increasing the branching of the dendritic structure enhanced the ability of CNTs to complex siRNA, most likely due to a higher water dispersibility of the conjugates and to the higher amount of positive charges on the surface of the nanotubes.

Another approach for the covalent linkage of amine polymers or dendrimers to the surface of CNTs is through amidation and/or peptide coupling chemistry. Carboxylic acid groups are introduced onto pristine CNTs through oxidation by acid treatment following established protocols, thereafter the carboxylic acid groups are transformed into acyl chloride groups or activated using carbodiimide chemistry. Subsequently, primary amine groups of the polyamine polymer, dendron, or dendrimer are covalently bonded to the CNTs surface. One of the advantages of the oxidation treatment is to increase the dispersibility of the CNTs. Moreover, the oxidation affords shortened and opened CNTs, creating defects on the sidewall of the material and introducing carboxylic functions, mainly at the nanotube tips.

Liu *et al.* published the first study using PEI 25 kDa grafted onto MWCNTs as pDNA carrier, showing the ability of this system to immobilize pDNA onto the CNTs surface and the capability of these complexes to enhance gene expression [102]. Following the amidation route, CNTs functionalized with cationic polymers, such as PEI [45], glycopolymers [103], PAMAM dendrimers [104], and ammonium and guanidinium dendrons [105], have shown efficiency in delivering DNA/siRNA.

An alternative route includes a noncovalent water-solubilization of the CNTs, since a stable aqueous suspension of short SWNTs can be achieved by noncovalent adsorption of phospholipid molecules.

Dai and coworkers [106] developed "smart" siRNA delivery systems based on SWC-NTs water-stabilized with phospholipids bearing PEG chains and terminal amine or maleimide groups. The phospholipid-PEG binds strongly to SWCNTs via van der Waals and hydrophobic interactions between two phospholipid alkyl chains and the SWCNT sidewall, with the PEG chain extending into the aqueous phase to impart solubility in water. siRNA cargo molecules are incorporated through biologically triggered cleavable disulfide bonds. Hence, in contrast to the approaches described above, siRNA cargos can be controllably released from the CNTs surface upon cellular uptake. Disulfide bonds can be cleaved by thiol reducing enzymes once the system is internalized into the endosomal or lysosomal compartments through endocytosis, thus releasing the cargos from the CNTs. Cellular studies performed on HeLa cells confirmed the occurrence of siRNA release and the resulting gene silencing. This system was also used to transfect human cells in order to silence the expression of HIV-specific cell surface receptors [107].

Biocompatible cationic cholesterol-based lipids also produce highly stable aqueous suspensions of SWCNTs. The hydrophobic cholesterol backbone of lipids wraps around the SWCNT surface, leaving their hydrophilic moiety exposed to water and leading to the formation of a stable SWCNT–lipid suspension which further interacts with phosphate residues of DNA. Enhanced transfection efficiency of these formulations was observed even in high serum conditions [108].

13.3.3.3   Graphene Oxide

Graphene, two dimensional (2D) single or a few layers of sp^2-bonded carbon nano-sheets, has extraordinary advantages, such as high electrical conductivity, large surface ratio, and remarkable mechanical strength. It has been extensively studied owning to its excellent physical, chemical, and mechanical properties, while exploration of its biomedical applications has just started (see Section 7.1.5).

Graphene is an optimal choice for aromatic drug delivery via $\pi-\pi$ stacking interactions, however, the poor distribution and stability of graphene sheets in aqueous solution is a challenge for its biomedical applications. To overcome this drawback, graphene oxide (GO), a derivative of graphene, is prepared via chemical oxidation. After oxidation or activation of graphene, the nano-sheets of GO possess hydroxy, epoxy, and carboxylic acid functional groups, which allow GO to be readily stable in aqueous solution and convenient for further modification or chemical functionalization.

Regarding the interaction of graphene with nucleic acids, graphene is able to strongly bind to ssDNA via $\pi-\pi$ stacking interactions with the DNA bases. However, the interaction of graphene with double-stranded DNA (dsDNA) molecules is rather weak since the bases are concealed within the double helix [109, 110]. Utilizing this mechanism and the effective fluorescence quenching ability of graphene, several groups have independently developed various grapheme-based DNA sensing platforms [110–113]. In addition, Lu *et al.* showed that graphene could protect oligonucleotides from enzymatic cleavage and was able to deliver ssDNA into cells [114].

GO decreases its binding capacity for both ssDNA and dsDNA, compared to graphene, due to electrostatic repulsion between the negative GO surface and the negatively charged phosphate groups of the DNA, which besides are externally oriented in the double helix of dsDNA. Thus, naked GO has limited applications as a gene or DNA vector. The functionalization of GO with cationic polyamine polymers or dendrimers can be achieved in two ways:

Via electrostatic interactions, since GO is negatively charged.

Via covalent functionalization, generating amide bonds via a conventional condensation reaction between the carboxylic acid groups of GO and the primary amines of the polymers, by using carbodiimide chemistry for activation of carboxylic acids in a first step.

In both cases, the negatively charged GO transforms into stable GO−polymer complexes highly enriched in positive charges, which allows effective loading of DNA/siRNA molecules, making possible the use of GO as a novel 2D nanovector for gene transfection.

In one of the first works in which GO has been used for intracellular deliver of a gene plasmid, it was functionalized with PEI polymers via noncovalent electrostatic interactions, therefore, the subsequent loading of the vector with DNA was performed *via* a layer by layer (LbL) assembly global process. In this work of Liu and coworkers [115] GO−PEI complexes were prepared with bPEI of two different molecular weights (1.2 and 10 kDa). The gene transfection efficiency was tested by using the model system of the pDNA encoding for EGFP. The complexes GO−PEI exhibited improved transfection efficiency over PEI-1.2k, with LMW, and lower toxicity than PEI-10k, with HMW. However, the transfection efficiency was only comparable to that of PEI-10k. The results of this work highlight the promise of GO as a novel nonviral nanovector for safe and efficient gene transfection. However, whether and how the structure of graphene, that is, size and thickness, would affect the gene transfection efficiency requires further investigation.

The following works make use of the covalent anchoring of the cationic polyamines PEI to afford polymer-GO hybrid materials as the gene vectors [116, 117]. In this way, the covalent bond ensures the vector integrity during the transfection process [118]. In summary, PEI-functionalized GO is able to effectively condense and transfer DNA into cells and the transfection efficiency of PEI−GO nanovectors is comparable to or even better than PEI under optimal conditions.

Zhang and coworkers [118] have developed a PEI-grafted GO nanocarrier for the delivery of Bcl-2-targeted siRNA and the anticancer chemical drug Dox. As previously explained, Bcl-2 protein, one of the main antiapoptotic defense proteins, is closely related to the MDR of cancer cells. Knockdown of the Bcl-2 protein expression level in cancer cells by Bcl-2-targeted siRNA would effectively overcome the MDR of cancer cells and sensitize them to anticancer drugs. For comparison, preparation of PEI−GO by noncovalent adsorption of PEI to GO by electrostatic and hydrogen-bonding interactions was also conducted. It was found that the thus-obtained PEI−GO was unstable in saline solution. It precipitated completely in 10% NaCl solution in 24 h, most likely due to the gradual removal of adsorbed PEI in the saline solution, thus limiting its applications in drug and RNA delivery. The PEI−GO nanocarrier system showed significantly lower cytotoxicity and substantially higher transfection efficacy of siRNA, at optimal siRNA/PEI−GO ratio, than PEI 25 kDa. Furthermore, the authors demonstrated that sequential delivery of Bcl-2-targeted siRNA and Dox by the PEI−GO nanocarrier leads to significantly enhanced anticancer efficacy due to a synergistic effect.

Kim and coworkers investigated the imaging properties of GO covalently functionalized with LMW bPEI acting as the gene transfection agent of a pDNA encoding for luciferase expression [119]. Reduced graphene exhibits photoluminescence (PL) and this optical property disappears in GO since, during oxidation of graphene, sp^2-hybridization of carbon atoms becomes sp^3 in the defects and functional groups. In this study, the PL of GO and GO/bPEI mixture (noncovalent) was negligible, but bPEI-GO displayed a significant increase in the PL (when excited at 480 nm). This is explained since during amidation and ring-opening amination of epoxy moieties there is a surface modification that leads to the reduction of GO after covalent conjugation of bPEI to GO, that is, the conjugation of bPEI to GO effectively restores the conjugated aromatic clusters. At the same time, the ionic interaction between bPEI and GO adversely affected the PL. However, combined effects gave rise to an overall increase in the PL intensity of bPEI–GO compared to that of GO. Moreover, a further increase in PL intensity was observed when bPEI–GO formed a complex with pDNA. After incorporation of pDNA, bPEI preferentially formed a complex with pDNA and induced recovery of the PL intensity of bPEI–GO that was partially lost due to the ionic interaction between GO and bPEI that adversely affected the PL. Therefore, these results illustrate the potential of the bPEI–GO nanosystem as a bioimaging reagent and gene delivery vector.

A step forward in this kind of multimodal systems is the design of stimuli-responsive gene vectors. The interesting feature of graphene and GO of having a strong NIR optical absorption ability has been explored for *in vitro* and *in vivo* photothermal therapy of cancer cells [120–124]. Furthermore, NIR light-controllable gene transfection has aroused significant interest owing to its easy operation, highly localized controlling, and deep-tissue penetration of NIR photons [125].

Very recently, Liu and coworkers have reported a dual functionalized nano-graphene oxide (NGO) for photothermally enhanced gene delivery. This study is the first to use photothermally enhanced intracellular trafficking of nanocarriers for light-controllable gene delivery. Two different polymers PEG and PEI have been covalently attached to GO via amide bonds, obtaining a physiologically stable dual-polymer-functionalized NGO (NGO-PEG-PEI) with ultra-small size [126]. Compared with free PEI and the GO–PEI conjugate without PEGylation, NGO–PEG–PEI shows superior gene transfection efficiency without serum interference, as well as reduced cytotoxicity. Utilizing the NIR optical absorbance of NGO, the cellular uptake of NGO–PEG–PEI is shown to be enhanced under a low power NIR laser irradiation, owing to the mild photothermally induced local heating, that increases the cell membrane permeability without significantly damaging cells, and accelerated intracellular trafficking of nanovectors. As a result, remarkably enhanced pDNA transfection efficiencies induced by the NIR laser are achieved using NGO–PEG–PEI as the light-responsive gene carrier. More importantly, it is shown that NGO–PEG–PEI is able to deliver siRNA into cells under the control of NIR light, resulting in obvious downregulation of the target gene, in the presence of laser irradiation. Then, NGO–PEG–PEI could also serve as a NIR-light-controllable siRNA delivery platform for potential applications in combined photothermal and gene therapies.

The latest study at the time of writing this chapter regarding photothermally controlled gene delivery has been carried out by the group of Kim [127]. For advanced design of the gene carrier LMW bPEI (1.8 kDa) is covalently grafted to GO through carbodiimide chemistry. Subsequently, bPEI–GO is chemically reduced by hydrazine to achieve reduced graphene oxide (rGO), since rGO sheets have more effective photothermal properties than GO. Then, bPEI–rGO is modified with PEG (5 kDa) attached to the PEI branches to enhance colloidal stability. The nanovector PEG–bPEI–rGO demonstrates an enhanced gene transfection efficiency upon NIR irradiation, which is attributed to accelerated endosomal escape of polyplexes augmented by locally induced heat. The endosomal escaping effect of the nanovector was investigated using a proton sponge effect inhibitor.

13.3.4   Magnetic Iron Oxide Nanoparticles

This ceramic material has not been introduced in this book until now so a brief description of magnetic nanoparticles (mNPs) and IONPs is made at this point, before tackling their use as nonviral vectors for gene magnetofection, and also their use in magnetic hyperthermia, which is considered for cancer treatment in the Chapter 14.

mNPs are an important class of nanomaterials that consist of typical magnetic elements, such as iron, nickel, cobalt, manganese, chromium, and gadolinium, as well as their chemical compounds. Furthermore, magnetic particles with controllable size in the nanometer range also possessing high values of saturation magnetization and magnetic susceptibility are applicable in biotechnology. Recently, these materials have attracted significant interest in biomedical research because they provide advanced therapeutic and diagnostic capabilities with dual-mode manipulation, controlled either by using a magnetic field or through surface ligand engineering [128, 129].

Colloidal nanoparticles that form a ferrofluid have superparamagnetic properties. They are mono-domain particles whose magnetic moments are randomly oriented. Under the action of an external magnetic field, these moments will rapidly rotate to be placed in the direction of the field, so as to increase the magnetic flux. Once the field is ceased, a return to the random initial arrangement is produced without keeping any magnetic remanence [130]. In addition, Brownian motion of the mNPs is sufficient to cause thermal energy. The aim is to design colloidal fluids stable at physiological pH and salinity, in which mNPs do not form aggregates able to impair circulation in the bloodstream. To accomplish this requirement, the dimension of the NPs should be sufficiently small to avoid precipitation due to gravitation forces, and the charge and surface chemistry must provide both steric and coulombic repulsions, which can be achieved by convenient functionalization or coating of the mNPs. These particles may achieve high values of magnetization under the action of body-tolerated magnetic field intensities and frequencies, eventually losing the magnetization on suspending the magnetic field.

From a therapeutic point of view, the suitability of magnetic nanomaterials is largely determined by three factors: toxicity, magnetic performance, and biocompatibility [131, 132]. The inorganic nanocrystals, based on nontoxic elements that offer a high potential for several uses in nanomedicine, are colloidal superparamagnetic IONPs, such as magnetite (Fe_3O_4) or its oxidized form maghemite (γ-Fe_2O_3). The characteristics and properties of IONPs in terms of magnetic susceptibility, narrow size distribution, superparamagnetic behavior, surface chemistry, and toxicity are designed in their synthesis [133, 134]. The synthetic methods for the preparation of IONPs must allow the preparation of particles of nearly uniform and customized size and shape. Some of the routes for the synthesis of superparamagnetic IONPs in colloidal solution are, for example, coprecipitation from Fe(II) and Fe(III) salts in aqueous solution by the addition of a base or reactions in constrained environments, that is, using synthetic or biological nanoreactors, such as microemulsions from reversed micellar structures. High temperature decomposition of organometallic precursors, or the use of polyols as solvents, such as PEG, are also solution techniques to obtain IONPs. Aerosol technologies, such as spray and laser pyrolysis, are also employed for continuous production of well-defined mNPs with a high production rate. The size, charge, and surface chemistry of the particles are very important factors that will strongly influence the properties of the mNPs to be used in bioapplications [135–137].

As aforementioned, therapeutic and diagnostic applications are the main biomedical uses of mNPs [138]. Remarkably, the superparamagnetic behavior provides the final biomaterial with multifunctionality through magnetic targeting in several biotechnological applications, such as bioseparation [139], magnetic guided drug delivery, hyperthermia treatment, magnetic resonance imaging (MRI), and magnetofection.

mNPs can be synthesized with controlled size from a few to tens of nanometers. This means that they can interact with biological material and can even be attached to biological molecules that may direct the mNPs to the site of interest in the body [140]. Secondly, the particle magnetism provides the chance for manipulation by an external magnetic field gradient. This feature allows either transportation or immobilization of nanoparticles and nanoparticle/biological entity composites, for the release of drugs, genes, or proteins [141]. Finally, mNPs can respond to the action of alternating magnetic fields, so there is a transfer of energy from the field to the particle. For example, the nanoparticle could be used to transmit certain amounts of thermal energy into tumor cells, which forms the basis of antitumor therapy by hyperthermia [142, 143] (see Section 14.4.4).

Regarding MRI [144], superparamagnetic nanoparticles represent a class of nuclear magnetic resonance (NMR) contrast agents that are referred to as T2 (transversal relaxation time) contrast agents, as opposed to T1 (longitudinal relaxation time) agents, such as paramagnetic gadolinium(III) chelates. A current trend regarding the use of mNPs is to integrate in a multifunctional nanosystem simultaneous imaging and therapy [145, 146], or to generate multimodal imaging probes that integrate into a single system a mNP, a radionuclide, an optical tag, and a targeting moiety [147].

When mNPs are combined in gene vectors it is possible to apply the same principles of magnetically targeted drug delivery to the gene transfection methods. The viral and nonviral alternatives of this technology were demonstrated by Scherer, Plank, and coworkers, who coined the term *magnetofection* [148, 149]. The magnetofection technique consists of the association of superparamagnetic NPs with compounds able to interact with DNA fragments, facilitating DNA binding and protection during extra- and intracellular trafficking. Then, the coupling of genetic material (DNA fragments, pDNA, or siRNA) to mNPs produces the magnetic vector, which is focused on the target site or cells via a magnetic field gradient. This technique improves the uptake and expression of DNA since the transfection into cells can be assisted by the application of an external magnetic field, which targets and reduces the duration of the gene delivery, enhancing the efficiency of the DNA vector.

The experimental procedures of this technology depend on whether the application is performed *in vitro* or *in vivo*. In the case of *in vitro* magnetofection, the DNA/mNPs complexes (magnetoplexes), normally in colloidal suspension, are introduced into the cell culture and then rare earth magnets are placed underneath the cell culture dish producing a field gradient. The magnetic field gradient pulls the mNPs toward the magnetic field source, increasing the sedimentation rate of the magnetic vectors onto the cells surface and thus increasing the speed of internalization via an endocytosis mechanism (Figure 13.4). *In vivo*, the magnetic fields are focused over the target site not only to enhance transfection,

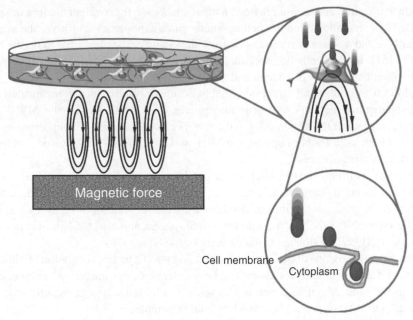

Figure 13.4 *Magnetofection technique: the magnet below the cell culture plate creates a gradient that pulls the DNA/mNPs complexes into contact with the cell membrane surface thus allowing endocytosis and speeding up the internalization rate of the magnetic vector*

but also to target the therapeutic gene to a specific organ or tissue within the body. Generally, the mNPs carrying the therapeutic gene are injected intravenously and high-gradient external magnets are used to attract the particles as they flow through the blood stream. Once captured by the field, the particles are held at the target, where they are taken up by the tissue.

Methods of surface modification to obtain magnetic hybrid composites are usually employed to provide the nanoparticles with functionality and biocompatibility, also stabilizing the colloid at the same time. Encapsulation of the IONPs in either organic, such as polymeric surfactants or inorganic matrices, mainly silica, may prevent the flocculation or the aggregation of the nanoscale particulate. Furthermore, the modified surface can impart non-toxicity and bear functional groups to allow the covalent grafting of biomolecules, therapeutic agents, or specific ligands for targeting.

Inorganic coatings present some advantages over polymeric matrices, such as chemical and mechanical stability and the lack of swelling or porosity changes with the pH. They also protect doped molecules (enzymes, drugs, etc.) against denaturalization induced by extreme pH and temperature. Many of the systems proposed in the literature are based on core–shell settings, in which species such as silica or gold [150] coat the magnetic nanocrystals.

The silica coating on a magnetic core leads to stable and biocompatible ferrofluids over a wide range of pH and at high electrolyte concentrations. This coating also prevents oxidation and degradation of the mNPs during and after synthesis. An advantage of having a surface enriched in silica is the presence of surface silanol groups that can easily react with silane coupling agents and provide an ideal anchorage for covalent binding of specific ligands. The preparation of core–shell magnetic silica nanoparticles involves the synthesis of the mNPs and a series of reactions to grow the coating silica layer based on the Stöber method [151]. In the synthetic procedure known as the Stöber method the silica species grow in alkaline mixtures of ethanol and water.

In addition to core–shell structures, magnetic nanocrystals can be encapsulated into the mesostructured network of silica nanospheres [152, 153]. Hence, the MSNPs with superparamagnetic IONPs embedded in the matrix can be externally manipulated by using magnetic fields, used for both optical and MRI, and used to store and deliver chemotherapeutic drugs into cancer cells.

Currently, the different possibilities for the combination of MSNPs and mNPs that further extend their functionalities are gaining great attraction in the field of nanomedicine [154, 155]. Other different possibilities of combinations between mesoporous silica and mNPs are rattle-type mNPs@hollow mesoporous silica spheres and also MSNPs decorated with mNPs, the mNPs being situated at the exterior of MSNPs.

As aforementioned, one of the advantages of using IONPs in the preparation of multifunctional gene vectors is that transfection can be accelerated, thus hindering the biochemical attack on the gene. Magnetofection is an excellent alternative to significantly reduce the transfection time from several hours to less than 60 minutes.

Magnetic IONPs are often covered with suitable coatings to improve their biocompatibility, colloidal stability at physiological pH, and to attain the possibility of functionalization. Moreover, plain IONPs do not effectively bind DNA. Furthermore, the combination of two or more nonviral components with DNA enables the combined vector to inherit the respective advantages of each component. In this sense, the coating of superparamagnetic iron oxide nanoparticles (SPIONs) with polycationic polymers such as PEI makes magnetofection feasible. Nowadays, there are some commercial products based on SPIONs coated with PEI of different molecular weights to perform magnetofection. For example, PolyMag (PEI 25 kDa, Chemicell, Germany) that has a hydrodynamic diameter of about 12 nm, as determined by dynamic light scattering analysis [156]. However, some of them are relatively large particles, such as transMAGPEI (PEI 800 kDa, Chemicell, Germany) with an average size of 200 nm [148] or Neuromag (Oz Biosciences, France) with an average particle size around 160 nm [157].

The size of the complexes is an important factor that may influence the behavior of transfection agents, since it affects the rate of uptake as well as toxicity. Endocytosis is more efficient with small particles of about 50 nm and also the velocity of cytoplasmatic movement is a function of particle size. Steitz *et al.* [158] described the coating of SPIONs with 25 kDa PEI, varying the PEI:Fe mass ratio among 2, 1, 0.4, 0.2, and 0.1. They also described the preparation of DNA/PEI:SPION complexes as well as their characterization according to their particle size, ζ-potential, morphology, DNA-complexing ability, magnetic sedimentation, and colloidal stability. In this study different sizes of DNA/PEI:SPION complexes were tested as transfection agents in COS cells. Only very low transfection rates of up to 0.6% could be achieved with the largest particles (about 200 nm), whereas the smallest particles (around 60 nm) resulted in much higher transfection rates (34.5%). The applied magnetic field reduced the free diffusion of the particles, which could have resulted in the higher uptake of nanoparticles.

A very thorough piece of research work regarding improvements in the design of nonviral vectors composed of mNPs and cationic polymers for optimal pDNA transfection has been recently carried out by Amal and coworkers [159]. In this work the effect of mNP, PEI, and DNA arrangement on transfection efficiency was investigated by varying the vector component mixing order. Four different vector configurations were prepared by systematically varying the mixing order of mNP, PEI, and DNA:

1. coupling mNP with premixed PEI/DNA (mNP + PEI/DNA),
2. coupling mNP precoated with PEI with PEI/DNA complex (mNP/PEI + PEI/DNA),
3. mixing PEI precoated mNP with naked DNA (mNP/PEI + DNA), and
4. adding extra PEI to mNP/PEI + DNA (mNP/PEI + DNA + PEI).

The authors demonstrate that the vector size, surface charge, cellular uptake of mNP and DNA, and the level of gene expression are highly dependent on the assembly, that is, mixing order, of the vector components. It was also studied how the presence of serum in culture media alters the size and surface properties of the vector. The vector dispersion is improved with the presence of serum, enhancing their stability, and the surface charge of all vectors is altered to negative charge, indicating serum protein adsorption. The gene expression efficiency of the four different mNP vectors after magnetofection was assessed in hamster kidney cells (BHK21). It was evidenced that all mNP vector configurations were taken up by the cells, despite the similar charge of mNP vectors with cell membrane. The highest mNP vector cellular uptake was observed for the largest mNP vector (mNP + PEI/DNA) and this was attributed to enhanced gravitational and magnetic aided sedimentation onto the cells. The highest gene expression was also observed for this mNP vector configuration. This fact is opposite to the findings of the above commented work, however, it must be highlighted that a high mNP uptake by the cells does not predictably lead to increased gene expression efficiency. Besides vector uptake, gene expression is affected by extracellular factors, such as premature DNA release from mNPs and DNA degradation by serum, as well as intracellular factors, such as vector endosomal degradation, inability of DNA to detach from mNPs, and cytotoxic effects of mNPs at high uptake.

Until the work of Kim and coworkers [46], publications regarding the combination of PEI and mNPs were based on mere mixing of the mNPs with the polymer to coat the mNPs. Since the combination of both nonviral vectors would achieve a synergistic effect, these authors suggest that the covalent conjugation of both components, PEI and mNPs, could provide a better level of consistency and reproducibility in the magnetofection efficiency. Hence, the hybrid vectors were prepared from thermally cross-linked SPIONPs [160], that are IONPs coated with an antibiofouling polymer so that the mNPs possess PEG as a protein resistant moiety and also carboxylic acid groups within the polymer coating. Thus, the -NH_2 groups of LMW PEI were covalently linked to the -COOH groups by means of carbodiimide chemistry (EDC/NHS (1-ethyl-3(3-dimethylamino propyl)carbodiimide/ N-hydroxysuccinimide) coupling), resulting in amide bonding. The ultimate aim of this work was to transfect therapeutic genes into intractable primary human umbilical vein endothelial cells (HUVECs) which play an important role in vascular-related diseases. The pDNA IL-10 gene regulates the expression of the plasminogen activator inhibitor-1 (PAI-1) that plays an important role in thrombus formation, causing adverse cardiovascular anomalies. Therefore, an efficient expression of IL-10 gene can regulate negatively the expression of PAI-1, thereby preventing endothelial cell-related vascular dysfunctions. However, the major concern of the assay arose from the extreme susceptibility of the HUVEC cells to the serum-free condition of the transfection experiment. So success of the strategy depended upon whether the magnetofection became rapid enough for survival of HUVEC during the experiment and on the overall efficiency of the transfection as well. The pDNA IL-10 was mixed with PEI-SPION and the formed DNA/PEI-SPION magnetoplexes were incubated with normal HUVEC for 15 min in serum-free conditions under the action of a magnet. A high transgene expression, that is, successful inhibition of expression of PAI-1, was achieved at very low vector dose within a very short incubation time. More importantly, the therapeutic gene was transfected into highly sensitive HUVEC cells efficiently with high cell viability in the serum-free media. Therefore, it further implied that magnetofection was efficient for therapeutic gene transfection into HUVEC in serum-free media, which is difficult to achieve by a standard transfection protocol.

As just outlined, a gene carrier composed of two or more vectors should possess a chemical stability to keep the integrity until reaching the intracellular level. Regarding this issue, the group of Vallet-Regí has recently reported a novel nonviral nanosystem for *in vitro* gene magnetofection [161]. The multifunctional vector consists of SPIONs functionalized with a low generation of PPI dendrimers, that is, a non-toxic generation. The nanosystem is chemically designed with a covalent bond between both components, thus ensuring the vector stability from the extracellular environment up to the cell interior. In this strategy the dendrimers are attached to the naked surface of maghemite nanoparticles (γ-Fe$_2$O$_3$ NPs) through covalent bonds via a sol–gel synthetic route. This approach allows a direct dendritic decoration of the IONPs without any additional surface modification. Furthermore, this strategy avoids the multistep procedures of dendritic growth onto solid surfaces, since the grafting is performed in one synthetic step. The *in vitro* transfection efficiency of the materials was evaluated as a function of parameters such as dendrimer generation, different plasmid/dendrimer-mNP weight ratios, incubation time, and the presence or absence of the magnetic field. A pDNA (pEGFP-N3) that codes for the EGFP was used as reporter gene. The highest transfection efficiency (about 12%), without cytotoxic reactions, was obtained with the third generation dendrimer (DNA/G3-mNPs) using a weight ratio of 1/5 and under magnetic stimulus. However, negligible values of transfection efficiency were obtained when using DNA/G1-mNPs and DNA/G2-mNPs complexes in the same conditions. These results could be explained in terms of the increasing amounts of primary and tertiary amine groups as the dendrimer generation increases. Comparing the three dendritic generations attached to the mNPs, it is observed that gene transfection occurs when the amine content is higher than a threshold amine value, which is the case for the third generation nanosystem. Both, DNA complexation for the endocytosis process and proton sponge behavior, would explain the efficiency found with the third generation system. Regarding the application of the magnetic field, this was essential for the gene delivery in short periods of time. Exposure to the magnet for up to 20 min produced significant transfection efficiency, and no transfection was observed for the same delivery time in the absence of the magnetic field. The biocompatibility of the DNA/G3-mNPs complex was evaluated by analysis of cell viability and apoptosis by flow cytometry after magnetofection. The results demonstrated that the treatment produced a slight decrease in viable cells (25%) in comparison with the control, but very low levels of apoptosis were detected (4%), thus indicating the absence of cytotoxicity of this material.

As described above, magnetic vectors are composed of two or more components, one being the SPIONs to provide the vector with the magnetic behavior and another component essential to facilitate DNA binding and protection during extra and intracellular trafficking. This component is usually an organic polyamine, either polymeric or dendritic. Nevertheless, a different approach would be the use of calcium phosphates to accomplish this issue.

Based on these facts, Spector and coworkers [162] designed a strategy to render superparamagnetic behavior to synthetic HAp NPs and natural bone mineral nanoparticles (NBM NPs) for their application in magnetofection of a pDNA encoding for a neurotrophic growth factor. For the preparation of the magnetic vectors, calcium phosphate nanoparticles (CaP NPs) were treated with iron ions using a wet chemical process, which consisted in the addition of an iron (II) chloride solution to dispersed suspensions of HAp and NBM nanoparticles, subsequent treatment with NH_4OH and aging. Characterization of the CaP NPs confirmed that the magnetic properties are the result of the heteroepitaxial growth of crystallites of magnetite on the individual HAp and NBM crystallites, that is, the result of the presence of adherent magnetite crystallites. The magnetic CaP NPs displayed enhanced *in vitro* gene transfection when used as nonviral vectors under the action of a magnetic field, the magnet being applied for only the first 15 min period that the cells were exposed to the magnetoplexes. It is also worth noting that the lactate dehydrogenase (LDH) assay showed no cytotoxicity of the magnetic CaP NPs. In addition, the present study demonstrates that unmodified IONPs have a poor binding affinity for DNA.

Three-dimensional (3D) cellular assays closely mimic the *in vivo* milieu, providing a rapid, inexpensive system for screening drug candidates for toxicity or efficacy in the early stages of drug discovery. The magnetofection technique has also been applied to cells grown in a 3D-culture environment [163]. Using PEI-coated magnetic IONPs, siRNA and pDNA have been delivered through a collagen matrix, allowing them to enter cells in a 3D-culture environment.

DNA vaccination is another possible application of the magnetofection technique, which has been recently investigated for the delivery of malaria DNA vaccine *in vitro* [164]. The transfection efficiency of PEI-coated iron oxide mNPs as carriers for malaria DNA vaccine into eukaryotic cells *in vitro* is significantly enhanced under the application of an external magnetic field, so the results have prompted the authors to extend the investigation to *in vivo* trials.

IONPs can also be used as vectors for gene delivery under standard conditions, that is, without the application of the magnetic field. In this sense, a dual-responsive (pH and temperature) core–shell nanosystem, the core being the magnetic IONPs, and the shell the double-responsive polymer, has been prepared for gene delivery and cell separation. After transfection with the hybrid NPs the cells acquire magnetic properties that can be used for selective isolation of transfected cells [165]. A multifunctional nanovector has been reported for the treatment of hepatitis C by delivering DNAzyme to induce knockdown of the hepatitis C virus gene. The IONPs have a magnetic core coated with dextran, and are conjugated to a synthetic DNAzyme targeting a gene of interest, a NIR fluorescent dye (Cy5.5), and a cell-penetrating peptide (CPP) that aids in membrane translocation. The conjugated Cy5.5 dye enables tracking of the therapeutic nanoparticle *in vitro* and *in vivo* using fluorescence imaging, and the iron oxide core can be used for tracking via noninvasive MRI [166].

References

1. Bielke, W. and Erbacher, C. (eds) (2010) *Nucleic Acid Transfection*, Topics in Current Chemistry, vol. **296**, Springer, Berlin, Heidelberg.
2. Friedmann, T. and Rossi, J.J. (eds) (2007) *Gene Transfer: Delivery and Expression of DNA and RNA, a Laboratory Manual*, CSHL Press.
3. Moazed, D. (2009) Small RNAs in transcriptional gene silencing and genome defense. *Nature*, **457**, 413–420.
4. Hannon, G.J. and Rossi, J.J. (2004) Unlocking the potential of the human genome with RNA interference. *Nature*, **431**, 371–378.
5. Kurreck, J. (2009) RNA interference: from basic research to therapeutic applications. *Angew. Chem. Int. Ed.*, **48**, 1378–1398.
6. Ozpolat, B., Sood, A.K. and Lopez-Berestein, G. (2009) Nanomedicine based approaches for the delivery of SiRNA in cancer. *J. Internal Med.*, **267**, 44–53.
7. Sibley, C.R., Seow, Y. and Wood, M.J.A. (2010) Novel RNA-based strategies for therapeutic gene silencing. *Mol. Ther.*, **18**, 466–476.
8. Schaffer, D.V. and Zhou, W. (eds) (2010) *Gene Therapy and Gene Delivery Systems*, Advances in Biochemical Engineering Biotechnology, vol. **99**, Springer.
9. Thomas, C.E., Ehrhardt, A. and Kay, M.A. (2003) Progress and problems with the use of viral vectors for gene therapy. *Nat. Rev. Genet.*, **4**, 346–358.
10. Glover, D.J., Lipps, H.J. and Jans, D.A. (2005) Towards safe, nonviral therapeutic gene expression in humans. *Nat. Rev. Genet.*, **6**, 299–310.
11. Pathak, A., Patnaik, S. and Gupta, K.C. (2009) Recent trends in nonviral vector-mediated gene delivery. *Biotechnol. J.*, **4**, 1559–1572.
12. Mintzer, M.A. and Simanek, E.E. (2009) Nonviral vectors for gene delivery. *Chem. Rev.*, **109**, 259–302.
13. Yu, H.J. and Wagner, E. (2009) Bioresponsive polymers for nonviral gene delivery. *Curr. Opin. Mol. Ther.*, **11**, 165–178.
14. Woodrow, K.A., Cu, Y., Booth, C.J. *et al.* (2009) Intravaginal gene silencing using biodegradable polymer nanoparticles densely loaded with small interfering RNA. *Nat. Mater.*, **8**, 526–533.
15. Tseng, Y.C., Mozumdar, S. and Huang, L. (2009) Lipid-based systemic delivery of siRNA. *Adv. Drug Delivery Rev.*, **61**, 721–731.
16. Dufès, C., Uchegbu, I.F. and Schätzlein, A.G. (2005) Dendrimers in gene delivery. *Adv. Drug Delivery Rev.*, **57**, 2177–2202.
17. Mintzer, M.A. and Grinstaff, M.W. (2011) Biomedical applications of dendrimers: a tutorial. *Chem. Soc. Rev.*, **40**, 173–190.
18. Rosi, N.L., Giljohann, D.A., Thaxton, C.S. *et al.* (2006) Oligonucleotide-modified gold nanoparticles for intracellular gene regulation. *Science*, **312**, 1027–1030.
19. Slowing, I.I., Vivero-Escoto, J.L., Wu, C.-W. and Lin, V.S.-Y. (2008) Mesoporous silica nanoparticles as controlled release drug delivery and gene transfection carriers. *Adv. Drug Delivery Rev.*, **60**, 1278–1288.
20. Sokolova, V. and Epple, M. (2008) Inorganic nanoparticles as carriers of nucleic acids into cells. *Angew. Chem. Int. Ed.*, **47**, 1382–1395.
21. De, M., Ghosh, P.S. and Rotello, V.M. (2008) Applications of nanoparticles in biology. *Adv. Mater.*, **20**, 4225–4241.

22. Ho, R.J.Y. and Gibaldi, M. (2004) *Biotechnology and Biopharmaceuticals: Transforming Proteins and Genes into Drugs*, John Wiley & Sons, Inc., Hoboken, NJ.

23. Sheridan, C. (2011) Gene therapy finds its niche. *Nat. Biotechnol.*, **29**, 121–128.

24. Miller, A.D. (1992) Human gene therapy comes of age. *Nature*, **357**, 455–460.

25. Schleef, M. (ed) (2005) *DNA-Pharmaceuticals: Formulation and Delivery in Gene Therapy, DNA Vaccination and Immunotherapy*, Wiley-VCH Verlag GmbH, Weinheim.

26. Templeton, N.S. (ed) (2009) *Gene and Cell Therapy: Therapeutic Mechanisms and Strategies*, CRC Press/Taylor & Francis Group, Boca Raton, FL.

27. Mancuso, K., Hauswirth, W.W., Li, Q. *et al.* (2009) Gene therapy for red-green colour blindness in adult primates. *Nature*, **461**, 784–787.

28. Feldman, A.L. and Libutti, S.K. (2000) Progress in antiangiogenic gene therapy of cancer. *Cancer*, **89**, 1181–1194.

29. Lechardeur, D., Verkman, A.S. and Lukacs, G.L. (2005) Intracellular routing of plasmid DNA during non-viral gene transfer. *Adv. Drug Delivery Rev.*, **57**, 755–767.

30. Bausinger, R., von Gersdorff, K., Braeckmans, K. *et al.* (2006) The transport of nanosized gene carriers unraveled by live-cell imaging. *Angew. Chem. Int. Ed.*, **45**, 1568–1572.

31. Reischl, D. and Zimmer, A. (2009) Drug delivery of siRNA therapeutics: potentials and limits of nanosystems. *Nanomedicine: NBM*, **5**, 8–20.

32. Elbashir, S.M., Harborth, J., Lendeckel, W. *et al.* (2001) Duplexes of 21-nucleotide RNAs mediate RNA interference in cultured mammalian cells. *Nature*, **411**, 494–498.

33. Niemeyer, C.M. and Mirkin, C.A. (2004) *Nanobiotechnology*, Wiley-VCH Verlag GmbH, Weinheim.

34. Sanchez, C., Shea, K.J. and Kitagawa, S. (2011) Recent progress in hybrid materials science. *Chem. Soc. Rev.*, **40**, 471–472.

35. Boussif, O., Lezoualc'h, F., Zanta, M.A. *et al.* (1995) A versatile vector for gene and oligonucleotide transfer into cells in culture and in vivo: polyethylenimine. *Proc. Natl. Acad. Sci. U.S.A.*, **92**, 7297–7301.

36. Dennig, J. and Duncan, E. (2002) Gene transfer into eukaryotic cells using activated polyamidoamine dendrimers. *Rev. Mol. Biotechnol.*, **90**, 339–347.

37. Zinselmeyer, B.H., Mackay, S.P., Schatzlein, A.G. and Uchegbu, I.F. (2002) The lower-generation polypropylenimine dendrimers are effective gene-transfer agents. *Pharm. Res.*, **19**, 960–967.

38. Fischer, D., Bieber, T., Li, Y. *et al.* (1999) A novel non-viral vector for DNA delivery based on low molecular weight, branched polyethylenimine: effect of molecular weight on transfection efficiency and cytotoxicity. *Pharm. Res.*, **16**, 1273–1279.

39. Xia, T., Kovochich, M., Liong, M. *et al.* (2008) Cationic polystyrene nanosphere toxicity depends on cell-specific endocytic and mitochondrial injury pathways. *ACS Nano*, **2**, 85–96.

40. Lv, H., Zhang, S., Wang, B. *et al.* (2006) Toxicity of cationic lipids and cationic polymers in gene delivery. *J. Controlled Release*, **114**, 100–109.

41. Xiong, M.P., Forrest, M.L., Ton, G. *et al.* (2007) Poly(aspartate-g-PEI800), a polyethylenimine analogue of low toxicity and high transfection efficiency for gene delivery. *Biomaterials*, **28**, 4889–4900.

42. Hu, C., Peng, Q., Chen, F. *et al.* (2010) Low molecular weight polyethyleneimine conjugated gold nanoparticles as efficient gene vectors. *Bioconjugate Chem.*, **21**, 836–843.

43. Namgung, R., Zhang, Y., Fang, Q.L. *et al.* (2010) Multifunctional silica nanotubes for dual-modality gene delivery and MR imaging. *Biomaterials*, **32**, 3042–3052.

44. Xia, T., Kovochich, M., Liong, M. *et al.* (2009) Polyethyleneimine coating enhances the cellular uptake of mesoporous silica nanoparticles and allows safe delivery of siRNA and DNA constructs. *ACS Nano*, **3**, 3273–3286.

45. Nunes, A., Amsharov, N., Guo, C. *et al.* (2010) Hybrid polymer-grafted multiwalled carbon nanotubes for *in vitro* gene delivery. *Small*, **6**, 2281–2291.

46. Namgung, R., Singha, K., Yu, M.K. *et al.* (2010) Hybrid superparamagnetic iron oxide nanoparticle-branched polyethylenimine magnetoplexes for gene transfection of vascular endothelial cells. *Biomaterials*, **31**, 4204–4213.

47. Larsen, A.K., Escargueil, A.E. and Skladanowski, A. (2000) Resistance mechanisms associated with altered intracellular distribution of anticancer agents. *Pharmacol. Ther.*, **85**, 217–229.

48. Eytan, G.D. (2005) Mechanism of multidrug resistance in relation to passive membrane permeation. *Biomed. Pharmacother.*, **59**, 90–97.

49. Pakunlu, R.I., Wang, Y., Tsao, W. *et al.* (2004) Enhancement of the efficacy of chemotherapy for lung cancer by simultaneous suppression of multidrug resistance and antiapoptotic cellular defense: novel multicomponent delivery system. *Cancer Res.*, **64**, 6214–6224.

50. Gergely, S., Katalin, J., Ferenc, A. and Balázs, S. (1998) Diagnostics of multidrug resistance in cancer. *Pathol. Oncol. Res.*, **4**, 251–257.

51. George, J., Banik, N.L. and Ray, S.K. (2009) Bcl-2 siRNA augments taxol mediated apoptotic death in human glioblastoma U138MG and U251MG cells. *Neurochem. Res.*, **34**, 66–78.

52. Beh, C.W., Seow, W.Y., Wang, Y. *et al.* (2008) Efficient delivery of Bcl-2-targeted siRNA using cationic polymer nanoparticles: downregulating mrna expression level and sensitizing cancer cells to anticancer drug. *Biomacromolecules*, **10**, 41–48.

53. Pakunlu, R.I., Cook, T.J. and Minko, T. (2003) Simultaneous modulation of multidrug resistance and antiapoptotic cellular defense by MDR1 and BCL-2 targeted antisense oligonucleotides enhances the anticancer efficacy of doxorubicin. *Pharm. Res.*, **20**, 351–359.

54. Kanda, T., Yokosuka, O., Imazeki, F. *et al.* (2005) Enhanced sensitivity of human hepatoma cells to 5-fluorouracil by small interfering RNA targeting Bcl-2. *DNA Cell Biol.*, **24**, 805–809.

55. Lima, R.T., Martins, L.M., Guimarães, J.E. *et al.* (2004) Specific downregulation of Bcl-2 and xIAP by RNAi enhances the effects of chemotherapeutic agents in MCF-7 human breast cancer cells. *Cancer Gene Ther.*, **11**, 309–316.

56. Hamilton, A.J. and Baulcombe, D.C. (1999) A species of small antisense RNA in posttranscriptional gene silencing in plants. *Science*, **286**, 950–952.

57. Wang, Y., Gao, S., Ye, W.H. *et al.* (2006) Co-delivery of drugs and DNA from cationic core-shell nanoparticles self-assembled from a biodegradable copolymer. *Nat. Mater.*, **5**, 791–796.

58. Maszewska, M., Leclaire, J., Cieslak, M. *et al.* (2003) Water-soluble polycationic dendrimers with a phosphoramidothioate backbone: preliminary studies of cytotoxicity and oligonucleotide/plasmid delivery in human cell culture. *Oligonucleotides*, **13**, 193–205.

59. Uskoković, V. and Uskoković, D.P. (2011) Nanosized hydroxyapatite and other calcium phosphates: chemistry of formation and application as drug and gene delivery agents. *J. Biomed. Mater. Res. Part B: Appl. Biomater.*, **96**, 152–191.

60. Lee, D., Upadhye, K. and Kumta, P.N. (2012) Nano-sized calcium phosphate (CaP) carriers for non-viral gene delivery. *Mater. Sci. Eng., B Adv.*, **177**, 289–302.

61. Maitra, A. (2005) Calcium phosphate nanoparticles: second-generation nonviral vectors in gene therapy. *Expert Rev. Mol. Diagn.*, **5**, 893–905.

62. Bisht, S., Bhakta, G., Mitra, S. and Maitra, A. (2005) pDNA loaded calcium phosphate nanoparticles: highly efficient non-viral vector for gene delivery. *Int. J. Pharm.*, **288**, 157–168.

63. Okazaki, M., Yoshida, Y., Yamaguchi, S. *et al.* (2001) Affinity binding phenomena of DNA onto apatite crystals. *Biomaterials*, **22**, 2459–2464.

64. Olton, D.Y.E., Close, J.M., Sfeir, C.S. and Kumta, P.N. (2011) Intracellular trafficking pathways involved in the gene transfer of nano-structured calcium phosphate-DNA particles. *Biomaterials*, **32**, 7662–7670.

65. Graham, F.L. and van der Eb, A.J. (1973) A new technique for the assay of infectivity of human adenovirus 5 DNA. *Virology*, **52**, 456–467.

66. Welzel, T., Radtke, I., Meyer-Zaika, W. *et al.* (2004) Transfection of cells with custom-made calcium phosphate nanoparticles coated with DNA. *J. Mater. Chem.*, **14**, 2213–2217.

67. Sokolova, V.V., Radtke, I., Heumann, R. and Epple, M. (2006) Effective transfection of cells with multi-shell calcium phosphate-DNA nanoparticles. *Biomaterials*, **27**, 3147–3153.

68. Sokolova, V., Neumann, S., Kovtun, A. *et al.* (2010) An outer shell of positively charged poly(ethyleneimine) strongly increases the transfection efficiency of calcium phosphate/DNA nanoparticles. *J. Mater. Sci.*, **45**, 4952–4957.

69. Klesing, J., Chernousova, S. and Epple, M. (2012) Freeze-dried cationic calcium phosphate nanorods as versatile carriers of nucleic acids (DNA, siRNA). *J. Mater. Chem.*, **22**, 199–204.

70. Olton, D., Li, J., Wilson, M.E. *et al.* (2007) Nanostructured calcium phosphates (NanoCaPs) for non-viral gene delivery: influence of the synthesis parameters on transfection efficiency. *Biomaterials*, **28**, 1267–1279.

71. Pedraza, C.E., Bassett, D.C., McKee, M.D. *et al.* (2008) The importance of particle size and DNA condensation salt for calcium phosphate nanoparticle transfection. *Biomaterials*, **29**, 3384–3392.

72. Giger, E.V., Puigmartí-Luis, J., Schlatter, R. *et al.* (2011) Gene delivery with bisphosphonate-stabilized calcium phosphate nanoparticles. *J. Controlled Release*, **150**, 87–93.

73. Jang, S., Lee, S., Kim, H. *et al.* (2012) Preparation of pH-sensitive CaP nanoparticles coated with a phosphate-based block copolymer for efficient gene delivery. *Polymer*, **53**, 4678–4685.

74. Kovtun, A., Neumann, S., Neumeier, M. *et al.* (2013) Nanoparticle-mediated gene transfer from electrophoretically coated metal surfaces. *J. Phys. Chem. B*, **117**, 1550–1555.

75. González, B., Colilla, M., López de Laorden, C. and Vallet-Regí, M. (2009) A novel synthetic strategy for covalently bonding dendrimers to ordered mesoporous silica: potential drug delivery applications. *J. Mater. Chem.*, **19**, 9012–9024.

76. Rosenholm, J.M., Meinander, A., Peuhu, E. *et al.* (2009) Targeting of porous hybrid silica nanoparticles to cancer cells. *ACS Nano*, **3**, 197–206.

77. Rosenholm, J.M., Sahlgren, C. and Lindén, M. (2010) Towards multifunctional, targeted drug delivery systems using mesoporous silica nanoparticles – opportunities and challenges. *Nanoscale*, **2**, 1870–1883.

78. Radu, D.R., Lai, C., Jeftinija, Y.K. *et al.* (2004) A polyamidoamine dendrimer-capped mesoporous silica nanosphere-based gene transfection reagent. *J. Am. Chem. Soc.*, **126**, 13216–13217.

79. Torney, F., Trewyn, B.G., Lin, V.S.Y. and Wang, K. (2007) Mesoporous silica nanoparticles deliver DNA and chemicals into plants. *Nat. Nanotechnol.*, **2**, 295–300.

80. Chen, A.M., Zhang, M., Wei, D. *et al.* (2009) Co-delivery of doxorubicin and Bcl-2 siRNA by mesoporous silica nanoparticles enhances the efficacy of chemotherapy in multidrug-resistant cancer cells. *Small*, **5**, 2673–2677.

81. Hom, C., Lu, J., Liong, M. *et al.* (2010) Mesoporous silica nanoparticles facilitate delivery of siRNA to shutdown signaling pathways in mammalian cells. *Small*, **6**, 1185–1190.

82. Meng, H., Liong, M., Xia, T. *et al.* (2010) Engineered design of mesoporous silica nanoparticles to deliver doxorubicin and p-glycoprotein siRNA to overcome drug resistance in a cancer cell line. *ACS Nano*, **4**, 4539–4550.

83. Bhattarai, S.R., Muthuswamy, E., Wani, A. *et al.* (2010) Enhanced gene and siRNA delivery by polycation-modified mesoporous silica nanoparticles loaded with chloroquine. *Pharm. Res.*, **27**, 2556–2568.

84. Kim, M.-H., Na, H.-K., Kim, Y.-K. *et al.* (2011) Facile synthesis of monodispersed mesoporous silica nanoparticles with ultralarge pores and their application in gene delivery. *ACS Nano*, **5**, 3568–3576.

85. Hartono, S.B., Gu, W., Kleitz, F. *et al.* (2012) Poly-L-lysine functionalized large pore cubic mesostructured silica nanoparticles as biocompatible carriers for gene delivery. *ACS Nano*, **6**, 2104–2117.

86. Gao, F., Botella, P., Corma, A. *et al.* (2009) Monodispersed mesoporous silica nanoparticles with very large pores for enhanced adsorption and release of DNA. *J. Phys. Chem. B*, **113**, 1796–1804.

87. Ashley, C.E., Carnes, E.C., Epler, K.E. *et al.* (2012) Delivery of small interfering RNA by peptide-targeted mesoporous silica nanoparticle-supported lipid bilayers. *ACS Nano*, **6**, 2174–2188.

88. Nakamura, E., Isobe, H., Tomita, N. *et al.* (2000) Functionalized fullerene as an artificial vector for transfection. *Angew. Chem. Int. Ed.*, **39**, 4254–4257.

89. Isobe, H., Nakanishi, W., Tomita, N. *et al.* (2006) Nonviral gene delivery by tetraamino fullerene. *Mol. Pharm.*, **3**, 124–134.

90. Isobe, H., Nakanishi, W., Tomita, N. *et al.* (2006) Gene delivery by aminofullerenes: structural requirements for efficient transfection. *Chem. Asian J.*, **1**, 167–175.

91. Klumpp, C., Lacerda, L., Chaloin, O. *et al.* (2007) Multifunctionalised cationic fullerene adducts for gene transfer: design, synthesis and DNA complexation. *Chem. Commun.*, 3762–3764.

92. Sitharaman, B., Zakharian, T.Y., Saraf, A. *et al.* (2008) Water-soluble fullerene (C60) derivatives as nonviral gene-delivery vectors. *Mol. Pharm.*, **5**, 567–578.

93. Sigwalt, D., Holler, M., Iehl, J. *et al.* (2011) Gene delivery with polycationic fullerene hexakis-adducts. *Chem. Commun.*, **47**, 4640–4642.

94. Hirsch, A. (2002) Functionalization of single-walled carbon nanotubes. *Angew. Chem. Int. Ed.*, **41**, 1853–1859.

95. Banerjee, S., Hemraj-Benny, T. and Wong, S. .S. (2005) Covalent surface chemistry of single-walled carbon nanotubes. *Adv. Mater.*, **17**, 17–29.

96. Cheung, W., Pontoriero, F., Taratula, O. *et al.* (2010) DNA and carbon nanotubes as medicine. *Adv. Drug Delivery Rev.*, **62**, 633–649.

97. Lacerda, L., Russier, J., Pastorin, G. *et al.* (2012) Translocation mechanisms of chemically functionalised carbon nanotubes across plasma membranes. *Biomaterials*, **33**, 3334–3343.

98. Pantarotto, D., Singh, R., McCarthy, D. *et al.* (2004) Functionalized carbon nanotubes for plasmid DNA gene delivery. *Angew. Chem. Int. Ed.*, **43**, 5242–5246.

99. Singh, R., Pantarotto, D., McCarthy, D. *et al.* (2005) Binding and condensation of plasmid DNA onto functionalized carbon nanotubes: toward the construction of nanotube-based gene delivery vectors. *J. Am. Chem. Soc.*, **127**, 4388–4396.

100. Podesta, J.E., Al-Jamal, K.T., Herrero, M.A. *et al.* (2009) Antitumor activity and prolonged survival by carbon nanotube-mediated therapeutic siRNA silencing in a human lung xenograft model. *Small*, **5**, 1176–1185.

101. Herrero, M.A., Toma, F.M., Al-Jamal, K.T. *et al.* (2009) Synthesis and characterization of a carbon nanotube-dendron series for efficient siRNA delivery. *J. Am. Chem. Soc.*, **131**, 9843–9848.

102. Liu, Y., Wu, D.-C., Zhang, W.-D. *et al.* (2005) Polyethylenimine-grafted multiwalled carbon nanotubes for secure noncovalent immobilization and efficient delivery of DNA. *Angew. Chem. Int. Ed.*, **44**, 4782–4785.

103. Ahmed, M., Jiang, X., Deng, Z. and Narain, R. (2009) Cationic glyco-functionalized single-walled carbon nanotubes as efficient gene delivery vehicles. *Bioconjugate Chem.*, **20**, 2017–2022.

104. Qin, W., Yang, K., Tang, H. *et al.* (2011) Improved GFP gene transfection mediated by polyamidoamine dendrimer-functionalized multi-walled carbon nanotubes with high biocompatibility. *Colloid Surf., B: Biointerfaces*, **84**, 206–213.

105. Battigelli, A., Wang, J.T.-W., Russier, J. *et al.* (2013) Ammonium and guanidinium dendron-carbon nanotubes by amidation and click chemistry and their use for siRNA delivery. *Small*, **9**, 3610–3619. doi: 10.1002/smll.201300264

106. Kam, N.W.S., Liu, Z. and Dai, H. (2005) Functionalization of carbon nanotubes *via* cleavable disulfide bonds for efficient intracellular delivery of siRNA and potent gene silencing. *J. Am. Chem. Soc.*, **127**, 12492–12493.

107. Liu, Z., Winters, M., Holodniy, M. and Dai, H. (2007) siRNA delivery into human T cells and primary cells with carbon-nanotube transporters. *Angew. Chem. Int. Ed.*, **46**, 2023–2027.

108. Misra, S.K., Moitra, P., Chhikara, B.S. *et al.* (2012) Loading of single-walled carbon nanotubes in cationic cholesterol suspensions significantly improves gene transfection efficiency in serum. *J. Mater. Chem.*, **22**, 7985–7998.

109. Tang, L.A.L., Wang, J. and Loh, K.P. (2010) Graphene-based SELDI probe with ultrahigh extraction and sensitivity for DNA oligomer. *J. Am. Chem. Soc.*, **132**, 10976–10977.

110. He, S., Song, B., Li, D. *et al.* (2010) A graphene nanoprobe for rapid, sensitive, and multicolor fluorescent DNA analysis. *Adv. Funct. Mater.*, **20**, 453–459.

111. Lu, C.-H., Yang, H.-H., Zhu, C.-L. *et al.* (2009) A graphene platform for sensing biomolecules. *Angew. Chem. Int. Ed.*, **48**, 4785–4787.

112. Liu, F., Choi, J.Y. and Seo, T.S. (2010) Graphene oxide arrays for detecting specific DNA hybridization by fluorescence resonance energy transfer. *Biosens. Bioelectron.*, **25**, 2361–2365.

113. Lu, C.-H., Li, J., Liu, J.-J. *et al.* (2010) Increasing the sensitivity and single-base mismatch selectivity of the molecular beacon using graphene oxide as the "nanoquencher". *Chem. Eur. J.*, **16**, 4889–4894.

114. Lu, C.-H., Zhu, C.-L., Li, J. *et al.* (2010) Using graphene to protect DNA from cleavage during cellular delivery. *Chem. Commun.*, **46**, 3116–3118.

115. Feng, L., Zhang, S. and Liu, Z. (2011) Graphene based gene transfection. *Nanoscale*, **3**, 1252–1257.

116. Chen, B., Liu, M., Zhang, L. *et al.* (2011) Polyethylenimine-functionalized graphene oxide as an efficient gene delivery vector. *J. Mater. Chem.*, **21**, 7736–7741.

117. Ren, T., Li, L., Cai, X. *et al.* (2012) Engineered polyethylenimine/graphene oxide nanocomposite for nuclear localized gene delivery. *Polym. Chem.*, **3**, 2561–2569.

118. Zhang, L., Lu, Z., Zhao, Q. *et al.* (2011) Enhanced chemotherapy efficacy by sequential delivery of siRNA and anticancer drugs using PEI-grafted graphene oxide. *Small*, **7**, 460–464.

119. Kim, H., Namgung, R., Singha, K. *et al.* (2011) Graphene oxide-polyethylenimine nanoconstruct as a gene delivery vector and bioimaging tool. *Bioconjugate Chem.*, **22**, 2558–2567.

120. Markovic, Z.M., Harhaji-Trajkovic, L.M., Todorovic-Markovic, B.M. *et al.* (2011) *In vitro* comparison of the photothermal anticancer activity of graphene nanoparticles and carbon nanotubes. *Biomaterials*, **32**, 1121–1129.

121. Yang, K., Hu, L., Ma, X. *et al.* (2012) Multimodal imaging guided photothermal therapy using functionalized graphene nanosheets anchored with magnetic nanoparticles. *Adv. Mater.*, **24**, 1868–1872.

122. Yang, K., Wan, J., Zhang, S. *et al.* (2012) The influence of surface chemistry and size of nanoscale graphene oxide on photothermal therapy of cancer using ultra-low laser power. *Biomaterials*, **33**, 2206–2214.

123. Yang, K., Zhang, S., Zhang, G. *et al.* (2010) Graphene in mice: ultrahigh in vivo tumor uptake and efficient photothermal therapy. *Nano Lett.*, **10**, 3318–3323.

124. Zhang, W., Guo, Z., Huang, D. *et al.* (2011) Synergistic effect of chemo-photothermal therapy using PEGylated graphene oxide. *Biomaterials*, **32**, 8555–8561.

125. Matsushita-Ishiodori, Y. and Ohtsuki, T. (2012) Photoinduced RNA interference. *Acc. Chem. Res.*, **45**, 1039–1047.
126. Feng, L., Yang, X., Shi, X. *et al.* (2013) Polyethylene glycol and polyethylenimine dual-functionalized nano-graphene oxide for photothermally enhanced gene delivery. *Small*, **9**, 1989–1997.
127. Kim, H. and Kim, W.J. (2013) Photothermally controlled gene delivery by reduced graphene oxide-polyethylenimine nanocomposite. *Small*. doi: 10.1002/smll.201202636
128. Xue, X., Wang, F. and Liu, X. (2011) Emerging functional nanomaterials for therapeutics. *J. Mater. Chem.*, **21**, 13107–13127.
129. Schladt, T.D., Schneider, K., Schild, H. and Tremel, W. (2011) Synthesis and biofunctionalization of magnetic nanoparticles for medical diagnosis and treatment. *Dalton Trans.*, **40**, 6315–6343.
130. Pankhurst, Q.A., Thanh, N.K.T., Jones, S.K. and Dobson, J. (2009) Progress in applications of magnetic nanoparticles in biomedicine. *J. Phys. D Appl. Phys.*, **42**, 224001.
131. Roca, A.G., Costo, R., Rebolledo, A.F. *et al.* (2009) Progress in the preparation of magnetic nanoparticles for applications in biomedicine. *J. Phys. D Appl. Phys.*, **42**, 224002.
132. Fang, C. and Zhang, M. (2009) Multifunctional magnetic nanoparticles for medical imaging applications. *J. Mater. Chem.*, **19**, 6258–6266.
133. Laurent, S., Forge, D., Port, M. *et al.* (2008) Magnetic iron oxide nanoparticles: synthesis, stabilization, vectorization, physicochemical characterizations, and biological applications. *Chem. Rev.*, **108**, 2064–2110.
134. Gupta, A.J. and Gupta, M. (2005) Synthesis and surface engineering of iron oxide nanoparticles for biomedical applications. *Biomaterials*, **26**, 3995–4021.
135. Polyak, B. and Friedman, G. (2009) Magnetic targeting for site-specific drug delivery: applications and clinical potential. *Expert Opin. Drug Deliv.*, **6**, 53–70.
136. Arruebo, M., Fernández-Pacheco, R., Ibarra, M.R. and Santamaría, J. (2007) Magnetic nanoparticles for drug delivery. *Nano Today*, **2**, 22–32.
137. Yu, M.K., Jeong, Y.Y., Park, J. *et al.* (2008) Drug-loaded superparamagnetic iron oxide nanoparticles for combined cancer imaging and therapy in vivo. *Angew. Chem. Int. Ed.*, **47**, 5362–5365.
138. Xie, J., Huang, J., Li, X. *et al.* (2009) Iron oxide nanoparticle platform for biomedical applications. *Curr. Med. Chem.*, **16**, 1278–1294.
139. Song, E.-Q., Hu, J., Wen, C.-Y. *et al.* (2011) Fluorescent-magnetic-biotargeting multifunctional nanobioprobes for detecting and isolating multiple types of tumor cells. *ACS Nano*, **5**, 761–770.
140. Bardhan, R., Chen, W., Bartels, M. *et al.* (2010) Tracking of multimodal therapeutic nanocomplexes targeting breast cancer in vivo. *Nano Lett.*, **10**, 4920–4928.
141. Foy, S.P., Manthe, R.L., Foy, S.T. *et al.* (2010) Optical imaging and magnetic field targeting of magnetic nanoparticles in tumors. *ACS Nano*, **4**, 5217–5224.
142. van Landeghem, F.K.H., Maier-Hauff, K., Jordan, A. *et al.* (2009) Post-mortem studies in glioblastoma patients treated with thermotherapy using magnetic nanoparticles. *Biomaterials*, **30**, 52–57.
143. Vauthier, C., Tsapis, N. and Couvreur, P. (2011) Nanoparticles: heating tumors to death? *Nanomedicine*, **6**, 99–109.

144. Khemtong, C., Kessinger, C.W. and Gao, J. (2009) Polymeric nanomedicine for cancer MR imaging and drug delivery. *Chem. Commun.*, 3497–3510.
145. McCarthy, J.R. and Weissleder, R. (2008) Multifunctional magnetic nanoparticles for targeted imaging and therapy. *Adv. Drug Delivery Rev.*, **60**, 1241–1251.
146. Kim, J., Piao, Y. and Hyeon, T. (2009) Multifunctional nanostructured materials for multimodal imaging, and simultaneous imaging and therapy. *Chem. Soc. Rev.*, **38**, 372–390.
147. Cheon, J. and Lee, J.-H. (2008) Synergistically integrated nanoparticles as multimodal probes for nanobiotechnology. *Acc. Chem. Res.*, **41**, 1630–1640.
148. Scherer, F., Anton, M., Schillinger, U. *et al.* (2002) Magnetofection: enhancing and targeting gene delivery by magnetic force *in vitro* and in vivo. *Gene Ther.*, **9**, 102–109.
149. Dobson, J. (2006) Gene therapy progress and prospects: magnetic nanoparticle-based gene delivery. *Gene Ther.*, **13**, 283–287.
150. Melancon, M.P., Elliott, A., Ji, X. *et al.* (2011) Theranostics with multifunctional magnetic gold nanoshells: photothermal therapy and T2* magnetic resonance imaging. *Invest. Radiol.*, **46**, 132–140.
151. Stöber, W., Fink, A. and Bohn, E. (1968) Controlled growth of monodisperse silica spheres in the micron size range. *J. Colloid Interface Sci.*, **26**, 62–69.
152. Liong, M., Angelos, S., Choi, E. *et al.* (2009) Mesostructured multifunctional nanoparticles for imaging and drug delivery. *J. Mater. Chem.*, **19**, 6251–6257.
153. Knežević, N.Ž., Ruiz-Hernández, E., Hennink, W.E. and Vallet-Regí, M. (2013) Magnetic mesoporous silica-based core/shell nanoparticles for biomedical applications. *RSC Adv.*, **3**, 9584–9593.
154. Lee, J.E., Lee, N., Kim, T. *et al.* (2011) Multifunctional mesoporous silica nanocomposite nanoparticles for theranostic applications. *Acc. Chem. Res.*, **44**, 893–902.
155. Colilla, M., González, B. and Vallet-Regí, M. (2013) Mesoporous silica nanoparticles for the design of smart delivery nanodevices. *Biomater. Sci.*, **1**, 114–134.
156. Song, H.P., Yang, J.Y., Lo, S.L. *et al.* (2010) Gene transfer using self-assembled ternary complexes of cationic magnetic nanoparticles, plasmid DNA and cell-penetrating Tat peptide. *Biomaterials*, **31**, 769–778.
157. Pickard, M.R., Barraud, P. and Chari, D.M. (2011) The transfection of multipotent neural precursor/stem cell transplant populations with magnetic nanoparticles. *Biomaterials*, **32**, 2274–2284.
158. Steitz, B., Hofmann, H., Kamau, S.W. *et al.* (2007) Characterization of PEI-coated superparamagnetic iron oxide nanoparticles for transfection: size distribution, colloidal properties and DNA interaction. *J. Magn. Magn. Mater.*, **311**, 300–305.
159. Arsianti, M., Lim, M., Marquis, C.P. and Amal, R. (2010) Assembly of polyethylenimine-based magnetic iron oxide vectors: insights into gene delivery. *Langmuir*, **26**, 7314–7326.
160. Lee, H., Lee, E., Kim, D.K. *et al.* (2006) Antibiofouling polymer-coated superparamagnetic iron oxide nanoparticles as potential magnetic resonance contrast agents for in vivo cancer imaging. *J. Am. Chem. Soc.*, **128**, 7383–7389.
161. González, B., Ruiz-Hernández, E., Feito, M.J. *et al.* (2011) Covalently bonded dendrimer-maghemite nanosystems: nonviral vectors for *in vitro* gene magnetofection. *J. Mater. Chem.*, **21**, 4598–4604.

162. Wu, H.-C., Wang, T.-W., Bohn, M.C. *et al.* (2010) Novel magnetic hydroxyapatite nanoparticles as non-viral vectors for the glial cell line-derived neurotrophic factor gene. *Adv. Funct. Mater.*, **20**, 67–77.

163. Zhang, H., Lee, M.-Y., Hogg, M.G. *et al.* (2010) Gene delivery in three-dimensional cell cultures by superparamagnetic nanoparticles. *ACS Nano*, **4**, 4733–4743.

164. Al-Deen, F.N., Ho, J., Selomulya, C. *et al.* (2011) Superparamagnetic nanoparticles for effective delivery of malaria DNA vaccine. *Langmuir*, **27**, 3703–3712.

165. Majewski, A.P., Schallon, A., Jérôme, V. *et al.* (2012) Dual-responsive magnetic core-shell nanoparticles for nonviral gene delivery and cell separation. *Biomacromolecules*, **13**, 857–866.

166. Ryoo, S.-R., Jang, H., Kim, K.-S. *et al.* (2012) Functional delivery of DNAzyme with iron oxide nanoparticles for hepatitis C virus gene knockdown. *Biomaterials*, **33**, 2754–2761.

14

Ceramic Nanoparticles for Cancer Treatment

Alejandro Baeza
Departamento de Química Inorgánica y Bioinorgánica, Facultad de Farmacia,
Universidad Complutense de Madrid, CIBER de Bioingeniería, Biomateriales y
Nanomedicina (CIBER-BBN), Spain

14.1 Delivery of Nanocarriers to Solid Tumors

Nanoparticles used as drug carriers are submicron-sized systems which should have a size below 100 nm in at least one orthogonal direction. In recent years, these nanocarriers have offered a promising alternative to conventional therapy in oncology and have been prepared using a wide range of materials, such as polymers, lipids, viruses, and ceramics [1]. Conventional antitumoral drugs usually present small size and they are removed from systemic circulation very fast, so higher dosages are needed. This raises the risk of severe side effects caused by the high toxicity of these antitumoral agents toward healthy tissues. The encapsulation of cytotoxic drugs in nanometric carriers improves their pharmacokinetic profile, protecting them against enzymatic degradation, and allows their selective accumulation into the diseased tissue owing to the enhance permeation and retention effect (EPR), as mentioned below. Moreover, using nanoparticles it may be possible to obtain other important advantages such as: to improve the delivery of hydrophobic drugs or macromolecular-type therapeutic agents (proteins, enzymes, or oligonucleotides), to transport and deliver multiple therapeutic agents at the same time, which is enormously important for combined therapy, to cross tight epithelial or endothelial barriers and even to visualize in real time the drug delivery process using imaging agents attached or encapsulated into the nanocarrier. The nanocarrier selectivity can be improved by attaching targeting agents onto its surface able to be specifically recognized by cancer cells or by the tumor microenvironment. Additionally, there is clear commercial interest in

Bioceramics with Clinical Applications, First Edition. Edited by María Vallet-Regí.
© 2014 John Wiley & Sons, Ltd. Published 2014 by John Wiley & Sons, Ltd.

the re-formulation of drugs which have not passed the clinical trials due to poor solubility or extensive side effects.

The maximum size of these nanocarriers is limited by the primary immunogenic system. Macrophages and other specialized cells are designed for the elimination of foreign entities and operate in all tissues. The clearance mechanisms orchestrated by these cells are generally not effective for objects with sizes less than 100–200 nm, although other factors such as particle shape or surface charge must be taken into account. On the other hand, particles or macromolecules with sizes smaller than 5–6 nm are rapidly excreted by the kidneys. Therefore, the optimal size for biomedical purposes of the nanocarriers should be between 10 and 200 nm.

14.1.1 Special Issues of Tumor Vasculature: Enhanced Permeation and Retention Effect (EPR)

The EPR effect was first reported by the Japanese researchers Matsumura and Maeda in 1986 [2]. They discovered that macromolecules larger than 40 kDa, which is the threshold of renal clearance, selectively leak out from tumor vessels and accumulate within tumor tissues [3]. This behavior is caused by the dramatic differences between tumoral and healthy blood vessels. Tumoral blood vessels are very irregular, highly heterogeneous, and tortuous, leaving unperfused spaces within the tumor. Moreover, their vessel-wall structure shows specific characteristics, such as wide interendothelial junctions, pericytes deficiency, aberrant basement membrane, large number of fenestrations, and transendothelial pores with diameters as large as several hundred nanometers. Consequently, vascular permeability is significantly higher in tumors than in healthy tissues. This enhanced permeability ensures enough supply of nutrients and oxygen to tumor tissues to allow their fast growth rate, but it could constitute their Achilles heel. Drug-loaded nanoparticles with sizes up to a few hundred nanometers would be able to extravasate selectively through the tumoral blood vessel fenestrations, releasing their payloads within solid tumors without affecting the rest of the tissues. On the other hand, macromolecules, waste products, or excess fluid in the interstitial space of healthy tissues are effectively recovered by the lymphatic system. By contrast, the rapidly growing tumoral cells compress the lymphatic vessels, especially at the center of the solid tumors, provoking their collapse. Thus, the clearance via lymphatic drainage is greatly compromised in neoplastic tissues, causing an additional retention of colloidal nanomedicines. The combination of high permeability and enhanced retention is the basis of the EPR effect and the reason why this phenomenon has become the "gold standard" in nanoparticle-based antitumoral therapy (Figure 14.1). It is important to remark that this particular vascular architecture has been observed even in tumor nodules smaller than 0.2 mm.

As has been mentioned above, nanoparticles exploit the EPR effect in order to be selectively accumulated into the tumoral tissue. However, it is necessary to take into account that this effect is a highly heterogeneous phenomenon which shows significant variations for each tumor, each patient, and even within a single tumor. Thus, in order to improve the therapeutic efficacy of nanomedicines, several strategies focused on the augmentation of the EPR effect can be employed. The administration of angiotensin II before the nanoparticle treatment produces systemic hypertension, which leads to a higher nanoparticle retention within the tumoral mass. The reason is that tumor blood vessels show very little response to

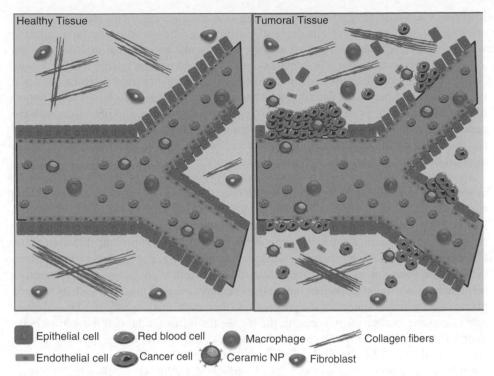

Epithelial cell   Red blood cell   Macrophage   Collagen fibers
Endothelial cell   Cancer cell   Ceramic NP   Fibroblast

Figure 14.1 *Differences between healthy and tumoral tissues (EPR effect) (See insert for color representation of the figure)*

this hormone due to the lack of smooth muscle layer or pericytes required for vasoconstriction, whereas healthy vessels show constriction. Therefore, the administered nanoparticles are pushed out into the tumoral space due to the different pressures between healthy or tumoral blood vessels. Nitric oxide (NO) is one of the main factors that sustain the EPR effect and is produced by the tumoral cells in higher amounts than normal cells in order to increase the blood flux and maintain their high nutrient requirements. Recently, it has been demonstrated that the administration of nitroglycerin and other NO-releasing agents, which can be converted to NO under hypoxic conditions, increase the efficacy of the nanomedicine therapy [4].

14.1.2 Tumor Microenvironment

The lack of an effective lymphatic system, particularly at the core of the tumor, in association with fluid leakage from the tumoral blood vessels produces interstitial hypertension, which constitutes an important barrier for the transport of nanoparticles within the tumor [5]. Due to the increased interstitial fluid pressure (IFP), the major mechanism of mass transport within a tumor is diffusion, a process highly dependent on the molecular weight and, therefore, very slow for nanoparticles. In some cases, the IFP inside tumors can exceed the vascular pressure, provoking the intravasation of the nanocarriers back to the blood system. This fact could originate systemic toxicity and poor efficacy of the therapy. Blood irrigation into tumors is very irregular, showing an average velocity of the red blood vessels one order of magnitude lower than the host vessels. Four different zones can be recognized within a solid tumor, based on perfusion rates: (i) a vascular necrotic region which exhibits a high cellular mortality due to low content of oxygen and nutrients, (ii) a transition zone called the seminecrotic region, (iii) a stabilized microcirculation region which present a normalized blood perfusion, and (iv) an advancing front [6]. The presence of these unperfused regions, which present low partial oxygen pressure and acidic pH, stimulates the appearance of multiresistant tumoral cells immune to radiotherapy and chemotherapeutic agents, due to evolution forces in this highly hostile microenvironment. Moreover, mutated tumoral cells able to migrate to different tissues usually show up in this hypoxic region.

An effective nanocarrier must be homogeneously distributed in all the affected tissue in order to eliminate all the tumoral cells before they acquire drug resistance and metastatic capacities. Thus, the nanoparticle should be able to diffuse into the tumoral extracellular matrix. This matrix is composed of a complex mixture of different proteins, mainly collagen, and glycoproteins, which form a dense network that hampers the diffusion of the nanocarriers. Particles with negative charges on their surface are strongly retained by the collagen fibers due to the positive charge of this protein at physiological pH. On other hand, tumors which exhibit higher amounts of negative charged proteins, such as sulfated glycosaminoglycans, can make the penetration of positive carriers difficult. One obvious rule is that the smaller the nanocarrier the higher the diffusion through the extracellular matrix (ECM). However, if the particle is too small, the discrimination between healthy and tumoral tissues by the EPR effect would be diminished. It is also compulsory to consider other aspects, such as the luminal surface of the blood vessels which presents a negative charge and, therefore, the positively charged particles can be nonspecifically attached on their surface suffering a rapid elimination from the bloodstream. For all these reasons, the nanocarrier should be carefully engineered with specific size, shape, and surface charge for each type of tumor.

14.2   Ceramic Nanoparticle Pharmacokinetics in Cancer Treatment

14.2.1   Biodistribution and Excretion/Clearance Pathways

The effective and safe application of a nanomaterial for drug delivery requires an understanding of its interactions with the biologic system [7]. The biocompatibility and biodistribution of each material depend not only on their characteristic properties (composition, size, shape, and surface charge) but also on other factors, such as the mode of administration (oral or intravenous), target body localization, or internalization route into the cell. Herein, general concepts about the introduction of nanoparticles into the human body will be briefly presented, considering the particularities of the ceramic particles. Although the nanoparticles can be introduced into the body by different routes, the most useful way in the case of antitumoral therapy is by peripheral intravenous injection and, therefore, only this way will be considered in this section.

Once the nanoparticles are injected into the bloodstream, some organs play an important role in relation to their circulation time in the body [8]. Among the body blood vessels, lung capillaries are the smallest ones, showing diameters between 2 and 13 μm. They constitute the first sieving constrains, especially in the case of rigid ceramic nanoparticles, which are not able to suffer any deformation of their structure in order to navigate through these tight channels. Therefore, ceramic particles with diameters bigger than 2 μm are rapidly trapped by the lungs and could constitute a serious risk for the integrity of the pulmonary circulation. The kidneys are in charge of the blood filtration. Macromolecules or particles whose size is less than 5–6 nm are freely removed from the bloodstream. The spleen is a lymphatic organ that participates in storage of spent blood components as well as in lymphocyte maturation. Particle retention in this organ is always undesired because it usually produces immunogenic reactions. The interaction of the nanoparticles with this organ is the responsible for the appearance of the accelerated blood clearance (ABC) effect [9]. This effect can appear after the administration of repeated doses of nanoparticles inducing a rapid elimination of these carriers. The ABC effect is divided into two phases, the induction phase or sensitization, originated by the interaction of the nanoparticles with B cells in the spleen. These cells recognize the particles as foreign entities and start the production of specific antibodies. Two to four days after the first injection, the amount of the specific antibodies against the nanoparticle increases. This is when the second phase, called the effectuation phase, begins. In this phase, the nanoparticles are rapidly covered by these antibodies being captured by the macrophages. Thus, the blood circulation times of the nanocarriers are shortened with each injection.

Finally, sooner or later, all the blood components of the blood, and consequently the injected nanoparticles, reach the liver. This organ contains almost 90% of the macrophage total body population. Once there, specific phagocytic cells called Kupffer cells engulf the nanoparticles and eliminate them by degradation or, if this is not possible, by retention in residual bodies within the cell. The interaction of smaller nanocarriers, less than 1 μm is less favorable and these particles can escape the liver retention. However, other characteristics of the carriers, such as shape or surface charge, can strongly influence the nanoparticle capture by these phagocytic cells and it is necessary to study in detail each type of particle. Thus, liver, spleen, and lungs usually retain the major amount of the injected nanoparticles and it is necessary to design these carriers in order to avoid their retention in these organs and also to avoid the renal clearance.

Another aspect that should be taken into account in order to study the fate and distribution of the nanomedicines is their interaction with the blood components. When nanoparticles are exposed to this biological fluid, they are in contact with more than 3700 different proteins and many other complex biomolecules, which can bind competitively with the nanocarrier surface. At a certain time after the injection, the particle surface will be covered by several proteins forming the *corona* of the nanoparticle, which constitutes a nanoparticle/biomolecule interface "readable" by the cells [10]. Among blood proteins, a very important type is the opsonin family, that are proteins designed to mark certain antigens or foreign bodies inducing the internalization and degradation of these entities by the mononuclear phagocytic system (MPS). This process is called opsonization. The macrophages of the MPS have the capacity to eliminate the nanocarriers from the bloodstream within seconds after the intravenous administration. However, these phagocytic cells cannot directly identify the nanoparticles by themselves, but they recognize the opsonins bound to the particle surface. Thus, the injected nanoparticles which are covered by opsonins become visible to the macrophages and are rapidly removed from the bloodstream. In order to avoid this detection, several methods have been reported. These methods

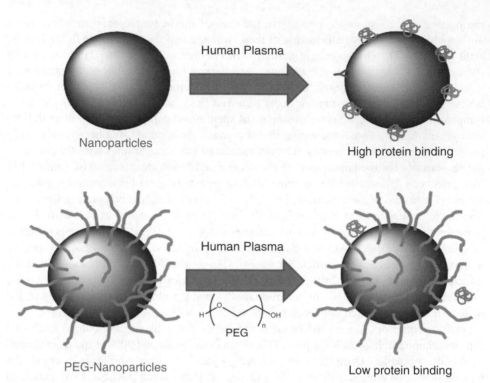

Figure 14.2   *PEGylation of nanoparticles*

usually employ the decoration of the external surface of the nanoparticle with molecules which interfere with the binding of opsonins. One of the most employed strategies is the grafting of shielding groups which block or make difficult the electrostatic or hydrophobic interactions that allow protein absorption on the nanoparticle surface. These groups are usually long hydrophilic polymeric chains, being the most famous polyethyleneglycol (PEG) chains of different lengths [11]. Anchoring PEG chains on the surface of the nanoparticles in order to increase the blood circulation half-life is a well-established procedure, called PEGylation. Thus, PEGylated nanoparticles are able to escape macrophage detection due to the lack of opsonin adhesion, avoiding the clearance from the blood stream for longer times and for this reason they are known as stealth particles (Figure 14.2).

14.2.2   Toxicity of the Ceramic Nanoparticles

According to the European Society of Biomaterials, the term biocompatibility can be defined as the "ability of a material to perform with an appropriate host response in a specific application". Regardless of the different nature of each type of inorganic nanoparticle (silica, calcium phosphates, carbon, etc.) they can cause several adverse effects due to their small size, in the range of the cellular organelles and compartments. Their characteristic high surface area over volume ratio greatly increases the interaction with biological entities. One of the most common effects caused by exposure to nanoparticles is the generation of reactive oxygen species (ROS) which can originate adverse side effects. In general, cells can tolerate small and transient increases of ROS species. However, under high or prolonged NP exposure, cells can suffer membrane damage or DNA alterations (genotoxicity). The ROS species can be generated as a consequence of exposure to the acidic endosomal environment, interaction of NP with the mitochondria and also through activation of certain intracellular signaling pathways. The kinetics of the ROS formation depends on the total surface area and the coating stability and it should be studied for each type of nanoparticle. In some cases, the surface functionalization plays an important role, as in the case of iron oxide nanoparticles (IONPs) coated with a dextran shell. These nanoparticles diminish the intracellular ROS levels, inducing cellular proliferation instead of adverse effects [12]. Also, cellular internalization of rigid nanoparticles can provoke disruption of the cytoskeleton network, which can originate secondary side effects.

Mesoporous silica nanoparticles (MSNs) (SBA-15 and MCM-41) exhibit low toxicity at low concentrations. Cellular uptake is strongly size dependent. Particles of 50 nm are the most capable of being internalized by human cells [13]. The capture of the mesoporous particles by macrophages is favored by increase in size, while smaller particles exhibit longer circulation times. Cytotoxicity is related to the uptake grade, thus, micrometric particles with size around 1 μm show lower toxicity than nanoparticles with sizes up to 200 nm. A general principle is that cationic nanoparticles induce more immunogenic response and cytotoxicity than their neutral or anionic counterparts. The silanol groups of the naked MSNs can interact with the lipids of the cellular membranes and some proteins altering their properties. Without any coating, these particles present negative surface charge at physiological pH. Hence, the particles are rapidly covered by opsonins, captured by the macrophages and are accumulated in the liver and spleen a short period of time after entering the blood-stream. As mentioned above, the use of different coatings (PEG or other polymers) can increase the circulation time and reduce the toxicity of this material. The degradability of these particles has been studied using simulated body fluid (SBF). The process takes place in three steps, fast bulk silica degradation (hour scale) followed by the deposition of magnesium/calcium silicate on the particle surface and, finally, slow dissolution of the particles (day scale) [14]. Regarding the *in vivo* toxicity, Kohane *et al.* have demonstrated that these materials exhibit high toxicity when bigger doses are employed (up to $1.2 \, g \, kg^{-1}$ in one single injection), whereas they did not show toxicity when the dose was below $40 \, mg \, kg^{-1}$ [15]. Recently, the research group of Tamanoi established $50 \, mg \, kg^{-1}$ as the safe dose for drug delivery purposes [16].

The intravenous administration of carbon nanotubes and fullerenes can activate the immune system, resulting in inflammation and granuloma formation [17]. On other hand, the biocompatibility and cytotoxicity of another carbon allotrope, graphene oxide (GO), have been recently studied in detail, showing that this material can be safely employed in the treatment of ocular tumors [18].

IONPs (mainly magnetite, Fe_2O_3, and its oxidized and more stable form maghemite, Fe_3O_4) are widely used as contrast agents in magnetic resonance imaging (MRI) and also in magnetic hyperthermia for cancer therapy. Once internalized into the cells, these particles are degraded into iron ions by several hydrolyzing enzymes [19]. As in the other cases, the biodistribution of these particles depends on their charge, size, shape, external coatings, and

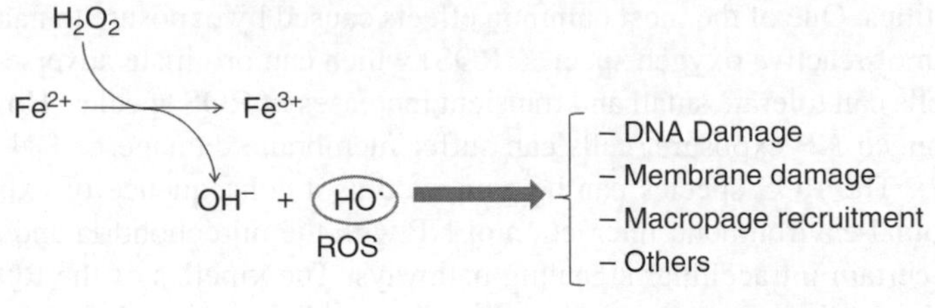

$$H_2O_2$$

$$Fe^{2+} \longrightarrow Fe^{3+}$$

$$OH^- + \boxed{HO^{\cdot}} \Longrightarrow$$
$$ROS$$

- DNA Damage
- Membrane damage
- Macropage recruitment
- Others

Scheme 14.1   *ROS generation by the Fenton reaction*

mode of administration. They are captured by the macrophages of the reticuloendothelial system (RES) when the nanoparticles are administered without any coating. IONPs are able to produce oxidative stress within the body by the Fenton reaction [20], which is one of the main sources of ROS in living systems (Scheme 14.1).

Moreover, the presence of IONPs can produce a homeostasis imbalance within the body which can affect different organs. It is known that the accumulation of redox-active metals in the brain induces multiple degenerative disorders, such as Parkinson's and Alzheimer's diseases. This fact is particularly important because magnetic nanoparticles are able to penetrate into the body through the skin and also to cross the blood-brain barrier (BBB) after exposure by inhalation [21].

It is important to remark that in all cases the toxicity of these materials is a dose-dependent phenomenon and the biocompatibility should be determined with each case and treatment.

14.3   Cancer-targeted Therapy

14.3.1   Endocytic Mechanism of Targeted Drug Delivery

In order to improve the efficacy of the antitumoral therapy, the nanocarrier must cross the cellular membrane and release its cargo in the cytosol. A nanoparticle can be internalized within a cell following a process called *endocytosis* [22]. Endocytosis can be divided into two major categories: phagocytosis, which is the uptake of large particles with sizes up to 20 µm, generally performed by specialized cells (macrophages and dendritic cells among others), and pinocytosis, which consist in the uptake of fluids and small substrates and is present in all types of cells. Endocytosis is a complex process which involves different stages:

1. The particle is engulfed in membrane invaginations that are pinched off to form membrane-bound vesicles which are known as endosomes, or phagosomes in the case of phagocytosis.
2. The endosomes circulate within the cells delivering the cargo toward different specialized compartments.
3. The particles are placed in different intracellular compartments for degradation and digestion or are expelled to the external milieu (transcytosis).

Among pinocytosis, different classifications can be performed according to the proteins that are involved in the process (Scheme 14.2).

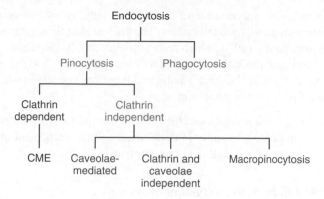

Scheme 14.2   *Classification of endocytosis*

- *Phagocytosis:* occurs mainly in specialized cells such as macrophages, dendritic cells, neutrophils and monocytes. The process starts with the opsonization of the material by immunoglobulins (IgG, IgM) and/or complement proteins present in the blood-stream. Then, the opsonized particle is recognized by a phagocytic cell through specific receptors. When certain receptors placed in the membrane of the phagocytic cell interact with the attached opsonins, a phagosome is formed by actin rearrangement leading to uptake of the particle. The formed phagosome follows a different maturation process which leads to the cargo transference to late endosomes or lysosomes where the digestion of the particle takes place.
- *Clathrin-mediated endocytosis (CME):* is the most common route of internalization of essential nutrients and also nanoparticles into practically all types of mammalian cells [23]. The endosome is produced by the polymerization of a cytosolic protein called clathrin-1. Then, the vesicle is pinched off for the cellular membrane by a GTPase called dynamin. Finally, the vesicles are sorted to late endosomes or lysosomes. The presence of certain ligands attached on the surface of the nanocarriers, such as mannose-6-phosphate, transferrin or nicotinic acid, among others, can direct the internalization of the particle to a CME process.
- *Caveolae-mediated endocytosis:* caveolae are abundant in muscle, endothelial cells and fibroblast, and it is produced by a membrane protein called caveolin-1. This protein directs the formation of the caveosome of diameter $60-80$ nm. One of the most important properties of these vesicles is that they present neutral pH and, in some cases, they evade fusion or conversion to lysosomes. Several pathogens, such as viruses or bacteria, use this mechanism in order to avoid lysosomal degradation. This internalization route has attracted the attention of researchers in nanomedicine because this property is very important for the delivery of sensitive molecules such as DNA, RNA, or proteins.
- *Clathrin- and caveolae-independent endocytosis:* these routes are mediated by multiple effectors (Arf6, flotillin, Cdc-42, and RhoA among others), and they require specific lipid composition in the membrane being dependent on cholesterol. There are only a few examples of nanomaterials which exploit this route, such as polymeric particles or liposomes conjugated with folate [24]. However, competition with caveolae- and clathrin-dependent mechanisms was observed using these particles, depending on the cell types [22].

- *Macropinocytosis:* is a special case of endocytosis initiated by transient activation of receptor tyrosine kinases by growth factors. The activation of these receptors leads to the formation of membrane ruffles which trap nutrients and fluids. Then, they fuse with the cellular membrane producing large intracellular vesicles (0.5–10 μm). This mechanism can serve as a non-specific entry point and it has been observed with nanoparticles able to use multiple internalization routes.

In order to develop efficient nanomedicines able to transport different therapeutic agents within cells it is compulsory to study in detail their uptake mechanism and also how to direct them toward specific intracellular compartments.

14.3.2   Specific Tumor Active Targeting

As has been mentioned above, nanoparticles administered intravenously tend to be accumulated in tumoral lesions due to the EPR effect. This effect constitutes a passive targeting of nanomedicines toward tumoral tissues and is one of the main reasons for their use in antitumoral therapy. However, solid tumors are composed of a heterogeneous mixture of tumoral cells and healthy cells and, therefore, it is very important to discriminate between them. The external surface of the nanoparticles can be decorated with molecules able to be specifically recognized by tumoral cells, providing the capacity to kill tumoral cells selectively in the presence of healthy ones. Therefore, these functionalized nanoparticles could attack specifically the tumoral cells present in the target tissue without affecting the neighboring cells, producing a strong decrease in the side effects usually associated with antitumoral chemotherapy. This strategy constitutes the active targeting and is another reason which nanoparticles are such promising materials for cancer treatment. Engineered nanoparticles can satisfy the requirements of the "magic bullet" principle postulated by Ehrlich in 1906 [25]. This concept is based on three principles: to find a proper target for a particular disease, to find the drug that effectively treats the disease and, finally, to find how to carry the drug to the desired place.

The basis of active targeting is the use of targeting agents, specifically selected for each pathology, placed on the periphery of the nanocarriers. In the case of cancer disease, the targeting ligand is usually chosen to bind to a receptor overexpressed by tumor cells or tumor vasculature and not expressed by healthy cells or vessels. This receptor should be homogeneously expressed by all diseased cells. One of the most useful families of targeting molecules is the antibodies. An antibody is a protein-based macromolecule, fabricated by living organisms, able to bind selectively to certain regions of foreign entities or pathogens. Once labeled by the antibody, the macrophages and other cells of the immune system are able to detect the foreign body and destroy it. Conjugation of antibodies on the surface of a nanoparticle could be performed randomly using chemical reagents such as carbodiimides, which create robust amide bonds between carboxylic acid groups placed on the nanoparticle surface and free amino groups of the antibody, provided by lysine or the *N*-terminal group. Using this strategy, the activity of the attached antibodies is usually lower because the antibody is bound to the surface in multiple ways, some of which can block the recognition region of the macromolecule. Alternatively, antibodies can be conjugated using maleimide-type cross-linkers, using thiol groups present in known regions of the antibody (provided by cysteine) or using engineered antibodies with thiol groups placed away from the recognition site. Other strategies for antibody grafting on inorganic nanoparticles are the use of biomolecules as cross-linkers, such as biotin-streptavidin bridges [26] or protein A [27]. The first involves the biotinylation of the antibody and the nanoparticle followed by the coupling to each other using streptavidin, as a bridge between them. Streptavidin is a 60 kDa protein isolated from the bacteria *Streptomyces avidinii* that presents a strong affinity to the biotin moieties, producing one of the strongest non-covalent intermolecular interactions in nature. This protein has four pockets able to bind a biotin molecule and it can act as a bridge between the antibody and the nanoparticle. The second strategy involves the use of protein A as linker. Protein A is a cell-wall associated protein, produced as a defensive mechanism by the bacteria *Staphylococcus aureus*. It binds to antibodies secreted by the host via their Fc region, avoiding the recognition of the pathogen. Thus, the use of this molecule for antibody grafting presents a very important advantage because the recognition zone of the antibody is not involved in the bond (Scheme 14.3).

Some antibodies conjugated with nanoparticles are currently used in clinical application, such as trastuzumab (anti-HER2) against breast cancer or panitumumab (anti-EGFR) against colorectal cancer, among others. It is also possible to employ small antibody fragments which contain only the recognition site [28]. The antibody fragment retains the capacity to recognize the antigen while it lacks the constant Fc effector region which is responsible for binding to the immune cells and can induce complementary activation, resulting in the premature clearance of the conjugated nanocarrier.

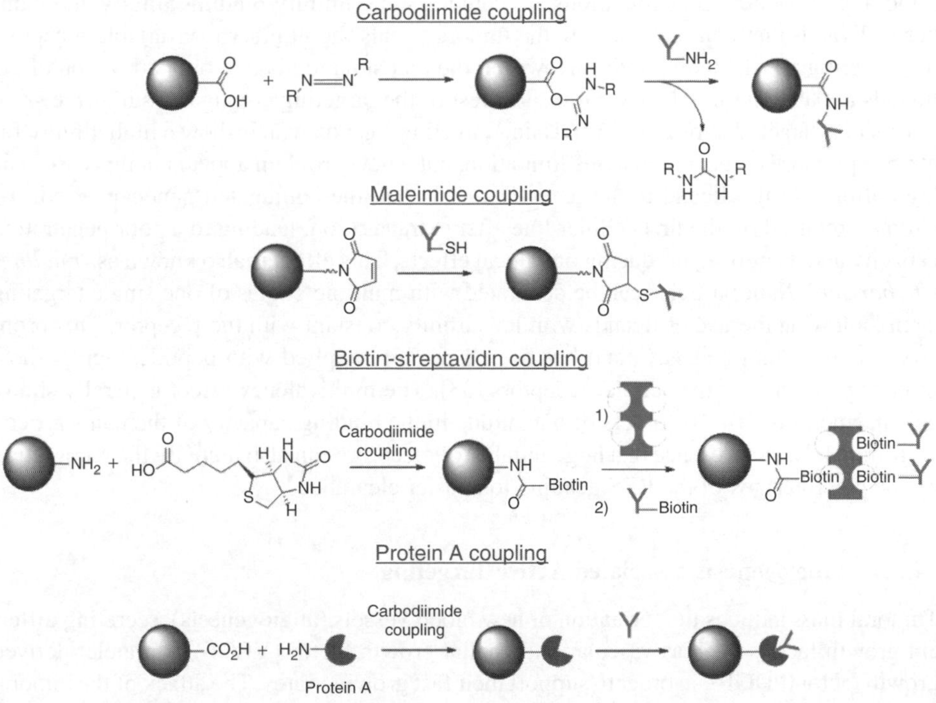

Scheme 14.3 *Antibody-grafting strategies*

Tumoral cells present a fast growing rate that requires major amounts of nutrients than normal cells, and for this reason they usually overexpress receptors to capture some essential molecules. Thus, the attachment of these molecules on the surface of the nanocarriers could improve the internalization grade and, therefore, the efficacy of the therapy. For instance, transferrin receptors are overexpressed 100-fold higher than the average expression in healthy cells due to the higher requirements of iron by the tumoral cells. Folic acid is an essential vitamin, required for the biosynthesis of DNA among other important roles. Thus, the folate receptor is usually upregulated in many tumoral cells (up to 40% of human cancers) in some cases by more than two orders of magnitude [29]. Due to the essential role of this molecule, the cell internalizes the folic acid into a vesicle and releases it into the cytosol. Folic conjugation is widely used in nanomedicines because this molecule is inexpensive, non-toxic, stable, non-immunogenic, and easy to conjugate with different substrates while retaining its binding capacity [30]. The cellular membrane is usually decorated with different glycoproteins which are a distinctive label of each type of cell. Tumoral cells present their own glycoproteins on their surface and this fact can be exploited in order to guide specifically the nanocarriers. Lectins are non-immunogenic proteins able to recognize the glycoproteins or glucides present in the cells and, therefore, they can be used as targeting agents. It is also possible to employ the opposite strategy that consists in the attachment of certain glucides on the nanoparticle surface capable of being recognized by the glycoproteins present in the tumoral cells [31].

This active targeting is limited by the receptor capacity. The numbers of receptors located on the cell surface limits the number of targeting molecules which can be bound to the tumoral cell. Under ideal conditions we could assume infinity binding affinity, the number of ligands that can be bound by the tumors equals the number of available receptors (considering a 1 : 1 binding ratio). However, the real situation is that only a fraction of the ligands are able to bind the receptors. The rest of the targeting conjugates suffer the same fate as non-targeted substrates [32]. Using targeting ligands which show a high affinity for the receptors we can overcome this limitation, but a new problem appears in this case. The high affinity of the ligand to the receptor means that the conjugated nanocarrier will be strongly retained by the first cellular line after extravasation, leading to a poor penetration capacity and, therefore, producing only local effects. This effect is also known as "*binding-site barrier*". Nanoparticles can be decorated with multiple copies of one single targeting agent, allowing the use of ligands with low affinity constant with the receptor. This property is called "multivalency" and has been especially applied with peptides which show moderate affinity for the cellular receptors [33]. The multivalency effect generally shows enhancements of $10-10^4$ orders of magnitude in the binding capacity of the nanocarriers. Unfortunately, the presence of a large number of receptors can also increase the recognition by the macrophages of the RES, leading to a faster clearance.

14.3.3   Angiogenesis-associated Active Targeting

Tumoral mass induces the formation of new blood vessels (angiogenesis), secreting different growth factors such as vascular endothelial growth factor (VEGF) or platelet-derived growth factor (PDGF), in order to support their fast growing rates. The attack of the tumoral vasculature can destroy the tumoral mass due to the lack of an efficient blood supply which provides nutrients and oxygen to the tumor. This strategy presents some interesting advantages:

1. Extravasation of the nanocarriers from the blood vessels is not required.
2. Tumoral blood vessels usually overexpress certain receptors which are easily accessible by the functionalized nanoparticles.
3. Endothelial cells, of which the tumoral vessels are composed, are less susceptible to suffering mutations due to their more stable environment, compared with the tumoral cells housed inside the solid tumoral mass which are exposed to hard conditions (lower pH values, low O_2 pressure, etc.) This fact reduces the risk of multidrug resistance under prolonged treatments.
4. The endothelial cell markers are common in different tumors.

The principal angiogenic markers which have been widely explored in nanomedicines are:

- *VEGF:* when the tumoral cells are exposed to hypoxic conditions they increase the production of VEGF. This results in an upregulation of the vascular endothelial growth factor receptors (VEGFR-1 and VEGFR-2) in the endothelial cells. Thus, it is possible to exploit the presence of these receptors using VEGF as targeting ligand inducing the endocytotic pathway. Also, the opposite strategy can be employed, using anti-VEGF attached on the nanocarrier as targeting ligands in order to inhibit ligand binding to VEGFR-2.
- $\alpha_v\beta_3$ *integrins:* are endothelial cell receptors for ECM proteins which are highly overexpressed in neovascular endothelial tumoral cells but are scarcely present in healthy cells. Oligopeptides harboring the RGD sequence (Arg-Gly-Asp) bind selectively to this receptor. Recently, Ruoslhati and coworkers have discovered that the use of a cyclic peptide with a hidden RGD motif (CRGDK/RGPD/EC) attached to the nanoparticle surface, allows not only targeting of tumoral blood vessels but also improves the penetration of the nanoparticle within the tumoral mass [34]. This peptide sequence binds to the $\alpha_v\beta_3$-integrins of the tumor endothelium and then, after a proteolytic cleavage, the RGD motif is exposed and interacts with neuropilin-1 receptors which promote the internalization of the nanoparticle.
- *Vascular cell adhesion molecule-1 (VCAM-1):* is an immunoglobulin transmembrane protein that is expressed only on the surface of tumoral blood vessels and inflammation. This protein is responsible for inducing cell-to-cell adhesion, an important requisite of the angiogenesis process. The attachment of antibodies specifically designed to bind to this molecule could improve the antitumoral activity of the nanocarrier, theoretically without side effects.
- *Matrix metalloproteinases (MMPs):* they are proteins that degrade the ECM, mainly hydrolyzing the collagen network. They play a determinant role in angiogenesis and metastasis. Some antibodies capable of binding selectively to these proteins have been conjugated to different nanocarriers, especially antibodies that recognize the membrane type-1 metallo-proteinase (MT-1 MMP) which is present on endothelial tumoral cells of a large number of malignancies.

14.4   Ceramic Nanoparticles for Cancer Treatment

14.4.1   Mesoporous Silica Nanoparticles

MSNs are one of the most promising materials for antitumoral purposes among the inorganic nanocarriers, and their application has reached *in vivo* trials, using mainly murine models [35]. Recently, the first silica-based nanoparticle has been FDA approved for first-in-human clinical trial in oncologic diagnosis, which constitutes a great step forward for the clinical acceptance of this material [36]. As has been mentioned above, MSNs present very interesting properties for biomedical uses, such as high specific surface area, robustness, easily tunable size, shape, and pore diameter, among others. These properties provide unique advantages for the encapsulation of different drugs, from small molecules to therapeutic macromolecules [37]. The production of mesoporous silica particles is simple, controllable, cost-effective, and easily scalable. Moreover, this material presents a high proportion of silanol groups on its surface which provide many options for the development of multifunctional materials through covalent grafting using the well-established silane chemistry. The most research efforts in the field of antitumoral drug delivery applications of these materials have been performed using MCM-41 and SBA-15 types, the former being the most studied because this material can be easily obtained in the form of nanoparticles smaller than 200 nm, which is a perfect size for drug delivery purposes [38]. Mesoporous silica particles with more exotic morphologies, such as hollow or rattle-type spheres have also received great attention [39].

The application of the MSNs as drug carriers in antitumoral therapy requires their effective suspension in biological solutions where high salt and protein contents are present. Bare mesoporous silica particles show a tendency to suffer aggregation due to formation of interparticle hydrogen bonds caused by the silanol groups. The particle aggregation is even more intense when they are exposed to biological fluids. In order to avoid this undesirable behavior, different strategies have been described which involve the attachment of hydrophilic moieties on the particle surface, such as phosphonates or polyethyleneglycol-derived phospholipids (PEG-lipid) [40]. The last method provides particles with significantly lower non-specific protein absorption due to the PEG coating. Small mesoporous nanoparticles (< 50 nm) with their surface decorated with polyethyleneimine polyethylene glycol copolymer (PEI-PEG) have been evaluated *in vivo* using a murine xenograft model of human squamous carcinoma, showing high passive accumulation (12%) within the tumoral mass and an improved therapeutic efficacy, higher than the free drug [41]. PEGylated silica nanorattles loaded with docetaxel, which is a potent hydrophobic antitumoral drug, have demonstrated an ability to inhibit the tumoral growth 15% higher than Taxotere, the clinical formulation of docetaxel, and also show less systemic toxicity [42]. MSNs can also be employed for the capture of important biomolecules within the tumoral cells provoking their cellular growth inhibition. Thus, Lin *et al.* have decorated the external surface of MSN with phenanthridinium molecules, which is a group able to bind to cytoplasmic oligonucleotides, as messenger RNAs, interfering with the normal development of the cell [43].

The shape of the particles is an important aspect regarding the cellular internalization and presents cell-type dependence. Lin *et al.* have reported that spherical MSN exhibited a faster endocytic rate than rod-shape particles when Chinese hamster ovarian (CHO) cells were employed, whereas the internalization rate for both was similar using human fibroblast cells [44]. Thus, the shape of the carrier has to be designed depending on each type of target cell.

Targeting molecules able to be selectively internalized by tumoral cells can be attached on the mesoporous silica surface in order to improve the efficacy of the therapy. Thus, the surface of MSNs has been conjugated with different ligands, such as sugars [45], folic acid [46, 47], transferrin or cyclic RGD-peptides [48], as well as antibodies [49] and DNA aptamers [50]. Instead of loading cytotoxic species, MSNs can transport photosensitizers, which are molecules commonly employed in photodynamic therapy (PDT). PDT is based on the use of a molecule (photosensitizer) which, upon light irradiation at specific wavelengths in the presence of oxygen, leads to the production of cytotoxic agents which provoke cell death [51]. One limitation of this strategy is that conventional photosensitizers require the use of high excitation powers which can provoke tissue damage, but this limitation can be overcome by combining PDT with two-photon excitation (TPE). In TPE, the photosensitizer is excited by simultaneous absorption of two photons with half the energy of one-photon excitation. The first one excites the sensitizer to a virtual intermediate state and then the second photon leads it to the singlet excited state, identical to the state achieved by PDT. This process occurs only at the focus of the light source (usually a laser beam) where the irradiation is more intense, due to the lower probability of the two-excitation event. Thus, TPE-PDT presents higher spatial resolution and causes less side damage. MSNs are capable of transporting high amounts of photosensitizers in one single particle and, therefore, can enhance the response to the laser irradiation. Gary-Bobo *et al.* employed mannose-functionalized MSNs with photosensitive molecules covalently attached to the particle in order to kill different human tumoral cell lines [52]. Moreover, the encapsulation of the photosensitizers within the silica matrix prevents the unwanted excitation of these molecules by UV/Vis light, which is one of the problems of PDT due to the presence of sunlight sensitivity in the patients treated with PDT agents. Finally, it can be possible to achieve a synergic effect combining this therapy with the release of cytotoxic compounds and thus improving the efficacy of the therapy [53].

Another interesting alternative is based on the conjugation of drug-loaded mesoporous silica particles with mesenchymal stem cells, using these cells as a Trojan horse due to their capacity to localize and penetrate into the tumors. Using fluorescently-labeled cells, it has been reported that human circulated bone marrow-derived mesenchymal stem cells (hMSCs) exhibit a tropism for different tumors and are even able to cross the BBB to reach brain tumors [54]. Thus, one recent research article shows that drug-loaded silica nanorattles can be attached on the hMSC surface, up to 1500 nanoparticles per cell, and they destroy effectively glioma tumoral cells in a xenograft murine model [55]. The use of human cells, able to navigate first through the blood stream and then into the target tissue, could provide a solution of one of the main problems of targeted nanotherapy, which is the poor penetration of the nanodevices within the tumoral mass due to the "*binding-site barrier*", as has been mentioned above. However, more research is necessary in order to test the suitability of this approach in further clinical developments.

14.4.1.1   *MSNs as Stimuli-Responsive Drug Delivery Systems*

MSNs (especially MCM-41 and SBA-15 types) present a very interesting advantage in drug delivery applications due to their ordered pore structure. The pores in these materials are independent parallel channels without interconnections between them and, therefore, each pore presents only two entrances. This peculiar structure allows the development of zero-release materials by placing at the pore openings different moieties, called gatekeepers or molecular gates, able to regulate the opening–closing process in response to the presence or not of certain stimuli. Thus, drug-loaded MSNs with molecular gates on the pore outlets are able to release the drug only in the presence of the specific stimuli, achieving a better control of the dosage of the therapeutic agents (Figure 14.3). In this section, the different stimuli which can be employed in order to trigger the drug release will be briefly described.

- *pH*, some tissues of the body shows alterations in their pH level when they are suffering a pathological state. Thus, tumors and inflamed tissues usually present lower pH values (up to pH 5.5–6) than blood or healthy tissues (at pH ≈ 7) [56]. This pH gradient can be employed as a trigger event with MSNs functionalized with pH sensitive gates attached to the pores. Moreover, the pH value of the late endosomes or lysosomes becomes even lower and, therefore, the internalized nanocarrier functionalized with these pH-gates can release the cargo once there, avoiding the premature release of the drug outside the cell. Different nanocaps, such as β-cyclodextrins (β-CDs) [57] or gold nanoparticles [58], can be attached to the pore outlets through acid cleavable linkers. Another strategy is based on coating the mesoporous surface with a polymeric shell which suffers a physicochemical transformation in acid media [59]. Inorganic pH-sensitive coatings, such as calcium phosphates, can also be applied on the MSN surface [60]. This shell is dissolved into non-toxic ions in acidic media (pH 5) allowing the drug release.
- *Temperature*, the artificial production of hyperthermia or hypothermia by physical means is a well-established procedure in clinical practice. Also, some tissues exhibit higher temperature values when they are affected by different diseases. One of the most common approaches to exploit this fact, in order to produce stimuli-responsive carriers, is the attachment of thermosensitive polymers, generally poly-N-isopropylacrylamide (PNIPAM) and derivatives, on the external surface of the MSNs [61, 62]. This polymer is in the hydrated form below the lower critical solution temperature (LCST) of 32 °C, which prevents the release of the drugs trapped inside, whereas it suffers a collapse if the temperature exceeds this value, producing the pore opening.
- *Magnetic field*, MSNs do not present magnetic properties and, therefore, it is necessary to place superparamagnetic moieties into their structure in order to make the material sensitive to magnetic fields. With exposure to alternative magnetic fields (AMFs), the

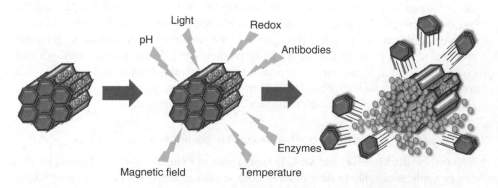

Figure 14.3 *Stimuli-responsive gatekeeper concept*

presence of these superparamagnetic nanocrystals provokes a temperature increase in the medium due to the rapid rotation of the magnetic nuclei (Brownian fluctuations) and the fluctuation of the magnetic moment (Néel fluctuations) [63]. The production of magnetic MSNs can be developed by trapping iron oxide nanocrystals into the silica matrix using aerosol-assisted [64] or modified Stöber methods [65]. These materials, under AMFs exposure, are able to generate enough heat to lead to the destruction of tumoral cells (magnetic hyperthermia) [66]. The external surface of these magnetic particles can be decorated with thermosensitive gates which open or close the pores in response to the temperature changes. A very interesting thermosensitive molecule is the polymer of deoxyribonucleic acid or DNA. A single strand of DNA is able to bind selectively with its complementary strand through hydrogen bonds between the nitrogenated bases; adenine (A) is complementary with thymine (T) and guanidine (G) is complementary with cytosine (C). The two strands will be separated if the temperature reaches a certain value, which is dependent on the G/C content and the length of the strand, in a process called dehybridization [67]. Thus, in one recent research work [68], the surface of MMSNs was decorated with a specific 15 base single pair oligonucleotide sequence, selected to display a melting temperature of 47 °C. After loading the particles with a drug model, the pores of these particles were capped with iron oxide nanocrystals (size around 5 nm) previously functionalized with complementary DNA strands. Under AMFs, the presence of the iron oxide particles produced a temperature increase up to the melting temperature of the DNA strands, with consequent drug release. Thus, the system showed a zero-release behavior at room temperature but released the drug when the temperature reached the melting temperature (Figure 14.4).

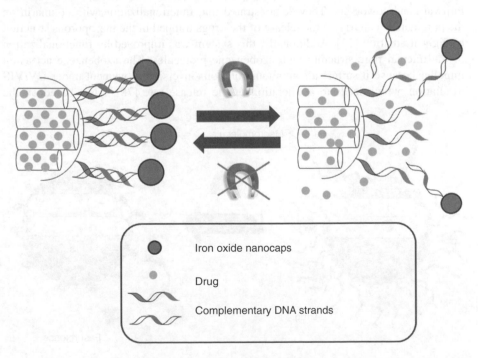

Iron oxide nanocaps

Drug

Complementary DNA strands

Figure 14.4   *Magnetic-responsive MSNs based on DNA gates*

Thermosensitive polymers have also been attached on the surface of magnetic MSNs in order to control the release. A thermoresponsive copolymer of poly(ethyleneimine)-*b*-poly(*N*-isopropylacrylamide) (PEI/NIPAM) has been used as a temperature-sensitive coating of MSNs which carried superparamagnetic iron oxide nanocrystals into their structure [69]. The reason for using this polymer coat is because it can combine two interesting properties: temperature-responsive gatekeeper behavior of the poly-*N*-isopropylacrylamide (NIPAM) and the protein or DNA retention capacity of the polyethyleneimine (PEI). This material has demonstrated its ability to transport and release two different molecules (small molecules retained within the porous channels and macromolecules trapped within the polymeric shell) in response to the application of an external magnetic field (Figure 14.5).

- *Redox*, there is a significant difference in the redox potential between the mildly oxidizing extracellular space and the reducing intracellular space which can be exploited in order to produce redox-responsive MSNs. The concentration of glutathione (GSH) is 1000 times higher in the cytosol than outside the cell. Thus, different moieties have been attached to the pore outlets through a redox-cleavable bond (disulfide) such as inorganic nanocaps [70, 71], biomolecules such as collagen [72], and also polymers which contain these breakable bonds [73]. However, when the carriers are taken up by the cells they are confined in endosomes vesicles where the presence of reductive species is lower than necessary, or even where mild oxidative agents are present [74]. Therefore, to design a system able to accelerate the endosomal escape, for instance disrupting the endosome membrane by irradiation of a photosensitizer attached to the carrier [75], is of paramount importance in this type of nanodevice [76].

- *Light*, pioneers in the use of light as a trigger stimulus in drug release with MSNs are Fujiwara and coworkers. They demonstrated that functionalization with coumarin on the pore outlets can trigger the release of the drugs trapped in the mesoporous structure by light irradiation [77]. Additionally, this system was improved by functionalization of the internal pore structure with azobenzene molecules. The azobenzene act as an impeller because it suffers a continuous rotation–inversion movement under UV/VIS irradiation, which leads to acceleration of the release rate [78]. Also, based on the

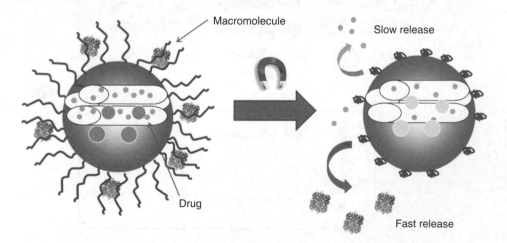

Figure 14.5 *Thermosensitive polymers as gatekeepers of magnetic MSNs*

photo-isomerization of azobenzene, Zink *et al.* have reported a mesoporous silica nanocarrier able to release its cargo in response to UV irradiation [79]. In this system, β-CDs are used as pore caps due to their binding affinity for the *trans*-azobenzene, which decorates the pore outlets. UV light irradiation at 351 nm causes azobenzene isomerization from the trans- to cis-conformation and consequently the β-CD caps are removed from the surface with subsequent drug release. As in the previous cases, it is also possible to use nanocaps attached through sensitive linkers. Thus, gold nanoparticles can be anchored to the mesoporous surface using thioundecyl-tetraethyleneglycoesteronitrobenzylethyldimethylammonium bromide (TUNA) which is converted to the negative form under UV irradiation. This fact leads to the gold nanocaps being repelled, causing the drug release [80]. All of these systems have been designed to respond in the presence of UV/VIS light which exhibits a poor penetration capacity in living tissues. The application of near-infrared (NIR) light could improve the efficacy of these devices because the transmission of this radiation through blood and soft tissue is optimal thanks to its low energy abosorption and, therefore, it allows deep penetration in body tissues. Recently, a mesoporous silica carrier has been designed to exploit this energy to trigger the release of an antitumoral drug [81]. In this system, a gold nanorod is trapped within the silica particle in order to provide the NIR-responsive behavior, because these gold rods are able to capture the NIR radiation transforming it into thermal energy. The pores are capped with a double DNA strand which suffers thermal dehybridization when the material is irradiated with NIR light. The DNA dehybridization produces the pore opening and drug release.

- *Enzymes*, some diseases are characterized by the overexpression of certain enzymes in levels higher than the normal values and therefore, this characteristic can be exploited to trigger the drug release [82]. Thus, as in the previous cases, different nanocaps can be covalently attached to the pore entrances using cleavable linkers sensitive to the presence of one particular enzyme, such as esterases [83] or amylases [84]. Dysregulation in the amount of proteases has been described in different pathologies and can be exploited as an internal triggering stimulus. Avidin molecules were attached to the MSNs surface, acting as caps using their well-known affinity for the biotin moieties which were previously grafted on the pore outlets [85]. These caps were eliminated by the action of a specific protease (trypsin) allowing the release of the loaded drug. Bhatia *et al.* [86] have covered the MSN surface with an enzyme-sensitive polymeric shell which prevents the premature drug release until certain proteases are present. In this case, the enzymes responsible for the polymer degradation are MMPs know to be highly overexpressed in tumoral environments.

- *Small molecules*, the presence of specific molecules in the target tissue can trigger the release in MSNs provided with their corresponding sensitive gatekeepers. Mártinez-Mañez *et al.* [87] have described MSNs capped with an antibody which recognizes selectively the presence of a certain molecule. When this molecule is present in the medium, the antibody is detached from the mesoporous surface leading to the drug release. Recently, MSNs have been capped with gold particles modified with aptamers able to bind to adenosine-triphosphate molecules (ATPs) [88]. Aptamers are single-stranded oligonucleotide polymers able to recognize certain molecules selectively. They are more stable to degradation and easier to prepare than antibodies. Thus, when these nanocarriers are exposed to the presence of ATP, the gold caps are removed by a displacement reaction.

14.4.2 Calcium Phosphates Nanoparticles

As has been mentioned extensively in previous sections, calcium phosphates (CaP) are the most important inorganic components of hard tissues and present excellent biocompatibility. This material is dissolved in acidic environments (pH $\approx 4-5$) which can be naturally present in lysosomes after internalization or in tumoral tissues. Therefore, cytotoxic drugs can be encapsulated during the formation of calcium phosphate nanoparticles (CaPNPs) and released once the nanocarrier is internalized by the target cell or when it reaches the tumoral environment. Hollow CaP nanoparticles have demonstrated their potential application as stimuli-responsive carriers able to release their cargo in response to ultrasounds [89]. Cis-platin has been trapped into CaPNPs showing a sustained release of the cytotoxic agent and an improved effectiveness against resistant A2780cis human ovarian cancer cell line [90]. Ceramide is a lipid-derived drug which induces apoptotic death of melanoma tumoral cells. Unfortunately, the potential application of this molecule is hampered by its low solubility in water. Recently, ceramide has been encapsulated into CaPNPs of 20 nm diameter stabilized by a PEG coat [91]. This material is easy to disperse in buffered solutions at physiological pH, remaining stable in solution for long times, which is an important issue for drug delivery purposes. The biological evaluation of this material was performed using different human tumoral cell lines, showing its efficacy to deliver hydrophobic drugs to these malignant cells. The same material has also demonstrated its ability to transport cytotoxic drugs and organic dyes in order to combine drug delivery and imaging in the same carrier [92]. Photosensitizers can also be transported by CaPNPs for PDT. Thus, Methylene Blue and 5,10,15,20-tetrakis(3-hydroxyphenyl)porphyrin (mTHPP) have been trapped into polymeric layers attached to the surface of CaPNPs [93]. These particles have demonstrated their capacity to destroy tumoral cells with high efficacy after light irradiation, showing very low toxicity under dark conditions [94]. Another strategy to destroy tumors is the encapsulation of superparamagnetic iron oxide nanocrystals into calcium phosphate particles. Thus, magnetic hydroxyapatite nanoparticles have been synthesized by the co-precipitation process using different concentrations of Fe^{2+} present in the medium [95]. The as-formed particles present round shapes of size $20-50$ nm and good magnetic properties, around $3-20$ emu g^{-1}. The ability to produce local hyperthermia of these particles under exposure to AMFs, with temperature increase from 38 to $40\,°C$, in mice models was confirmed, showing a significant tumor volume reduction. At the same time, magnetic nuclei have also been trapped in dicalcium phosphate particles, showing excellent biocompatibility and promising capacity to destroy tumoral cells under magnetic stimulus without affecting healthy cells [96].

14.4.3 Carbon Allotropes

The development of novel graphene-based drug delivery nanocarriers has received great attention in recent years, due to their ultra-high surface area which is available for drug loading [97]. The application of these types of materials in drug delivery applications requires their surface functionalization with organic groups, polymers, or biomolecules, which allow their suitable dispersion in aqueous solutions. Thus, PEG has been attached to the graphene surface in order to avoid agglomeration in serum media [98]. The delocalized electron of the graphene sheets allows retention of different cytotoxic drugs by π-stacking, such as in this case, doxorubicin. Surfactants as Pluronic F127 have been employed for the fabrication of water-stable graphene carriers which present a high loading capacity of doxorubicin (up to 280% w/w) [99]. Moreover, this carrier is able to release its cargo in response to pH changes, achieving higher release in mild-acidic conditions (pH 5). Camptothecin is a quinolone alkaloid which shows a very potent cytotoxic activity against several human tumoral cell lines. However, due to its low solubility in water and high toxicity, the clinical application of this agent is really compromised. Nanoparticles of graphene oxide (NGO) covered with chitosan have demonstrated their ability to retain this drug by intermolecular forces guided by π-stacking attractions, acting as a drug-delivery carrier to different tumoral cell lines [100]. One additional advantage of this type of material is its intrinsic photoluminescence in the visible and infrared regions, which can be used in order to localize the particle along its path into the body.

As mentioned above, functional groups present in GO allow the attachment of targeting molecules able to guide the graphene-carriers to the tumoral cells. As in the previous materials, folic acid has been attached on the NGO surface in order to provide selectivity [101]. NGO functionalized with folic acid on the surface can also deliver multiple drugs at the same time, such as doxorubicin and camptothecin, increasing the efficacy of the nanodevice by the synergistic effect [102]. As in the case of mesoporous silica particles, a very interesting approach is the development of NGO devices able to release their cargo in response to external or internal stimuli. A redox-responsive NGO carrier has been synthesized attaching PEG on the graphene surface using disulfide crosslinkers [103]. The presence of a PEG-layer hampers the release of the loaded drugs and, therefore, the release rate is higher when the polymeric layers are detached from the surface, which happens if the nanoparticle reaches the intracellular space. Finally, pH has also been exploited in order to create NGO-responsive systems. Thus, NGO combined with superparamagnetic iron oxide nanoparticles (SPIONs) has been synthesized by a chemical precipitation method [104]. Targeting molecules, in this case folic acid, were attached to the iron oxide surface and doxorubicin was loaded on the graphene, showing a very high loading capacity, as high as $0.387 \, mg \, mg^{-1}$. This material exhibits strong pH dependence, presenting a higher release in a mild acidic environment (pH 5).

Single-walled carbon nanotubes (SWNTs) have also been employed in drug delivery for cancer treatment using similar strategies as with GO, due to their very similar chemical nature [105]. As in the previous case, they require external functionalization with hydrophilic moieties in order to increase their dispersability in water solutions. PEG-functionalized SWNT has demonstrated its ability to transport and release doxorubicin in a pH-dependent manner [106]. Although SWNT present the potential to encapsulate doxorubicin or other small drugs, recent studies indicate that the drugs are usually attached on the surface by π-stacking forces, because the average diameter of the SWNT increases after loading [107]. The release rate of doxorubicin is related to the diameter and surface charge of the SWNT, being slow when the SWNT has a smallerer diameter and surface charges. Drug release is also accelerated with higher temperatures due to the increase in molecular motion and, therefore, heat can also be employed as a trigger stimulus. Other cytotoxic drugs have been loaded on the SWNT and studied using *in vivo* models. Paclitaxel was chemically conjugated via a cleavable ester bond to branched PEG chains placed on the SWNT surface [108]. This material has demonstrated high efficiency in suppressing tumor growth in a murine 4T1 breast cancer model.

Both SWNT and NGO can be employed in photothermal therapy (PTT) against cancer because they are able to generate heat under light irradiation. The *in vivo* application of PEGylated NGO labeled with a fluorescent group was studied using different xenograft tumor mouse models [109]. This material showed a high passive tumor accumulation and relatively low retention by the RESs. Under laser irradiation in the NIR region, the particles were able to generate enough heat to destroy the tumoral cells, which indicates the suitability of these materials in this therapy. On the other hand, SWNT functionalized with PEGylated phospholipids has the capacity to destroy tumors under irradiation of NIR using low doses up to $3.6\,mg\,kg^{-1}$ without toxic side effects [110].

14.4.4   Iron Oxide Nanoparticles and Hyperthermia

In this section we will describe briefly the use of IONPs for cancer treatment, mainly focused on the use of these systems as magnetic hyperthermia inducers. The use of IONPs for this purpose was first described by Gilchrist in 1957 [111]. IONPs, mainly Fe_3O_4 or γ-Fe_2O_3 nanoparticles, with sizes around 5–20 nm exhibit unique magnetic properties since each nanoparticle can be considered as a single magnetic domain [112]. When these nanomaterials are exposed to AMFs, they are able to produce thermal energy by both the fast rotation of the particle itself (Brownian fluctuation) and the fluctuation of the magnetic moment within the particle (Néel fluctuations) [63]. There are two different approaches for the use of IONPs in local magnetic hyperthermia depending of the reached temperature range:

1. *Magnetic hyperthermia*, which involves the generation of temperatures up to 45–47 °C. This strategy is generally applied in combination with conventional chemotherapy or radiotherapy because the presence of higher temperatures in the media improves the efficacy of the other therapies by a synergistic effect [113].
2. *Magnetic thermoablation*, if the reached temperature is between 43 and 55 °C. This is a more drastic strategy because it is based on the destruction of tumoral cells by the higher temperatures.

The administration of aminosilane-coated IONPs for hyperthermia treatment in a rat model of intracerebral glioblastoma multiforme has achieved promising results, showing 4.5-fold prolongation of survival time over controls after a single injection of nanoparticles and two cycles of magnetic exposure [114]. Recently, these nanoparticles have been applied in human clinical trials with 59 patients showing significant improvements in the overall survival [115]. In order to enhance the biodistribution, biocompatibility, and toxicity of IONPs, different coating strategies have been employed, using polymers [116], silica [117, 118], and gold [119], among others. Moreover, IONPs hosted by albumin nanospheres have shown their capacity to reach the brain because they are internalized by the erythrocytes and transported by them across the brain blood barrier [120].

The coating shell can also be employed as a carrier of different cytotoxic compounds attached by adsorption or covalently bound. Thus, doxorubicin can be transported by PEGylated [121] or polyamidoamine-coated IONPs [122]. In this last case, the polymer layer is able to retain doxorubicin in basic media (pH 11) while it releases the drug when the pH drops to more acidic values (pH 7 to 5), showing pH-sensitive drug release behavior. β-CD can be conjugated with IONPs, improving the stability of the nanoparticles at the same time that is used as drug carrier due to the inclusion properties of the CD [123]. This material releases the trapped drug under the application of an external AMF due to the temperature increase. IONPs can also be encapsulated within polymeric micelles in order to combine the tunable biodegradability and controlled release properties of the micelles, by polymer composition changes, with the capacity to produce hyperthermia. To this end, IONPs have been encapsulated inside thermosensitive and biodegradable micelles composed of poly(ethylene glycol)-b-poly[*N*-(2-hydroxypropyl) methacrylamide dilactate] (mPEGb p(HPMAm-Lac2)) [124]. Inorganic coatings can also be employed in order to improve its properties [125].

The external surface of the particles or also the organic and inorganic coatings can be decorated with targeting agents, as in the previous materials, in order to provide the selectivity to attack tumoral cells without affecting the healthy ones. Thus, folic acid has been conjugated to the surface of poly (ethylene oxide)-coated IONPs loaded with doxorubicin to target liver cancer in rat models, showing a significantly higher efficiency than the free drug [126]. Antibodies against human prostate cancer cells (scAbPSCA) [127] or human breast cancer cells (HER2) [128] can also be conjugated to this type of drug-loaded particles with promising results. The cyclic peptide sequence (cRGD) which encodes the RGD sequence has been attached to IONPs, showing high preference to attack glioblastoma cells due to these cells expressing approximately 10^5 $\alpha v \beta 3$ receptors per cell [129]. The selectivity was confirmed using receptor-deficient breast cancer cells (MCF-7) as control because they present lower expression of these receptors. The use of IONPs coated with glycol chitosan/heparin has demonstrated the capacity to target tumoral tissues [130]. In contrast with the healthy tissues, tumors exhibit a mesh of clotted plasma proteins (fibrinogen-derived product) in the tumor stroma and the vessel walls. Heparin is able to interact with these fibrinogen-derived compounds and can direct the nanoparticles to the tumoral tissue. The micelles with IONPs trapped into their structure can also be functionalized with targeting agents. In one recent example, the specific cancer cell monoclonal antibody (mAb2C5), able to recognize breast cancer cells, has been conjugated on the surface of iron oxide/polyethylene glycol phosphatidylethanolamine (PEG-PE) micelles [131].

An additional advantage of the use of IONPs for cancer treatment is that these particles can combine their ability to transport and release different therapeutic compounds with their capacity to be used as contrast agents in MRI and, therefore, they can be used as theranostic agents, as will be briefly discussed in the following section.

14.5 Imaging and Theranostic Applications

As has been mentioned before, the external surface of the nanoparticles can be decorated with different molecules or polymers in order to provide interesting properties, such as targeting capacity, improved solubility in water solutions, and stimuli-responsive properties. Additionally, fluorescent dyes, superparamagnetic crystals, or radioactive compounds can be incorporated, making them multifunctional nanodevices which combine drug delivery capacity with imaging properties, receiving the name of theranostics (therapeutics and diagnostics devices) [132]. If both drugs and imaging agents are incorporated within the same carrier, it is possible to consider that they present the same biodistribution and targeted accumulation, and the efficacy of the therapy can be evaluated in real time.

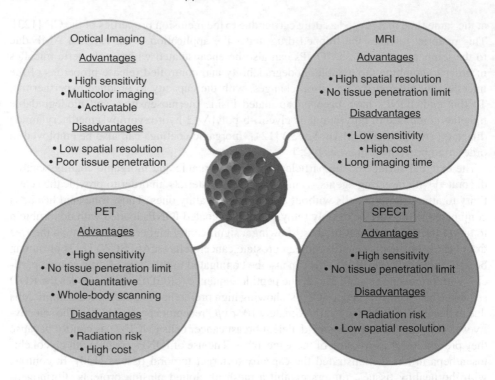

Figure 14.6 *Imaging techniques with nanoparticles*

Nowadays, there are some techniques for imaging applied in oncology in which nanoparticles can play an important role, such as optical imaging, MRI, positron emission tomography (PET), and single photon emission computed tomography (SPECT), each of them with their own advantages and disadvantages (Figure 14.6). These techniques employ different agents (dyes, magnetic particles or radioactive compounds) that are preferentially accumulated into the target organ. One single nanoparticle can transport large amounts of these agents, thus obtaining enhanced signals. Also, the targeting capacity of the nanoparticles usually leads to better selectivity toward the target tissue. Moreover, different imaging agents can be incorporated on the same nanocarrier in order to overcome the limitations of each technique, obtaining an improved diagnosis agent. For instance, it can be possible to combine PET with fluorescence or PET with MRI in the same particle. These multimodal imaging agents can provide more reliable and accurate detection of disease sites. Finally, if the nanocarrier is also loaded with therapeutic compounds, it can be possible to treat the disease while getting information on the pathology at the same time. In this section, a brief summary of the recent advances in the use of ceramic nanoparticles for theranostic applications will be described.

- *MSNs:* different dyes can be encapsulated into silica particles. Thus, fluorescein has been covalently attached to the surface of mesoporous silica particles previously functionalized with PEI [133]. The presence of the fluorescein was employed in order to track the fate of the carriers by fluorescence microscopy. The external surface of this carrier was also functionalized with folic acid and the resulting system was able to be selectively recognized by human tumoral cells (HeLa). Fluorescein-labeled hollow MSNs have been employed as carriers of doxorubicin and PDT agents in order to improve the therapeutic response by a synergic effect and to trace the particle during the process [134]. The use of conventional fluorophores, such as fluorescein or rhodamine, in fluorescence microscopy is limited due to the attenuation of photon propagation and poor signal to noise ratio caused by living tissue autofluorescence. In order to overcome these limitations, fluorophores excited with NIR radiation (usually between 750 and 900 nm) can be employed as this radiation presents higher penetration in living tissues [135]. An additional advantage is that MSNs protect the dyes from degradation. Lo *et al.* [136] have described a MSN-based theranostic device composed of NIR dyes (ATTO 647N) and photosensitizers (*meso*-tetratolylporphyrin-Pd) attached within the silica matrix, and cyclic RGD moieties anchored to PEG chains placed on the external surface of the particles. This device is able to destroy human tumoral cells (breast and glioblastoma) selectively under irradiation at 552 nm. Recently, Shi *et al.* [137] have developed a novel mesoporous silica material based on a post-calcination and heating process, which presents luminescent properties without the addition of fluorophores.

MRI is a very useful non-invasive diagnostic technique which provides high resolution images using highly penetrating magnetic fields. Gadolinium is one of the most well-known agents for this imaging tool but usually presents low selectivity and unsatisfactory image contrast enhancement due to its small size. This molecule can be incorporated in MSNs using chelates covalently anchored to their external surface. This fact improves the sensitivity due to the enhanced relaxivity as a result of reduced tumbling rates and large payloads of active magnetic nuclei [138]. One of the main limitations of Gd-conjugated MSNs is their long-term tissue accumulation that can produce excessive toxicity. In order to reduce the toxicity, the Gd-chelate can be attached using a redox-cleavable linker which allows the rapid excretion of the metal after imaging [139]. Iron oxide nanocrystals can be encapsulated within fluorescein-labeled MSNs, obtaining multifunctional nanodevices which combine imaging by MRI and fluorescence microscopy, with the capacity to transport and release therapeutic compounds housed inside the pores [140]. Gold nanorods have also been trapped into MSNs due to this metal surface plasmon resonance property, which allows its use as an imaging probe combined with the capacity to produce heat under laser irradiation [141].

- *Carbon allotropes:* carbon nanotubes show interesting properties for imaging applications, such as high absorption in the NIR region, strong Raman shift, or photoacoustic properties [142]. Different radioisotopes, such as ^{64}Cu, ^{125}I, and ^{66}Ga, can be attached to the CNT surface in order to obtain *in vivo* information about their distribution by PET [143–145]. The *ex vivo* organ distribution of the particles can be evaluated by Raman spectroscopy thanks to the intrinsic Raman shift of the CNT. As in the previous material, fluorescent labels can be attached to its surface in order to trace the material by fluorescence imaging [146]. Graphene can produce fluorescence under an excitation of 400 nm. Thus, the cell internalization of drug-loaded graphene sheets can be easily monitored by fluorescence microscopy [98]. However, the visible fluorescence of graphene competes with the intrinsic auto-fluorescence of living tissue which makes its application in animal models difficult. As an alternative, graphene can be decorated with NIR fluorophores, such as Cy7, in order to avoid this fluorescence quenching [109]. Also, gold nanoparticles with NIR photoluminescence [147] and quantum dots (QDs) with strong fluorescence

[148] have been attached to the graphene surface in order to improve the detection of these systems. Finally, superparamagnetic GO has been produced by growing IONPs on the GO surface in order to employ this system as an MRI contrast agent [149].

- *IONPs:* IONPs provide a large T_2 relaxation effect and are widely used as contrast agents in MRI [150]. Hyeon and coworkers have developed a simple methodology to synthesize water dispersible iron oxide particles for dual imaging (MR and optical imaging) using a polyethyleneglycol-derivatized phosphine oxide (PO-PEG) moiety with fluorescein attached on the end [151]. The surface of these particles can also be decorated with NIR dyes, such as Cy.5.5, in order to obtain *in vivo* information combining two imaging techniques [152]. The combination of IONPs with PET or SPECT probes is receiving special attention nowadays. The most common approach to creating these multifunctional devices consists in the conjugation of radionucleotides such as ^{64}Cu, ^{111}In, and ^{124}I, with different chelates (DOTA or DTPA) anchored on the iron oxide surface. Thus, ^{64}Cu-DOTA complexes conjugated to IONPs with RGD-functionalized IONPs allow non-invasive imaging of tumor targeting using PET and MRI [153]. Trimodal imaging probes have been synthesized by the attachment of PET probes to dual MRI/optical probes [154]. The attachment of therapeutic compounds on the surface of IONPs by physical adsorption or covalent grafting, enables the combination of imaging with drug delivery. Thus, methotrexate, an antitumoral drug, can be anchored to the iron oxide surface through an amide bond. Once the particle is internalized by the cancer cells, the drug is released due to the presence of intracellular proteases [155]. Other antitumoral drugs, such as doxorubicin [121], or glucocorticoid prednisolone [156] have been loaded on the surface of polymer-coated-IONPs achieving excellent antitumoral activities. Proteins, such as human serum albumin, have been absorbed on the drug-loaded IONPs in order to improve the biocompatibility and provide tumor target capacity [157]. The cytotoxic drugs can be attached through sensitive bonds which suffer hydrolysis or other chemical transformations in the presence of certain stimuli. Thus, doxorubicin has been grafted to PEI-coated-IONP via a pH-sensitive hydrazone bond [158]. In mild-acid conditions, which are usually present in the endosomal or lysosomal intracellular compartments, this hydrazine bond undergoes hydrolysis, producing the release of the drug. These types of devices combine the capacity to release drugs in a controlled manner with their imaging properties.

References

1. Riehemann, K., Schneider, S.W., Luger, T.A. *et al.* (2009) Nanomedicine-challenge and perspectives. *Angew. Chem. Int. Ed.*, **48**, 872–897.
2. Matsumura, Y. and Maeda, H. (1986) A new concept for macromolecular therapeutics in cancer chemotherapy: mechanism of tumoritropic accumulation of proteins and the antitumor agent SMANCS. *Cancer Res.*, **46**, 6387–6392.
3. Maeda, H., Nakamura, H. and Fang, J. (2013) The EPR effect for macromolecular drug delivery to solid tumors: improvement of tumor uptake, lowering of systemic toxicity, and distinct tumor imaging in vivo. *Adv. Drug Deliv. Rev.*, **65**, 71–79.
4. Fang, J., Nakamura, H. and Maeda, H. (2011) The EPR effect: unique features of tumor blood vessels for drug delivery, factors involved, and limitations and augmentation of the effect. *Adv. Drug Deliv. Rev.*, **63**, 136–151.

5. Jain, R.J. and Stylianopoulos, T. (2010) Delivering nanomedicine to solid tumors. *Nat. Rev. Clin. Oncol.*, **7**, 653–664.
6. Jain, R.K. (2001) Delivery of molecular and cellular medicine to solid tumors. *Adv. Drug Deliv. Rev.*, **46**, 149–168.
7. Naahidi, S., Jafari, M., Edalat, F. *et al.* (2013) Biocompatibility of engineered nanoparticles for drug delivery. *J. Controlled Release*, **16**, 182–194.
8. Bertrand, N. and Leroux, J.C. (2012) The journey of a drug carrier in the body: an anatomo-physiological perspective. *J. Controlled Release*, **161**, 152–163.
9. Ishihara, T., Takeda, M., Sakamoto, H. *et al.* (2009) Accelerated blood clearance phenomenon upon repeated injection of PEG-modified PLA-nanoparticles. *Pharm. Res.*, **26**, 2270–2279.
10. Lundqvist, M., Stigler, J., Cedervall, T. *et al.* (2011) The evolution of the protein corona around nanoparticles: a test study. *ACS Nano*, **5**, 7503–7509.
11. Karakoti, A.S., Das, S., Thevuthasan, S. and Seal, S. (2011) PEGylated inorganic nanoparticles. *Angew. Chem. Int. Ed.*, **50**, 1980–1994.
12. Huang, D.M., Hsiao, J.K., Chen, Y.C. *et al.* (2009) The promotion of human mesenchymal stem cell proliferation by superparamagnetic iron oxide nanoparticles. *Biomaterials*, **30**, 3645–3651.
13. Chithrani, B.D., Ghazani, A.A. and Chan, W.C.W. (2006) Determining the size and shape dependence of gold nanoparticle uptake into mammalian cells. *Nano Lett.*, **6**, 662–668.
14. He, Q., Shi, J., Zhu, M. *et al.* (2010) The three-stage in vitro degradation behavior of mesoporous silica in simulated body fluid. *Microporous Mesoporous Mater.*, **131**, 314–320.
15. Hudson, S.P., Padera, R.F., Langer, R. and Kohane, D.S. (2008) The biocompatibility of mesoporous silicates. *Biomaterials*, **29**, 4045–4055.
16. Lu, J., Liong, M., Zink, J.I. and Tamanoi, F. (2010) Biocompatibility, biodistribution, and drug delivery efficiency of mesoporous silica nanoparticles for cancer therapy in animals. *Small*, **6**, 1794–1805.
17. Slavador-Morales, C., Flahaut, E., Sim, E. *et al.* (2006) Complement activation and protein absorption by carbon nanotubes. *Mol. Immunol.*, **43**, 193–201.
18. Yan, L., Wang, Y., Xu, X. *et al.* (2012) Can grapheme oxide cause damage to eyesight? *Chem. Res. Toxicol.*, **25**, 1265–1270.
19. Gupta, A.K., Naregalkar, R.R., Vaidya, V.D. and Gupta, M. (2007) Recent advances on surface engineering of magnetic iron oxide nanoparticles and their biomedical applications. *Nanomedicine (London)*, **2**, 23–39.
20. Nel, A., Xia, T., Mädler, L. and Li, N. (2006) Toxic potential of materials at the nanolevel. *Science*, **311**, 622–627.
21. Shubayeb, V.I., Pisanic, T.R. II, and Jin, S. (2009) Magnetic nanoparticles for theragnostics. *Adv. Drug Deliv. Rev.*, **61**, 467–477.
22. Sahay, G., Alakhova, D.Y. and Kavanov, A.V. (2010) Endocytosis of nanomedicines. *J. Controlled Release*, **145**, 182–195.
23. Bareford, L.M. and Swaan, P.W. (2007) Endocytic mechanisms for targeted drug delivery. *Adv. Drug Deliv. Rev.*, **59**, 748–758.
24. Lu, Y. and Low, P.S. (2002) Folate-mediated delivery of macromolecular anticancer therapeutic agents. *Adv. Drug Deliv. Rev.*, **54**, 675–693.

25. Ehrlich, P. (1960) *The Collected Papers of Paul Ehrlich*, Pergamon, London, p. 3.
26. Narain, R., Gonzales, M., Hoffman, A.S. *et al.* (2007) Synthesis of monodisperse biotinylated p(NIPAAm)-coated iron oxide magnetic nanoparticles and their bioconjugation to streptavidin. *Langmuir*, **23**, 6299–6304.
27. Mazzucchelli, S., Colombo, M., De Palma, C. *et al.* (2010) Single-domain protein A-engineered magnetic nanoparticles: toward a universal strategy to site-specific labeling of antibodies for targeted detection of tumor cells. *ACS Nano*, **4**, 5693–5702.
28. Vigor, K.L., Kyrtatos, P.G., Minogue, S. *et al.* (2010) Nanoparticles functionalised with recombinant single chain Fv antibody fragments (scFv) for the magnetic resonance imaging of cancer cells. *Biomaterials*, **31**, 1307–1315.
29. Danhier, F., Feron, O. and Préat, V. (2010) To exploit the tumor microenvironment: passive and active tumor targeting of nanocarriers for anti-cancer drug delivery. *J. Controlled Release*, **148**, 135–146.
30. Low, P.S. and Antony, A.C. (2004) Folate receptor-targeted drugs for cancer and inflammatory diseases. *Adv. Drug Deliv. Rev.*, **56**, 1055–1058.
31. Minko, T. (2004) Drug targeting to the colon with lectins and neoglycoconjugates. *Adv. Drug Deliv. Rev.*, **56**, 491–509.
32. Ruoslahti, E., Bathia, S.N. and Sailor, M.J. (2010) Targeting of drugs and nanoparticles to tumors. *J. Cell Biol.*, **188**, 759–768.
33. Reulen, S.W., Dankers, P.Y., Bomans, P.H. *et al.* (2009) Collagen targeting using protein-functionalized micelles: the strength of multiple weak interactions. *J. Am. Chem. Soc.*, **131**, 7304–7312.
34. Sugahara, K.N., Teesalu, T., Karmali, P.P. *et al.* (2009) Tissue-penetrating delivery of compounds and nanoparticles into tumors. *Cancer Cell*, **16**, 510–520.
35. Mamaeva, V., Sahlgren, C. and Lindén, M. (2013) Mesoporous silica nanoparticles in medicine-recent advances. *Adv. Drug Deliv. Rev.*, **65**, 689–702.
36. Benezra, M., Penate-Medina, O., Zanzonico, P.B. *et al.* (2011) Multimodal silica nanoparticles are effective cancer-targeted probes in a model of human melanoma. *J. Clin. Invest.*, **121**, 2768–2780.
37. Li, Z., Barnes, J.C., Bosoy, A. *et al.* (2012) Mesoporous silica nanoparticles in biomedical applications. *Chem. Soc. Rev.*, **41**, 2590–2605.
38. Trewyn, B.G., Giri, S., Slowing, I.I. and Lin, V.S.-Y. (2007) Mesoporous silica nanoparticle based controlled release, drug delivery, and biosensor systems. *Chem. Commun.*, 3236–3245.
39. Tang, F., Li, L. and Chen, D. (2012) Mesoporous silica nanoparticles: synthesis, biocompatibility and drug delivery. *Adv. Mater.*, **24**, 1504–1534.
40. Liong, M., Lu, J., Kovochich, M. *et al.* (2008) Multifunctional inorganic nanoparticles for imaging, targeting, and drug delivery. *ACS Nano*, **2**, 889–896.
41. Meng, H., Xue, M., Xia, T. *et al.* (2011) Use of size and a copolymer design feature to improve the biodistribution and the enhanced permeability and retention effect of doxorubicin-loaded mesoporous silica nanoparticles in a murine xenograft tumor model. *ACS Nano*, **5**, 4131–4144.
42. Li, L., Tang, F., Liu, H. *et al.* (2010) In vivo delivery of silica nanorattle encapsulated docetaxel for liver cancer therapy with low toxicity and high efficacy. *ACS Nano*, **4**, 6874–6882.

43. Vivero-Escoto, J.L., Slowing, I.I. and Lin, V.S.-Y. (2010) Tuning the cellular uptake and cytotoxicity properties of oligonucleotide intercalator-functionalized mesoporous silica nanoparticles with human cervical cancer cells HeLa. *Biomaterials*, **31**, 1325–1333.

44. Trewyn, B.G., Nieweg, J.A., Zhao, Y. and Lin, V.S.-Y. (2008) Biocompatible mesoporous silica nanoparticles with different morphologies for animal cell membrane penetration. *Chem. Eng. J.*, **137**, 23–29.

45. Brevet, D., Gary-Bobo, M., Raehm, L. *et al.* (2009) Mannose-targeted mesoporous silica nanoparticles for photodynamic therapy. *Chem. Commun.*, 1475–1477.

46. Rosenholm, J.M., Meinander, A., Peuhu, E. *et al.* (2009) Targeting of porous hybrid silica nanoparticles to cancer cells. *ACS Nano*, **3**, 197–206.

47. Lu, J., Li, Z., Zink, J.I. and Tamanoi, F. (2012) In vivo tumor suppression efficacy of mesoporous silica nanoparticles-based drug-delivery system: enhanced efficacy by folate modification. *Nanomedicine*, **8**, 212–220.

48. Ferris, D.P., Lu, J., Gothard, C. *et al. et al.* (2011) Synthesis of biomolecule-modified mesoporous silica nanoparticles for targeted hydrophobic drug delivery to cancer cells. *Small*, **7**, 1816–1826.

49. Tsai, C.P., Chen, C.Y., Hung, Y. *et al.* (2009) Monoclonal antibody-functionalized mesoporous silica nanoparticles (MSN) for selective targeting breast cancer cells. *J. Mater. Chem.*, **19**, 5737–5743.

50. He, X., Zhao, Y., He, D. *et al.* (2012) ATP-responsive controlled release system using aptamer-functionalized mesoporous silica nanoparticles. *Langmuir*, **28**, 12909–12915.

51. Robertson, C.A., Evans, D.H. and Abraharnse, H. (2009) Photodynamic therapy (PDT): a short review on cellular mechanisms and cancer research applications for PDT. *J. Photochem. Photobiol. B*, **96**, 1–8.

52. Gary-Bobo, M., Mir, Y., Rouxel, C. *et al.* (2011) Mannose-functionalized mesoporous silica nanoparticles for efficient two-photon photodynamic therapy of solid tumors. *Angew. Chem. Int. Ed.*, **50**, 11425–11429.

53. Gary-Bobo, M., Hocinea, O., Brevet, D. *et al.* (2012) Cancer therapy improvement with mesoporous silica nanoparticles combining targeting, drug delivery and PDT. *Int. J. Pharm.*, **423**, 509–515.

54. Nakamizo, A., Marini, F., Amano, T. *et al.* (2005) Human bone marrow–derived mesenchymal stem cells in the treatment of gliomas. *Cancer Res.*, **65**, 3307–3318.

55. Li, L., Guan, Y., Liu, H. *et al.* (2011) Silica nanorattle doxorubicin-anchored mesenchymal stem cells for tumor-tropic therapy. *ACS Nano*, **5**, 7462–7470.

56. Lee, E.S., Gao, Z. and Bae, Y.H. (2008) Recent progress in tumor pH targeting nanotechnology. *J. Controlled Release*, **132**, 164–170.

57. Zhao, Y.L., Li, Z., Kabehie, S. *et al.* (2010) pH-operated nanopistons on the surfaces of mesoporous silica nanoparticles. *J. Am. Chem. Soc.*, **132**, 13016–13025.

58. Liu, R., Zhang, Y., Zhao, X. *et al.* (2010) pH-responsive nanogated ensemble based on gold-capped mesoporous silica through an acid-labile acetal linker. *J. Am. Chem. Soc.*, **132**, 1500–1501.

59. Liu, R., Liao, P., Liu, J. and Feng, P. (2011) Responsive polymer-coated mesoporous silica as a pH-sensitive nanocarrier for controlled release. *Langmuir*, **27**, 3095–3099.

60. Rim, H.P., Min, K.H., Lee, H.-J. *et al.* (2011) pH-tunable calcium phosphate covered mesoporous silica nanocontainers for intracellular controlled release of guest drugs. *Angew. Chem. Int. Ed.*, **50**, 8853–8857.
61. Fu, Q., Rama, R., Ward, T.L. *et al.* (2007) Thermoresponsive transport through ordered mesoporous silica/PNIPAm copolymer membranes and microspheres. *Langmuir*, **23**, 170–174.
62. Park, J.H., Lee, Y.-H. and Oh, S.-G. (2007) Preparation of thermosensitive PNIPAm-grafted mesoporous silica particles. *Macromol. Chem. Phys.*, **208**, 2419–2427.
63. Laurent, S., Dutz, S., Häfeli, U.O. and Mahmoudi, M. (2011) Magnetic fluid hyperthermia: focus on superparamagnetic iron oxide nanoparticles. *Adv. Colloid Interface Sci.*, **166**, 8–23.
64. Ruiz-Hernández, E., López-Noriega, A., Arcos, D. *et al.* (2007) Aerosol-assisted synthesis of magnetic mesoporous silica spheres for drug targeting. *Chem. Mater.*, **19**, 3455–3463.
65. Arcos, D., Fal-Miyar, V., Ruiz-Hernández, E. *et al.* (2012) Supramolecular mechanisms in the synthesis of mesoporous magnetic nanospheres for hyperthermia. *J. Mater. Chem.*, **22**, 64–72.
66. Martín-Saavedra, F.M., Ruíz-Hernández, E., Boré, A. *et al.* (2010) Magnetic mesoporous silica spheres for hyperthermia therapy. *Acta Biomater.*, **6**, 4522–4531.
67. Jin, R., Wu, G., Li, Z. *et al.* (2003) What controls the melting properties of DNA-linked gold nanoparticle assemblies? *J. Am. Chem. Soc.*, **125**, 1643–1654.
68. Ruiz-Hernández, E., Baeza, A. and Vallet-Regí, M. (2011) Smart drug delivery through DNA/magnetic nanoparticle gates. *ACS Nano*, **5**, 1259–1266.
69. Baeza, A., Guisasola, E., Ruiz-Hernández, E. and Vallet-Regí, M. (2012) Magnetically triggered multidrug release by hybrid mesoporous silica nanoparticles. *Chem. Mater.*, **24**, 517–524.
70. Lai, C.-Y., Trewyn, B.G., Jeftinija, D.M. *et al.* (2003) A mesoporous silica nanosphere-based carrier system with chemically removable CdS nanoparticle caps for stimuli-responsive controlled release of neurotransmitters and drug molecules. *J. Am. Chem. Soc.*, **125**, 4451–4459.
71. Giri, S., Trewyn, B.G., Stellmaker, M.P. and Lin, V.S.-Y. (2005) Stimuli-responsive controlled-release delivery system based on mesoporous silica nanorods capped with magnetic nanoparticles. *Angew. Chem. Int. Ed.*, **44**, 5038–5044.
72. Luo, Z., Cai, K., Hu, Y. *et al.* (2011) Mesoporous silica nanoparticles end-capped with collagen: redox-responsive nanoreservoirs for targeted drug delivery. *Angew. Chem. Int. Ed.*, **50**, 640–643.
73. Liu, R., Zhao, X., Wu, T. and Feng, P. (2008) Tunable redox-responsive hybrid nanogated ensembles. *J. Am. Chem. Soc.*, **130**, 14418–14419.
74. Austin, C.D., Wen, X., Gazzard, L. *et al.* (2005) Oxidizing potential of endosomes and lysosomes limits intracellular cleavage of disulfide-based antibody-drug conjugates. *Proc. Natl. Acad. Sci. U.S.A.*, **102**, 17987–17992.
75. de Bruin, K.G., Fella, C., Ogris, M. *et al.* (2008) Dynamics of photoinduced endosomal release of polyplexes. *J. Controlled Release*, **130**, 175–182.
76. Shauer, A.M., Schlossbauer, A., Ruthardt, N. *et al.* (2010) Role of endosomal escape for disulfide-based drug delivery from colloidal mesoporous silica evaluated by live-cell imaging. *Nano Lett.*, **10**, 3684–3691.

77. Mal, N.K., Fujiwara, M. and Tanaka, Y. (2003) Photocontrolled reversible release of guest molecules from coumarin modified mesoporous silica. *Nature*, **421**, 350–353.
78. Zhu, Y. and Fujiwara, M. (2007) Installing dynamic molecular photomechanics in mesopores: a multifunctional controlled-release nanosystem. *Angew. Chem. Int. Ed.*, **46**, 2241–2244.
79. Ferris, D.P., Zhao, Y.-L., Khashab, N.M. *et al.* (2009) Light-operated mechanized nanoparticles. *J. Am. Chem. Soc.*, **131**, 1686–1688.
80. Vivero-Escoto, J.L., Slowing, I.I., Wu, C.W. and Lin, V.S.-Y. (2009) Photoinduced intracellular controlled release drug delivery in human cells by gold-capped mesoporous silica nanosphere. *J. Am. Chem. Soc.*, **131**, 3462–3463.
81. Chang, Y.-T., Liao, P.-Y., Sheu, H.-S. *et al.* (2012) Near-infrared light-responsive intracellular drug and siRNA release using Au nanoensembles with oligonucleotide-capped silica shell. *Adv. Mater.*, **24**, 3309–3314.
82. de la Rica, R., Aili, D. and Stevens, M.M. (2012) Enzyme-responsive nanoparticles for drug release and diagnostics. *Adv. Drug Deliv. Rev.*, **64**, 967–978.
83. Patel, K., Angelos, S., Dichtel, W.R. *et al.* (2008) Enzyme-responsive snap-top covered silica nanocontainers. *J. Am. Chem. Soc.*, **130**, 2382–2383.
84. Park, C., Kim, H., Kim, S. and Kim, C. (2009) Enzyme responsive nanocontainers with cyclodextrin gatekeepers and synergistic effects in release of guests. *J. Am. Chem. Soc.*, **131**, 16614–16615.
85. Schlossbauer, A., Kecht, J. and Bein, T. (2009) Biotin-avidin as a protease responsive cap system for controlled guest release from colloidal mesoporous silica. *Angew. Chem. Int. Ed.*, **48**, 3092–3095.
86. Singh, N., Karambelkar, A., Gu, L. *et al.* (2011) Bioresponsive mesoporous silica nanoparticles for triggered drug release. *J. Am. Chem. Soc.*, **133**, 19582–19585.
87. Climent, E., Bernados, A., Martínez-Máñez, R. *et al.* (2009) Controlled delivery systems using antibody-capped mesoporous nanocontainers. *J. Am. Chem. Soc.*, **131**, 14075–14080.
88. Zhu, C.-L., Lu, C.-H., Song, X.-Y. *et al.* (2011) Bioresponsive controlled release using mesoporous silica nanoparticles capped with aptamer-based molecular gate. *J. Am. Chem. Soc.*, **133**, 1278–1281.
89. Cai, Y., Pan, H., Xu, X. *et al.* (2007) Ultrasonic controlled morphology transformation of hollow calcium phosphate nanospheres: a smart and biocompatible drug release system. *Chem. Mater.*, **19**, 3081–3083.
90. Cheng, X. and Kuhn, L. (2007) Chemotherapy drug delivery from calcium phosphate nanoparticles. *Int. J. Nanomed.*, **2**, 667–674.
91. Kester, M., Heakal, Y., Fox, T. *et al.* (2008) Calcium phosphate nanocomposite particles for in vitro imaging and encapsulated chemotherapeutic drug delivery to cancer cells. *Nano Lett.*, **8**, 4116–4121.
92. Morgan, T.T., Muddana, H.S., Altinoglu, E.I. *et al.* (2008) Encapsulation of organic molecules in calcium phosphate nanocomposite particles for intracellular imaging and drug delivery. *Nano Lett.*, **8**, 4108–4115.
93. Schwiertz, J., Wiehe, A., Gräfe, S. *et al.* (2009) Calcium phosphate nanoparticles as efficient carriers for photodynamic therapy against cells and bacteria. *Biomaterials*, **30**, 3324–3331.

94. Klesing, J., Wiehe, A., Gitter, B. *et al.* (2010) Positively charged calcium phosphate/polymer nanoparticles for photodynamic therapy. *J. Mater. Sci. Mater. Med.*, **21**, 887–892.

95. Hou, C.-H., Hou, S.-M., Hsueh, Y.-S. *et al.* (2009) The in vivo performance of biomagnetic hydroxyapatite nanoparticles in cancer hyperthermia therapy. *Biomaterials*, **30**, 3956–3960.

96. Hou, C.-H., Chen, C.-W., Hou, S.-M. *et al.* (2009) The fabrication and characterization of dicalcium phosphate dihydrate-modified magnetic nanoparticles and their performance in hyperthermia processes in vitro. *Biomaterials*, **30**, 4700–4707.

97. Yang, K., Feng, L., Shi, X. and Liu, Z. (2013) Nano-graphene in biomedicine: theranostic applications. *Chem. Soc. Rev.*, **42**, 530–547.

98. Sun, X., Liu, Z., Welsher, K. *et al.* (2008) Nano-graphene oxide for cellular imaging and drug delivery. *Nano Res.*, **1**, 203–212.

99. Hu, H., Yu, J., Li, Y. *et al.* (2012) Engineering of a novel pluronic F127/graphene nanohybrid for pH responsive drug delivery. *J. Biomed. Mater. Res. Part A*, **100A**, 141–148.

100. Bao, H., Pan, Y., Ping, Y. *et al.* (2011) Chitosan-functionalized graphene oxide as a nanocarrier for drug and gene delivery. *Small*, **7**, 1569–1578.

101. Yang, Y., Zhang, Y.-M., Chen, Y. *et al.* (2012) Construction of a graphene oxide noncovalent multiple nanosupramolecular assembly as a scaffold for drug delivery. *Chem. Eur. J.*, **18**, 4208–4215.

102. Zhang, L., Xia, J., Zhao, Q. *et al.* (2010) Functional grapheme oxide as a nanocarrier for controlled loading and targeted delivery of mixed anticancer drugs. *Small*, **6**, 537–544.

103. Wen, H., Dong, C., Dong, H. *et al.* (2012) Engineered redox-responsive PEG detachment mechanism in PEGylated nano-graphene oxide for intramolecular drug delivery. *Small*, **8**, 760–769.

104. Yang, X., Wang, Y., Huang, X. *et al.* (2011) Multi-functionalized graphene oxide based anticancer drug-carrier with dual-targeting function and pH-sensitivity. *J. Mater. Chem.*, **21**, 3448–3454.

105. Meng, L., Zhang, X., Lu, Q. *et al.* (2012) Single walled carbon nanotubes as drug delivery vehicles: targeting doxorubicin to tumors. *Biomaterials*, **33**, 1689–1698.

106. Liu, Z., Sun, X., Nakayama-Ratchford, N. and Dai, H. (2007) Supramolecular chemistry on water-soluble carbon nanotubes for drug loading and delivery. *ACS Nano*, **1**, 50–56.

107. Zhang, X.-K., Meng, L.-J., Lu, Q.-H. *et al.* (2009) Targeted delivery and controlled release of doxorubicin to cancer cells using modified single wall carbon nanotubes. *Biomaterials*, **30**, 6041–6047.

108. Liu, Z., Chen, K., Davis, C. *et al.* (2008) Drug delivery with carbon nanotubes for in vivo cancer treatment. *Cancer Res.*, **68**, 6652–6660.

109. Yang, K., Zhang, S., Zhang, G. *et al.* (2010) Graphene in mice: ultrahigh in vivo tumor uptake and efficient photothermal therapy. *Nano Lett.*, **10**, 3318–3323.

110. Robinson, J.Y., Welsher, K., Tabakman, S.M. *et al.* (2010) High performance in vivo near-IR (>1 μm) imaging and photothermal cancer therapy with carbon nanotubes. *Nano Res.*, **3**, 779–793.

111. Gilchrist, R.K., Medal, R., Shorey, W.D. *et al.* (1957) Selective inductive heating of lymph nodes. *Ann. Surg.*, **146**, 596–606.

112. Goya, G.F., Berquo, T.S. and Fonseca, F.C. (2003) Static and dynamic magnetic properties of spherical magnetite nanoparticles. *J. Appl. Phys.*, **94**, 3520–3528.

113. Hilger, I., Fruhauf, K., Andra, W. *et al.* (2002) Heating potential of iron oxides for therapeutic purposes in interventional radiology. *Acad. Radiol.*, **9**, 198–202.

114. Jordan, A., Scholz, R., Maier-Hauff, K. *et al.* (2006) The effect of thermotherapy using magnetic nanoparticles on rat malignant glioma. *J. Neuro-Oncol.*, **78**, 7–14.

115. Maier-Hauff, K., Ulrich, F., Nestler, D. *et al.* (2011) Efficacy and safety of intratumoral thermotherapy using magnetic iron-oxide nanoparticles combined with external beam radiotherapy on patients with recurrent glioblastoma multiforme. *J. Neuro-Oncol.*, **103**, 317–324.

116. Hervé, K., Douziech-Eyrolles, L., Munnier, E. *et al.* (2008) The development of stable aqueous suspensions of PEGylated SPIONs for biomedical applications. *Nanotechnology*, **19**, 465608.

117. Baeza, A., Arcos, D. and Vallet-Regí, M. (2013) Thermoseeds for interstitial magnetic hyperthermia: from bioceramics to nanoparticles. *J. Phys. Condens. Matter*, **25**, 484003.

118. Colilla, M., González, B. and Vallet-Regí, M. (2013) Mesoporous silica nanoparticles for the design of smart delivery nanodevices. *Biomater. Sci.*, **1**, 114–134.

119. Goon, I.Y., Lai, L.M.H., Lim, M. *et al.* (2009) Fabrication and dispersion of gold shell-protected magnetite nanoparticles: systematic control using polyethyleneimine. *Chem. Mater.*, **21**, 673–681.

120. Cintra-Silva, D.O., Estevanato, L.L.C., Simioni, A.R. *et al.* (2012) Successful strategy for targeting the central nervous system using magnetic albumin nanospheres. *J. Biomed. Nanotechnol.*, **8**, 182–189.

121. Yu, M.-K., Jeong, Y.-Y., Park, J. *et al.* (2008) Drug-loaded superparamagnetic iron oxide nanoparticles for combined cancer imaging and therapy in vivo. *Angew. Chem. Int. Ed.*, **47**, 5362–5365.

122. He, X., Wu, X., Cai, X. *et al.* (2012) Functionalization of magnetic nanoparticles with dendritic – linear – brush-like triblock copolymers and their drug release properties. *Langmuir*, **28**, 11929–11938.

123. Hayashi, K., Ono, K., Suzuki, H. *et al.* (2010) High-frequency, magnetic-field-responsive drug release from magnetic nanoparticle/organic hybrid based on hyperthermic effect. *Appl. Mater. Interfaces*, **2**, 1903–1911.

124. Talelli, M., Rijcken, C.J.F., Lammers, T. *et al.* (2009) Superparamagnetic iron oxide nanoparticles encapsulated in biodegradable thermosensitive polymeric micelles: toward a targeted nanomedicine suitable for image-guided drug delivery. *Langmuir*, **25**, 2060–2067.

125. Knezevic, N.Z., Ruiz-Hernandez, E., Hennink, W.E. and Vallet-Regi, M. (2013) Magnetic mesoporous silica-based core/shell nanoparticles for biomedical applications. *RSC Adv.*, **3**, 9584–9593.

126. Maeng, J.H., Lee, D.-H., Jung, K.H. *et al.* (2010) Multifunctional doxorubicin loaded superparamagnetic iron oxide nanoparticles for chemotherapy and magnetic resonance imaging in liver cancer. *Biomaterials*, **31**, 4995–5006.

127. Ling, Y., Wei, K., Luo, Y. *et al.* (2011) Dual docetaxel/superparamagnetic iron oxide loaded nanoparticles for both targeting magnetic resonance imaging and cancer therapy. *Biomaterials*, **32**, 7139–7150.
128. Dilnawaz, F., Singh, A., Mohanty, C. and Sahoo, S.K. (2010) Dual drug loaded superparamagnetic iron oxide nanoparticles for targeted cancer therapy. *Biomaterials*, **31**, 3694–3706.
129. Yu, M.K., Park, J., Jeong, Y.Y. *et al.* (2010) Integrin-targeting thermally cross-linked superparamagnetic iron oxide nanoparticles for combined cancer imaging and drug delivery. *Nanotechnology*, **21**, 415102.
130. Yuk, S.H., Oh, K.S., Cho, S.H. *et al.* (2011) Glycol chitosan/heparin immobilized iron oxide nanoparticles with a tumor-targeting characteristic for magnetic resonance imaging. *Biomacromolecules*, **12**, 2335–2343.
131. Sawant, R.M., Sawant, R.R., Gultepe, E. *et al.* (2009) Nanosized cancer cell-targeted polymeric immunomicelles loaded with superparamagnetic iron oxide nanoparticles. *J. Nanopart. Res.*, **11**, 1777–1785.
132. Lee, D.E., Koo, H., Sun, I.C. *et al.* (2012) Multifunctional nanoparticles for multimodal imaging and theragnosis. *Chem. Soc. Rev.*, **41**, 2656–2672.
133. Rosenholm, J.M., Peuhu, E., Eriksson, J.E. *et al.* (2009) Targeted Intracellular delivery of hydrophobic agents using mesoporous hybrid silica nanoparticles as carrier systems. *Nano Lett.*, **9**, 3308–3311.
134. Wang, T., Zhang, L., Su, Z. *et al.* (2011) Multifunctional hollow mesoporous silica nanocages for cancer cell detection and the combined chemotherapy and photodynamic therapy. *ACS Appl. Mater. Interfaces*, **3**, 2479–2486.
135. Lee, C.-H., Cheng, S.-H., Wang, Y.-J. *et al.* (2009) Near-infrared mesoporous silica nanoparticles for optical imaging: characterization and in vivo biodistribution. *Adv. Funct. Mater.*, **19**, 215–222.
136. Cheng, S.-H., Lee, C.-H., Chen, M.-C. *et al.* (2010) Tri-functionalization of mesoporous silica nanoparticles for comprehensive cancer theranostics-the trio of imaging, targeting and therapy. *J. Mater. Chem.*, **20**, 6149–6157.
137. He, Q., Shi, J., Cui, X. *et al.* (2011) Synthesis of oxygen-deficient luminescent mesoporous silica nanoparticles for synchronous drug delivery and imaging. *Chem. Commun.*, **47**, 7947–7949.
138. Taylor, K.M.L., Kim, J.S., Rieter, W.J. *et al.* (2008) Mesoporous silica nanospheres as highly efficient MRI contrast agents. *J. Am. Chem. Soc.*, **130**, 2154–2155.
139. Vivero-Escoto, J.L., Taylor-Pashow, K.M.L., Huxford, R.C. *et al.* (2011) Multifunctional mesoporous silica nanospheres with cleavable Gd(III) chelates as MRI contrast agents: synthesis, characterization, target-specificity, and renal clearance. *Small*, **7**, 3519–3528.
140. Kim, J., Kim, H.S., Lee, N. *et al.* (2008) Multifunctional uniform nanoparticles composed of a magnetite nanocrystal core and a mesoporous silica shell for magnetic resonance and fluorescence imaging and for drug delivery. *Angew. Chem. Int. Ed.*, **47**, 8438–8441.
141. Zhang, Z., Wang, L., Wang, J. *et al.* (2012) Mesoporous silica-coated gold nanorods as a light-mediated multifunctional theranostic platform for cancer treatment. *Adv. Mater.*, **24**, 1418–1423.

142. Ji, S.-R., Liu, C., Zhang, B. *et al.* (2010) Carbon nanotubes in cancer diagnosis and therapy. *Biochim. Biophys. Acta, Rev. Cancer*, **1806**, 29–35.

143. Liu, Z., Cai, W., He, L. *et al.* (2007) In vivo biodistribution and highly efficient tumour targeting of carbon nanotubes in mice. *Nat. Nanotechnol.*, **2**, 47–52.

144. Yang, K., Wan, J.M., Zhang, S.-A. *et al.* (2011) In vivo pharmacokinetics, long-term biodistribution, and toxicology of PEGylated graphene in mice. *ACS Nano*, **5**, 516–522.

145. Hong, H., Zhang, Y., Engle, J.W. *et al.* (2012) In vivo targeting and positron emission tomography imaging of tumor vasculature with ^{66}Ga-labeled nano-graphene. *Biomaterials*, **33**, 4147–4156.

146. Bhirde, A.A., Patel, V., Gavard, J. *et al.* (2009) Targeted killing of cancer cells in vivo and in vitro with EGF-directed carbon nanotube-based drug delivery. *ACS Nano*, **3**, 307–316.

147. Wang, C., Li, J., Amatore, C. *et al.* (2011) Gold nanoclusters and graphene nanocomposites for drug delivery and imaging of cancer cells. *Angew. Chem. Int. Ed.*, **50**, 11644–11648.

148. Hu, S.-H., Chen, Y.-W., Hung, W.-T. *et al.* (2012) Quantum-dot-tagged reduced graphene oxide nanocomposites for bright fluorescence bioimaging and photothermal therapy monitored in situ. *Adv. Mater.*, **24**, 1748–1754.

149. Chen, W., Yi, P., Zhang, Y. *et al.* (2011) Composites of aminodextran-coated Fe_3O_4 nanoparticles and graphene oxide for cellular magnetic resonance imaging. *ACS Appl. Mater. Interfaces*, **3**, 4085–4091.

150. Kim, J., Piao, Y. and Hyeon, T. (2009) Multifunctional nanostructured materials for multimodal imaging, and simultaneous imaging and therapy. *Chem. Soc. Rev.*, **38**, 372–390.

151. Na, H.B., Lee, I.S., Seo, H. *et al.* (2007) Versatile PEG-derivatized phosphine oxide ligands for water-dispersible metal oxide nanocrystals. *Chem. Commun.*, 5167–5169.

152. Moore, A., Medarova, Z., Potthast, A. and Dai, G. (2004) In vivo targeting of under-glycosylated MUC-1 tumor antigen using a multimodal imaging probe. *Cancer Res.*, **64**, 1821–1827.

153. Lee, H.Y., Li, Z., Chen, K. *et al.* (2008) PET/MRI dual-modality tumor imaging using arginine-glycine-aspartic (RGD)–conjugated radiolabeled iron oxide nanoparticles. *J. Nucl. Med.*, **49**, 1371–1379.

154. Xie, J., Chen, K., Huang, J. *et al.* (2010) PET/NIRF/MRI triple functional iron oxide nanoparticles. *Biomaterials*, **31**, 3016–3022.

155. Kohler, N., Sun, C., Fichtenholtz, A. *et al.* (2006) Methotrexate-immobilized poly(ethylene glycol) magnetic nanoparticles for MR imaging and drug delivery. *Small*, **2**, 785–792.

156. Gianella, A., Jarzyna, P.A., Mani, V. *et al.* (2011) Multifunctional nanoemulsion platform for imaging guided therapy evaluated in experimental cancer. *ACS Nano*, **5**, 4422–4433.

157. Quan, Q., Xie, J., Gao, H. *et al.* (2011) HSA coated iron oxide nanoparticles as drug delivery vehicles for cancer therapy. *Mol. Pharmaceutics*, **8**, 1669–1676.

158. Kievit, F.M., Wang, F.Y., Fang, C. *et al.* (2011) Doxorubicin loaded iron oxide nanoparticles overcome multidrug resistance in cancer in vitro. *J. Controlled Release*, **152**, 76–83.

Index

References to figures are given in italic type. References to tables are given in bold type.

Bioceramics with Clinical Applications, First Edition. Edited by María Vallet-Regí.
© 2014 John Wiley & Sons, Ltd. Published 2014 by John Wiley & Sons, Ltd.